THE GEOTOURISM INDUSTRY IN THE 21ST CENTURY

The Origin, Principles, and Futuristic Approach

THE GEOTOURISM INDUSTRY IN THE 21ST CENTURY

The Origin, Principles, and Futuristic Approach

Edited by
Bahram Nekouie Sadry

Apple Academic Press Inc.
4164 Lakeshore Road
Burlington ON L7L 1A4, Canada

Apple Academic Press Inc.
1265 Goldenrod Circle NE
Palm Bay, Florida 32905, USA

© 2021 by Apple Academic Press, Inc.

Exclusive co-publishing with CRC Press, a Taylor & Francis Group

No claim to original U.S. Government works

International Standard Book Number-13: 978-1-77188-826-4 (Hardcover)
International Standard Book Number-13: 978-0-42929-279-8 (eBook)

All rights reserved. No part of this work may be reprinted or reproduced or utilized in any form or by any electric, mechanical or other means, now known or hereafter invented, including photocopying and recording, or in any information storage or retrieval system, without permission in writing from the publisher or its distributor, except in the case of brief excerpts or quotations for use in reviews or critical articles.

This book contains information obtained from authentic and highly regarded sources. Reprinted material is quoted with permission and sources are indicated. Copyright for individual articles remains with the authors as indicated. A wide variety of references are listed. Reasonable efforts have been made to publish reliable data and information, but the authors, editors, and the publisher cannot assume responsibility for the validity of all materials or the consequences of their use. The authors, editors, and the publisher have attempted to trace the copyright holders of all material reproduced in this publication and apologize to copyright holders if permission to publish in this form has not been obtained. If any copyright material has not been acknowledged, please write and let us know so we may rectify in any future reprint.

Trademark Notice: Registered trademark of products or corporate names are used only for explanation and identification without intent to infringe.

Library and Archives Canada Cataloguing in Publication

Title: The geotourism industry in the 21st century : the origin, principles, and futuristic approach / edited by Bahram Nekouie Sadry.

Names: Nikū'ī Ṣadrī, Bahrām, 1973 or 1974- author.

Description: Includes bibliographical references and index.

Identifiers: Canadiana (print) 20200154311 | Canadiana (ebook) 20200154338 | ISBN 9781771888264 (hardcover) | ISBN 9780429292798 (PDF)

Subjects: LCSH: Ecotourism. | LCSH: Geotourism. | LCSH: Geoparks.

Classification: LCC G156.5.E26 N55 2020 | DDC 333.78/3—dc23

Library of Congress Cataloging-in-Publication Data

Names: Nikū'ī Ṣadrī, Bahrām, 1973 or 1974- editor.

Title: The geotourism industry in the 21st century : the origin, principles, and futuristic approach / edited by Bahram Nekouie Sadry.

Description: Burlington, ON ; Palm Bay, Florida : Apple Academic Press, 2020. | Includes bibliographical references and index. | Summary: "Here is an engaging overview of the development of, definition of, and approach to modern geotourism, a growing movement to help sustain and showcase the distinctive geographical characteristics of many places around the world. This volume, The Geotourism Industry in the 21st Century: The Origin, Principles, and Futuristic Approach, provides a clear conceptual framework with illustrative examples from all corners of the world to better understand abiotic nature-based tourism. The volume looks at the establishment and effective management of the over 130 UNESCO geoparks around the world and other travel and tourism destinations of interest for their significant historical, cultural, and frequently stunning physical attributes. With studies from a selection of geotourist areas in Poland, Japan, Turkey, Brazil, Albania, California, Mexico, Peru, and other places, the volume explores urban geotourism, mining heritage, geomorphological landforms, geoheritage (based on cultural and historical interest), roadside geology of the U. S., community engagement and volunteer management programs, and much more. There is even a chapter on space and celestial geotourism. The volume encourages academics, practitioners, and students in the fields of tourism, geology, geography, and also environmental and conservation science to learn more about the geopark movement and this arising new discipline. Key features: Provides guidance for all aspects of geotourism as it relates to the establishment and effective management of geoparks Offers specific information on the geo-conservation and effective management of geotourism in geoparks Identifies significant geological and mining heritage areas that could be formally reserved as national geoparks or geosites by nations Provides a model and schematic mechanism for integrating geodiversity into all relevant geotourism activities and also to geoheritage stakeholders, such as UNESCO, the mining industry, and others"-- Provided by publisher.

Identifiers: LCCN 2019057666 (print) | LCCN 2019057667 (ebook) | ISBN 9781771888264 (hardcover) | ISBN 9780429292798 (ebook)

Subjects: LCSH: Geotourism--Case studies.

Classification: LCC G155.A1 G426 2020 (print) | LCC G155.A1 (ebook) | DDC 338.4/791--dc23

LC record available at https://lccn.loc.gov/2019057666

LC ebook record available at https://lccn.loc.gov/2019057667

Apple Academic Press also publishes its books in a variety of electronic formats. Some content that appears in print may not be available in electronic format. For information about Apple Academic Press products, visit our website at **www.appleacademicpress.com** and the CRC Press website at **www.crcpress.com**

DEDICATION

This book is dedicated to the Nekouie Sadry family;

And to:
All my mentors, who inspired me through their books and speeches;
My kind friends who encouraged and supported me;
My keen students who were a source of motivation to me;

And:
To all hard-working researchers, executives, and managers who endeavor to alleviate poverty and who popularize geoscience through developing geotourism and geoparks around the world.

About the Editor

Bahram Nekouie Sadry, PhD
Adjunct Senior Lecturer and Geotourism Consultant

Bahram Nekouie Sadry, PhD, is an Adjunct Senior Lecturer and a Geotourism Consultant. Dr. Sadry conducts research in the fields of geotourism, ecotourism, and wildlife tourism and heritage interpretation. He has published several books and textbooks on geotourism and has undertaken geotourism consultancy projects for the private sector. He is also an education consultant and a curriculum developer on tourism and geotourism in higher education and has conducted geo-tour guide training courses for his national government. Dr. Sadry is deeply involved in the development of geotourism and is a passionate advocate for the creation of national geoparks and UNESCO Global Geoparks around the world. He holds a BE degree (Mining Eng.), MSc (Geography), and PhD (Educational Administration).

Contents

Contributors .... *xiii*
Preface .... *xxiii*
Foreword 1 by Thomas A. Hose .... *xxv*
Foreword 2 by William (Bill) Witherspoon .... *xxix*
Acknowledgments .... *xxxiii*

Part I: Geotourism Concepts in the 21st Century 1

1. **The Scope and Nature of Geotourism in the 21st Century** 3
   Bahram Nekouie Sadry

2. **Historical Viewpoints on the Geotourism Concept in the 21st Century** 23
   Thomas A. Hose

3. **Urban Geotourism in Poland** 93
   Krzysztof Gaidzik

4. **Mining Heritage as Geotourism Attractions in Brazil** 117
   Antonio Liccardo, Virginio Mantesso-Neto, and Marcos Antonio Leite do Nascimento

Part II: Geoassessments: Geoheritage Assessments for Geotourism 145

5. **Geomorphosites: Esthetic Landscape Features or Earth History Heritage?** 147
   Emmanuel Reynard

6. **Geoheritage and Geotourism in Albania** 169
   Afat Serjani

7. **Establishing an Appropriate Methodology for the Management of Geological Heritage for Geotourism Development in the Azores UNESCO Global Geopark** 189
   Eva Almeida Lima and Marisa Machado

8. **"Simply the Best": The Search for the World's Top Geotourism Destinations** 207
   Murray Gray

Part III: Geointerpretation: Interpreting Geological and Mining Heritage 227

9. Interpreting Mining: A Case Study of a Coal Mine Exhibit 229
   Ted T. Cable

10. Geotrails 247
    Thomas A. Hose

11. Interpreting Geological and Mining Heritage 277
    Ross Dowling

12. Evolving Geological Interpretation Writings about a Well-Traveled Part of California, 1878–2016 299
    William (Bill) Witherspoon

13. Commercially Successful Books for Place-Based Geology: Roadside Geology Covers the US 325
    William (Bill) Witherspoon and John Rimel

Part IV: Geoparks and Community Developments: A Base for Geotourism Promotion 355

14. Community Engagement in Japanese Geoparks 357
    Kazem Vafadari and Malcolm J. M. Cooper

15. The Role of Volunteer Management Programs in Geotourism Development 375
    Cristian Ciobanu and Alexandru Andrășanu

16. Geotourism and Proposed Geopark Projects in Turkey 387
    Gülpinar Akbulut Özpay

17. Geotourism Development in Latin American UNESCO Global Geoparks: Brazil, Uruguay, Mexico, and Peru 421
    José Luis Palacio Prieto, César Gosso, Diego Irazábal, José Patrício Pereira Melo, Francisco do O´de Lima, Júnior, Carles Canet, Miguel A. Cruz-Pérez, Erika Salgado-Martínez, Juan Carlos Mora-Chaparro, Krzysztof Gaidzik, Jerzy Żaba, and Justyna Ciesielczuk

Part V: Globalization and the Future of Geological Attraction Destinations 447

18. Dinosaur Geotourism: A World-Wide Growing Tourism Niche 449
    Nathalie Cayla

Contents

19. **Accessible Geotourism: Constraints and Implications** 473
    Mamoon Allan
20. **Space and Celestial Geotourism** 481
    Bahram Nekouie Sadry
21. **Post-Mining Objects as Geotourist Attractions: Upper Silesian Coal Basin (Poland)** 507
    Krzysztof Gaidzik and Marta Chmielewska
22. **Geotourism vs Mass Tourism: An Overview of the Langkawi UNESCO Global Geopark** 527
    Kamarulzaman Abdul Ghani
23. **The Future of Geotourism** 535
    Bahram Nekouie Sadry

Index *553*

Contributors

Mamoon Allan is currently an associate professor of tourism marketing at the Department of Tourism Management, the University of Jordan. Dr. Allan has worked in the tourism and hospitality fields in three countries: Australia, Jordan, and Libya. He completed his PhD in Tourism and Leisure Marketing from Edith Cowan University in Perth, Western Australia. Dr. Allan conducts international research and projects in the fields of geotourism, accessible tourism, and tourism motivation. E-mail: mamoon1073@yahoo.com

Alexandru Andrășanu is on the teaching staff at the University of Bucharest (Romania). Dr. Andrășanu was involved in the development of educational and training structures and programs, and in the management of more than 40 national and international projects. He is coordinator of the master program on Applied Geo-Biology for Natural and Cultural Heritage Conservation, Director of the Hațeg Country UNESCO Global Geopark, initiator of the Buzau Land Geopark project, member of the Coordination Committee of the European Geoparks Network, and member of the European Association ProGEO. He is experienced in geoconservation, geopark development, geoeducation and, geotourism and is a UNESCO expert for Global Geoparks, with missions in China, France, Greece, Morocco, Belgium, Japan, Indonesia, and Italy. E-mail: alex-andru.andrasanu@gg.unibuc.ro

Ted T. Cable is Professor Emeritus in the Park Management and Conservation program at Kansas State University. He worked as a heritage interpreter for Cook County Forest Preserve District (Illinois) and for Lake County Parks and Recreation Department (Indiana) before receiving his Ph.D. at Purdue University. Professor Cable has received the "Master of Interpretation" award from the National Association for Interpretation. He also is a Founder and Fellow of the National Association for Interpretation. In addition to numerous university honors and awards from state and local organizations, he has been honored by the U.S. Environmental Protection Agency and the U.S. Department of Agriculture for outstanding teaching in the fields of environmental education and heritage interpretation. E-mail: tcable@ksu.edu

Carles Canet holds a PhD in geology from the Universidad de Barcelona, Spain. He is a Senior Researcher in the Institute of Geophysics, Universidad Nacional Autónoma de México (UNAM). His research interests include mineralogy and geochemistry applied to the study of mineral deposits. He is the Scientific Coordinator of the Comarca Minera UNESCO Global Geopark, México, and President of the Latin American and Caribbean Geoparks Network. E-mail: ccanet@atmosfera.unam.mx

Nathalie Cayla currently works at the Laboratory of Environment Dynamics and Territories of the Mountain (EDYTEM), University of Savoie Mont-Blanc. The subject of her PhD thesis was the valorization of the geological heritage in the alpine mountain range. Dr. Cayla is involved in the development of geotourism working closely with two French geoparks: the Massif des Bauges Geopark and the Chablais geopark. In recent years, she has studied geotourism around dinosaur sites (mass mortalities, nesting areas or trackways) and carries out area analyses of the networks of actors involved and the socioeconomic effects of this tourist activity. E-mail: Nathalie.cayla@univ-savoie.fr

Marta Chmielewska holds a PhD in urban geography and is an adjunct associate professor in the Department of Economic Geography, Faculty of Earth Sciences, University of Silesia in Poland. Her research focuses mainly on urban morphology, evolution of the urban space, revitalization of postindustrial sites, as well as conservation of the postindustrial heritage and postindustrial tourism. E-mail: marta.chmielewska@us.edu.pl

Justyna Ciesielczuk holds a PhD in Geology from the University of Silesia, where she is acting as director of the Department of Fundamental Geology. Her fields of interest include mineralogy, petrology of the hydrothermal alterations of granites and transformations occurring on burning waste dumps after coal-mining exploitations. E-mail: justyna.ciesielczuk@us.edu.pl

Cristian Ciobanu has a PhD in Geography with the paper Study of Mental Geography in Bucharest, which created a trend of geographies of the mind in Romania. Since 2011, he teaches Heritage Management at the Geobiology masters. He is a member of the Geology and Geophysics Faculty, University of Bucharest. He is a member of the Caleidoscop team, an informal group for the multidisciplinary research of space. Since 2012, he lives in Hațeg region, where he works at the Hațeg Country UNESCO Global Geopark. His scientific preoccupations include heritage interpretation using space perception, sacred geography of the Hațeg region and a whole series of the geographies of perception: imaginative geography, geoteology, geocritics, and others. E-mail: cristian.ciobanu@unibuc.ro

Malcolm J. M. Cooper is Emeritus Professor at Ritsumeikan Asia Pacific University, Beppu, Japan, and is a former Vice President of that university. He is a specialist in education, economic development, tourism and hospitality management and law, environmental and water resources management, and environmental law and has published widely in these fields. He has held previous appointments at the Open University (UK), the Universities of New England, Adelaide, and Southern Queensland (Australia), and the Waiariki Institute of Technology (New Zealand), and has worked in the educational policy, environmental planning, and tourism policy areas for federal, state, and local governments in Australia. He has also been both a private consultant and an education consultant to the governments of New Zealand, Sri Lanka, China, and Vietnam. He is a visiting professor at Sabaragamuwa, Rajarata, and Ruhuna Universities in Sri Lanka, and at Nagasaki University in Japan. He is a recipient of the Australian Centennial Medal and has published over 150 books and papers. He currently edits manuscripts for a range of organizations in the education, economic, environment, and tourism fields. E-mail: cooperm@apu.ac.jp

Miguel A. Cruz-Pérez works for the Comarca Minera UNESCO Global Geopark and does research in geology and geochemistry applied to mineral systems. He also works with geoheritage perspectives and strategies for this geopark at the Institute of Geophysics, Universidad Nacional Autónoma de México (UNAM); Member of the Scientific Committee of the Comarca Minera UNESCO Global Geopark, México. E-mail: macruz@geofisica.unam.mx

Ross Dowling is an Honorary Professor of Tourism in the School of Business and Law, Edith Cowan University, Australia. He has degrees in geology, geography, and environmental science. Professor Dowling conducts international research in the fields of geotourism, ecotourism, and cruise ship tourism and has over 200 publications in these fields, including 16 books. In 2011, he was awarded the Medal of the Order of Australia for his contributions to tourism education and development. He has a deep interest in the development of geotourism and geoparks. He has coedited three books on geotourism and convened the first three global conferences on the subject. E-mail: r.dowling@ecu.edu.au

Krzysztof Gaidzik is an associate professor in the Department of Fundamental Geology, Faculty of Earth Sciences, University of Silesia, Poland. His research focuses mainly on structural geology, active tectonics, and tectonic geomorphology, as well as geodiversity, geoheritage protection, and geotourism. E-mail: k.gaidzik@gmail.com

Kamarulzaman Abdul Ghani has been a retired officer of the Malaysian Government officer for over 30 years, having served in various key agencies such as the Economic Planning Unit of the Prime Ministers Department and the Ministry of Finance. He earned his Bachelor of Arts degree in Economics from the National University of Malaysia (UKM) in 1977 and a Master's degree in Development Economics from Vanderbilt University, Nashville, USA, in 1985. He was bestowed a honorary doctorate in 2008 from UKM in recognition of his commitment in the area of sustainable development during his tenure as the General Manager of Langkawi, which brought about the establishment of Langkawi UNESCO Global Geopark. Upon retirement, he was enrolled as a visiting scholar in the Institute for Evironment and Development (LESTARI) UKM. He is now the Founder and President of FLAG—Friends of Langkawi Geopark, an NGO dedicated to popularizing and supporting the Geopark agenda. He is currently a member of the Geopark Implementation Committee of Malaysia. E-mail: kzamang@gmail.com

César Gosso is a geologist and an associate professor of the Institute of Geological Sciences at the Universidad de la República, Uruguay. His research interests include paleontology, geology and mineralogy, and geological heritage. He is the Scientific Coordinator of the Grutasdel Palacio UNESCO Global Geopark, Uruguay. E-mail: cesar.goso@gmail.com

Murray Gray is the Emeritus Reader in the School of Geography, Queen Mary University of London, England, and is a visiting professor at the University of Minho, Portugal. Trained as a glacial geomorphologist, he is the author of the book *Geodiversity*, the first globally scoped book about the importance of the geological diversity of our planet. His current research interests include the principles of geodiversity and geoconservation and their relationship with geotourism. E-mail: j.m.gray@qmul.ac.uk

Thomas A. Hose is an Honorary Research Associate in the School of Earth Sciences, University of Bristol, United Kingdom and visiting professor in the Department of Geography, Tourism and Hotel Management, University of Novi Sad, Serbia. He is an honors graduate (London University) in geology and geography and has a Postgraduate Certificate in Education (Liverpool University), together with a Master's (City University) in museum and gallery administration. He was formerly a museum professional (holding natural science and education roles), school teacher (in geography and science, finally as a head of geology), and latterly a university principal lecturer (in countryside management, heritage management, and tourism). His professional and research interests in landscape history, geoconservation, and environmental interpretation underpinned his ground-breaking doctoral thesis (Birmingham University) on geotourism, a paradigm first defined and pioneered by Tom in the early 1990s. He convened the 2012 Geological Society of London conference "Appreciating Physical Landscapes: Geotourism 1670–1970" and edited the subsequent conference volume; he had previously authored geotourism chapters for four previous Geological Society volumes and another five books, together with some 50 journal papers and reports. He edited and was the major contributor to *Geoheritage and Geotourism: A European Perspective*, published in 2016, the first volume to summarize Europe's historical engagement in geoheritage protection, management, and promotion. His current geotourism research foci are geological literature and cyclists' geotrails. E-mail: gltah@bristol.ac.uk

Diego Irazábal is the Director General of Development, Tourism and Environment of the Municipality of Flores, Uruguay. He has a degree in political sciences from the Universidad Católicadel Uruguay "Dámaso Antonio Larranaga." He is the Coordinator of the Grutasdel Palacio UNESCO Global Geopark, Uruguay. E-mail: diegoirazabal@gmail.com

Francisco do O'de Lima, Júnior, is an associate professor at the Economics Department of URCA, Brazil. He holds a Master's in economics from the Federal University of Uberlândia (UFU), and a PhD in Economic Development from the State University of Campinas (UNICAMP). He is a leader of the Research Group on Economic Territories and Regional and Urban Development. His research interest is economic development. E-mail: francisco.lima@urca.br

Antonio Liccardo is a professor at the State University of Ponta Grossa, Brazil. As a geologist, he works with general geology, gemstone mineralogy, mining heritage, geoheritage, geodiversity, and geotourism. His research into these new areas of geology emphasizes geoscience divulgation and informal education, with interface actions with different knowledge. Dr. Liccardo is the author of 16 books, including *Geotourism in Curitiba, Mining History* *of Paraná, Geodiversity in Education*, and *Gemstones of Minas Gerais*. His current research focuses on wide access of geological information by community people (informal education) and cultural landscapes relating to mining heritage. E-mail: aliccardo@uepg.br

Eva Almeida Lima is responsible for the Geo-conservation and Environmental Planning of Azores UNESCO Global Geopark, Portugal. She is also the Director of the TURGEO project "Definition of carrying capacity for touristic use of geosites: a tool for the sustainability and tourism valuing of the natural resources of the Azores." E-mail: evalima@azoresgeopark.com

Marisa Machado graduated as a Nature Guide from the Azores University, Portugal. She is currently a collaborator of the TURGEO project "Definition of carrying capacity for touristic use of geosites: a tool for the sustainability and tourism valuing of the natural resources of the Azores." She is also responsible for the Tourism, Communication, and Marketing sector of the Azores UNESCO Global Geopark. E-mail: marisamachado@azoresgeopark.com

Virginio Mantesso-Neto holds a degree in geology (University of São Paulo, 1968) and a degree in history (University of São Paulo, 1994). He has the title of Tourist Guide (SENAC/Ministry of Tourism of Brazil, 2015). Since 2002, he has been studying the history of geosciences in Brazil, including geodiversity, geoheritage, and geotourism and promoting these themes to his more conventional colleagues and to the general public. He has coauthored the first Brazilian book about these subjects, *Geodiversidade, Geoconservação e Geoturismo: Trinômioimportantepara a proteção do patrimôniogeológico*, and organized Festschrift books about two masters of Brazilian geosciences: *Geologia do Continente Sul-Americano—Evolução da Obra de Fernando Flávio Marques de Almeida*, and *A Obra de Aziz NacibAb'Sáber*. E-mail: virginio@imigracaoitaliana.com.br

José Patrício Pereira Melo graduated in law from the Regional University of Cariri, Brazil, and holds Master is in constitutional law from the Federal University of Ceará and PhD in Economic and Socioambiental Law of the Graduate Program in Law at Puc Curitiba. He is currently Rector of the Regional University of Cariri and UNESCO Adviser for the UNESCO World Geoparks Program. E-mail: patricio.melo@urca.br

Juan Carlos Mora-Chaparro holds a PhD in geology from the Universidad Nacional Autónoma de México (UNAM). He is a researcher in the Institute of Geophysics and member of the Scientific Committee of the Comarca Minera UNESCO Global Geopark, México. His research interests include geological hazards and vulnerability. E-mail: jcmora@geofisica.unam.mx

Marcos Antonio Leite do Nascimento is a professor at the Federal University of Rio Grande do Norte, in Brazil. As a geologist, he works with igneous petrology, field geology, geodiversity, geoheritage, geotourism, and geoparks. He is the author of several national and international scientific articles; he published the first Brazilian book dedicated to the subject, entitled *Geodiversity, Geoconservation and Geotourism: important trinomial for the protection of the geological heritage*, beyond *Geodiversity in the Rock Art in the Potiguar Seridó and Geodiversity of Natal City/RN*. E-mail: marcos@geologia.ufrn.br

Gülpinar Akbulut Özpay is a geographer and an associate professor at the Sivas Cumhuriyet University in Turkey. Her fields of interest include geoheritage, geopark, and geotourism. She is a scientific coordinator of the proposed Upper Kızılırmak Geopark, Sivas geoheritage inventory map, Sivas geotourism potential areas, and Upper Kızılırmak Culture and Natural Route projects in Turkey. She is an academic advisor of the geopark student community, which was established to educate people including school students with the aim of raising awareness on geoheritage protection and the history of Earth. E-mail: gakbulut58@gmail.com

José Luis Palacio Prieto is a senior researcher and lecturer in the Institute of Geography, Universidad Nacional Autónoma de México (UNAM); he has a PhD in geography from UNAM and completed courses on watershed management, remote sensing, and geographic information systems in the International Institute for Aerospace Survey and Earth Sciences in the Netherlands. He is also the Scientific Coordinator of the Mixteca Alta UNESCO Global Geopark, México. E-mail: palacio@unam.mx

Emmanuel Reynard is a professor of physical geography at the University of Lausanne (since September 2005). He has been the Director of the Institute of Geography (IGUL) from September 2008 to July 2012 and of the Institute of Geography and Sustainability (IGD) from August 2012 to December 2016. He has also been chairman of the Working Group on Geomorphosites of the International Association of Geomorphologists

(IAG) from 2001 to 2013. Since 2013, he has been a member of the Executive Committee of the IAG, as Publication Officer. Emmanuel Reynard is also president of the Working Group on Geotopes (Scnat) and of the Association "Mémoires du Rhône." E-mail: emmanuel.reynard@unil.ch

John Rimel is the publisher at Mountain Press Publishing in Missoula, Montana, an independent press that has achieved national recognition for its books for both young and old on natural history and history. Mountain Press is perhaps best known as the publisher of the popular *Roadside Geology and Geology Underfoot* series. He served as the president of the Rocky Mountain Book Publishers Association and as a board member of the Publisher's Association of the West. A native Montanan, John is active locally, serving as a trustee of the Historical Museum at Fort Missoula, and formerly on the Missoula County Open Lands Committee. E-mail: jrimel@mtnpress.com

Erika Salgado-Martínez is a geologist from the Universidad Autónoma de Guerreo, Mexico, and works in the Institute of Geophysics, Universidad Nacional Autónoma de México. She is a member of the Scientific Committee and Coordinator of Dissemination and Communication of the Comarca Minera UNESCO Global Geopark, México. E-mail: eri_10_50@hotmail.com

Afat Serjani was educated (1957–1961) at the Mining Institute of Saint Petersburg (Leningrad of the former Soviet Union). He has been a researcher on prospecting and publications on chromite, phosphorite ores, and also industrial minerals. In 1986, he completed his doctoral thesis on phosphorites in the Ionian Zone, working full time in the Geological Research Institute in Tirana, Albania. In 1994, the Albanian Higher Commission granted him the title "Leader of Researches" (professor). He has more than 150 scientific publications, including books, papers, and presentations. Also, he has been involved in international scientific activities such as UNESCO IGCP Projects. For the first time in Albania, he initiated studies on the geological heritage, and since 1995, he has been the national representative of ProGEO (The European Association for the Conservation of the Geological Heritage) in Albania. E-mail: afatserjani@gmail.com

Kazem Vafadari is the Director of the International Center for Asia Pacific Tourism (iCAPt); the academic director of Kunisaki City Research Center for World Agriculture Heritage, and Division Head, Graduate School of Ritsumeikan Asia Pacific University Tourism and Hospitality Program (THP) Beppu, Japan. He has worked with the United Nations University and Kanazawa University in Japan before joining APU in 2011. He is an expert on tourism and the natural resource management and health-oriented tourism applications of agricultural heritage landscapes. He worked with the United Nations Food and Agriculture Organization as an advisor and scientific committee member for the Globally Important Agriculture Heritage Systems (GIAHS) program. His research interests also include the development of health and wellness destinations through promoting edible and medicinal plant food tourism in agriculture heritage landscapes. E-mail: kazemv@gmail.com, kazem@apu.ac.jp

William (Bill) Witherspoon is the co-author of *Roadside Geology of Georgia*. His PhD research at the University of Tennessee concerned the geological structure of the Great Smoky Mountains National Park. From 1997 until his retirement in 2014, he was an instructor at Fernbank Science Center, guiding thousands of students and their teachers at local geology sites. He served as president of the Georgia Geological Society and was named an Outstanding Earth Science Teacher by the National Association of Geoscience Teachers. Since retirement, Bill has presented walks, talks, and workshops in parks, libraries, and museums throughout Georgia. He offers resources for geotourism exploration at georgiarocks.us. E-mail: bill@georgiarocks.us

Jerzy Żaba is the director of the Department of Fundamental Geology at the University of Silesia. His research interests are focused on the structural evolution of the Bohemian Massif, tectonic processes on the boundary between Małopolska and Upper Silesian Blocks, tectonics of the crystalline basement of Tuareg Shield, volcanic processes in the Great Rift Valley in Africa, and in Kamchatka, as well as the tectonic and volcanic activity of the South American subduction zone. E-mail: jzaba@interia.pl., jerzy.zaba@us.edu.pl

Preface

The first comprehensive English geotourism book was published by Ross Dowling and David Newsome, titled *Geotourism* (Elsevier 2006). In different countries, there are atlases and books integrating geology and tourism before and after 2006 in various languages and at national scale or as textbooks. However, despite the fact that geotourism and geoparks are developing strikingly, there are very few international comprehensive books in English that can assist in the subject combination and promotion.

As yet there have not been many geotourism studies accomplished, and there are many issues, until now unarticulated, that should be shared and discussed among researchers and practitioners.

The present book is a continuity of publications and studies in the 21st century by international activists, which aims to provide up-to-the-minute illustration of the progress made.

This book illustrates the direct association between abiotic nature attributes and the tourism industry. Geotourism attractions and relevant abiotic nature attributes involve scales from microscopic to the global and astronomical and from hours to eons, such as a story behind a grain of sand to narrate to public people in an edutainment manner. We should remember here that the pioneer 19th-century evolutionist Thomas Henry Huxley introduced the people to the story of "A piece of Chalk." Every geo-object and geoheritage site has its own wonderful story and a sense of wonder to develop in individuals. Michel Walland in his book, *Sand: The Never Ending Story* (published by University of California Press, 2010), utters that just in a grain of sand, there are worlds to see! Everybody has been always enchanted by the earth's geological wonders; geotourism is a considerably soaring subject, and the geotourism industry helps people to see geoworlds.

This editor is passionate about geotourism and devotes himself to this science. I believe that the geotourism and geoparks development potential in order to alleviate poverty has not yet been recognized correctly in poor and far-off places among managers and policymakers. This subsegment of sustainable tourism has not yet been addressed correctly in countries where their biological diversity is limited and which have been accounted as part of the dry or semidry areas that comprise the less-developed countries of the world. On the other hand, in the 21st century, as the destinations are expanding vertically and horizontally (in both developed and developing countries), environmental protection and conservation needs to be advancing vertically and horizontally and the need for recognition of abiotic nature in tourism sustainable development and its influence on life and growth on Earth is a must. Geotourism scientific–experimental progress facilitates this issue.

This book has been compiled for a broad audience including tourism professionals, planners and managers, government and business decision makers, and for students of different courses intent upon geotourism development. Compiled with the introduction of 23 chapters altogether contributed by 35 authors from 18 countries include: Albania, Australia, Brazil, England, France, Iran, Japan, Jordan, Malaysia, Mexico, Peru, Poland, Portuguese, Romania, Switzerland, Turkey, Uruguay, and USA, across four continents; the book has been published as a reference book in geotourism subject. I hope this book may assist in geotourism experimental development internationally and that all readers will be delighted.

Finally, it is the custom of an author to indicate the place of writing. Indeed this book is being written in my hometown, Tabriz.

—**Bahram Nekouie Sadry**
Tabriz, Iran
November 2019

Foreword 1

It is less than 25 years since the first formal definition of geologically/geomorphologically based modern geotourism was published. Geotourism provision essentially gives travelers (or geotourists) the opportunity to acquire knowledge and understanding of a destination's Earth history, geological elements, and landscapes. Indeed, across most of the world, geotourism has long been considered a geologically focused form of tourism. Perhaps, surprisingly, its antecedents can be traced back to at least the 17th century. By promoting and imparting something of the wonder of geoscientific inquiry and its outputs, geotourism can engender amongst its participants sufficient empathy to motivate their support for the protection and conservation of geodiversity and geoheritage, that is, "geoconservation." Both modern geotourism and geoconservation, alongside practitioner reportage, have from the early 1990s attracted academic study and consequently evolving managerial and theoretical underpinnings.

The seminal 1995 geotourism definition, published in *Environmental Interpretation* magazine, following further research and reflection, was subsequently revised several times by its author. That author has freely acknowledged that others—particularly in Australia, China, and Europe—had either mentioned "tourism geology" or something similar but had generally not indicated any specific meaning of their terms; their mentions with the associated studies really helped to lay the groundwork for modern geotourism's widespread acceptance as a new paradigm.

The first dedicated national geotourism conference, *Tourism in Geological Landscapes*, was held in Belfast at the Ulster Museum in 1998. *The Inaugural Global Geotourism Conference*, much practitioner focused, was held only a decade ago in Fremantle, Australia. The first international conference on the history of geotourism, *The Appreciating Physical Landscapes: Geotourism 1670–1970* conference, was held in London at the Geological Society as recently as 2012. Meanwhile, *The First International Conference on Geoparks* was held in Beijing in 2004.

The emergence of modern geotourism and the provision of geosites and geomorphosites interpreted for tourists both predate, by at least a couple of decades, the designation of the first geoparks; the latter can be standalone ensembles of geosites and geomorphosites or part of some national or international designation. The first geoparks were established in Europe in 2000. Significantly, UNESCO's original Geoparks Programme Feasibility Study report of that year included the seminal geotourism definition and its major concepts. Formal designation as a member of Global UNESCO Network of Geoparks network has been available since 2004. Although geoparks were initially a European development, it is in Asia that they have particularly expanded in numbers and popularity. For example, the People's Republic of China has more than a quarter (31 of 111) and the region had over a third (39 of 111) of the UNESCO Global Geoparks designated by the mid-2010s; additionally, that country had by then designated 185, and recognized another potential 56, National Geoparks. Geoparks have done much, and more significantly than any other single initiative, to promote and develop geology-based tourism. Indeed, to maintain their UNESCO geopark membership, they must offer interpretative services. Probably the greatest contribution of geoparks is their requirement to engage with the broader, especially local and business, than just the Earth science communities. They are one of the success stories of 21st-century sustainable tourism, something for which their proponents are to be congratulated.

Modern geotourism provision meets geotourists' needs by encouraging them to visit localities with spectacular or readily appreciated, and usually (on-site and/or off-site) interpreted, geological/geomorphological features. These features are often more readily, at least in the marked seasonal climates of Europe and North America, seen outside of the major vegetation growth period; hence, potentially it can extend the tourism season in some coastal and upland areas. Of course, the appreciation of physical landscapes and the extraction of their mineral resources has been a pragmatic human activity, long before the recognition and practice of geotourism, especially for the purposes of agriculture, construction, and metallurgy; evidence for this can be found in both the archaeological and historical records.

The breadth of geotourism's encompass is clear from the preceding paragraphs. They suggest that any attempt to summarize this breadth of geoscience and its tourism component in a single book is a major and challenging undertaking. This book's 23 chapters are spread across five sections covering geotourism's concepts, assessment, interpretative provision, geoparks, and its global future. They have been contributed by an international assemblage of 35 authors, each contributing from their own perspectives and experiences. The various authors explore the spectrum of modern geotourism provision, practice, and development. The range of topics covered range from urban tourism to the world's best geosites and from mining geoheritage to geotrails and interpretative writing. The included case studies, geographically spread from Albania to the Azores and the Americas to Japan, indicate the variety and wide distribution, geographically and by type, of modern geotourism provision. The diversity of views expressed in these chapters, helpfully summarized by the Editor, challenges readers to engage in studies to further understanding and disseminating geotourism beyond its current strongholds.

This book is a timely contribution to studies on the status and practice of modern geotourism. Its publication would not have been realized without the personal vision, wide connections, organizational skills and perseverance of the Editor, Bahram Nekouie Sadry. Fellow students of geotourism and the wider readership of geography and tourism specialists owe him, for this sterling effort, a considerable debt of gratitude—one which I am most happy to wholeheartedly express. As Bahram suggests in the book's opening sentence "Geotourism is an emerging and promising field for enjoyable and meaningful experiences in contemporary tourism." Similarly, this book is an enjoyable read and will add understanding and meaning to its readers' own geotourism experiences!

—Thomas A. Hose, PhD
Honorary Research Associate
School of Earth Sciences
University of Bristol
United Kingdom
&
Visiting Professor
Faculty of Science
University of Novi Sad
Serbia

March 2019

Foreword 2

This is an important book because it brings together the interpretive and management aspects of geotourism on a global scale. It reveals geotourism's potential to popularize the fundamentals of geoscience and open up new pathways of sustainable economic development. Through his own chapters and those of the authors he has invited, Dr. Sadry has conveyed his global knowledge of the geopark movement and the art of geoscience interpretation.

For me as an American geologist, geoscience educator, *Roadside Geology* coauthor, and chapter author in this book, it is my first introduction to the terms geopark, geosite, or geomorphosite. This may be an example of the cross-fertilization a work of this scope can achieve, for example, inspiring its US readers to advocate for our country joining UNESCO's geopark system.

People know some aspects of nature in their bones. Scanning the night sky, they know that the Moon is Earth's neighbor and the stars are far, far away. In contrast, awareness is rare that shapes of hills area snapshot result of continuing Earth processes, while the bedrock is a legible record of many moments in the unimaginably distant past. This book offers the hope that in a well-interpreted geopark, visitors will get "deep time" and Earth processes into their bones, the way we geologists experience them.

I first heard the word geotourism in the late 1970s from my PhD advisor, Professor Dietrich Roeder. He is credited with, in 1969, introducing the word "subduction" from the Alpine literature into its present plate-tectonic sense. Though German was his native language, you only knew that because his English was "too good." He was playful in his use of it. Geotourism for him described the pleasure of blending geologic insights along the roadside with awareness of the geology's connections to local nature and culture.

The geotourism concept inspired me in 2001 to take some Georgia teachers to the Grand Canyon. I wanted my group to internalize the canyon's lessons on Earth processes and deep time. Was it possible to build the story up piece by piece on the drive from Phoenix Airport to our destination? Though I did not know the word "geosites" until I read this book, such sites were my answer. In *Roadside Geology of Arizona* by Halka Chronic, I learned that nearly the whole sequence of layers seen in the Grand Canyon is exposed in road cuts heading north from Payson, a town about 90 miles northeast of Phoenix. Four miles from Payson, in a county park and nearby road cut, you can lay a hand on the Great Unconformity, with your thumb on a 1.6-billion-year-old rock that represents the roots of a mountain chain planed flat by erosion. Your fingers touch sandstone layers from the Cambrian Period, when life with shells first appeared, more than a billion years later. A perfect geosite to create a memorable experience, but with neither sign nor park brochure to mention it.

In the 21st century, the need for bone-deep awareness of Earth processes is greater every day. Without it, people are prone to imagining that scientists are only guessing when they say that burning coal is raising global temperatures, or that mostly gradual changes to landscapes and climate enabled the evolution of whales, tree frogs, and people. Admittedly, much progress could be made if politicians would finally yield to expert advice that geoscience in school needs equal emphasis with chemistry, physics, and biology. But this book offers an additional vision. Imagine putting the "Great Unconformity" in a starring role in a geopark promoted as a tourist attraction. Imagine, too, the local economic benefits of opening this new front of popular attention.

In 2012, I wanted to help teachers and others internalize plate boundary processes in a trip to California to be called "Geology on the Edge." I thought to promote it using Professor Roeder's word "geotourism." Having the internet by then, I searched for the term. To my surprise, Wikipedia informed me that a team with the blessing of the *National Geographic Society* had lassoed it as a synonym for sustainable tourism, with only incidental Earth science significance. Fortunately, I persisted to find a blog by Dr. Sadry that identified geotourism as a parallel concept to ecotourism, distinguished as "abiotic nature-based tourism."

The phrase "abiotic nature" seems strange at first. From the outer edges of Earth's atmosphere to the deepest mines and drill-holes, it is now clear that some form of life is ever present. Even the first tourist in space, Dennis Tito, whom Dr. Sadry mentions in his intriguing "Space and Celestial Geotourism" chapter, brought biota with him. But ecologists use "abiotic" abstractly for the nonliving components of the larger Earth system. Without them, life could not exist. If we fail to preserve Earth's favorable abiotic features, its special atmosphere, hydrosphere, and geosphere, we fail to preserve ourselves.

Having found his blog, I emailed Dr. Sadry to ask how I could help support his meaning of geotourism. One eventual answer was his invitation to write a chapter (which became two) for this book. Being further asked to write this Foreword gave me the opportunity to review his entire labor of love, this book, filling in numerous gaps in my knowledge of geotourism.

In these chapters, you will read about successes and challenges in several countries with setting aside localities for geotourism. Whether carved out

Foreword 2 xxxi

by glaciers and rivers, or miners' dynamite and pickaxes, rock outcroppings attract visitors, and tourism can help local economies.

Reading on, you will learn of the tradeoffs between signage, leaflets, and digital media in conveying geoscience concepts. You will experience the tension between the often-lucrative enterprise of selling access to nature's joyrides, and the challenging and sometimes costly work of imparting wisdom in the ways of the planet. And you will see that like its sister activity, ecotourism, geotourism includes both keeping local people invested as volunteers and beneficiaries, and managing resources, so that we visitors do not love a place to death.

Regarding education by firsthand experience, in the 19th century, American geologist James McFarlane summed up geotourism's potential (his italics):

> to teach persons not versed in geology ... not as in a textbook, but by pointing to the things themselves ... *There are some kinds of knowledge too that cannot be obtained from books, but must be gathered by actual observation.*

This book will help geopark managers, geoscience educators, and all researchers and students of geotourism to use those "things themselves" to instill bone-deep awareness of how the planet works. It will also help them use geotourism to create economic value for alleviating poverty while protecting resources for future generations. Thank you, Dr. Sadry, for having the knowledge, vision, and energy to bring it all together.

—**William (Bill) Witherspoon, PhD, PG**
Retired Geologist Leading Walks and Talks
Co-author, Roadside Geology of Georgia
March 2019

Acknowledgments

First, I would like to thank the contributors, all thirty-four of them. They are from various countries all over the world and are all professionals in their fields. I am grateful to them for entrusting their work to me in the following 23 chapters, as well as being patient as the project moved through its various stages. The contributors are to be thanked for their willing participation in this project and for bestowing on me the joy of working with such an outstanding cadre of authors. While some of the contributors are young researchers, others are august professors, global geotourism pioneers, and/or well-known writers. Though some of them are from private business and administration firms, most are academics, and some have become friends while writing this book. I would like to express my deep thanks to them all for addressing my requests. This is actually your book.

Secondly, I would like to thank the publishing staff, the staff of Apple Academic Press Inc. USA, who worked with me. It was impossible for this book to be accomplished without Sandra Jones Sickels assistance (VP, Editorial and Marketing of Apple Academic Press Inc.); she reviewed the book proposal after it had been declined. I wish specially to thank Ashish Kumar, AAP publisher, as well as Rakesh Kumar (pre-production Manager for AAP) and the whole managerial and production teams. Thank you also to my colleague Dr. Neda T. Farsani for her effective guidance during the publishing contract phase at the outset of the project.

I particularly wish to thank my friend, Noah A. Razmara, for his support and encouragement. It is highly appreciated. I also wish to thank my Polish friend and colleague, Dr. Krzysztof Gaidzik, who constantly applied the important pressure of asking how this project was going even before a publishing contract. His following persuaded me to make efforts and discuss with different publishers in order to sign a contract in the most burdensome period of my private life.

I also am grateful for initial acceptance of chapter writing by Dr. Malcolm Cooper, a great professor of APU in Japan, who was the first scholar to welcome my suggestion for contribution (before a publishing contract). Meeting at a particular governmental session of geotourism in Tehran, he undoubtedly asked for the deadline of my geotourism book project. His trust in me and his enthusiasm for the work, encouraged me to continue my efforts toward accomplishing this work.

I also wish to thank my students who have participated in my postgraduate geotourism and ecotourism classes. Their research tasks were based on topics where I have learned a lot about the industry. Also, initial thanks go to Professor Ross Dowling and Dr. Patricia Erfort who were my inspiration to compile such a work. Also, I would like to thank Dr. Susan Turner, Dr. Bill Witherspoon, and Mr. John Macadam for precious help to improve the work. In addition, I want to thank my two geo-colleagues, Mr. John Pint for putting me in contact with Latin American colleagues, and Mr. John Macadam for introducing me to Malaysian and Romanian geopark managers, Dr. Kamarulzaman Abdul Ghani, Founder/President FLAG-Friends of Langkawi Geopark and Dr. Cristian Ciobanu at the Hațeg Country UNESCO Global Geopark.

I also express my wholehearted gratitude for assistance of Mrs. Fatemeh Fehrest, in a certain period of time to help meeting the deadline of submitting the whole manuscript to the publisher and to improve the work. I am particularly appreciative of the kindness of two dear friends and colleagues, Thomas A. Hose, mentioned as the "father" of geotourism in this book, who first defined the word geotourism in the modern era, and William (Bill) Witherspoon, coauthor of *Roadside Geology of Georgia,* in agreeing to write the Forewords for this book.

Finally, I wish to thank my parents for their support throughout the work on this book in the last year during a complex time; I could not have achieved this without them. I also wish to acknowledge my daughter *Newsha*, who has always given me encouragement. This book is partly for you all.

PART I
Geotourism Concepts in the 21st Century

CHAPTER 1

The Scope and Nature of Geotourism in the 21ˢᵗ Century

BAHRAM NEKOUIE SADRY

Adjunct Senior Lecturer and Geotourism Consultant

E-mail: Bahram.Sadry@gmail.com

ABSTRACT

A brief introduction to the history of geological tourism and the definition of modern geotourism are given. The boundaries as well as the theoretical and experimental definitions of geotourism are discussed in this chapter. The theoretical framework about geotourism principles and its development are proposed. The structure and contents of the new book are reviewed.

1.1 INTRODUCTION

Geotourism is an emerging and promising field for enjoyable and meaningful experiences in contemporary tourism. Geological tourism has a history of over a hundred years in some countries, including England, Australia, Iceland, and so on. To illustrate, geology-based tourism in England is said to be rooted in the last years of the 17th century (Hose, 2008), and many visitors to Iceland during the 18th and 19th century mentioned the geological features as their primary reason to visit the country (Ólafsdóttir and Tverijonaite, 2018). So, the emergence of geological tourism in other countries was probably around the same time (e.g., Gray, 2004; Turner, 2013). The co-author of *Roadside Geology of Georgia*, William Witherspoon (2012) has uttered that "[in the USA], I have been hearing the term geotourism since the 1970's, in the sense of geological tourism" (W. Witherspoon, pers. comm., July 16, 2012), but modern "geotourism" only appeared for the first time discussed by Hose (1994) in the first focus of university research and the academic literature in 1994, and was first then defined by Hose (1995), gaining credibility with its acceptance by UNESCO's Earth Sciences division in 2000 and with the 2009 launch of the "peer-reviewed Journal of Geoheritage" (Hose, 2012).

The *Geoheritage journal* is an international journal exploring all aspects of our global geoheritage, both in situ and portable. According to Coratza et al. (2018 and all references therein) two terms, which are being utilized more and more regularly have been made known in scientific terminology: "Geoheritage" and "Geodiversity"; More recently, Geotourism as tourism related to geoheritage and geodiversity and all in all, tourism related to abiotic nature appeared (e.g., Hose, 1995, 2016; Gray, 2004; Sadry, 2009; Reynard et al., 2009; Newsome and Dowling, 2010; Farsani et al., 2012; Reynard and Brilha, 2018; Dowling and Newsome, 2006, 2010, 2018). Therefore, over the past 15 years, geotourism has developed from an unknown niche trend to an approach in tourism planning and for abiotic nature conservation, especially with highlighting the UNESCO sponsored global geopark movement in the 21st century. Geotourism is somehow selling the story and beauty of a country's rocks, which is noticeable in quite extensive subsectors, such as "Rural geotourism" (Farsani et al., 2013), Urban geotourism (Hose, 2006; Sadry, 2009; Riganti and Johnston, 2018; Del Lama, 2018; Gaidzik (Chapter3)), Celestial geotourism (Chapter 20 in this book) and Space geotourism (Chapter 20 in this book), Roadside geotourism (e.g., Sadry, 2009; Strba et al., 2016; Witherspoon and Rimel (Chapter 13 in this book)), Community-based Geotourism (Turner, 2005; Mukwada and Sekhele, 2015), Underground and Cave geotourism (Garofano, 2018), Submarine geotourism (e.g., Lima et al., 2014), Dinosaur Geotourism (Turner, 2004; Cayla (Chapter 18)) even Meteorite geotourism (Chapter 20); Mining geotourism: one particular type of tourism related to mining heritage and mine sites (e.g., Hose, 2006; Sadry, 2009; Garofano and Govoni, 2012; Mata-Perelló et al., 2018; Gaidzik and Chmielewska (Chapter 21)); Adventure geotourism (e.g., Sadry, 2009; Farsani et al., 2013; Newsome and Dowling, 2018: 477), Volcanic Geotourism (e.g., Erfurt-Cooper and Cooper, 2009; Woo et al., 2010; Hose, 2010; Dowling, 2010; Cooper and Eades, 2010; Turner, 2013; Erfurt, 2018) and, according to Erfurt-Cooper and Cooper (2009), connected geotourism with volcanic geotourism phenomena which consists of thermal springs (the springs of natural hot water, which are extensively used in health tourism, e.g., Turner, 2005; incorporating such hot springs to regular events held annually is also quite common in some countries (e.g., Japan and Indonesia). "Many resorts benefit from these spas for health and wellbeing reasons, which is especially common in Iceland, Japan and New Zealand, where there are numerous active volcanoes offering an amazing scenery for visitors experiencing the spas. In this respect, the preservation, sustainability and education of various geosites is particularly based on volcanic heritage" (Erfurt-Cooper and Cooper, 2009, 2014). Therefore, in the 21st century, geotourism subsectors are unique and broad enough so that they deserve direct management, planning and marketing by specialized institutions as well as purposeful research. It is sometimes convenient to utilize one of these products which are highly specialized, at a larger geotourism or tourism scale; however, it would depend on the aims of the related companies. To accomplish this, the perfect essence or three key elements of geotourism (Sadry, 2013) are:

1. Abiotic nature as the main attraction;
2. Geological heritage interpretation[1]; and
3. Positive outcomes for nature (and locals).

Also, in a parallel manner, according to Weaver (2011) the core ideals of ecotourism are: meaningful participant learning and maximization of positive ecological and sociocultural impacts.

Hence, the core ideals of geotourism can also be developed in a similar manner comprising: meaningful participant learning and maximization of positive geological and sociocultural impacts. The present chapter intends to enhance the understanding of the geotourism industry as an outstanding subsector of the wider tourism industry, which truly deserves special attention from academics and related organizations. Meanwhile, geotourism activities should control tourism's rising interactions with abiotic nature and especially geological heritage.

The first section below defines the term "geotourism" and proposes a theoretical frame for geotourism principles. Following this, a review of the structure and contents of the book is discussed.

1.2   TOWARD A DEFINITION OF GEOTOURISM

According to Newsome and Dowling (2010: 4) "a more specific definition of geotourism helps to develop a focused strategy, which is fundamental in accomplishing the objects of geotourism. It includes geodiversity conservation, visitor education, and empowering local communities by providing knowledge about their geological resources and employment opportunities." There are five elements of the definition of geotourism in more specific ways (see Fig. 1.1).

---

[1] Also, see Chapters 9 (based on the 15 principles in interpretation), 11 (ABC approaches in geointerpretation), and 23 of this book.

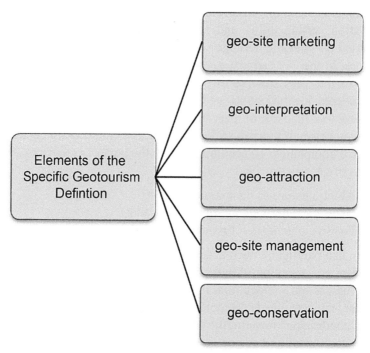

FIGURE 1.1 The five elements of the definition of geotourism in more specific ways.
Source: Materials adapted from Newsome and Dowling (2010: 4).

1.2.1 TRUE GEOTOURISM

"Modern tourism originated after the 1840s. It is just 60–70 years since geographers began to study modern tourism from a geographical perspective, and *Relationship between Recreational Activities and Land Use* by K.C. McMurry was generally recognized as the first work of geographers studying modern tourism" (Chen et al., 2015). Although it has long been recognized that tourism and geological background of our environment are inextricably interwoven, it is generally accepted that the first widely published definition of geotourism in the modern era was coined and began, hence, by the "father" of geotourism, Thomas A. Hose in 1995 as

The provision of interpretative and service facilities to enable tourists to acquire knowledge and understanding of the geology and geomorphology of a site (including its contribution to the development of the Earth sciences) beyond the level of mere esthetic appreciation (Hose, 2006: 221) (see Fig. 1.2).

Modern geotourism's aforementioned original definition along with some of its associated concepts was integrated within the UNESCO 2000 Geoparks Programme Feasibility Study (Patzak and Eder, 1998 cited by Hose, 2010: 268) that effected the movement of global geoparks in the 21st century. A UNESCO geopark is territory with well-defined limits that has a large enough surface area for it to serve local economic development. The geopark comprises a number of geological heritage sites of special scientific importance, rarity, or beauty; it may not be solely of geological significance but also of archaeological, ecological, historical, or cultural value (UNESCO, 2000).

Therefore, in a philosophical manner, geotourism with the suffix (-ism) is an umbrella for geoparks and "geoparks are obviously a major subsector of geotourism provision" (T.A. Hose, Pers. Comm., March 23, 2019) but from the managerial view point (exactly for practitioners) geotourism is just one of the tourism activities (but the main one) within geoparks among other "geopark tourism" activities (see Fig. 1.3).

1.2.1.1 THE BASIC PRINCIPLES OF GEOTOURISM

The basic principles of geotourism consist of tourism promotion, geological conservation measures, community benefits, and stakeholders' involvement. For a planning strategy to be effective, the aforementioned four components must be used correctly (see Fig. 1.2).

1.2.2 "TOURTELLOTIC GEOTOURISM": A TAG FOR DESTINATIONS

Witherspoon (2018) states that "the geologist I clearly remember using this word [geotourism] in the 1970s was my mentor, Dietrich Roeder, who coincidentally also had brought the term 'subduction' into the plate tectonics lexicon" (W. Witherspoon, pers. comm., Dec. 15, 2018). Also he previously mentioned that "the 1995 definition you cite [from Thomas A. Hose in 1995 who first defined the term geotourism in the globe] is older than the very different definition that the Wikipedia article cites [Jonathan Tourtellot's geographical sustainable tourism charter & concept] dating back to 2002. And taking over a useful term [Geotourism] purely as a brand name is a *shame*" (W. Witherspoon, pers. comm., July 16, 2012). Unfortunately, recent usage of the same coined word "geotourism" by Jonathan Tourtellot, senior editor of National Geographic Traveler is not a new subsector of tourism industry in the 21st century, "with no focus on geodiversity" (Neches and Erdeli, 2014); and according to Ólafsdóttir and Tverijonaite (2018), National Geographic and its so-called Geotourism Map Guides just seek to inform visitors concerning the more sustainable choices provided in each area, thereby helping to enhance the region's geographical character and contributing to the well-being of local people.

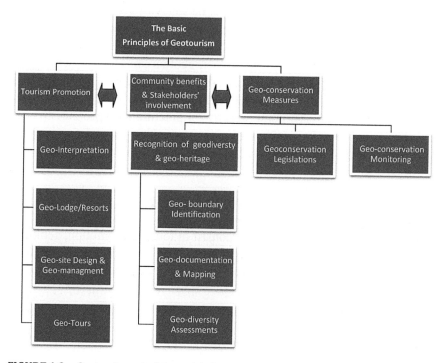

FIGURE 1.2 Geotourism principles and their promotion codes within geoparks and individual geosites (as geotourism sites).

Source: Own construction; some materials adapted from Gray, 2004; Dowling and Newsome, 2006, p 5; Hose, 2006; Sadry, 2012, 2013.

Fortunately it is sustainable tourism principles and a new comprehensive strategy for sustainable tourism for any destinations and a related "charter and concept promotes behaviors such as buying local products, improving employee well-being, and donating to local causes" (Jorgenson and Nickerson, 2016) and this is just a *broad* term that covers all *tourism*. Also it is an excellent strategical device to change mass tourism to sustainable forms in the 21st century. Indeed, "*National Geographic* has spread use of the word 'geotourism' as shorthand for 'geographical sustainable tourism,' which is outlined in a charter document and in more recent interviews, Jonathan Tourtellot increasingly uses 'destination stewardship,' which would be a far more appropriate tag for the charter." Destination stewardship is indeed of critical importance, when so many of the world's special places are being 'loved to death.' Use of "geotourism" to mean "destination stewardship" ought to be ended, because it causes confusion with a prior use of the term, as a category of nonliving nature-based tourism. Geotourism is a sister category to ecotourism, which informs the traveler about a destination's ecology of living systems. Geotourism teaches the traveler about a location's Earth history, rock types, and landscapes. By increasing the visitors' sense of wonder, both ecotourism and geotourism can motivate them to practice destination stewardship. This must not be swept aside by obscuring the term "geotourism" with a less appropriate use (W. Witherspoon, pers. comm., Dec. 15, 2018). We can't change the name of the tourism industry, which is internationally well-known, to the geotourism industry on the pretext of integrating a sustainability charter to the tourism industry. The tourism industry would entail its own name and a brilliant sustainable charter.

Tourism science would develop its independent subsegments for interpretation, attractions, marketing, management, and nature conservation. More recently, adding "geology" to the holistic concept and definition of the approach taken by National Geographic in the Arouca declaration have been reiterated to take this approach by José Luis Palacio Prieto and his colleagues (Chapter 17) in this book. As they noted that: "More recently, geotourism (as a modality of geographical sustainable tourism) was defined as 'tourism which sustains and enhances the identity of a territory, taking into consideration its geology, environment, culture, esthetics, heritage and the well-being of its residents' (Arouca, 2011). Undoubtedly, geological heritage is a key component of the geotourism offer of the geoparks referred to herein, with the relationship between geological heritage with other types of heritage—both natural and cultural—as major factors." And added that "This chapter does not provide an in-depth discussion of these two perspectives, although geotourism activities in Latin American geoparks are compatible with both."

According to Reynard (Chapter 5 in this book), "some authors (e.g., Hose, 2016: 6) consider that the Declaration has added confusion more than solved the problem," that emerged from National Geographic side (see Hose (Chapter 2 in this book, for more discussion, to avoid confusion and more probably to close the debate)).

1.2.3 CLARIFYING THE SITUATION

1.2.3.1 COMMON CONFUSION: ASSUMING GEOTOURISM AND GEOPARK AS EQUIVALENT

Following the correspondence with geopark executives, and also considering the previous educational and consulting experiences of the author, as is mentioned in Chapter 17 of the book, considering geoparks and geotourism as equivalents among geopark executives and even sometimes among researchers, is quite prevalent. This may cause confusion in written articles, book chapters, and in using different concepts. For instance, when geotourism and geopark are taken as synonyms, a holistic definition of geotourism covers more attractions. So, geoparks would be more successful and diverse. This is a common mistake by geopark executives intending to be more successful in managing a geopark. While geotourism is an independent scientific and research subsector in itself and by considering the fact that geoheritage and local community are essential in order for the geopark to be registered, the development of other forms of tourism in geoparks is not only permitted but also would add to the diversity of the geopark, the attractions, and the revenues. Another benefit of incorporating other forms of tourism into geoparks would be enhancing the attractiveness and economic success of such areas. Geotourism is a central part of geoparks but in addition to geotourism other types of sustainable tourism using other diversity and heritage of a region are applicable. In most geoparks around the world, other forms of tourism such as ecotourism (e.g., The Langkawi UNESCO Global Geopark in Malaysia; see Chapter 22), agricultural tourism (e.g., a "GIAHS/Geopark" site like Kunisaki Peninsula in Japan; see Chapter 14), wildlife tourism (e.g., whale watching in the Azores UNESCO Global Geopark in Portugal; see Chapter 7), cultural tourism, and so on alongside geotourism as foundation to a geopark, are promoted and related products are supplied for the visitors, which has added to the diversity of the geopark. Also, "geotourism is the vehicle for geoparks leading to the important UNESCO recognized global geopark criterion of sustainable development" (Newsome and Dowling, 2018: 478). As a result, geotourism is one of the tourism components of the UNESCO recognized global geopark criterion of sustainable development (see Fig. 1.3). Since geotourism has its own conceptualization and its own geoproducts, the definition of geotourism should be narrowed down and become more specific like abiotic nature-based tourism.

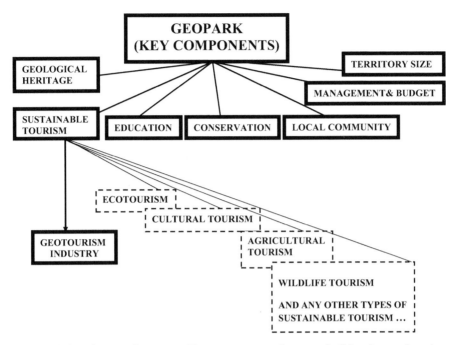

FIGURE 1.3 The core elements and key components of a geopark. Other forms of tourism within geoparks, as noncentral components, dashed.
Source: Materials adapted from UNESCO, 1999, 2000 and GGN Guidelines, 2010.

1.2.3.2 GEOTOURISM DEFINITION IN THE 21ST CENTURY

According to Ólafsdóttir and Tverijonaite (2018) researchers emphasize the importance of narrowing the definition of geotourism to geology and its conservation, pointing out that applying a broader approach to geotourism might reduce the impact of the concept. To clarify the situation, I propose the following definitions as conceptual and practical definitions:

    a.   A Conceptual Definition (Sadry, 2009: 17):
        It is knowledge-based tourism, an interdisciplinary integration of the tourism industry with conservation and interpretation of "abiotic nature" attributes, besides considering related cultural issues, within the geosites for the general public.

b. A Practical Definition (Newsome and Dowling, 2010: 4):
Geotourism is a form of natural area tourism that specifically focuses on geology and landscape. It promotes tourism to geosites and the conservation of geodiversity and an understanding of Earth sciences through appreciation and learning. This is achieved through in dependent visits to geological features, use of geotrails and view points, guided tours, geoactivities, and patronage of geosite visitor centers.

1.3 STRUCTURE AND CONTENTS OF THIS BOOK

This book comprises 23 chapters written by 35 authors from different countries around the world. Topics of different areas, including various samples and case studies, are covered so it enhances our understanding of geotourism as a distinct form of tourism, which is becoming a rising issue all over the world.

Chapter 2 (Thomas Hose) examines historical viewpoints on the geotourism concept. He fully discusses the history surrounding the definition of geotourism from Asian "definitions," Chinese thoughts on geology-related tourism, Australasian and European definitions to historical aspects of geointerpretation in detail. Hose's emphasis on the historical roots of geotourism includes geosites and geomorphosites and artistic outputs, and so on. He finishes by suggesting that recognition of geotourism, recent development, and history can and should be placed within a broader tourism history context.

In Chapter 3, Krzysztof Gaidzik outlines the opportunities and challenges for urban geotourism in the Polish small-sized towns to large cities like Kraków, Kielce, and Wrocław. He argues that the combination of urban tourism with geo-education are recognized by local governments as a great idea for promoting tourism in the town. He suggests that the geology of cities and town provides geo-educational elements for development of urban geotourism to visitors.

From Poland to Brazil, where Antonio Liccardo, Virginio Mantesso-Neto, and Marcos Antonio Leite do Nascimento (Chapter 4) discuss the relevance of Mining Heritage Attractions to promotion of geotourism. They present six cases chosen throughout Brazil that considered as cultural heritage. These sites are in the OuroPreto, Diamantina, Chapada Diamantina, Ametista do Sul, Pedro II, CurraisNovos as well as Itu, which allow geotourism for the general public to be developed.

Geomorphosites—Esthetic Landscape Features or Earth History Heritage? is the focus of Chapter 5 by Emmanuel Reynard. He describes the esthetic character of geomorphosites as heritage sites and argues that geotourism development depends on the geohistorical value of geomorphosites and explaining and interpretation of Earth's history and processes. The link between the esthetic dimension of geomorphosites and geotourism are illustrated with two examples: glacier tourism and geocultural sites.

In Chapter 6, Afat Serjani describes Albania's geoheritage and geotourism through the pioneering work of writing the first geotourism book in Albania. This work presented the main chapters of the first book of the series: "Geoheritage and Geotourism in Albania" for the selection of important Earth science sites, and the classification and categorization of geological sites for geotourism development.

Further promotion of geotourism in a fascinating island archipelago belonging to Portugal is outlined by Eva Almeida Lima and Marisa Machado in the Azores UNESCO Global Geopark (Chapter 7). The chapter highlights a number of planning and management efforts to promote the development of a sustainable tourism industry by integrating geotourism into its overall destination development strategy. They argue that the main challenge to this new form of tourism, geotourism, is to keep the balance between the enjoyment into the geolandscapes, geological heritage, and the geoconservation. Additional key points are outlined including the integration of the existing services and infrastructures with new interpretative services and products that allow the implementation of a high-quality geotourism in this archipelago.

The topic of the "Search for the World's Top Geotourism Destinations" is the topic of Chapter 8 by Murray Gray. The list presents sites that are both internally geodiverse and represents the world's geodiversity. Gray listed all the wonderful landscapes he has been to and considers as eye-catching and amazing. He proposed that the world's top 10 geotourism destinations are: Brazil/Argentina, Zimbabwe/Zambia, the United States, Australia, China/Vietnam, Switzerland, Bolivia, Iceland, and South Africa. However, these places are neither necessarily the best place on the Earth nor a perfect one for geological heritage interpretation, but the examples are a combination of outstanding natural beauty presenting geological features. Seemingly some of these spots are an ideal area for practicing ecotourism and probably perfect for establishing new geoparks simultaneously. He finishes by suggesting that these sites can be used to raise the public profile of geology, geotourism and geoheritage and also the list should also attempt to include sites from all the continents.

Moving from the Azores Islands to the United States, Ted T. Cable in Chapter 9 considers Interpreting Mining through a case study of a coal mine exhibit. The chapter presents definitions and principles of heritage interpretation and discusses the application of interpretation approaches at the famous coal mine exhibit in the Museum of Science and Industry in Chicago, Illinois (USA). Cable presents potential benefits of using heritage interpretation in interpreting mining and promoting geotourism, and he encourages the application of professional interpretation practices to interpret mining. The chapter presents potential benefits of heritage interpretation to both the visiting public and to the sponsoring institutions.

Geotrails is the focus of Chapter 10 by Thomas A. Hose. This chapter provides a summary of the history, development, and nature of geotrails within the United Kingdom. The chapter necessarily explores the different types of geo-interpretative media, together with a brief consideration of their efficacy, associated with geotrails. It also examines the nature and needs of those persons who access geotrails. Hose suggests that, and indicates some of the ways in which, new geotrails might better appeal and be accessible to wider and younger audiences than those usually targeted by geotrail providers.

Geological and Mining Heritage Interpretation is the focus of Chapter 11 by Ross Dowling. The chapter identifies a number of characteristics of sound geo-interpretation. Dowling describes interpretation before characterizing geological interpretation (geo-interpretation) through a variety of interpretation methods. He then proposes the "ABC" Interpretation Method for geo-interpretation and argues that the key elements of geoheritage are the interpretation of the geological heritage which a visitor is viewing. This is referred to as geo-interpretation which places geology within the environment's **A**biotic (nonliving), **B**iotic (living), and **C**ultural (human) attributes. This chapter illustrates geo-interpretation by showcasing interpretation for three geotourism attractions and three mine sites from six countries around the world.

In Chapter 12, William Witherspoon examines the Evolving Geological Interpretation Writings about a Well-Traveled Part of California, 1878–2016, USA according to the previous nine outstanding geology publications published between 1878 and 2011. These books cover place and routes connected to geology and give a clear explanation of maps designed to be used along the routes. California has a central role in the importance of geoscience concepts including early uses of plate's tectonics to interpret outcrops and landscapes. The significant role of digital tools, particularly Google Earth (R) and Google MyMaps (R) to apply plate tectonics, location markers, and planning out routes with driving time and also the possibility of social media to build on this foundation are discussed in this chapter.

This theme of the roadside geology is continued in Chapter 13 by William Witherspoon and John Rimel and the authors of this chapter describe Commercially Successful Books For Place-Based Geology: Roadside Geology Covers the U.S.; The *Roadside Geology* series books present interesting geological features and are widely available all over the Unites States and parts of Canada. The books are considered a valuable source of information for travelers and also a textbook at universities. It is a beginning point for its authors to tie geology to landscape, gold rush and settlement history, fossil collecting, climate history, and specific travel destinations.

From Community Engagement to the impact of Geoparks on the community, Malcolm Cooper and Kazem Vafadari share their experiences with Japanese Geoparks (Chapter 14). This chapter assesses the processes and patterns to be found in the Japanese community's engagement with Geoparks. It is based on trends identified by the authors over the past 15 years and longer, but is a commentary based on the activities of actual communities rather than surveys of tourists. The themes covered are the threads that make up the current community interest in geoconservation, rural decline and revitalization, globally important agricultural and other forms of heritage conservation (Geoparks, GIAHS and cultural heritage, Satoyama), and geotourism in Japan. These themes help us identify the impact of geoparks on the community, and the Japanese community's engagement with them.

In Chapter 15, Cristian Conibulo and Alexandru Andrășanu examine the role of volunteer management programs in geotourism development. The authors suggest that youth is a reservoir of creativity and energy for any community. If properly engaged and empowered it can have a very high impact on the growth of a geopark. Since 2013, Hațeg UNESCO Global Geopark has developed a Volunteer Program to use and inspire the young people. The chapter outlines the success story of the Volunteer Group in Romania. This chapter shows why volunteers are needed, how a volunteer program works in a geopark and what results were reached for local development, geoconservation, and geotourism promotion.

Chapter 16 presents a case study on the Geotourism and Proposed Geopark Projects in Turkey (Gülpinar Akbulut Özpay). This chapter includes general information about the geographical and geological characteristics of Turkey. At the same time, this chapter is aimed at determining the proposed geopark projects and the perception of geotourism of Turkey.

Geotourism development in Latin American UNESCO Global Geoparks (Chapter 17, by José Luis Palacio Prieto, César Gosso, Diego Irazábal, José

Patrício Pereira Melo, Francisco do O´de Lima Júnior, Carles Canet, Miguel A. Cruz-Pérez, Erika Salgado-Martínez, Juan Carlos Mora-Chaparro, Krzysztof Gaidzik, Jerzy Żaba, and Justyna Ciesielczuk) is outlined through the developing four UNESCO global Geoparks, establishing the Latin American and Caribbean Geopark Network (GeoLAC) and the development of geotourism in Latin America. The chapter present five case studies from Brazil (Araripe), Uruguay (Grutas el Palacio), and Mexico (Comarca Miner and Mixteca Alta) as well as one aspiring geopark in Peru (Colca y Volcanes de Andagua). The chapter focuses on geoparks registered with UNESCO in Latin America. The authors argue that geotourism displays a growing dynamism and acceptance in various Latin American territories, although there are striking differences in their development. Latin American Geoparks are contrasting territories, some with a long tourism tradition, extensive infrastructure while others are socially and economically marginalized. They describe that geotourism is representing a true development strategy in rural territories in various Latin American territories currently characterized by limited social and economic progress that have led to emigration.

A worldwide growing Dinosaur Geotourism development is the focus of Chapter 18 by Nathalie Cayla. She argues that dinosaurs are not only a very popular science topic but also paleontological tourism highlighting dinosaur paleontological sites is booming. This chapter aims to understand how, over time, the attractiveness of dinosaur geotourism destinations will be elaborated. After having presented the scope of the diversity of dinosaurian outcrops, several case studies illustrate the geotourism trajectories of these sites.

In Chapter 19, Mamoon Allan promotes the concept of "Geotourism for all" through discussing constraints and implications for Accessible Geotourism. He argues that despite the developments in the nature and scope of geotourism experiences in the last decade, people with special needs are generally excluded from different types of geological tourism activities, and to date, geological tours for people with special needs are still very scant. He suggests that it is critical to promote the concept of "Geotourism for all" in different geological tourism sites and to enhance the ethical practices in geotourism business. The chapter identifies the implications for such concepts in the geotourism context.

In Chapter 20, further development and promotion of geotourism is outlined by Bahram Nekouie Sadry. He highlights the development of celestial geotourism and space geotourism through emphasizing the abiotic aspects of environment. He also proposes a categorization for these activities as a subsegment of geotourism. Managerial efforts to promote the development of a Space, considering parts of the universe as geotourism attractions which will be used as space geological environment in the future, and Celestial Geotourism industry by integrating geotourism into its vertical destination development strategy is also discussed in this chapter. The author argues that the development of celestial geotourism, in terrestrial activities, can be used as a factor that contributes to poverty alleviation and income generation for local communities in remote areas—especially in developing countries, a component resulting in environmental sustainability, and as a platform to promote public space geotourism in developed countries in the future.

In Chapter 21, Krzysztof Gaidzik and Marta Chmielewska outline the opportunities for geotourism in the Post-Mining Objects through a case study of Upper Silesian Coal Basin (USCB) in Poland. They argue that postmining tourism is a very intensively developing branch of geotourism industry and includes tourism of stone quarries, mine sites, exploitation hollows, and postexploitation areas with specific buildings, infrastructure, and culture, that brings out their geo-educational, cognitive, and esthetic values. They suggest that the area of the USCB is a perfect spot for the development of postmining geotourism. The chapter presents the main attractions of the region and aims to present some of the so-called hidden potential of this region, that is, smaller or lesser-known post-mining sites, waste dumps, and objects with different functions. The authors believe that appropriate regional policy and promotion, long-term sustainable development, and geoprotection of the present geoheritage will lead to geotourism development and economic growth of the USCB postmining area.

The penultimate chapter (Chapter 22) is on the Langkawi UNESCO Global Geopark. Kamarulzaman Abdul Ghani outlines the popularity of Langkawi as an island holiday destination and its sizeable number of visitors. He addresses the fact that an increasing number of uncontrolled and unregulated visitors would cause serious sustainability issues regarding over development in various areas of the ecosystem, namely environment, waste disposal, electricity, and water supply. This chapter seeks to discuss the issues of conflict and balance between mass tourism, overtourism, and geopark tourism, specifically geotourism and what the outlook of Langkawi UNESCO Global Geopark could preferably be.

In Chapter 23, the final chapter of the book, the editor synthesizes the topics discussed in the book. While reviewing the topics covered by contributors, I manage to introduce some new matters. In order to expand concepts, various theoretical and practical observations are brought together; I also develop some policies and strategies essential for geotourism planning and expansion. In this way, the chapter's main effort is to provide a deep insight into geology and tourism and leaves a list of items to be discussed in the future.

ACKNOWLEDGMENTS

I am very grateful for help from Fatemeh Fehrest and Susan Turner, which has improved the chapter.

KEYWORDS

- geotourism definitions
- geotourism industry
- the basic principles of geotourism
- Tourtellotic geotourism
- true geotourism
- key components of geoparks

REFERENCES

Chen, A.; Yunting, L. N.; Young, C. Y. *The Principles of Geotourism*, Springer Geography, Springer-Verlag Berlin Heidelberg and Science Press Ltd., 2015.
Cooper, M.; Eades, J. The Auvergne—Centre of Volcanic Tourism in France. In *Volcano & Geothermal Tourism: Sustainable Geo-Resources for Leisure and Recreation;* Erfurt-Cooper, P., Cooper, M., Eds.; Earthscan Publications: the UK, 2010; pp 247–258.
Coratza, P.; Emmanuel, R.; Zbigniew, Z. Geodiversity and Geoheritage: Crossing Disciplines and Approaches. *Geoheritage J.* **2018,** *10,* 525–526.
Del Lama, E. A. Urban Geotourism with an Emphasis on the City of Sao Paulo, Brazil. In *Handbook of Geotourism*; Dowling, R. K., Newsome, D., Eds.; Edward Elgar Publishing: Cheltenham, Gloucestershire, 2018; pp 210–223.
Dowling, R. K. Emerging Volcano and Geothermal related Tourism in Iceland. In *Volcano and Geothermal Tourism: Sustainable Geo-Resources for Leisure and Recreation*; Erfurt-Cooper, P., Cooper, M., Eds.; Earthscan: London, 2010; pp 209–220.
Dowling, R. K.; Newsome, D., Eds.; *Geotourism*; Elsevier/Heineman: Oxford, 2006.
Dowling, R.; Newsome, D., Eds.; *Global Geotourism Perspectives*; Goodfellow Publishers Limited: Oxford, UK, 2010.

Dowling, R. K.; Newsome, D., Eds.; *Handbook of Geotourism*; Edward Elgar Publishing: Cheltenham, Gloucestershire, 2018.

Erfurt-Cooper, P., Cooper, M. *Health and Wellness Tourism: Spas and Hot Springs*; Channel View Publications: Bristol, UK, 2009.

Erfurt-Cooper, P., Ed.; *Volcanic Tourist Destinations*; Springer-Verlag: Berlin Heidelberg, 2014.

Erfurt, P. Geotourism Development and Management in Volcanic Regions. In *Handbook of Geotourism*; Dowling, R., Newsome, D., Eds.; Edward Elgar Publishing: Cheltenham, Gloucestershire, 2018; pp 152–167.

Farsani, N. T.; Coelho, C.; Costa, C., Eds.; *Geoparks and Geotourism: New Approaches to Sustainability for the 21st Century*; Brown Walker Press, 2012.

Farsani, N. T.; Coelho, C.; Costa, C. Rural Geotourism: A new Tourism Product. *Acta Geoturistica* **2013**, *4* (2), 1–10.

Garofano, M.; Govoni, D. Underground Geotourism: A Historic and Economic Overview of Show Caves and Show Mines in Italy. *Geoheritage* **2012**, *4* (1–2), 79–92.

Garofano, M. Developing and Managing Show Caves in Italy. In *Handbook of Geotourism*; Dowling, R. K., Newsome, D., Eds.; Edward Elgar Publishing: Cheltenham, Gloucestershire, 2018; pp 126–138.

GGN Guidelines. Guidelines and Criteria for National Geoparks Seeking UNESCO's Assistance to Join the Global Geoparks Network (GGN). 2010. Available at: http://www.unesco.org/new/fileadmin/MULTIMEDIA/HQ/SC/pdf/sc_geoparcs_2010guidelines.pdf (accessed 1 Dec. 2018).

Gray, M. *Geodiversity. Valuing and Conserving Abiotic Nature*. John Wiley & Sons: Chichester, 2004.

Gray, M. "Simply the Best": The Search for the World's Top Geotourism Destinations. In *The Geotourism Industry in the 21st Century*; Sadry, B. N. Ed.; The Apple Academic Publishers Inc., 2020; pp 207–226.

Hose, T. A. *Rockhounds Welcome!—Assessing the United Kingdom Client Base for Site-Specific Geological Interpretation*. Unpublished Paper to the Visitor Studies Association Annual Conference, 26–30 July, Raleigh, North Carolina, USA, 1994.

Hose, T. A. Selling the Story of Britain's Stone. *Environ. Interpret.* **1995**, *10* (2), 16–17.

Hose T. A. Geotourism and Interpretation. In *Geotourism*; Dowling, R., Newsome, D, Eds.; Elsevier Butterworth-Heinemann: Oxford-Burlington, 2006; pp 221–241.

Hose, T. A. Towards a History of Geotourism: Definitions, Antecedents and the Future. *Geol. Soc. Lond. Spec. Publ.* **2008**, *300*, 37–60.

Hose, T. A. Volcanic Geotourism in West Coast Scotland. In *Volcano and Geothermal Tourism: Sustainable Geo-Resources for Leisure and Recreation*; Erfurt-Cooper, P., Cooper, M., Eds.; London: Earthscan, 2010; pp 259–271.

Hose, T. A. 3G's for Modern Geotourism. *Geoheritage* **2012**, *4*, 7–24

Hose, T. A., Ed.; *Geoheritage and Geotourism: A European Perspective*; Boydell and Brewer: Woodbridge, 2016.

Hose, T. A. Email Personal Communication with Dr Thomas (Tom) Alfred Hose, 2019.

Hose, T. A. Historical Viewpoints on the Geotourism Concept in the 21st Century. In *The Geotourism Industry in the 21st Century*; Sadry, B. N., Ed.; The Apple Academic Publishers Inc., 2020; pp 23–92. International Congress of Geotourism. Arouca Declaration; International Congress of Geotourism, 2011. https://dl.dropboxusercontent.com/u/36358978/News/Declaration_Arouca_%5BEN%5D.pdf (accessed Dec 1, 2015).

Jorgenson, J.; Nickerson, N. Geotourism and Sustainability as a Business Mindset. *J. Hosp. Mark. Manage.* **2016**, *25* (3), 270–290.

Lima, E.; João Carlos, N.; Manuel, C. A Presentation on Azore Geoparks: Experience the Azores's Geotourism, 2014. https://www.visitazores.com/en/experience-the-azores/geotourism.

Mata-Perelló, J.; Carrión, P.; Molina, M.; Villas-Boas, R. Geomining Heritage as a Tool to Promote the Social Development of Rural Communities. In *Geoheritage: Assessment, Protection and Management*; Reynard, E., Brilha, J., Eds.; Elsevier Inc.: Amsterdam, 2018, pp 167–177.

Mukwada, G; Sekhele, N. The Potential of Community-based Geotourism in Rural Development in South Africa: The Case of Witsie Cave Project. *J. Asian Afr. Stud.* **2015**, *52* (4), 471–483. https://doi.org/10.1177/0021909615595991.

Necheș, I. M.; Erdeli, G. Geolandscapes and Geotourism: Integrating Nature and Culture in the Bucegi Mountains of Romania, Landscape Research, 2014. DOI: 10.1080/01426397.2014.939616.

Newsome, D.; Dowling, R. K., Eds. *Geotourism: The Tourism of Geology and Landscapes*; Goodfellow: Oxford, 2010.

Newsome, D.; Dowling, R. K.; Setting an Agenda for Geotourism. In *Geotourism: The Tourism of Geology and Landscape*; Newsome, D., Dowling, R., Eds.; Goodfellow Publishers Limited: Oxford, UK, 2010; pp. 1–12.

Newsome, D.; Dowling, R. K. Conclusions: Thinking about the Future. In *Handbook of Geotourism*; Dowling, R. K., Newsome, D., Eds.; Edward Elgar Publishing: Cheltenham, Gloucestershire, 2018; pp 475–482.

Ólafsdóttir, R.; Tverijonaite, E. Geotourism: A Systematic Literature Review. *Geosciences* **2018**, *8*, 234.

Patzak, M.; Eder, W. 'UNESCO GEOPARK, a New Programme—A New UNESCO label'. *Geol. Balcan.* **1998**, *28*, (3–4), 33–35.

Palacio-Prieto, J. L.; Gosso, C.; Irazábal, D.; Melo, J. P. P. ; Lima Júnior, F. D. D.; Canet, C.; Cruz-Pérez, M. A.; Salgado-Martínez, E.; Mora-Chaparro, J. C.; Gaidzik, K.; Żaba, J.; Ciesielczuk, J. Geotourism Development in Latin American UNESCO Global Geoparks: Brazil, Uruguay, Mexico, and Peru. In *The Geotourism Industry in the 21st Century*; Sadry, B. N., Ed.; The Apple Academic Publishers Inc., 2020; pp 419–443.

Reynard, E.; Coratza, P.; Regolini-Bissig, G., Eds. *Geomorphosites*; Pfeil: München, 2009.

Reynard, E.; Brilha, J., Eds. *Geoheritage: Assessment, Protection and Management*; Elsevier Inc.: Amsterdam, 2018.

Reynard, E. Geomorphosites: Esthetic Landscape Features or Earth History Heritage?. In *The Geotourism Industry in the 21st Century*; Sadry, B. N., Ed.; The Apple Academic Publishers Inc., 2020; pp 147–167.

Riganti, A.; Johnston, J. Geotourism—a Focus on the Urban Environment. In *Handbook of Geotourism*; Dowling, R. K.; Newsome, D., Eds.; Edward Elgar Publishing: Cheltenham, Gloucestershire, 2018; pp 192–201.

Sadry, B. N. *Fundamentals of Geotourism: With Special Emphasis on Iran*; SAMT Publishers: Tehran, 2009 (English summary available online at: http://physio-geo.revues.org/4873?file=1; Retrieved: Jan. 01, 2020) (in Persian).

Sadry, B. N. An Introduction to Geomorphosites Studies in Iran. Proceedings of the first Congress (IRAG) of Iranian Association of Geomorphology; In *Geomorphology and Human Settlements,* Yamani, M., Negahban, S., Eds.; Tehran, 2012; pp 136–134 (in Persian).

Sadry, B. N. Geotourism Concepts. Presented at the APO's Intensive Training Course on Planning and Management of Ecotourism, National Training & Productivity Centre of the Fiji National University, Tanoa Hotel, Nadi, Fiji, 25–29 Nov. 2013.

Sadry, B.N. Space and Celestial Geotourism. In *The Geotourism Industry in the 21st Century*; Sadry, B. N., Ed.; The Apple Academic Publishers Inc., 2020; pp 479–503.

Strba, L.; Bartolomew, B.; Marián, L. Roadside Geotourism—An alternative Approach to Geotourism. *e-Rev. Tour. Res.* **2016,** *13* (5–6), 598–609.

Turner, S. Recommendation for National Geopark Initiatives seeking UNESCO's Assistance in Geological Heritage Promotion. The Australian Geologist No. 131, June 30, 2004, pp 37–38.

Turner, S. Geoheritage and Geoparks: One (Australian) Woman's Point of View. *Geoheritage* **2013,** *5,* 249–264. DOI: 10.1007/s12371-013-0085-5.

Turner, S. "Geotourism": Where Wise Birds Stay Awhile. *GeoExpro Mag.*, **2005,** 56–58, 60 (An online version published in 2008, available at: https://www.geoexpro.com/articles/2016/08/where-wise-birds-stay-awhile) (retrieved: Jan. 01, 2020).

UNESCO. UNESCO Geoparks Programme—a New Initiative to Promote a Global Network of Geoparks Safeguarding and Developing Selected Areas having Significant Geological Features, 1999, Document 156.

UNESCO. *UNESCO Geoparks Programme Feasibility Study*, UNESCO, Paris, 2000.

Weaver, D. Celestial Ecotourism: New Horizons in Nature-Based Tourism. *J. Ecotour.* **2011,** *10* (1), 38–45.

Witherspoon, W. Email Personal Communication with Dr William (Bill) Witherspoon, 2012.

Witherspoon, W. Email Personal Communication with Dr William (Bill) Witherspoon, 2018.

Woo, Kyung-Sik; Young Kwan, S.; Lisa, M. King. Jeju: South Korea's Premier Island Geotourism Destination. In *Volcano & Geothermal Tourism: Sustainable Geo-Resources for Leisure And Recreation*; Erfurt-Cooper, P., Cooper, M., Eds.; Earth Scan Publications: the UK, 2010; pp 170–179.

CHAPTER 2

Historical Viewpoints on the Geotourism Concept in the 21st Century

THOMAS A. HOSE

Honorary Research Associate, School of Earth Sciences, University of Bristol, Wills Memorial Building, Queens Road, Clifton, Bristol, BS8 1RJ, UK

E-mail: gltah@bristol.ac.uk

ABSTRACT

Geotourism is a late 20th century recognized form of niche or special interest tourism with considerable global growth potential in the 21st century. Initially researched and defined in the UK and Europe, it is an emerging and growing field of academic study and publication. It is focused on the usage of geosites and geomorphosites, often subsumed within the concept of natural landscapes, for tourism purposes. As such, it is also a form of natural heritage marketing. Although relatively recently defined, the paradigm is already undergoing redefinition and refinement, sometimes departing from the original earth science basis, and this benefits from a new appreciation of its historical roots. Its resource base includes geosites and geomorphosites, museum collections and exhibitions, and library archives and artistic outputs. Although the term passed into general usage only in the early 1990s, its antecedents can be traced to at least the 17th century in Europe and possibly earlier elsewhere. It has significant underpinnings within social history and industrial archeology. Its recognition, recent development, and history can and should be placed within a broader, if niche, tourism history context.

2.1 ON CONTEMPLATING A HISTORY OF GEOTOURISM

Anyone contemplating writing a history of geotourism might initially think it to be a straightforward task, founded upon an examination of quite recent events and literature associated with geology and tourism. This actually is far from the case for, although as defined in the closing decade of the 20th century, geotourism is a new form of tourism, many of the activities it encompasses have long and occasionally tenuous antecedents. Geotourism provision is underpinned by the protection and management of the locations in which it takes place, that is, geoconservation. Any competent consideration of geotourism's history must necessarily examine the development of tourism that can be directly attributed to the promotion of landscapes and geology, coupled with the history of their recognition and protection, for whatever reasons. Landscape promotion is an older activity than geological promotion, although both were initially done, from around the middle of the 18th century, for and at the behest of an educated discerning elite. It is when we consider the promotion of the two for the mass audience, from somewhere around the middle of the 19th century onwards, that the concept of modern geotourism emerges. Consequently, geotourism allied to geo-interpretative provision (Hose, 1995a) for geoconservation purposes can only be truly recognized from the late 20th century. Indeed, it is only from the closing decade of that century that it burgeoned to the extent that it has become a recognized strand within tourism development strategies, especially following the development of geoparks. Therefore, "Given the volume of published material and general usage (especially in policy documents) and practice, it could be argued that "geotourism" has sufficient theoretical and conceptual status to qualify as a paradigm" (Hose, 2016a). As such it can be set alongside another contemporaneous geological paradigm, geodiversity (Gray, 2004, 2008); this all-embracing term has evolved to become for the English Geodiversity Forum (EGF) "…the variety of rocks, minerals, fossils, landforms and soils, together with the natural processes that shape them. Geodiversity is a foundation for life and our society. It influences landscape, habitats and species as well as our economy, historical and cultural heritage, education, health and well-being" (EGF, 2014: 2). The means by which the vast compass of landscape-focused tourism development, underpinned by an extraordinary range of geological and geomorphological phenomena, over at least some 300 years (Hose, 2016b) and incorporating defined European esthetic movements (such as the "sublime," the "picturesque," and the "romantic"), building upon much earlier travel motivators, can be tackled in an introductory chapter, without losing the helpful detail of key events, localities, publications, and personalities poses a considerable challenge. So too does the search for material likely to be unfamiliar to readers to avoid the excessive repetition evident in far too many other accounts of the topics covered herein. Hopefully, this is met in some measure in this chapter through its various summaries of the key interrelated elements of geotourism laid out in a sequence that coherently explains (and provides informative and contextualizing snippets) the underpinning theoretical, practical, and historical elements, which have culminated in the emergence of the global geopark movement.

2.2 A TOURIST IS...

2.2.1 TOURISM AND THE TOURIST

At the outset of examining geotourism, it is worth noting that the concepts of "tourism" and "the tourist," somewhat central to studies of its modern form, considerably postdate the recognition of travel and the traveler in Europe, and probably elsewhere in the world. Indeed, "Observers set out armed with courage and notebooks. Certain lands beckoned. Long ago the Chinese were going to India. Arabs went to Africa and, to complete the circle, Italians set out for China" (Foss, 1989: 9) long before specific tourist routes and areas had been formalized in guidebooks and infrastructure specifically provided for tourists had been developed. Significantly, tourism and the tourist were first recognized when travel in Europe for leisure purposes was emerging, at the beginning of the 19th century, as a reason for travel among the rising mercantile or middle classes. Previously, such leisure travel in Europe and North America had been the preserve of a generally rich and aristocratic elite, although they might be accompanied by clerics, tutors, or some other hired (for their knowledge of the places to be visited) assistant or servant. Of course, there have always been some much-traveled individuals from across society's social spectrum whose prime motivation was not in pursuit of leisure or pleasure—such as pilgrims, migrant workers, soldiers, and entertainers. Indeed, "If early travellers had been consulted, private curiosity would have rated rather low as a reason for travelling" (Foss, 1989: 9).

2.2.2 PILGRIMS AND TRAVELERS

Generally, pilgrimage is a journey undertaken by anyone, even of modest means, as an act of religious contrition or devotion. In fact, pilgrim, in use as early as 1200 in England, is derived from the Latin "peregrinus" meaning a foreigner; effectively, someone who travels abroad. It had such dual use for both a foreign traveler and a religious traveler into the 16th century. Christian pilgrims, from the 12th century onward, traveled afar in and across Europe to reach holy shrines as distant as the Levant of the eastern Mediterranean

(Hindley, 1983). Such Christian pilgrimage had begun in the 4th century Roman Empire, somewhat aided by the Empire's transport infrastructure and political control over the territory. Indeed, as Christian pilgrimage became established along certain routes accommodation and dining, together with souvenir industries were established. Hospitals were also built, often like some of the accommodation, linked to monasteries and convents; something of a precursor of the later tourist provision. In the Muslim world, pilgrimage has been a key feature of religious observance since the 7th century, although it was considered a continuation of such religious travel from the time of Abraham (Ibrahim) some 2500 years earlier. Further, "Abraham and Buddha, Alexander and Julius Caesar, though remembered for other reasons, were also famous for their journeys. Indeed, a person can hardly get a name in history without many busy travels or a grand and perilous expedition" (Foss, 1989: 9). While pilgrim and pilgrimage are then recognized from 12th century Middle English, *The Oxford English Dictionary* advises that the term "tourist" first made its appearance in English in only the late 18th century, as yet another synonym for "traveler." Among the earliest entries in *The Oxford English Dictionary* for "traveler" is Samuel Pegge's statement, in a book of 1800 on new English usages, that "A Traveller is now-a-days called a Tour-ist." Today, the term is widely employed, especially by many scholars, in a derogatory sense. It conjures up for those scholars a personality profile, life-style, social class identification, and scenarios in which "the tourist" performs some characteristic banal act; "the tourist" just blindly follows where the few authentic travelers have gone before, directed there by actual guides or guidebooks and often on a tour bus. Conversely, it has been asserted that the traveler "…exhibits boldness and gritty endurance under all conditions (being true to the etymology of "travel" in the word "travail"); the tourist is the cautious, pampered unit of a leisure industry. Where tourists go, they go *en masse*, remaking whole regions in their homogeneous image." The original author's motivation in developing geotourism (Hose, 1995a, 2011) was that it would provide both a value- added activity for mass tourists at popular holiday destinations and a focus for special interest travel and tourism for niche adventure and special interest tourists; a necessary by-product being funding and support for geoconservation.

2.2.3 THE TOURISM INDUSTRY

Whatever the image or perceptions of tourists, and the industry by which they are supported, held by host communities, researchers, and stakeholders;

in general, it is globally acknowledged that the tourism industry is one of the world's fastest growing and most important economic activities. It generates around a tenth of global gross domestic product, employs around 200 million people, and involves some 700 million international, let alone, many more domestic, travelers. Hence, geotourism is a relatively minor but developing form of niche or special tourism with much growth potential. As an aspect of the burgeoning "heritage industry" (Hewison, 1987; Hunter, 1996), in which geotourism provision can be an element, of the United Kingdom (UK), it is worth noting that it is a "…multi-million pound industry and its contribution to the success of tourism in the United Kingdom is undoubted…there is a powerful argument for protecting the heritage because it earns its own living" (Ross, 1991: 175). Thus, in addition to contributing to economic growth and regeneration, it could provide the rationale for, and some funding in support of, geoconservation.

The latter was the main reason that the modern geotourism paradigm was recognized and developed in England (Hose, 1995a) in the closing decade of the 20th century. It followed the late-1980s recognition by school, university, and museum geologists, of which the author—having worked in museums and schools—was one, in the UK of the increasing number and range of geosites and geomorphosites becoming unavailable for study by geology field parties. Although this was mainly due to their closure as they were worked out, it was also noted that many were becoming waste depositaries, whilst others were closed to visitors because of security and health and safety issues. Also identified was the issue of managing the impact of field, particularly student, parties at vulnerable geosites in popular areas of geological inquiry (Toghill, 1972: 1994). The means to tackle such threats to the geoheritage were explored and formed the basis of geoconservation approaches in the UK, Europe, and wider afield from the middle of the 20th century.

2.3 THE GEOHERITAGE: THREATS AND PROTECTION

2.3.1 THE NEED FOR GEOCONSERVATION

2.3.1.1 THE NATURE OF THE THREATS

Geoconservation has been particularly well advanced in practice and theoretical underpinning in Europe (Bentivenga and Geremia, 2015; Burek and Prosser, 2008; Hose, 2012b; Last et al., 2013; O'Halloran et al., 1994; McKirdy, 1991; Wimbledon and Meyer, 2012), Australia (Sharples, 1993,

2002), and to a limited extent in the United States (Santucci, 2005) and Canada (Maini and Carlisle, 1974). Its history, particularly in the UK (Burek and Prosser, 2008), has also been well documented. Hence, this chapter will particularly focus on the UK and Europe, but will also consider countries further afield, such as Australia and China, when they add something worthwhile to the specific matters being considered. Europe's geoheritage of both primary (Crowther and Wimbledon, 1988) and secondary (Doughty, 1981) geosites was increasingly threatened with neglect and destruction from the last quarter of the 20th century. This was due to the enduring ignorance by the public, politicians, and planners of their value and future potential for scientific and economic developments. Quaternary sites were then (Bridgland, 1994), and still are, especially threatened because of their particular nature (Bridgland, 2013). Unlike hard rock exposures they require much maintenance (because of issues around slope instability, vegetation growth, and flooding) and research input to maintain their geoscientific interest. This was when it was noted that the UK's geoheritage "...is being damaged and slowly destroyed by new "developments," waste disposal, and afforestation and people erosion. The restoration of derelict land and the building of new houses, roads, motorways and other civil engineering projects can remove, sometimes forever, vital clues to the geological history of an area...infilling of quarries may obscure important and rare exposures, and many sites may be "reduced" by the continual tread of tourists' feet... or "removed" by the collection of horizons of limited extent by hundreds of field parties..." (Stanley, 1993: 17). A significant threat was posed by a new generation of large-scale civil engineering works for coastal defense (Hooke, 1998; McKirdy, 1987), flood prevention, and roadside slope stability (Ager, 1986: 85; Larwood and Markham, 1995). In the UK, economically worked out hard-rock and soft-rock quarries were filled in with domestic and commercial waste.

2.3.1.2 SOME EXAMPLES OF UK LOSSES

This happened even at scientifically significant and supposedly protected geosites, Webster's claypit SSSI in Coventry. A former brick pit, its geological significance was such that it was included in the *Geological Conservation Review* volume on British Upper Carboniferous Stratigraphy (Cleal and Thomas, 1996). Despite a local and national campaign to raise awareness of the claypit's significance, the local authority obliterated the site. It was clearly, given that "...keen to infill the quarry to create recreational green space in a relatively deprived urban area. It regarded this need as much greater than the need to retain a geological SSSI...a compromise scheme allowing both green space and a geological face to be retained...was not accepted..." (Prosser, 2002), these issues that prompted Hose (2011: 345) to develop geotourism. As reported in the UK's chief geoconservation magazine, "The total destruction of a geological SSSI is thankfully a very rare occurrence" (Prosser, 2002), but it is not an isolated example of a determined landowner damaging or destroying a geological SSSI. Many landowners, particularly arable farmers, see SSSI designation as something that interferes with the right to manage their own land, at their own expense, as they see fit. This is, as the recent abandonment of a geopark proposal in Queensland, Australia exemplifies, not a uniquely UK or past issue.

Soft-rock quarries, usually working Quaternary sands and gravels, were also allowed to flood; they were then turned into wildlife reserves usually managed by the Royal Society for the Protection of Birds (at over a million members, the largest conservation body in Europe) or one of the county wildlife trusts, and leisure water parks. Similarly, the landscaping of redundant coal mining industry sites, following its rapid decline across the UK in the mid-1980s, obliterated both geological and usually any industrial heritage interest. By the 1990s, this was also an issue with the worked-out Jurassic limestone iron ore mines of the UK's English Midlands (Knell, 1990). Roadside exposures at the time were routinely covered with soil and grassed over, encased in netting or concreted over (Baird, 1994). Natural geosites and geomorphosites were also being increasingly lost because of planning decisions permitting, for example, the construction of hard (essentially concreted) coastal defenses (Leafe, 1998). New warehouse and retail developments built on the sites of former quarries and mines often obliterated or severely limited access to any remaining geological interest. The major driving force behind most of these losses was the economic downturn of the 1980s that had led to reduced extractive industry activity, together with the restructuring and downsizing of heavy industries such as iron and steel. Finally, increased emphasis on workplace health and safety, particularly in quarries and mines, with concomitant insurance and security concerns much reduced permitted access to geology field parties and researchers.

2.3.1.3 VARIOUS MINOR THREATS

A minor threat to geosites is posed by students and amateur geologists (who are usually attracted to geology by the joy of casual collecting) collecting material indiscriminately from sensitive locations—especially hammering out in situ fossils and minerals. Indeed, "...visitor pressure, particularly excessive hammering and sample and fossil collection, which can have a major impact, if not properly controlled, and result in the degradation of key sites" (Anon., 1994: 5). At a number of major UK geosites much visited by amateur collectors, such as the Isle of Wight (Anon., nd), attempts were made to educate visitors about the associated geoconservation issues of their activities. The Isle of Wight, along with the adjacent Dorset coast (nowadays marketed to geotourists as the "Jurassic Coast"), is one of the earliest localities to be much visited by amateur collectors, no doubt due to several significant early geological guides such as the large library volume by Englefield (1816) and the highly pocketable guide by Mantell (1847, 1851, 1854); the latter, *Geological Excursion Round the Isle of Wight and Along the Adjacent Coast of Dorsetshire*, is considered (Hose, 2008b) to be the antecedent of the modern illustrated geology fieldguide (Hose, 2006a).

The most visually obvious intrusions on hard-rock exposures have been by postgraduate students unwilling to follow, or ignorant of, professional codes of practice for rock-coring (Robinson, 1989) and collecting, compounded by their inadequate supervision by university staff. All too often it has led to the despoliation of fieldguide exposures (MacFadyen, 2010), as noted by the author in the Malvern Hills and the Lake District in England where rock-faces fronting onto roads and footpaths were despoiled when samples could easily have been almost invisibly collected from their backs. Another, and more serious, threat is from a very few unscrupulous commercial collectors. There are several reported cases of foreign commercial collectors raiding classic UK sites (Rolfe, 1977; Taylor, 1988) to sell material to unprincipled dealers abroad. Such vandalism by a British collector occurred in 2009; three illegally collected dinosaur footprints, cut out from a dinosaur track-way at Bendrick Rock in South Wales, were apparently found for sale in a Dorset fossil shop. However, the legal commercial extraction and sale of geological specimens has a long history (Taylor, 1988). Indeed, some eminent professional geologists of their day were involved (Taylor, 1992) in the activity; for example, there was "...A.J. Jukes-Browne, the noted author of many of the Geological Survey's maps and memoirs, who seems to have had a sideline in acquiring...fossils...and selling them to museums" (Taylor, 1992: 30).

Most modern museum collections house material bought mainly in the 19th century from commercial collectors and dealers (Rolfe et al., 1988; Taylor, 1987). Unfortunately, this past investment is commonly undervalued by modern managers of natural science collections. Taylor (1987) noted that today's auction market for fossils is insignificant compared with that for *objects d'art* and antiquarian books. Consequently, but of significance to geoconservation in secondary geosites, most museum manager falsely assumed that fossils had no real financial value. Therefore, there was "... little to lose by trumpeting the fact that fossils are worth good money. Unless we do this, no museum will want to spend money on its fossils. The only way to convince a committee to allocate £400 to conserving an ichthyosaur will, all too often, be to emphasise its value of £10,000, not just its scientific or historical importance" (Taylor, 1987: 60). During the 19th century, the sale of fossils was a real tourist attraction; this was especially so at Lyme Regis and Charmouth in Dorset where this had begun in the latter part of the 18th century (Woodward, 1907: 115) and continues today (Taylor, 1992). An occasional, but generally minor, threat is caused when normally permitted access to geosites is threatened, often by attempts to charge fees for access. There was such a notable attempt on the Isle of Arran (Robinson, 1990) where the landowner (whose land accounted for four-fifth of the localities in its popular fieldguide) proposed a *per capita* weekly charge (initially £10-00 and subsequently reduced to £4-00) for geology field parties. The National Trust had made similar proposals, ostensibly as a geoconservation measure, in the preceding decade for parts of Dorset (Rea, 1982) and later the Isle of Wight. It was the National Trust that had been instrumental, following its implementation from the mid-1930s of its country house acquisition scheme (Mandler, 1996: 101–107), in popularizing a rather romanticized (instead of a workmanlike scientific or practical) approach to countryside visits.

2.3.2 APPROACHES TO GEOCONSERVATION

2.3.2.1 UK RECOGNITION AND DEFINITIONS

Geoconservation has been succinctly defined as "the act of protecting geosites and geomorphosites from damage, deterioration or loss through the implementation of protection and management measures" (Hose, 2012), and this approach has been employed by several authors (e.g., Dong et al., 2013). Pressures on the UK's and Europe's geosites increased in the 1980s. Particularly for the UK, this was partly a result of the abandonment of domestic metalliferous mining and the virtual demise of deep coal mining; these had contributed to the actual and perceived dereliction of their former major areas. Hewison (1987) argued that this contributed to the rise of the heritage sector in the UK, on the back of which industrial archeology flourished. Meanwhile, concomitant land reclamation and amenity landscaping (Culshaw et al., 1987) supposedly intended to remove the "blight of industrial dereliction" (Oxenham, 1966; Kotler et al., 1993) increased apace. Such amenity landscaping of these old disturbed industrial landscapes (Fairbrother, 1970: Chapter 6) was often for the benefit of urban developers and occasionally the romantic urban-fringe country-goer. Intended to hide the scars of past extraction and exploitation, their emphasis was on masking and burying the old industrial landscapes under remodeled (but to the experienced observer never quite naturally contoured) hills, new retail/warehouse developments, and housing estates. Their industrial and scientific significance was largely ignored at the time. Little thought was given to their conservation and potential as future tourism assets. The geological heritage suffered much more than its associated industrial heritage.

Consequently, there are now numerous cases of well-preserved and restored mine buildings and associated domestic structures standing in splendid isolation from their adjacent remodeled or even removed spoil heaps; for example, Snailbeach Mine (Richards, 1992) in rural Shropshire. This is like the fate, before the recognition and promotion of their industrial archeology significance, of redundant textile mills, potteries, metal refineries, and workshops (Alfrey and Putnam, 1992). This is well exemplified by a comment on England's major industrial archeology complex at Ironbridge in Shropshire that was "…terribly spoilt by the forges and foundries, the banks of slag and refuse that run down to the water's edge. Tiers of dirty cottages rise on the hill-side, which is very steep" (Sillitoe, 1995: 232).

Rather belatedly in the 1990s, the value of such landscapes, their individual features, and geology were recognized in one of the UK's oldest coal mining areas (Durham County Council, 1994). In the 2000s, the UK began to adopt an action planning strategy process for managing geosites and geomorphosites; for example, Local Geodiversity Action Plans (LGAPs) (Burek and Potter, 2006; Burek, 2008: 84–86) set out actions to conserve, enhance, and promote the geoheritage of a particular area, usually based on local authority boundaries, such as a county, or some specified area—such as an area of outstanding natural beauty (AONB)—for example, for Lincolnshire (Lincolnshire Geodiversity Group, 2010) and the North Wessex Downs AONB (Oxfordshire Geology Trust, 2007) respectively. These plans are often preceded by a separate "geodiversity audit," which describes the geology of the area; for example, that for North Pennines AONB and geopark (North Pennines AONB Partnership, 2010). A century earlier, the UK and mainland Europe were at the forefront of practical geoconservation, although its origins can be traced to a century earlier in Ireland (Doughty, 2008). In the mid-nineteenth century UK, the Cheesewring (a granite tor above a quarry in Cornwall was saved by restrictive clauses in quarrying leases signed in 1845 and 1865; this was the first such recorded geoconservation measure.

2.3.2.2 GERMANY AND "GEOTOPES"

In Germany, in the early 19th century, the need to protect, especially from quarrying, significant geosites was recognized. The first nature reserve in Germany, the Drachenfels (formed of the remnants of a long-extinct volcano) in the Siebengebirge uplands, notified in 1836 was of some geological value; likewise, the Teufelsmauer, notified in 1852 (Grube and Albrechts, 1992: 16). The Drachenfels is the closest of all the hills in the Siebengebirge to the River Rhine, which facilitated the use of barges to transport its quarried rock. It has been quarried since Roman times, and much later; its stone was used in the construction of the Cologne Cathedral. Quarrying ended when the Prussian government bought the quarry in 1836. It was declared a national park in 1956. The Drachenfels and its ruined castle was popularized by the Romantic Movement. The poems by Lord Byron (1788–1824), Edward Bulwer-Lytton (1803–1873), and Heinrich Heine (1797–1856) ensured its pan-European recognition. Most noteworthy was its mention in Byron's *Childe Harold's Pilgrimage* (published 1812–1828 in four parts), which established the poet's career. Its popularity with tourists was also much aided in 1883 when a railway to its summit was opened; it is still a very popular tourist attraction. The Teufelsmauer (or "Devil's Wall"), a rock formation of hard Cretaceous age sandstones in the Harz Mountains, was protected in 1832 and 1852 from destruction due to the extension of a major quarry. Its protection was extended with its designation as a nature reserve in 1935.

In Germany, geosites are termed "geotopes." The modern approach, employing geotopes, to geoconservation in Germany has been summarized by Röhling and Schmidt-Thomé (2004). The role of geotopes in geoconservation has been examined by Krieg (1996). In Germany, because it is a federal state, the statutory protection afforded to geotopes is not uniform. Geotope protection at any scale can be initiated by government agencies or individuals. Nature conservation agencies are normally responsible for statutorily notifying geotopes; prior to that the geotope proposal is examined by the respective states' geological survey. The country's national Nature Conservation Law and the Natural Monument Law (covering archeological sites and fossil localities) are the main provisions for implementing geotope conservation. Within the framework of the former, each state enacts its own specific legislation. Such laws have some provisions compatible with the needs of geotope conservation by including legal protection for geological exposures as natural monuments; these may be fossiliferous or mineral rich, and other geological features, such as landforms and springs, are included. In some states (Baden-Wurttemberg, Brandenburg, Hesse, North Rhine-Westphalia, Rhineland-Palatine, and Thuringia), fossils and fossil localities are included under natural monuments in the legislation. The latter can be supplemented by designations such as protected natural feature, or for large areas nature reserve. However, in some states, regional land use planning measures may provide some protection for geotopes.

2.3.2.3 THE RELICT OTTOMAN EMPIRE AND GEOCONSERVATION

The complexities of geoconservation on Europe's extreme eastern edge, in Turkey and Albania, have been recently examined (Hose and Vasiljevic, 2016a) in the context of southeast Europe. This has, perhaps unexpectedly, shown that the legacy of the influence of the Ottoman Empire still binds much of their two approaches to geoheritage management and protection. While Turkey was at the center of the particularly bureaucratic Empire, Albania was at its periphery. After the Great War (1914–1918), Turkey moved toward a model of secular statehood with a tolerant religious policy, and Albania initially retained a monarchy; in the second half of the 20th century, the latter was overtaken by a strictly secular communist regime with an emphasis on standardization in all things. The frequently changing nature of the Turkish governmental conservation bodies from the mid-twentieth century is unmatched in the rest of Europe. This was partly as a consequence of governmental fears that conservation measures might prevent mineral exploitation and negatively impact on the country's economic development. Geoconservation interest only developed gradually from the late-1960s (Ketin, 1970; Ongur, 1976) before which a 1961 constitutional Article about the maintenance of forest boundaries had the unintended consequence of protecting some geosites. The first populist Turkish geo-journal, *Yeryuvarı veİnsan*, was released by the Geological Society of Turkey and drew attention to "natural monuments" (Arpat, 1976; Arpat and Guner, 1976). Then, due to the difficult domestic political situation, until the 1990s, cultural and nature protection activities were curtailed.

From 1961, two Turkish Ministry of Culture national bodies authorized nature conservation site selection, registration, and conservation; these were the Superior Council for the Conservation of Natural and Cultural Property and the Regional Conservation Council, respectively as advisory and technical committees. The Environmental Protection Agency for Special Areas was established in 1989 to examine and protect important landscape areas. From 2003, natural heritage protection was then entrusted to the Ministry of the Environment and Forestry under whose control, until 2011, was the Environmental Protection Agency for Special Areas. The latter was renamed in 2011 as the General Directorate for the Conservation of Natural Properties and placed under the Ministry of Environment and Urban Planning. However, the General Directorate of Culture and Museums, which replaced the General Directorate for the Conservation of Cultural and Natural Property in 2004, has direct responsibility for natural resources conservation. In 2004, the General Directorate of Nature Conservation and National Parks was formed under the then Ministry of the Environment and Forestry (now the Ministry of Water and Forestry). The Geological Society of Turkey (1945–1985) and the Chamber of Geological Engineers (founded in 1974), from 1947, organized annual congresses promoting geoconservation. However, the staff of the governmental Turkish Petroleum Corporation have discovered and published most of the geosites and geo-monuments. Some 26 university geology departments, along with 50 variously sized NGOs, have also contributed to nature conservation and geoconservation. Only the Turkish Association for the Protection of Geological Heritage (JEMIRKO), an offshoot of ProGEO, established in 2000, is now specifically working on geoheritage and geoconservation. JEMİRKO identified eight potential geoparks and over 450 geosites. Only about 70 geosites on its list have been conserved by the state to international standards. The various NGOs have been significant in bridging the gulf between Turkish society and state bureaucracy to enact nature conservation. It is through them that nature and landscape tourism, particularly geotourism, have developed. Turkey has 41 national parks (covering 8977 km^2), 34 nature parks (covering 7905 km^2), 31 nature conservation areas (covering 4658 km^2), and 105 natural monuments (covering 529 km^2). All protected sites, except some natural monuments, are owned by the state. Only two geosites, at Pamukkale (carbonate terraces) and Cappadocia (erosional earth features), are officially protected and included on the UNESCO World Heritage List. The Kızılcahamam-Camlıdere Geopark in Ankara, comprises 23 geosites—some of which are endangered (Kazancı, 2012). It was Turkey's first and the result of a joint project of the state bureaux and several NGOs. Privatization and the hiring out of caves and other interesting sites now seem to be the emerging government strategy to service geoconservation. However, there is still no national systematic management or maintenance of geosites; these also mostly lack management plans focused on their geological interest. The abundant laws, ministries (now Culture and Tourism, Water and Forestry, and Urban and Environment), and NGOs, together with the enforcement agencies with their different remits, has probably created an unwieldy and potentially ineffective nature conservation, let alone geoconservation, bureaucracy; perhaps one rather reminiscent of its Ottoman forebears.

Albania, the other ex-Ottoman Empire country, has no specific geoheritage or geoconservation statutes. However, some geosite protection is indirectly provided by laws pertaining to the rational exploration of mineral ores and raw materials; for example, "The Protection of Cultural Monuments and Rare Natural Wealth." Several state institutions have natural monuments and environmental protection within their remit; for example, the Committee of Environment Preservation and Protection, the Centre of Geographical Studies of the Academy of Sciences of Albania, and the Tourist Committee of Albania (Serjani et al., 2005). In the mid-1990s the official geological survey and ProGEO-Albania presented a "Project-Proposition on Geosites and Geoparks in Albania" to the Council of Ministers; a provisional list of geosites proposed for protection had already been prepared (Serjani and Cara, 1996) and their basis partly outlined (Serjani, 1996). The geosites were selected on the basis of a desktop study of numerous publications on Balkan geology and the eastern Mediterranean, plus presentations from various ProGEO meetings. For each selected geosite, the draft documentation included its index and number in the First National Inventory; topographic coordinates; status; a short description; and photo-illustrations. In 1998–1999, both bodies undertook a geoheritage inventory and the 1:50,000 *Map of Geological Sites of Albania* was published in 2001.

The 2002 Decision of Council of Ministers No 676 (2002) protected 669 areas and sites as Natural Monuments, of which almost 300 have geoheritage interest: 100 karst caves; 58 cold-water springs, mineral-water, and thermal springs; 28 wetlands (mainly karst and glacial lakes and marshes; and 195 geological sites. Surprisingly, the list of National Monuments omitted many geosites of local, national, and international importance already listed; it neither mentions nor defines geosites and geoparks. The "Regional, National and Local Lists of Geological Sites of Albania" study was completed in 2006 (Hallaci and Serjani, 2007). By the early 21st century Albania's geo-monuments, geosites, and geomorphosites had been inventoried and digitized in GIS format, facilitating recognition of potential geoparks. Albania's geoparks are areas with special biodiversity as well as geoheritage. Their definition was reviewed to include geo-monuments, so they could then (where appropriate) be proclaimed as National Geoparks. Following support by governmental bodies to complete their documentation, management, and infrastructure, some might eventually be registered with the European Geoparks Network. What Albania perhaps shows best is the influence of a specific, if relatively short-lived (compared to its mixed Ottoman and Habsburg roots), centralized political system. It has influenced the country's modern focus on landscapes as an economic resource, particularly for mineral extraction, which has, in turn, influenced emergent geoheritage and geoconservation matters.

2.3.2.4 GEOCONSERVATION IN CHINA

The various European (and UK) experiences and timeframes were, somewhat surprisingly, part mirrored by the situation in China. In 1956, its Government included the country's important geosites in a National Nature Reserve. It was another 30 years, following the country's economic reforms, before much more geoconservation work was achieved. In 1985, the First National Geological Natural Reserve (NGNR) was designated for its Middle Upper Proterozoic geology. In 1987, the Ministry of Geology and Mineral Resources (MGMR), incorporated into the Ministry of Land and Resources in 1998, issued a six-page circular (MGMR, 1987) proposing the establishment of Geological Nature Reserves; this also suggested that geoparks are part of GNRs. It was the first time the conservation of geosites was proposed in the form of ministerial regulations (Zhao and Zhao, 2003). The Nature Reserve regulation was issued, in 1994, as a basic framework for protecting China's geoheritage. The MGMR published *Regulations on the Protection and Management of Geosites* (MGMR, 1995) in 1995, in which geoparks were considered a means of protecting such geoheritage. These various 1990s' regulations documented the principles of geoconservation, identified appraisal criteria for the different levels of GNRs (e.g., county, provincial, and national) and outlined a classification system for their protection. These were swiftly acted upon and 86 GNRs (including 12 NGNRs) were established by 1999 as geological nature reserves rather than geoparks (Dong et al., 2013: 2).

2.3.2.5 THE GEOSITES PROGRAM

Obviously, geoconservation and geotourism, coupled with visitor management strategies must be interrelated activities if geotourism is to be globally accepted as a form of sustainable tourism. When visitor numbers and activities are well managed and matched to the physical resource of primary geosites and geomorphosites, seemingly there is little conflict between the two. Research in the UK by English Nature found that ordinary visitors' damage to geosites was then actually rare (Badman, 1994: 430), potentially supporting geotourism's future sustainable tourism credentials. The UK's Geological Conservation Review indicated that "Achieving recognition of a site with regard to its importance to conservation is possible through education and site publicity. This is also part of conservation, as is encouraging the 'use' of the site for scientific research or education and training" (Ellis, 1996: 99). Research and education were at the heart of a major European geoconservation initiative, later applied in Australia (Joyce, 2010).

This was the "Geosites" program, operated under the auspices of the International Union of Geological Sciences (IUGS) until 2004 (Dingwall et al., 2005) which began in 1995 (Wimbledon, 1996). Its aim was to develop an international database of the global geoheritage based on an internationally applied systematic approach to inventorying. Key selection criteria were: Representativeness; Uniqueness; Suitability for correlation; Complexity and geodiversity; Degree of research/study; and Site availability and Potential. In the selection of geosites, rather than enforcing a rigid framework, individual countries were encouraged to adopt their specific stratigraphic, tectonic, fossil, or other appropriate geological frameworks for the purpose. These would then identify potential "global geosites" that provided the evidence of major geological events and processes. Its methodology and application in the UK, where frameworks such as Carboniferous palaeobotany and the igneous history of the Caledonian orogeny were employed, has been discussed by Cleal et al. (1999). The program's primary purpose was to provide a sound factual basis for national and international initiatives to protect the geoheritage for the purposes of research and education. The database was intended to provide advice to the IUGS, and other bodies such as UNESCO, on priorities for geoconservation in a global context (Dingwall et al., 2005); therefore, it was of potential benefit to World Heritage Program. An initial Geosites list for Europe with that potential was published in 1998 (Wimbledon et al., 1998).

As the Geosites program developed, it was noted that "The need to preserve the planet's biological heritage touches a wide audience and continues to enjoy a high profile…Geological conservation on the other hand, does not enjoy this profile. The need for conserving examples of our geological and geomorphological heritage, remains an academic debate" (Bastion, 1994: 392). This is partly in recognition that "Science in fact is a cultural exercise and the strong links between geological features and the development of the science, raises the status of sites important in the history of the geology, to a status of cultural importance" (Page, 1998: 206) but this was not widely recognized. It follows then that the potential for geoconservation and geotourism promotion focused on both their inherent scientific worth and socio-cultural significance needs to be developed and that everyone involved should "…be mindful of the key role that education plays in shaping knowledge and appreciation of the environment; and those working at the chalkface (in education) should understand that the knowledge and attitudes that are inculcated can often formulate an understanding of the needs and value of conservation. What is true for the conservation movement at large is also true for geological conservation, perhaps even more strikingly as students of earth science are immediate consumers through fieldwork activities" (Hawley, 1994: 26). Of course, fieldwork is the key element of both geological training and professional practice. Getting out into the field is one of the main features of geotourism. Typically, but not exclusively, it takes place in non-urban areas, and rural landscapes are the preferred setting.

2.4 GEOTOURISM

2.4.1 DEFINITIONS AND APPROACHES

2.4.1.1 A DISCRETE TERM

Although in the last quarter of the 20th century, a few authors in Europe and Asia had fleetingly co-mentioned tourism and geology (Bastion, 1994; Jenkins, 1992; Komoo, 1997; Maini and Carlisle, 1974; Martini, 1994; Page 1998; Spiteri, 1994), until the mid-1990s' "geotourism" as a discrete term encompassing both was neither published nor defined; likewise, for "geotourists." The first contemporaneous published accessible account of research on geology and tourism (Jenkins, 1992) employed the terms "fossicking" and "fossickers"—both originated from Australia's 1850s' "gold rush" (Wilkes, 1978). In North America, "rockhounding" and "rockhounds" has been employed for similar activities and participants from at least the early 20th century. In the UK, "amateur geology" and "amateur geologists" has been, sometimes disparagingly, used from the mid-nineteenth century. In the UK, but intended for global usage, two geotourist groups were recognized, "casual" and "dedicated" (Hose, 2000), to overcome any potentially disparaging epithet. Casual geotourists occasionally visit geosites mainly for recreation, pleasure, and some limited intellectual stimulation; provision for them in the form of populist guides, trails, and visitor center is relatively recent. Dedicated geotourists intentionally visit geosites for the purposes of personal educational or intellectual improvement and enjoyment; provision for them in the form of field guides (Hose, 2006a) and journal papers is long-standing. Geotourists can also be split into "recreational" and "educational" geotourists (Hose, 1997).

2.4.1.2   AUSTRALASIAN AND EUROPEAN DEFINITIONS

In Australia, an early mention of geology and tourism suggested that "Geology has a basic role in ecotourism…educated lay people undertaking ecotours are clamouring for well presented explanations of landscape form" (Casey and Stephenson, 1996). This and another paper (Mayer, 1996) presented to the 1996 Geological Society of Australia (GSA) national conference are seemingly the earliest mentions of tourism geology (but not called geotourism) and geotourism in that country. Mayer (1996) rather followed the other paper's ecotourism approach, suggesting that geology and tourism, in areas such as the Great Barrier Reef and Kakadu, could work together for ecotourism. He argued for "small, compact, but well-illustrated guidebooks" that might be produced by the GSA. A 2002 review (Sharples, 2002) of Tasmania's geoconservation adopted a similar ecotourism approach. It noted that "The direct values of geological, landform and soil systems to humans, as our 'geoheritage,' are the reasons most frequently cited to justify geoconservation, and these are indeed important, albeit not the only, reasons to value geodiversity" (Sharples, 2002: 11); further, "Geoheritage may be of value to humans as…features which inspire us because of their esthetic qualities…; features of recreational or tourism significance (e.g., mountains, cliffs, caves, beaches, etc)…; features which form the basis of landscapes that have contributed to the 'sense of place' of particular human communities…" (Sharples, 2002: 11). However, the document made no mention of tourism geology or geotourism but indicates that "…the ecological value of geodiversity is also of direct value to humans in maintaining the amenity of the environment in which we live" (ibid).

Interestingly, in Tasmania, the approach adopted to promote geo-conservation was unusually to focus on its geomorphology. This was because of the recognized need for a better fit with nature conservation strategies than could be achieved with the traditional geosite approaches employed on mainland Australia (Household and Sharples, 2008). Finally, the Geological

Society of Australia's latest adopted definition, on its website, of geotourism is "...tourism which focuses on an area's geology and landscape as the basis for providing visitor engagement, learning and enjoyment."

Meanwhile, a review of the application of the European geosites initiative in Australia recorded that "Geotourism is a relatively new term, and does not yet appear in dictionaries. It can be seen as an extension of tourism generally, and a part of ecotourism in particular. A working definition of geotourism could be "people going to a place to look at and learn about one or more aspects of geology and geomorphology" (Joyce, 2006: 2); not dissimilar to a slightly later version from England (Larwood and Prosser, 2008). In a review paper, examining the history of the study of Australia's geoheritage, Joyce (2010) reiterated his earlier definition and suggested that "Geotourism, or tourism related to geological sites and features, including geomorphological sites and landscapes, can be seen as a new phenomenon and also a subset of geology and tourism" (Joyce, 2010: 53). However, "Australia's interest in geology for the public goes back to the mid-19th century. In 1866, the Jenolan Caves area in the Blue Mountains of New South Wales was specifically set aside for the use of tourists" (Joyce, 2007: 25).

Pullin (2016), however, reports an early Australian geotourist. The Austrian-born landscape painter Eugene von Guerard (1811–1901) traveled to and around Australia and New Zealand between 1852 and 1882. He was convinced that professional landscape painters needed to approach their work from a scientifically informed perspective. This informed approach is evident (Pullin, 2011) in his large body of work, often developed under arduous conditions. Indeed, as can be gauged from an 1870 letter extract "I had to put thousands of miles behind me on horseback, on foot and over the water, defeat troubles of every kind, endure many months of privation in the wilderness, to unite those few sheets in one volume, which now can be leafed through in a few minutes in a drawing room" (Heger, 1884: 156), he was truly a dedicated geotourist!

2.4.1.3 ASIAN "DEFINITIONS"

In Asia, specifically Malaysia, the term "tourism geology" was employed to "...place conservation geology at the same level of importance as the widely recognized conservation biology and will push geology to the fore" (Komoo, 1997: 2973) as a new form of applied geology that could support ecotourism's growth. Although "geotourism" had also been discretely used by Komoo and Deas (1993), they did not define the term. In China, the similar term "tourism earth science geology" (but not geotourism) was seemingly employed sometime in the 1980s for tourism focused on geology and its interactions with ecology and culture. It was defined in 1985 as "Tourism earth-science is a comprehensive marginal discipline which is aimed to find, evaluate, plan and protect natural landscapes and cultural relics with tourism value, and discuss their formation causes and evolution history on the basis of earth scientific theories and methods and in combination with the knowledge of other disciplines, with a view to promoting the development of tourism" (Chen and Li, 1985); regrettably, there appears to be no on-line access for scholars to this early journal paper. Tellingly, the definition makes no mention of geosites/geomorphosites or their rocks, minerals, and fossils and thus it is seemingly not, strictly speaking, true geotourism. Later, it was suggested that tourism earth-science has a branch, "tourism geology" that "…studies the distributions, types, characteristics, causes of formation and changes of varied scenic spots by geological theories, methods, technologies, and results. The comprehensive survey and evaluation of basic geology, karst geology, dynamic geology, and environmental geology of scenic areas and spots are conducted to organize targeted earth scientific travels; reasonably select travel routes and supporting facilities; maximally display the esthetic, cultural, and scientific values of scenic areas and spots; and integrate scientificity and interest, which constitute an emerging interdisciplinary science with unique research objects and methods" (Chen et al., 2015: 4); this, in the broadest sense, is landscape tourism and practiced since at least the 17th century in Europe (Hose, 2008b), if not in China.

The claim has been made that "Tourism earth-science overseas dates back to the 1930s at the earliest, while in China, it emerged in the late 1970s to the early 1980s…The term "tourism earth-science" was first put forward by Chinese scholars in 1985" (Chen et al., 2015: 2), but it was not termed geotourism (Hose, 2008b, 2016b). Further, in the same volume, it is suggested that this 2015 published English version, *The Principles of Geotourism*, is based on an earlier 1991 text; its Forward written by Jiqing Huang in 1989 opens with the statement that *"The Principles of Geotourism"* (1991), compiled by Anze Chen, Yunting Lu, et al. of China Tourism Earth-science Research Association, is a summary of research findings of several 100 members of the Association in five years since its establishment, as well as a pioneering move of China's earth-science workers to serve tourism and apply the theories and methods of earth-science to tourism "…as the honorary chairman of the Association, feel very happy about so many research findings in just several years after the establishment of the Association, and feel it necessary for me to recommend this monograph to the earth-science, tourist and academic circles and say a few words about the birth of this new discipline." However, this is seemingly a convenient modern retranslation of the original Chinese title, *The Principles of Tourism Earth-Science*, published in 1991 to unjustifiably claim a priori recognition of geotourism to those in UK and Australia (Hose, 1994b, 1995a). In a section in the same volume entitled *Research History of Tourism Earth-Science Abroad* (Chen et al., 2015: 13–16), that effectively purports to be a comprehensive overview, the seminal works on geotourism—at least the section was not entitled "geotourism" —by Dowling and Newsome (2006), Newsome and Dowling (2010), Hose (1994b, 1995a, 1997, 2000), and other European researchers are not mentioned; presumably because the chapter was not fully revised from the original 1991 version. However, setting aside such scholarly qualms, the volume is to be welcomed as a most readable overview of Chinese thoughts on geology-related tourism; one to be critically read by others tempted to be too domestically parochial in their literature reviews.

2.4.1.4 THE INITIAL UK AND FIRST EVER DEFINITIONS

It is generally acknowledged by most geotourism scholars that the first published definition of geotourism was in the UK in 1995 as "The provision of interpretive and service facilities to enable tourists to acquire knowledge and understanding of the geology and geomorphology of a site (including its contribution to the development of the Earth sciences) beyond the level of mere esthetic appreciation" (Hose, 1995a: 17). It was in a specially commissioned article in a professional interpretation magazine themed issue. The definition, although developed earlier, had only first been publicly included in a 1994 presentation (Hose, 1994b) at that year's USA Visitor Studies Association annual conference. That first published definition with some of its associated concepts was included within the *Geoparks Programme Feasibility Study* (Patzak and Eder, 1998; UNESCO, 2000). Also included were the essential elements of a later redefinition to "The provision of interpretative facilities and services to promote the value and societal benefit of geologic and geomorphologic sites and their materials, and ensure their conservation, for the use of students, tourists and other recreationalists" (Hose, 2000: 136).

The initial definition was developed from one for earlier research informally undertaken for English Nature (Hose 1994a, 1995b) on "Site-Specific Geological Interpretation," defined as "The promotion and explanation to a non-specialist audience of the geologic features and/or significance of a delimited area by either a fixed facility and/or populist publication"

(Hose, 1994a: 2). Following much subsequent research, it has been continuously refined, most recently as "The provision of interpretative and service facilities for geosites and geomorphosites and their encompassing topography, together with their associated in situ and ex situ artefacts, to constituency-build for their conservation by generating appreciation, learning, and research by and for current and future generations" (Hose, 2012a: 11). Therefore, it reinforces the initial geoconservation rationale and provides a succinct summary, employing an easily translatable vocabulary, of the nature, focus, and location of geology-based geotourism. These various definitions by Hose explicitly encompass an examination and understanding of geosites' physical bases, together with an examination of their interpretative media and promotion. Further, they encompass the life, work, collections, publications, artwork and field notes, personal papers, workplace, residences, and even the final resting places of geoscientists. The original term (sometimes hyphenated as "geotourism"), with its focus on geology, gained widespread recognition within the UK's geoscience community following the first dedicated national conference at the Ulster Museum in 1998 (Hose, 1998; Robinson, 1998).

2.4.1.5 OTHER EUROPEAN DEFINITIONS

However, other Europeans have employed broader geotourism definitions, but they are arguably less appropriate in terms of supporting and promoting geoconservation. In England, Larwood and Prosser defined it as "travelling in order to experience, learn from and enjoy our Earth heritage" (Larwood and Prosser, 1998: 98); at least the authors made the apposite assertion that "Geotourism is therefore, in part, a consequence of successful Earth heritage conservation as this ensures the presence of a resource to "experience and learn from and enjoy" (Larwood and Prosser, 1998: 98). In Germany, Frey considered "Geotourism means interdisciplinary cooperation within an economic, success-orientated and fast moving discipline that speaks its own language. Geotourism is a new occupational and business sector. The main tasks of geotourism are the transfer and communication of geoscientific knowledge and ideas to the general public" (Frey et al., 2008: 97–98). Her approach was used in the contemporary management of the Vulkaneifel Geopark where geological, economic, and political considerations combined to develop commercially orientated geotourism. In Poland, the inaugural issue of *Geoturystyka* had a paper that considered geotourism an "… offshoot of cognitive tourism and/or adventure tourism based upon visits to geological objects (geosites) and recognition of geological processes integrated with esthetic experiences gained by the contact with a geosite" (Slomka and Kicinska-Swiderska, 2004: 6).

2.4.1.6 BEYOND EUROPE

Outside of Europe, and the disciplines of geology and geomorphology, geotourism has been given a somewhat broader usage. In Australia, building upon their earlier collaborative text (Newsome et al., 2002), Dowling and Newsome's (2006) book was the world's first to be entitled *Geotourism* and clearly promoted geotourism's geological approach, something for which the authors are to be commended. In their later text, *Geotourism: The Tourism of Geology and Landscape* (Newsome and Dowling, 2010: 4), they stated that "Geotourism is a form of natural area tourism that specifically focuses on geology and landscape. It promotes tourism to geosites and the conservation of geo-diversity and an understanding of earth sciences through appreciation and learning. This is achieved through independent visits to geological features, use of geo-trails and view points, guided tours, geoactivities and patronage of geosite visitor centres" (Newsome and Dowling, 2010: 4). Dowling (2011) also recognized that "…the character of geotourism is such that it is geologically based and can occur in a range of environments from natural to built, it fosters geoheritage conservation through appropriate sustainability measures, it advances sound geological understanding through interpretation and education, and finally it generates tourist or visitor satisfaction" (Dowling, 2011: 1), thus, he reinforced the geoconservation value of geotourism.

2.4.1.7 CONFUSION BY NATIONAL GEOGRAPHIC

In the United States, National Geographic ignored previously published and widely available work in the UK and Europe and claimed to have actually coined the term itself for "tourism that sustains or enhances the geographical character of the place being visited. By that we mean its environment, its heritage, its esthetics, its culture and the well-being of its citizens." In short, geotourism is a "…destination's geographic character–the entire combination of natural and human attributes that make one place distinct from another…" (Stueve et al., 2002: 1). National Geographic also correctly initially considered that "Geography is, of course, about place. And tourism is also about place, but not necessarily all tourism, as we will see. In order to highlight this basic relationship between tourism and "sense of place" we've introduced a new term, geotourism, which derives from "geographical character." Of course, by then "geotourism" was most definitely not a new term having been in usage for around 10 years. National Geographic further differentiated their approach by indicating that ecotourism in focusing only on nature is a niche activity, while "Geotourism talks about everything that goes into making a place a place. Without question, it is sustainable tourism, but it focuses…on recognizing that there are opportunities to build on character of place, and so enrich both the travel experience and the quality of the locale." National Geographic's approach is remarkably akin to sustainable tourism. Indeed, this is reinforced by their definition in its preamble by "The concept of sustainable tourism is not new to the travel industry. In the past, its primary concern was to sustain balance with the ecological environment and minimize the impact upon it by mass-market tourism. The term "Geotourism" is closely related, but is concerned instead with preserving a destination's geographic character—the entire combination of natural and human attributes that make one place distinct from another. Geotourism encompasses both cultural and environmental concerns regarding travel, as well as the local impact tourism has upon communities and their individual economies and lifestyles." It has been argued, in support of their approach, that "Attracting the geotourist means focusing attention in a holistic way on all of the natural and human attributes that make a place worth visiting. That, of course, includes flora and fauna, historic structures and archeological sites, scenic landscapes, traditional architecture, and all of the things that contribute to culture, like local music, cuisine as well as the agriculture traditions that support the cuisine, local crafts, dances, arts, and so forth" (Buckley, 2003: 79). Proponents of the National Geographic approach have dismissed geology-focused tourism as a minor activity and suggest "The older use of the term is as shorthand for geological tourism, travelling to see rocks. This is a rather small specialist subsector!" (Buckley, 2003: 78–79). Conversely, the seminal Australian study, although it did not specifically refer to geotourism, noted "Fossicking is a popular special interest recreational and tourist activity…it comprises one of the world's largest single hobby groups" (Jenkins, 1992: 129). It has been erroneously suggested that National Geographic's approach singularly led to the rapid acceptance of the term, but given from their viewpoint "… If you travel to see particular scenery or wildlife or experience a particular local culture, climb a particular mountain or kayak a particular river, then in this sense you would be a geotourists" (Buckley, 2003: 79) it is clearly a re-branding of already recognized tourism activities. Besides which, others in the UK, Europe, and Australia had already done much more to promote the acceptance of the geological meaning of geotourism through their conference presentations, professional workshops, and publications.

Despite this overwhelming acceptance of the purely geological basis of geotourism, in 2011, the European Geoparks Network proposed (without recourse to any other interested stakeholders in the UK and Europe) acceptance of the National Geographic approach, but with a loose geoheritage emphasis. The Organizing Committee of the 11th European Geoparks Congress, held in Portugal, somewhat erroneously stated in their published "Arouca Declaration" that "...there is a need to clarify the concept of geotourism. We therefore believe that geotourism should be defined as tourism which sustains and enhances the identity of a territory, taking into consideration its geology, environment, culture, esthetics, heritage and the well-being of its residents. Geological tourism is one of the multiple components of geotourism" (International Congress of Geotourism, 2011). The Committee had really embraced "ecotourism," rather than true geotourism, which the United Nations World Tourism Organisation (UNWTO) defined as "tourism which leads to management of all resources in such a way that economic, social and esthetic needs can be fulfilled while maintaining cultural integrity, essential ecological processes, biological diversity and life support systems" (UNWTO, 1997). The Committee had also failed to understand that ecotourism might not be completely environmentally benign; it could really consume the very landscapes within which it is based, that is "geo-exploitation" (Hose 2008a, 2011). For example, tourists might be "...encouraged to visit spectacular remote mountain regions but be provided with western standards of comfort and accommodation resulting in associated environmental problems" (Acott et al., 1998: 238). The Committee's disputed break with the strictly geological approach adopted by most European governmental agencies, NGOs, and authorities involved in geotourism is at best unhelpful and at worst divisive for its stakeholders and confusing to governments and funding bodies.

Accepting its geological focus, knowledgeable practitioners and researchers of true geotourism do not preclude the benefits of working closely with other nature conservation and industrial heritage promotion interests; incorporation of, especially esthetic, landscape considerations (Hose, 2010a) would seem to make this obvious. The chief criticism of the Committee's and other such general and geographically focused definitions is that they do not prominently include geoconservation. Thus, they leave open the possibility that geotourism could be misused for "geo-exploitation" purposes. Martini (2000) was quite explicit geotourism had commercial potential "...for geological heritage, its economic value is revealed..." and clearly indicated that this was an attempt, because Europe's governments were unwilling to make available the necessary monies for geoconservation programs. Similarly, it has been noted in the UK, as stated in the original approach (Hose, 1995a), that geotourism and funding for geoconservation have tangible links; for example, "In today's economically stretched climate, tourists are a valuable source of local income. The encouragement of the tourist industry to include geodiversity within its remit is therefore high…" (Burek, 2012: 45).

2.4.2 GEOTOURISM IS...

2.4.2.1 "RECREATIONAL GEOLOGY"

Geotourism at the participant level, whether for mass or niche tourists, is essentially "recreational geology." This, unlike many other forms of mainly countryside recreation, is not limited in the UK, Europe, and other temperate lands by the seasons. This is because "…the weeds have died back and the leaves fallen from the trees making it easier to find and see rocks and landforms, when the days are short and cool" (Hose, 1995a: 17). Therefore, it could extend the tourism season and provide the means for developing alternative employment and regeneration strategies in places such as summer coastal and mountain winter resorts; likewise, in defunct mining and industrial areas. Geotourism's success depends upon identifying and promoting its physical basis, knowing about and understanding its user base, together with developing and promoting effective interpretative materials, all underpinned by geoconservation measures. In its present and strictest application, as tourism associated with geological and geomorphological sites (primary geosites and geomorphosites) and collections (secondary geosites), it is a relatively new and emerging form of niche (Novelli, 2005) or special interest (Weiler and Hall, 1992) tourism. Accepting geotourism's strict geological, which necessarily encompass geomorphology (Reynard et al., 2011), and geoconservation focused definition, it can be argued that many of the activities now associated with it can be traced back to earlier natural history-based movements and their activities.

2.4.2.2 A FORM OF NICHE TOURISM

In the strictest sense, geotourism is both a form of "niche tourism" (Hose, 2005) and of "special interest tourism." It has some overlap with other emerging forms of tourism such as "ecotourism" (Boo, 1990), "sustainable tourism" and "alternative tourism" (Cohen, 1987; Gonsalves, 1987) and potentially much overlap with "educational travel" (O'Rourke, 1990), "environmental tourism," "nature-based" (Richter 1987, 1989; Kutay, 1989) and "heritage" (Boniface and Fowler, 1993) tourism. The latter was invoked in a seminal USA geotourism, if not so named, study (O'Halloran, 1996) in which Brown's (1991) definition of heritage tourism was summarized as "...visits by persons from outside the host community that are motivated entirely or in part by interest in the natural and historical offerings of a community, region, group or institution." Further, the same study noted that natural heritage tourism "...embodies a belief that one can learn something new and interesting from fossils and other artifacts from earlier eras and that the experience of viewing fossils and artifacts can be quite exciting" (Behrensmeyer, 1994). The activities undertaken by, and probable motivations of, special interest tourists have been succinctly considered by Hall and Weiler (1992: 3) and similarly for special interest travel which involves "... people who are going somewhere because they have a particular interest that can be pursued in a particular region or at a particular destination" (Read, 1980: 195). The pioneering published Australian, if not so named as such, geotourism study reported that "The quality of the rural landscape with regard to features such as peace and quiet and scenery are almost as essential to on-site fossickers as the activity itself..." (Jenkins, 1992: 134). Special interest tourism is essentially "...tourism involving group or individual tours by people who wish to develop certain interests and visit sites and places connected with a specific subject. Generally speaking, the people concerned exercise the same profession or have a common hobby" (UNWTO, 1985: 3). It relies upon when the "...traveler's motivation and decision-making are primarily determined by a particular special interest. Therefore, the term... implies 'active' or 'experiential' travel"(Hall and Weiler, 1992: 5).

Some tourism studies indicate that the tourism vogue for active and participative elements suit tourists inclined toward conservation, scholarship, science, and environmental awareness (Heywood, 1990, 46). This would indicate it has a limited market relying upon well-educated and wealthy tourists. Despite the limited published research on special interest tourists (Hall and Weiler, 1992: 4), it is probable that "...many, though not all, broadly correspond to the allocentric category of Plog's (1974) psychographic continuum." Special interest tourism has been a recognized aspect of tourism studies from the 1980s, following major changes in the type of tourists, tourism product development and consumption, and research into these (Eadington and Redman, 1991; Krippendorf, 1986, 1987); it is still a growing (Hall, 1989; Read, 1980; UNWTO, 1985) tourism market segment.

2.4.3 THE PRECURSORS OF MODERN GEOTOURISM

2.4.3.1 LANDSCAPES AS A SOCIOCULTURAL CONSTRUCT

The landscapes visited by geotourists are as much sociocultural constructs as they are physical entities. Since the Renaissance, UK and European leisure travelers have chosen landscapes to visit that matched their quest for the novel, exotic, and authentic, initially on the bases of their esthetic preferences. Of course, outside of Europe, there must have been, if generally unreported in the limited geotourism historical literature, travelers elsewhere with similar motivations. Three major esthetic movements are generally considered, especially by literary and art historian scholars, in Europe to underpin people's changing perception of and subsequent relationship with the natural world, especially as it is observed in assumed Europe's "wild places."

Bunce (1994) has explored two of them, the Romantic and the Picturesque, in relation to Anglo-American perceptions of landscape. The movements subsequently impacted on the emergence of geology as an accepted pursuit, especially in the field, by those from the social elite—aristocrats and the wealthier mercantile class. They reflected three intertwined threads from the late-17th through to the mid-nineteenth centuries. These were: the nature of the traveler or visitor; the meanings ascribed to, and understandings of, the natural or assumed "wild" phenomena observed; and the shift from a mainly rural to an industrial society (in the UK after 1840, but later elsewhere in Europe).

2.4.3.2 THREE ESTHETIC MOVEMENTS

The first of the movements, the pursuit of the "Sublime," was overlapped and followed by the "Picturesque" before both were overtaken by the "Romantic." The Romantic poet William Wordsworth's (1820) *Guide to the Lakes*—which had first appeared as the anonymous introduction to a set of engravings of the region (Wilkinson, 1810)—distinguished two landscape formation phases that contributed separate but linked elements. "Sublimity" was the result of Nature's production of the Earth's initial rugged surface that was followed by the general tendency of subsequent processes to produce the more rounded beauty of fields and hills that created a whole unified landscape. Hence, the "Sublime" was concerned with Nature's first great underpinnings (the masses of rock, hills, and lakes) and their ability to solicit from the spectator a feeling of wonder and awe about their ruggedness and wildness. Such an approach to observing landscapes was adopted by the earliest of travelers, from around the late 17th century, into the British countryside. Wordsworth's *The River Duddon, A Series of Sonnets...* of 1835 was a key publication in promoting romantic landscapes, but generally for the social elite of the country. His 1810 *Guide to the Lakes* has been described, because of its approach to the description and analysis of landscape, as one of the first systematic geographical studies of a specific UK region (Whyte, 2000); unlike the few earlier guidebooks and travelogues to the Lake District, for example by Gilpin (1786), Green (1819), Hutchinson (1774), and West (1778). Gilpin's book with its promotion of the "Picturesque" had really kick-started the Lake District as a tourist destination; it gave rise to the contemporary epithet of "Lakers" for its tourists. Later guidebooks, such as Mackay's (1846) *The Scenery and Poetry of the English Lakes* and Hudson's (1842) *Complete Guide to the Lakes*, continued the theme of literature to guide and be read by tourists at specific locations in the region. It was also the first of the United Kingdom's mountain regions to have a specifically geology-focused guidebook (Otley, 1823).

Wordsworth along with some of the other romantic poets was familiar with the geological concepts of his day (Wyatt, 1995). Similarly, in north America, the Hudson River School of painters were familiar with geological concepts (Bedell, 2009); principal among those influenced by geology, like Eugene von Guerard in Australia, was John Frederick Kensett (1816–1872) who, like Wordsworth, had undertaken a Grand Tour and was much influenced by the Romantic movement. The 19th century American confluence of geology, tourism, and landscape painting in the mountains has been examined by Myers (1987). Interestingly, it was in America that from the last quarter of the 19th century where landscape photography began to supplant painting as an art form, particularly for the then remoter locations such as the Grand Canyon (Trimble, 2006). In the 20th century, this was taken to new levels of technical excellence and popularization by Ansell Adams (1902–1984), especially his work in the USA's national parks, later published in collected volumes such as *Yosemite and the Range of Light* (Adams, 1979) and *Our National Parks* (Adams, 1992) but beginning with the eighteen-print portfolio *Parmelian Prints of the High Sierras* (Adams, 1927).

Returning to the English Lake District, its geotourism and early tourism literature has been examined by Hose (2008b, 45–48) in the light of its role as an early geotourism destination. The "Picturesque" was concerned with the softer effects stemming from Nature's subsequent operations that produced the variegation and harmony expressed by the meandering curve of river or lake shore, the grouping of the rocks and trees which flanked them, the interplay of light and shade over these features, and the subtle color gradations that melded such scenes into one complete view. Such an approach was adopted, from the late 18th century, by the later groundbreaking travelers into the UK's less pastoral countryside. Waterfalls, with their obvious display of the power of Nature, were especially liked by such travelers; the epithet "cataractist," to describe those who particularly chose to visit waterfalls (Hudson, 2016), briefly passed into uncommon English usage. Many of the waterfalls they visited are the focus of modern, especially coach-based, tourists.

The overlap of the Picturesque and Romantic movements seems clear from the definition, in the 1801 *Supplement to Dr Johnson's Dictionary*, of the "Picturesque" as "What pleases the eye." By the beginning of the 19th century, it was in popular usage sufficient enough to overlap with the notion of the "Sublime." The "Romantic" was a movement that saw the expression of the feeling of landscape and its evocation in art and literature, but especially poetry. Romantic tourism was especially developed in the UK at the time of the closure of Europe to British tourists due to the French Revolution and the Napoleonic wars (1789–1815). This meant that the wealthy leisured British had to re-evaluate and explore their own, particularly upland, landscapes. Later, after the 1850s, the railway companies encouraged such domestic tourism and promoted these landscapes, especially with their own posters and guidebooks. These landscapes were so established by the close of the 19th century that the 20th century motor tourist followed in their tracks and continue to do so today.

2.4.3.3 THE GRAND TOUR

However, in the UK, most people—except perhaps the aristocracy (and later the offspring of wealthy merchants), some professionals (such as clerics and doctors), and the military—stayed quite close to their birthplace until after the Great War (1914–1918). Travel in Europe up to the 19th century had mainly been for commerce, employment, military service, or pilgrimage. Journeys were also undertaken by officials, those commissioned by government agencies, and researchers as a necessary means to undertake surveys; for example, Arthur Young (1741–1820), an agricultural and travel writer was one such individual. His best-known travel writings are *Tour in Ireland* (Young, 1892) and *Travels in France* (Young, 1792). His comprehensive accounts of UK farming include *The Farmer's Tour through the East of England* (Young, 1771). All contain some, albeit very limited, mention of geological matters. Those who did have the time to pursue pleasure travel were the aristocratic and upper middle class who ventured abroad on the "Grand Tour" (Black, 1992; Dolan, 2001; Hibbert, 1969) from the 17th up to the 19th centuries. Landscape tourism dating from the Grand Tour explored the spiritual elements of landscape, together with an examination of antiquities, significant (often ecclesiastical) architecture, and art works. However, there was also some interest shown by the tourists in natural history. For example, Norton Nicholls wrote in 1771 from Zurich that "Here I have... passed three days of seeing cabinets of natural history, and with the learned people of the place, such as they are; principally with Mr. Gesner..." (Black, 1992: 35); Solomon Gesner (1730–1788) was a poet and landscape painter held in some esteem at the time.

The Grand Tour landscape tourism was generally to safe and prepared destinations in the Alps and along the Mediterranean coast as far afield as Greece and sometimes Turkey. The Grand Tour itself was a feature of early adulthood from the 17th century for the offspring of the aristocracy and later also wealthy merchants. John Bacon Sawrey Morritt (1772–1843) was one of the more adventurous, farthest traveled, and almost last of the young "grand tourists" when he undertook his tour (Morritt, 1914) in the late 18th century when he was 22 years old. He made quite detailed observations on the countryside and antiquities he encountered on his journeys; for example, writing about the Peloponnese peninsula, then known as Morea, of southern Greece he observed in letter of April 1795 that "The scenery is a suite of little retired valleys, with clear streams or rocky rivers down them, the sides ornamented with wood, which only opened to discover glades covered with flowering shrubs and verdure" (Morritt, 1914: 196). Most grand tourists, as Morritt did, wrote letters home and kept a diary or journal. Some of these were published and then used by later tourists. However, the most useful guidebook was *The Grand Tour Containing an Exact Description of Most of the Cities, Towns and Remarkable Places of Europe*, originally published as four-volume set in 1743, by Thomas Nugent. Later, in the mid-nineteenth century, young, wealthy, and journeymen painter Americans began in increasing numbers to undertake the Grand Tour. The painters went on to use the experience to re-imagine and re-image their own countries. For example, John F. Kensett, following his European sojourn, painted and sketched landscapes for sale to tourists, much as he had done on his Grand Tour, and to illustrate a travel guide, *Lotus Eating*, by George William Curtis (1852).

As well as Continental grand tours, not so grand domestic tours within the UK, became popular from the end of the 17th century (Ousby, 1990;

Trench, 1990). From the late 18th century, the middle classes particularly "discovered" the English Lake District (Bray, 1793; Hose, 2008b) and the Scottish Highlands (Murray, 1799; Hose, 2010b) on extended tours. One of the earliest recorded of these tours lay as an unpublished journal (Fiennes, 1949) for over 200 years; it is an account of the horseback perambulations of Celia Fiennes (1662–1741) whom Hose (2008b) considered one of England's earliest geotourists because she visited a number of working mines. Such tours were particularly popular, with numerous published accounts, in the late 18th and early 19th centuries; the former is represented by a tour of the English midlands (Bray, 1783) and the latter a tour of northern England (Warner, 1802), both undertaken by reverends. The Hon. Mrs. Sarah Murray (1744–1811) undertook an extensive tour of northern England and Scotland (Murray, 1799). Scotland was particularly popularized for such early travelers by James Boswell (1740–1795) and Samuel Johnson (1709–1784). In the late summer of 1773, they went on a three-month tour from Edinburgh, taking in the islands of Skye, Raasay, Coll, Mull, and Iona on the west coast, as well as Inverness on the east coast—the Scottish Highlands and Islands. Both used these travels as the basis for their books *A Journey to the Western Islands of Scotland* (Johnson, 1775) and *The Journal of a Tour to the Hebrides, with Samuel Johnson, LL.D* (Boswell, 1786); the former, despite its unfavorable opinions of the Scots and distaste for Gaelic culture, surprisingly encouraged subsequent generations of travelers to follow their route. Boswell's volume included several geological observations.

2.4.3.4 HISTORICAL GEOTOURISTS

Many, now recognized, historical and essentially casual geotourists in the UK, such as Boswell, were also mainly traveling for other reasons, but they made sufficient geological and geomorphological observations to qualify in the modern sense as geotourists. Such an individual was Celia Fiennes (1662–1741) who mainly traveled for health reasons. Celia's contemporary, Daniel Defoe (1660–1731), published the widely read *Tour Through England and Wales Divided into Circuits* (Defoe, n.d.); however, supposedly based on his direct observations, there is some evidence that he plagiarized, as was common at the time, older travel writings. Both Fiennes and Defoe could have availed themselves of the UK's earliest, of a sort, regional tourist guides (Cotton, 1681; Hobbes, 1678) for the Peak District that mention geosites still visited today. These are the seven "wonders" described by Cotton and Hobbes: *"two fonts"*—the wells of Tideswell and St Ann;

"two caves"—Poole's Hole and Peak Cavern; *"one palace"*—Chatsworth House; *"one mount"*-Mam Tor; and *"a pit"*-Eldon Hole pothole. The original "wonders" are all well described by Ward (1827) in a volume also containing geological information useful to 19th century geotourists visiting the region. Much farther afield in China, Xu Xiake (1587–1641), whose considerably more extensive travels preceded those of Celia Fiennes and Daniel Defoe by a good century, was a geographer and renowned travel writer (compiled later into volumes such as the *Travel Notes of Xu Xiake*) at the time of the Ming Dynasty (1368–1644). Because he documented such geological features as river gorges, mineral beds, and the volcanic rocks of Tengchong mountain (noting the shape and type of its red pumice, with some scientific explanation of its formation), he surely qualifies as one of the earliest published Chinese, and indeed of any nationality, geotourists; it would be good to see his writings translated into English so that they might reach the wider audience they deserve. As the 19th century progressed, an increasing, but still comparatively small, number of the UK's middles classes visited the near continent (Mullen and Munson, 2009), mainly Belgium, France, Germany, and the Netherlands. Such continental perambulations were much informed by the burgeoning publication of travel guides aimed at the middle classes. Indeed, as such guides "…expanded in size, travelers used them not just as guides but as aids to plan their trips. To a large extent, they not only determined where people went and what they saw but often what they thought" (Mullen and Munson, 2009: 106). As Hose (2008b: 52) has already considered, such continental guides, in part precursors of the geology fieldguide (Hose, 2006a), were mainly produced by two publishers, Baedeker and Murray. Baedeker's were printed on poorer quality paper and always the cheaper of the two. Black of Edinburgh published good-quality regional guides for the domestic tourist from 1826. More popular and cheaper, as the eponymous title suggests, were Ward Lock's "Red Shilling Guides." Many of the continental and domestic guides had some limited mention of geology, usually in relation to some scenic attraction. In the 19th century, however, most of the British middle class took their holidays within the UK, which at the time politically encompassed the whole of Ireland. Such domestic tourists related mainly to their local or near holiday area because many lacked much leisure time and distant travel was difficult and expensive. This fostered an interest in local and regional landscape studies. Field excursions became a major attractive feature of local philosophical and natural history society programs. The few geological societies also organized field trips. Chief among these in the number and longevity of its field excursion program must be the Geologists' Association. Founded

in London in 1858, it encouraged fieldwork by amateur and professional/ academic geologists. Unusually, for the time, women were as welcome on its field excursions as men. The Association was innovative in developing its excursion program in terms of their locations and transport. Hose (2018) has recently reviewed the Association's Edwardian period excursions that were made by bicycle, a machine then at the cutting edge of personal transport. The Association advertised in advance its, mostly half- or full-day Saturday, excursions alongside its ordinary paper reading meetings and other business in the *Monthly Circular of the Geologists' Association* (hereafter, the *Circular*). Excursion reports were then published within three months in the *Proceedings of the Geologists' Association*. These usually repeated a lot of the information in the original *Circular* notices supplemented by additional illustrations and accounts of observations made on the excursions. They also often included social (commonly luncheon and refreshments arrangements, and any discussions) and weather information. Just occasionally, they include the number and type of excursionists, together with their mode(s) of transport. The accounts of its 19th century field excursions were twice published (Holmes and Sherborn, 1891; Monckton and Herries, 1910) as compendium volumes. These various accounts provide an invaluable record of not only the excursions, but also the field practices and contemporary attitudes to geoconservation (Green, 1989; Himus, 1954).

2.4.3.5 MANUFACTURIES

Although initially focused on exploring art collections, major buildings, and classic archeological sites, the Grand Tour, Continental or domestic, did not preclude visits to "manufacturies" or places where souvenirs were made, and raw materials extracted and refined. Such places would later become the focus of industrial heritage tourists' visits. The UK's industrial heritage, partly owing its genesis to geology as well as engineering, attracted early geotourists who compared these anthropogenic sites with "wild" landscapes; hence, Romantic landscape tourism is the antecedent of modern geotourism (Hose, 2008b). The social upheavals of the Great War led to increasing political pressure, with the emergence of hiking and cycling as major leisure activities, for access to the countryside, especially in the "wild" places jealously guarded by the major landowners. The public was at the time interested in the countryside for esthetic recreation and escapism (especially from the overcrowded towns and cities) rather than for scientific enquiry. In summary, the 19th century "romantic gaze," by which landscape tourists derived esthetic pleasure from the informed contemplation of beautiful and sublime scenes, was the product of esthetic and social influences that affected European society at a particular period of history (Urry, 1990), characterized by the move from a mainly rural to an increasingly urban economy and life. Of course, this does not imply that other societies in other periods of history did not develop esthetic responses, not necessarily the same as Europeans, to their own landscapes that also featured mountains, hills, lakes, rivers, and waterfalls. Immediately after the Second World War (1939–1945), the creation of National Parks and National Nature Reserves in the UK and similar designations across Europe probably seemed a natural culmination of the activities initiated by the small and mainly male social elite of late Georgian times in the UK.

2.4.4 EARLY INFLUENCES ON THE GROWTH OF MODERN GEOTOURISM

2.4.4.1 PHILOSOPHICAL AND NATURALIST SOCIETIES

The role of local philosophical societies, naturalists' societies, field clubs, and geological societies was crucial to the development of early modern geotourism in the UK and Europe. Their history in the UK, beginning in late Georgian times, is mirrored on a comparatively minor scale in the rest of Europe. In that context, it is worth noting that from the mid-18th century, scientific geological study in the field was mainly a British occupation, albeit with significant French and German contributions. The various natural science societies were founded, in the late 18th and early 19th centuries, from the close of the Age of Enlightenment. Just a very few were founded as late as the last quarter of the 19th century and exceptionally in the 20th century; the Woodstock Field Naturalists' Club founded in 1934 was unusually late. The first philosophical society, founded in Derby in 1783, served as something of a regional forum for male professionals (especially the gentry, clerics, doctors, industrialists, and manufacturers); it also established a large scientific library, a tool for research and dissemination emulated by many later kindred bodies. The society was founded by a leading figure of the day, Erasmus Darwin, something also emulated by many later kindred bodies.

The first local geological society, the Royal Geological Society of Cornwall, was founded in 1814. It was followed by the Newcastle Geological Society in 1829 and the Edinburgh Geological Society in 1834 (Burek, 2008). A philosophical society in Newcastle-upon-Tyne, founded in 1793, was superseded in 1829 by the Newcastle Geological Society. The Yorkshire Geological Society and The Liverpool Geological Society, respectively founded in 1837 and 1859, still exist and continue to publish their journals. Just like the other natural science societies, geological societies had mixed fortunes with closures and mergers. For example, The Liverpool Geological Association founded in 1880 eventually merged with the Liverpool Geological Society in 1910 (Tresise, 2013). The very short-lived Dudley and Midland Geological Society was founded in 1842, seemingly defunct after 1843, was replaced in 1862 by the Dudley and Midland Geological and Scientific Society and Field Club, which only just survived into the 20th century. Post 1830, these natural science societies were generally named naturalists' societies or field clubs rather than philosophical societies. For example, the Darlington Naturalists' Society was founded in 1860. This was belatedly after the demise of the short-lived Darlington Society for Promoting the Study of General and Natural History and Antiquities (founded at the latest in 1793 and then unusual in its choice of name). In 1891, the Darlington Naturalists' Society transferred its assets to the newly formed Darlington Naturalists' Field Club, renamed, in 1896, to the Darlington and Teesdale Naturalists' Field Club, which still exists. The Belfast Field Naturalists' Club (now the Belfast Naturalists' Field Club) was the first such formed club, in response to the increasing interest in natural sciences, in 19th-century Ireland. Its inaugural field-trip, on April 6, 1863, had nearly 90 of its members collecting fossils. This was a popular activity at a time because geology was then at the leading edge of scientific interest in the popular imagination. In fact, all these bodies organized both lecture and field-trip programs.

2.4.4.2 NATURAL HISTORY SOCIETY MUSEUMS

By the 1850s, many natural history societies had also begun to found museums to house their burgeoning collections (Knell, 2000) and libraries and in which to deliver their lectures. Both The Woolhope Naturalists' Field Club and The Chester Society for Natural Science (Robinson, 1971), respectively founded in 1851 and 1871, followed this pattern. Both of their buildings and much of their collections remain, albeit now in local authority management. By the 1920s, with a considerable demise in interest in the natural sciences but an increasing interest in archeology, most of the museums and libraries had passed into the care of local authorities. However, this was not always to the benefit of their natural science collections or libraries.

Sadly, this situation continued into the 1990s. The closure of the Passmore Edwards Museum (opened in 1990 as the Essex Museum of Natural History) in east London in 1995 is a case in point. Its natural history collections were removed to a redundant fire station, hardly conducive to their safe long-term storage, but at least there is now a voluntary effort, by members of the Essex Field Club, in the absence of a dedicated natural sciences curator to manage what remains of the natural history material; this work is undertaken at its Centre for Biodiversity and Geology in the Wat Tyler Country Park at Pitsea in Essex. The original purpose-built museum building is now used by the University of East London's Student Union. Quite similar histories of natural history society formation and decline, although on a much-reduced level when compared with the UK, are found in Europe.

2.4.5 LATER INFLUENCES ON THE GROWTH OF MODERN GEOTOURISM

2.4.5.1 GEOLOGY IN THE SCHOOL CURRICULUM

Twentieth and 21st century geotourism's growth is partly related to the public's exposure to geology during their schooldays and as adults in the mass media. The situation in the UK with regard to geology education matters reflects much of what is found across and outside of Europe. Although, for much of the 19th century, geology had enjoyed widespread interest and support, especially with the founding of numerous societies (Burek and Hose, 2016) in the UK, Europe, and North America, by the outbreak of the Great War (1914–1918), its popularity, like natural science society membership, had considerably dwindled (Allen, 1978). It had also much declined in interest to society at large, when compared to its popularity in the 19th century, to be supplanted particularly by archeology after the Great War with the renewed European interest in Egyptology. Following the Second World War (1939–1945) renewed general-public interest in dinosaurs, environmental concerns, and the emergence of the North Sea oil industry, developed a need for increased geological education in the UK in its universities in the 1960s and 1970s.

At least the practical value of school-based geological training was actually recognized in the UK in the 19th century. In 1890, the first national syllabus for geology was published by the British Association (Hamilton, 1976: 105–107) but it failed to have any influence on school curricula. School field work had been promoted, particularly for the teaching of physical geography (geomorphology) and science (although geology was not specifically mentioned), at around the same time (Cowham, 1900). By the 1930s, the UK's universities were annually only producing some 30 geology graduates (Hamilton, 1976: 105). Setting aside a few early teaching posts, most such appointments in England's universities were only introduced from around 1920. Indeed, "The first occupants of these [geography] posts could not, therefore, hold degrees in geography. They came into a new subject from outside, bringing with them the techniques of geology, botany, history and many other subjects" (Appleton, 1996: 6).

Finally, by the mid-1950s, in the UK's post-war period of educational innovation, geology "...had an important role in scientific education in schools in Britain. Experiments in integrated sciences, field sciences, and rural and environmental studies developed in the non-examination secondary schools. Geology's role in the school curriculum was examined in some detail" (Schools Council Geology Curriculum Review Group, 1977). Its role as an observational science with required fieldwork was accepted with some enthusiasm (Hamilton, 1976: 110). However, the underlying assumption was that geology, perceived perhaps as not a true science by many in the educational establishment, was unsuitable for the academically able pupils in the nation's grammar schools but was fine for those in its secondary modern schools. In schools, it was an adjunct to geography rather than a recognized component of science (Hamilton, 1976: 105). Until recently, when much of its content was spread across the science curriculum, it still occupied a similar position in the nation's comprehensive schools in which it was mainly "...taught by geographers in a way which was in accord with their needs and methods...geology or geomorphology had a largely service role...functional in understanding geographical issues" (Fisher, 1994: 477).

Further, teachers "...particularly those who came from scientific disciplines, found themselves more interested in the physical environment and others, especially those with a historical training, concentrated on the man-made landscape, they were forced without exception to direct their... teaching to the whole range of landscape, emphasizing particularly the interaction of its various components" (Fisher, 1994: 477). The quality of those teaching elements of geology and geomorphology in the geography curriculum is crucial to the positive engagement of pupils with the subjects. However, many were considered to be poorly prepared (Hawley, 1996, 1998), probably because of the increasing undergraduate emphasis from the 1980s on human and economic geography studies, in the 1990s. Today, the same issues continue but with science teachers forced to teach geological concepts in their respective disciplines. To compound such inter-curricular issue, schools were then, and especially so today, increasingly reluctant to

Historical Viewpoints on the Geotourism Concept 61

permit fieldwork (Robinson and McCall, 1996: 239), reducing the appeal of geology to most pupils, due to cost and purported health and safety issues.

2.4.5.2 INTERNATIONAL SCHOOL GEOLOGY EDUCATION

The First International Geoscience Education Conference was surprisingly held as late as 1993 (Stow and McCall, 1996). It revealed a wide range of national approaches to the inclusion of geology in the school curricula (King, Orion, and Thompson, 1995). A review of 1990s school curriculum changes in England and Wales, Israel and the USA showed that there had been a marked move away from discrete discipline-based science, including geology, toward integrated approaches to science teaching (Orion et al., 1999a, 1999b). Generally, the present European situation of geology in the school curriculum can be summarized (King, 2008) as either a small compulsory element of the national science curriculum (as in the UK) or a small compulsory element of the national geography curriculum (as in Germany and other northern European countries). Elsewhere in the world, such as Australasia and south-east Asia, it is generally an element within the science curriculum. In most of Africa, little geology appears in the school curricula of any country.

Unsurprisingly, the international demand, as exemplified by the UK (King, 1993; Smith, 1997), for discrete geological education at school examination level has always been extremely limited. By the close of the 20th century, geology's annual examination uptake in England and Wales (Hawley, 1998) was only around 2000 GCSE entries and 2500 "A" level entries; geography had around 265,000 GCSE entries and 45,000 "A" level entries. The "A" level entries between 1971 and 2004 (King and Jones, 2006) showed a big drop in the late 1980s and a steady fall in the second half of the 1990s, with entries stabilizing at around 1750 until 2005 when they dropped to 1689. In some ways, more significant than such pure statistics is they mask a steady decline in entries from comprehensive schools and further education colleges, while the small number of grammar and private school entries remained steady. Further, the gender balance has remained roughly constant at two-thirds male and one-third female. While there was also a geological element in "A" level Environmental Science, it too showed a decline. Likewise, GCSE geology entries continued their steady fall from more than 8000 in 1988, when GCSEs were first examined, to a mere 709 in 2004. The situation of school examination geology in the present decade is no better. Indeed, by the mid-2000s, there was even a suggestion to remove it entirely as an examination subject at GCSE when the only examination board offering geology proposed to withdraw it in 2010 (Anon., 2007). Of course, such past examination entry statistics to some extent correlate with the uptake of geology on university courses.

2.4.5.3   GEOLOGY EDUCATION AND THE UNIVERSITIES

From the 1960s (Hamilton, 1976) to the late-1980s, university geology provision in the UK increased, probably because of its popularization through plate tectonics and the North Sea oil industry. This saw university departments, "…having to expand to an unprecedented size and with more honours graduates in geology completing their training in any one of the many universities in one year than were produced for the whole country per year before World War II." (Hamilton, 1976: 105). However, by the 1980s, reduced demand for geology education in the UK is probably due to fewer opportunities for direct employment in its domestic coal mining and oil exploration industries and reduced "A" level geology entries, all coupled with the reluctance of many universities (even for places on geology courses!) and employers to recognize it as a science on par with chemistry and physics; behind all of this was the economic downturn of the 1980s. Up to half of university geology departments were forcibly merged with other departments or were closed down. Fortunately, where it is still offered in today's universities, such as in environmental science and some archeology courses, geology is a popular first-year undergraduate subject. However, only about 5% of undergraduates then specialize in geology. Postgraduate geology course provision has also markedly declined since the late-1980s. Hence, in many specialist fields, such as micro-palaeontology, no university-based courses were available for some time. Consequently, at present, specialist input to geological education and geo-interpretation for the public is increasingly from an ageing and decreasing pool of geologists. Undergraduate science communication and public engagement courses are unlikely to provide the geologically literate and skilled geo-interpreters that previously came from within the discipline.

2.5   THE DEVELOPMENT OF GEO-INTERPRETATION

2.5.1   HISTORICAL ASPECTS OF GEO-INTERPRETATION

Modern geotourism was developed in order to promote and help fund access to geosites and geoconservation by developing geology-focused sustainable tourism provision such as leaflets, geotrails, and visitor centers—that is, "geo-interpretation." The Geoconservation Review suggested this could be achieved, at the time, for geosites by "...erecting information sign boards on site, media publicity and producing books and leaflets..." (Ellis, 1996: 99). Of course, the available interpretative and informational media have increased since the mid-1990s. In reviewing the 1990s' launched Regionally Important Geological and Geomorphological Sites (RIGS) scheme (Harley and Robinson, 1991) in the UK, its chief protagonist, when summarizing its geosite selection criteria (Harley, 1996: 727), included three with direct geotourism application. These were the: value of the site for educational fieldwork; historic value of a site in terms of important advances in geological or geomorphological knowledge; and esthetic and cultural value of a site in the landscape, particularly in relation to promoting public awareness and appreciation of geology and geomorphology, its links with society, and the need for conservation. He further suggested its member groups could undertake geosite management (including the creation of stockpiles of spoil material for collectors) educational and promotional work (Harley, 1996: 729) as a new approach to UK geoconservation. Its member groups have been one of the major producers of inexpensive site-specific geotourism publications. Yet, much of this work, alongside practical geosite management measures (such as clearing scrub and litter, path maintenance, and fencing), has been undertaken by its mainly amateur (in the sense of unremunerated) workforce (Burek, 2011). Such RIGS-based interpretative provision has some remarkably early antecedents in the UK.

From the late-18th century onwards, there were several innovations, particularly in the UK but building upon Mediterranean influences seen on the Grand Tour, in the presentation and promotion of geology to the public. Many 17th and 18th century ornamental buildings and structures with no practical purpose (known as "follies") had considerable geological interest. For example, the grotto at Oatlands Park (built between 1760 and 1778 and demolished in 1948), near Weybridge (Barton and Delair, 1982) was encrusted with fossils and modern shells, together with real and artificial cave features. Similar grottos could be seen in Europe, especially in Italy, at around the same time. An interesting survivor of these is Pope's Grotto (Henry, 2007), originally in the grounds of a Palladian manor built for the poet Alexander Pope (1688–1744) on the banks of the Thames in 1720. In this house and its gardens, he incorporated some of the classical elements he had personally seen on his Grand Tours, particularly to Italy, including the grotto. It originally provided a passage for Pope's visitors, who had arrived by river, to enter the large formal garden. Having visited Hot wells thermal spa in Bristol and the Avon gorge in 1739, Pope was determined to create a mineral cave in the shell grotto he had originally built. To create the impression of a mine, the tunnel-like galleries had their walls and ceilings encrusted with the thousands of rocks, minerals, and fossils. Pope asked and/or cajoled numerous wealthy friends to send him the materials he needed from their estates and collections, and some 50 letters survive documenting the shipment of around 30 tons, to reline his grotto. Pope's house was demolished in 1808 by a later owner, irritated by souvenir hunters visiting the empty property. At least the cellar survived. Since the early 1900s, it has formed part of the foundations of a private school, now Radnor House Independent School. The grotto is now an unusual Grade 2 listed and protected building, but in the "Heritage at Risk" category, currently undergoing conservation. The School is interested in developing its Pope association and organizes visits on some weekends and during school holidays. The world's first attempt to create a geological theme park, albeit with admirable educational aims, was the construction, and location on accurately rendered geological sections, of three-dimensional re-constructions of prehistoric animals and plants at Crystal Palace (Doyle, 1993, 1995; Doyle and Robinson, 1993; Hawkins, 1854; McCarthy and Gilbert, 1994). These now survive as one of the few geological features to have Grade 1 Listed status, the highest legal protection, in the UK.

The world's first constructed geology trail (Baldwin and Alderson, 1996: 227), urban in character, was established in 1881, in a Rochdale (near Manchester) churchyard. It consists of 30 individually identified stone pillars. Each pillar is engraved with the rock type and a Biblical quotation; for example, "He made the Earth by His power. He hath established the world by his wisdom. The series of stones commencing here Boulder Stones, and in the descending order terminating with Lava elucidates the arrangement of the Strata of the Earth's crust in the order they were formed by the creator. Speak to the Earth and it will teach thee." This all aligned with the moral educational role ascribed to the study of the natural sciences, which was expected to lead to understanding and respect for the Creator's work.

The preservation of spectacular fossil finds where they originally grew or lived in the remote past was first undertaken in the 19th century. The first recorded was when, in the grounds of Sheffield's Yorkshire County Lunatic Asylum (the present-day Middle wood Hospital) the remains of a "fossil forest" were unearthed in 1873 (Sorby, 1875: 458). Several large fossil tree stumps were subsequently protected by two small locked sheds, the keys to which were available on request from the Asylum keeper; the sheds have long since gone but the original and newly discovered in 2004 (Boon, 2004) fossils, now surrounded by a housing estate, were preserved under a covering of earth (Cleal and Thomas, 1995a). A second preserved "fossil forest" was discovered in Edinburgh in 1887 (Kidston, 1888). Following the erection of a large covering building in 1888, "Fossil Grove" became the first publicly available and adequately interpreted (Black, 1988) fossil site; it is the world's first and longest open geological visitor center (Cleal and Thomas, 1995b). However, in recent years, its survival has been threatened by cuts in Local Authority expenditure and it has been passed into the management of charitable trust. Two other "fossil forests" (Buckland, 1840; Young, 1868) had earlier been discovered in the Glasgow area but they were not preserved (Thomas and Warren, 2008, 20–21) and the sites have long been built over. After all these various promising historic innovations, the UK's early lead in geo-interpretative provision was lost as the discipline itself declined in popularity. Little was then achieved, excepting in some museums, until the 1960s, but by the 1990s, such provision had multiplied and was again leading the way. However, in Europe, much was being achieved in novel interpretative provision, often employing the then new media, as in Andalusia in Spain (Hose, 2008c: 273–274).

2.5.2 MODERN ASPECTS OF GEO-INTERPRETATION

2.5.2.1 UK COUNTRY PARKS

In the 1990s, modern geotourism was thought, by those involved in geoconservation, to be a way to promote and help fund access to geosites and their geoconservation; geology-focused sustainable tourism provision was created such as leaflets, geotrails, and visitor centers. The relationship between geotourism and geoconservation, promoted by geo-interpretation, is underscored by noting that generally in the countryside "Site interpretation has much to offer the conservation profession and will continue to play a significant role in countryside recreation planning and management" (Barrow, 1993: 278). The increasing public demand for countryside-based recreation and interpretative provision that developed in the 1990s had its UK and European origins in the 1960s with the spectacular rise in motor vehicle ownership and leisure time, coupled with newly promoted countryside access measures. The creation in the UK of Country Parks (CPs), following the Countryside Act (1968), was significant in attracting, as had been their aim to relieve visitor pressure on the nation's National Parks (NPs) and Areas of Outstanding Natural Beauty (AONBs), people into the urban fringe for activity-based learning and recreation.

Many CPs were established on the farmlands and especially woodlands of old estates with their large country houses, such as Normanby Hall, near Scunthorpe in North Lincolnshire. Some, such as Park Hill Country Park in Staffordshire developed on the site of old coal workings, were established on areas reclaimed from extractive industry sites. Most of the parks had in place interpretative schemes, particularly nature trails, and visitor centers to enhance visitors' enjoyment and understanding. Indeed, "Virtually every countryside management plan has an interpretation element. Virtually all ranger services offer some interpretation to visitors…its integration into site planning and management in the countryside is perhaps the greatest single achievement of the past 20 years. You will now find interpretation provided at literally thousands of countryside localities…" (Phillips, 1989: 124). Such management and interpretation (Aldridge, 1975) was commonly based upon USA practice where wildlife recreation, especially field-sport based, and interpretation were much promoted from the 1920s onward. This was later much influenced by Tilden's (1957) seminal publication *Interpreting our Heritage*; an influence that has been revived, despite misgivings research (Hose, 2016a, 6) based upon modern visitor studies, in Europe by some geoparks following the 2011 "Arouca Declaration" (International Congress of Geotourism, 2011). The interpretation schemes recognized that people in ever increasing numbers usually visit the countryside in a social and pleasure-seeking context seeking experiences different to their daily lives "…from seeing the countryside, wildlife habitat and native birds and animals in their natural setting where man is clearly an observer of a system which is apparently controlled by something else. This is a phenomenon not restricted to an educated elite minority, but is of mass appeal" (Barrow, 1993: 271). An element of escapism coupled with estheticism is a prominent motivational factor for countryside visitors. Geologists in the UK were slow to recognize such potential to communicate to the public the significance of even its most significant geological localities.

2.5.2.2 THE WREN'S NEST AND LUDFORD CORNER

Among these was the Wren's Nest National Nature Reserve after its designation as the UK's first such reserve designated on purely geological grounds in 1956; it was designated a geological SSSI in 1990 and a Scheduled Monument (along with the nearby Castle Hill) in 2004 because of its contribution to its region's significant lime industry during the Industrial Revolution. Up to 90,000 tons of rock were removed annually from the Wren's Nest during the height of the industrial revolution and used in the many local iron-making blast-furnaces. Mining and quarrying ceased in 1924, leaving the hills honeycombed with a network of underground workings and caverns, and the area was left derelict as an informal recreation area. Wren's Nest is composed of Silurian Wenlock limestone packed with beautifully preserved marine fossils. Over 600 fossil species are known from around the Wren's Nest area, which was the first place in the world where a third of them were found. Internationally, it is famous, and its fossils can be found in museums throughout the world. It was something of a challenge in the 1950s to the then statutory conservation agency, the Nature Conservancy. Indeed, for "...the three geologists who then covered the nation in the interests of conservation...the concerns grew that there was a danger from over-collecting, and the approach changed to one of look-and-see rather than hammer-and-take" (Robinson, 1996a: 211). Several versions of the trail route (modified over time for safety reasons as parts of the site's underground workings deteriorated) and compact guidebook were supplemented by an on-site display in 1996.

Another one of these significant localities is at Ludlow (Lawson and White, 1989) in Shropshire. It is also an area rich in Silurian marine, as well as uncommon Devonian fish and terrestrial invertebrate, fossils. It is home, at Ludford Corner, to the internationally famous Ludlow Bone Bed and the overlying Downton Castle Sandstones, internationally famous for their rich fauna of acanthodian spines, the lodont denticles, and up to 14 species of Late Silurian fishes; it is the type locality for five of these species. An on-site plaque, rather well hidden by an inappropriately positioned bench, at Ludford Corner commemorates its global geological significance. It was where the UK's first purposely established educational geology trail, The Mortimer Forest Geology Trail (Hose, 2006b: 229–237), was opened in 1973 after the Forestry Commission (specifically, its Forest Enterprises division) and the Nature Conservancy Council cooperated on the project. The trail guide (Lawson, 1973, 1977) describes 13 sites (identified by numbered short wooden posts), of which all but two are old small quarries, along a 4-km route adjacent to a minor road. Another geology trail, also of 13 localities, was published in 2000 (Allbutt et al., 2002); its 10-km route is entirely along forestry tracks and suitable for both pedestrians and mountain-bikers.

2.5.2.3 GEOLOGY VISITOR CENTERS

In the late 1980s, several geology-focused heritage and visitor centers were opened; for example, the National Stone Centre (Hose, 1994c; Thomas and

Hughes, 1993; Thomas and Prentice, 1984) and the Charmouth Heritage Coast Centre (Edmonds, 1996). A range of interpretative activities such as talks, identification services, and guided walks were provided by both centers. The Geochrom project (Anon., 1990) within Dudley Zoo (near Birmingham) was a short-lived novel approach, integrating geology and wildlife conservation issues; the building and its displays has now reverted to a typical zoo type. A dual approach incorporating interpretative and geoconservation measures, although more really concerned with the latter, is the provision of on-site collecting facilities for recreational geologists. Such facilities can be provided by clearing rock faces (to make them both safe and their features more visible) to improve geological access (Anon., 1994: 16) and providing collection opportunities from mine and quarry spoil (noncommercial waste) material. For example, leveling and re-contouring work in the 1980s on colliery conical waste tips (Robinson, 1993a: 20) revealed a then unrecognized Upper Carboniferous land arthropod fauna (and the UK's richest insect fauna). Subsequently, the Writhlington National Nature Reserve was established (Jarzembowski, 1989: 219). These facilities offer little or no on-site interpretative media provision; they usually rely on published trail guides (Duff et al., 1985: 61–64). Sometimes collaboration between commercial collectors and public agencies, such as museums and national conservation agencies, to provide interpretative provision during fossil excavations are developed. For example, in the 1980s, the discovery of the world's oldest known amphibian (Rolfe, 1988), which was excavated in the grounds of a Glasgow social housing estate, capitalized on its potential for good public relations and income generation. Souvenirs were sold and conducted tours (Wood, 1983) were led.

2.5.2.4 SOME UK GEOTRAILS

Other interpretative possibilities are provided by urban geology and building stones trails (Burek and France, 1998). Sometimes, geology-themed art and sculptural works are created. One such large sculptural work is a modern henge, the European Community of Stone (Robinson, 1993b); its monumental stones represented the geology of the 12 then member states. Erected in 1992 at Frome in Gloucestershire, it was one of the few successful pan-European and EEC-funded geological interpretative projects. Commemorative plaques for significant geosites are more common; for example, that for the Silurian (especially the Ludlow Bone Bed) at Ludford Corner and Louis Aggisiz's glacial site in Edinburgh; the former is included in the Mortimer

Forest geotrail (Hose, 2006b: 229–237). More commonly, monuments or plaques commemorating the residence or workplace of some famous geologist are installed; for example, those in central London and Churchill in Oxfordshire for William Smith, the "Father of British Geology" and author of the world's first modern national geological map.

2.5.2.5 MAJOR UK GEOLOGY MUSEUMS

The UK's first official dedicated national geology museum, established by the Geological Survey, opened to the public in 1841. This was the Museum of Economic Geology, which displayed useful minerals and stones, together with manufactured items; it also had a laboratory in which the public could have samples of rocks and soils analyzed (Bailey, 1952). It moved to other premises in 1851 and, after closing in 1923, to its present home in 1935. Its last building, opened during the Geological Survey's centenary year, was purpose-built and was soon attracting over a quarter of a million annual visitors. Until the late 1960s, its displays, mainly three-dimensional representations of the Survey's regional geology guides, were virtually unaltered from the 1930s. Concurrently, most provincial museums also adopted and retained static specimen-rich displays that were essentially aimed at helping visitors identify their local geological finds. There was, however, some concern about the quality of provincial museum displays because "Poor displays serve only to reinforce negative attitudes to geology. Better displays in local museums concentrate on aspects of local geology but are almost always put together on a shoestring budget. Textbook in style and jargon-riddled - even the keen amateur would have difficulty in understanding some geological displays" (Knell and Taylor, 1991: 24). The national museums, with considerably larger budgets and access to research staff, began in the 1970s to move away from the type of traditional display with which their Victorian visitors would have been all too familiar.

The Geological Museum's ground-breaking multimedia exhibition, "The Story of the Earth," opened in 1973 (Dunning, 1974, 1975; Tresise, 1973; Walter and Hart, 1975) was the world's first permanent exhibition dedicated to plate tectonics. A number of other equally innovative exhibitions, such as "British Fossils" and "Dinosaurs," followed in the late 1970s and 1980s. A then £12 million (some £26 million in today's money) program to replace and renew displays (Clarke, 1991) and exhibitions (Dagnall, 1995; Robinson, 1996b; Sharpe et al., 1998) started in the 1990s; the first two of the new galleries, "The Power Within" (fundamentally about geology) and

"Restless Earth" (fundamentally about geomorphology), opened in 1996. Similar gallery approaches were adopted, on much smaller budgets, in the provinces; for example, at Liverpool Museum where "Earth Before Man"—a major revamp of a 1960s exhibition—opened in 1973 (Tresise, 1966). Spectacular multi-media exhibitions opened at the National Museum of Wales (Johnston and Sharpe, 1997; Kelly, 1994) and the National Galleries of Scotland in the mid-1990s. Unsurprisingly, due to financial constraints, most provincial museums did not adopt multimedia approaches for their geological exhibitions (Knell, 1993), but some have attempted innovative immersive exhibitions such as the "Time Trail" (Reid, 1994) at the now-closed Dudley Museum and Art Gallery. The trend in provincial museums, or at least those which have survived recent funding cuts by UK Local Authorities, is toward more holistic local community exhibitions rich in contemporary graphics with a social history bias in virtually all displayed disciplines.

2.5.2.6   GEOLOGY AND THE MASS MEDIA

Many of the new geology museum and heritage/visitor center exhibition techniques were pioneered by the mass media. Indeed, from the early 1970s, geology exhibitions in major museums employed computer graphics, supported by video displays and lighting systems imported from television and movie productions. The latter innovations had begun in the mid-20th century outside of Europe. From the 1940s, in the USA and in the UK, the communicative potential of the mass media, such as radio, to educate and inform the public on scientific matters, was recognized. Almost certainly, the first of these focussd on geology was in the USA. In 1954, the WGBH (Boston) radio station broadcast a 20-hour classroom geology course (Lyons, Robertson and Milton, 1993: 170). The BBC in the UK eventually adopted a similar format, in 1998, for its first series of six radio geology programs specifically aimed at schools. However, the use of radio for schools' programs in the UK had been much developed in the 1950s, but generally for the humanities, particularly music and dance.

Geology has had a varied but higher television exposure. One of the first television programs about the then new geology of plate tectonics, which emerged in the 1960s, was the BBC's 1972 "The Restless Earth" (Calder, 1972). The BBC broadcast in 1974 its first major television series about geology, "On the Rocks," albeit for a further education audience for which a supporting book was published (Wood, 1978). The 1974 program "The Weather Machine" (Calder, 1974) was the first populist program to examine ancient and recent ice ages for a mass audience. Sometimes elements of geology and geomorphology are included within the content of wildlife or astronomy programs. For example, the then imminent return of Haley's Comet was covered by a 1981 BBC program, which included an account of planetary impacts and mass extinctions; the inevitable BBC book (Calder, 1980) followed. In 1998, geology had a specific BBC program slot with the broadcast of "Rock Solid" (Grayson, 1988a, 1988b), a series of magazine-style programs. In the same year, the BBC broadcast "Earth Story" (Lamb and Singleton, 1998), a spectacular geology series, but narrated by an eminent biologist! The UK's sole major contemporary commercial television offering at the time was the 1990 series "Landshapes." Sir David Attenborough (1952-), the eminent British natural history broadcaster, has several times examined geological topics; for example, in the 1989 series "Lost Worlds, Vanished Lives" and the 2010 series "First Life." The emergence of computer generated-imagery (CGI) in the movie industry enabled the production of novel television programs on fossils that "brought to life" the bare bones in the rocks; for example, the BBC broadcast "Walking with Dinosaurs" (Haines, 1999), "Walking with Beasts" (Haines, 2001), and "Walking with Cavemen" (Lynch and Barrett, 2002) programs. Educational crossover with television and video recorded (tape and then CD/DVD) media (Booy, 1976), and now on the internet, have been promoted to disseminate virtual field-trips (Argles et al., 2015). However, whatever their attractions and educational value, it is difficult to believe they will, especially in the training of professional geologists, replace actual field-trips in the outdoors. Indeed, one of the attractions of the various geoparks movements is their commitment to providing on-site educational facilities.

2.6 GEOPARKS

2.6.1 GEOPARKS AND GEOTOURISM

Geoparks are essentially areas that include particularly significant geoheritage, within clearly defined boundaries of sufficient surface area for true territorial economic development and having a sustainable territorial development strategy. They have been the geotourism growth phenomenon of this century. Barring Antarctica and North America, they have either been established or their designation is actively being pursued on every continent as either national or UNESCO-recognized geoparks. Prior to their development, UNESCO international recognition, but not really formal protection, had been available since 1972 for a variety of landscapes and sites; the most significant recognition being their citation within the World Heritage List, which currently has some 630 sites, 7% of which are inscribed primarily for their geological interest. Because the "outstanding universal value" criterion, required by the World Heritage Convention, could not be met by many significant geosites and geomorphosites, an alternative recognition was sought. Hence, geoparks were proposed in order to promote landscapes on a holistic, rather than on a purely geological, basis. The initial geoparks program proposal documentation (UNESCO, 2000) incorporated, acknowledged, and quoted from the original geotourism approach adopted and defined by Hose (1995a). It envisaged that geoparks would recognize the relationships between people and geology and help to foster economic development within them.

The program's other major community benefit was to be to focus attention on geoconservation together with the related matter of sustainable development. Ideally, the geoscientific interest of geoparks should be allied to some archeological, cultural, historical, or ecological interest. To maintain their UNESCO Geoparks membership, parks must provide adequate educational provision. Significantly, for local and international geoconservation, the sale of geological material, whether from within or outside the geopark, is prohibited within them. Highly laudable as this intended sustainability measure is, it does not seem that the impact on pre-existing jewelry and geological souvenir businesses was considered. Consequently, the ban has created issues at classic localities (such as the "Jurassic Coast" of southern England) with a long history of collecting, preparing, and retailing fossils and minerals to their visitors, and from which many specimens have already found their way into museums worldwide, that might otherwise have been ideal geoparks. However, quarried or mined material for industrial or domestic use that extracted under national legislation is permitted within geoparks.

Sustainable geotourism within geoparks has been considered by Farsani et al. (2012). It is this author's contention that it is most likely to be promoted by the: sale of local products and souvenirs; creation of new products with geological connotations; reinforcement of local hotel and restaurant business through tourists and visitors; creation of new jobs linked to geology such as guides, technicians and craftspeople; and support for local public transport. However, for geoconservation purposes, it is regrettable that geopark designation of any type does not generate any national or international statutory protection. Whatever protection geoparks have is through other designations for their whole territory or its most significant elements. In the UK, this is provided for a landscape by either Area of Outstanding Natural Beauty (AONB) or National Park (NP) designation; specific elements of significant biological or geological importance are protected as a Site of Special Scientific Interest (SSSI) or a National Nature Reserve (NNR).

2.6.2 EUROPEAN GEOPARKS

In 2000, UNESCO supported the establishment of the European Geoparks Network (EGN) for sites with significant geoheritage and sustainable development strategies (Zouros, 2006). The development and rationale of Europe's geoparks has been recently documented (Zouros, 2013) and need not covered in detail herein. The EGN had four founding geoparks that had signed a convention to share information and expertise. One of these was at Aliaga (located within the Cultural Park of Maestrazgo) in Spain. Spain now has more than ten other geoparks: Cabo de Gata in Almería (Hose, 2008c); Sobrarbe in the central Pyrenees; and the Subbetic mountains in Córdoba and so on. Spain has a similar but lesser category of geotourism promotion, with "Geological Parks" such as the Chera Geological Park (in Valencia) and the Isona Cretaceous Park (in Lérida). In 2001, the EGN signed a collaborative agreement with, and placing it under the auspices of, UNESCO's Division of Earth Sciences. In 2004, the EGN agreed the "Madonie Declaration," recognizing the EGN as the official branch of the UNESCO Global Geoparks Network for Europe; it currently has over 75 geopark members in 26 countries.

Although there is no EGN membership fee, attendance at its twice yearly meetings is a requirement of membership. There is also a €1000 annual promotional fee for the publication and distribution cost of the EGN magazine, website, and its other corporate promotional activities. The evaluation of the application (and re-evaluation every four years) is undertaken by two experts appointed by the EGN. These experts may be from outside Europe because the EGN is part of the Global Geopark Network; hence, because their travel, accommodation, and ancillary expenses must be borne by the applicant, these costs can be significant. Some EGN geoparks began as national geoparks before becoming candidate EGN geoparks. For example, the Swabian Alb Geopark (based on Jurassic geology with typical karst features, Tertiary volcanic activity, and a meteorite crater) in Germany was established in 2002 as a national geopark; it joined the EGN in 2005. However, not all European or UK geoparks are members of the EGN. This is because they have either not applied for, or have withdrawn from, EGN membership. In the UK, the Abberley and Malvern Hills Geopark and the Lochaber Geopark withdrew, citing financial reasons, from the EGN in 2008 and 2011 respectively; the latter has indicated its intention to reapply once its finances improve. It is worth noting that any area, because it is not a statutory designation, may name itself a geopark. Consequently, Abberley and Malvern Hills and Lochaber, both famous for their Precambrian geology, continue to market themselves as geoparks. Lochaber, is unique among Europe's geoparks in having a record that involves both ancient tectonic plate collisions and the later rifting-apart of plates; in some ways, its geology complements that of Canada's Tumbler Ridge geopark.

The first UK geopark, designated in 2001, was the (now UNESCO global) Marble Arch Caves in Northern Ireland. In Eire, two UNESCO global geoparks, the Copper Coast and the Burren and Cliffs of Moher, were established 2001 and 2011, respectively as EGNs; both became UNESCO global geoparks in 2015. Interestingly, Ireland's best-known geological attraction, the Giant's Causeway and Antrim Coast with its Tertiary volcanic interest, is unlikely to be designated a geopark because it was given the higher accolade of a UNESCO World Heritage Site in 1986, having also been a NNR designated by the Department of the Environment for Northern Ireland in 1987. It is also Ireland's earliest recorded, from the end of the 17th century, geotourism attraction (Doughty, 2008). The first geopark in England, designated in 2003, was the North Pennines, based on Carboniferous geology and a major ore field. The designation in 2006 of the Forest Fawr Geopark (Davies et al., 2006) based on Devonian and Carboniferous geology, in Wales, within the Brecon Beacons NP, suggests that the two regions (the Peak District and Lake District) that were most significant in geotourism's early development (Hose, 2008b) could be similarly recognized. Indeed, there are plans for the former to make an application to the EGN (Benghiat, 2015); these include information centers, signed viewpoints, self-guided walks, cycle trails, and a Peak District GeoPark Way as five of its 16 projected outcomes. Both NPs are already the UK's most popular. Geopark designation would see the strong relationship between their geology and scenic beauty better promoted.

2.6.3 AUSTRALASIAN, CANADIAN, AND CHINESE GEOPARKS

Outside of Europe, the geoparks movement has seen considerable expansion, especially in China. In Australia, attempts to establish geoparks have been ongoing since the mid-2000s. However, the country already employs an alternative designation with some commonalities with the geopark approach. This was the Australian National Landscape (ANL) government initiative led until recently by a partnership of Parks Australia and Tourism Australia. The program was a national long-term strategic approach to tourism and conservation aimed at highlighting the nation's natural and cultural environments as tourism assets, by improving the visitor experience, consequently increasing support for their conservation; geotourism with geoconservation in all but name! Indeed, there are now 16 designated ANLs noted for their integrated focus on landscapes in which the development of geotourism aligns with the core focus and sustainable development. However, ANLs do not focus on geoconservation, but on broader natural heritage conservation. Further, any ANL may seek geopark branding if it considers that such global branding would enhance its geoheritage attractiveness to international visitors and/or enhance regional development opportunities for their state/territory governments. Australia's first UNESCO global geopark, Kanawinka (encompassing 26,910 km^2) was established in 2008. Its name was derived from the language of the Buandik aboriginal people, the traditional owners of the land, and perhaps fittingly means "Land of Tomorrow." It spreads across two States (South Australia and Victoria) and nine local government areas. It encompasses a significant portion of the Otway Basin (Douglas et al., 1988) with some 374 volcanic sites.

Today it is merely a national geopark because it was deregistered in 2010 as a UNESCO global geopark. This followed the decision, perhaps not helped by the complex political interrelationships across the geopark's area, of the Australian Government not to support its UNESCO status. Most recently, plans for a UNESCO geopark in Queensland stalled after concerns over its impact on the local cattle and mining industries, rather reminiscent of the situation in Turkey. The Etheridge Shire Council had planned to make an application in 2017 for a UNESCO global geopark (encompassing 40,000 km^2), which would have included the spectacular sandstone Gobbold Gorge. However, local landowners, especially ranchers, expressed concerns that if they were forced to work within a conservation area, such as the geopark, there was a level of implied regulation, which they would find uncomfortable.

This might well explain in part why there have been no geoparks established or proposed in the USA. Canada, however, has two: Stonehammer in New Brunswick and Tumbler Ridge in British Columbia, established in 2010 and 2014 respectively. The former, a UNESCO global geopark, has a geological record spanning a billion years. It has been examined, as one of Canada's oldest such research localities, by geologists for some 200 years. Tumbler Ridge is noteworthy for its sheer remote, wilderness location. It is the first to represent the plate tectonics phenomenon underpinning the formation of the Rocky Mountains. Its geology, seen in 41 geosites and its Dinosaur Discovery Centre building, thus spans the Precambrian to Cretaceous, as well as the Quaternary. It is probably best known for its Cretaceous dinosaur track-ways (many of global significance) and dinosaur bone bed. It also has important fossil fishes and marine reptiles from the Triassic.

Since 2000, China's provinces, autonomous regions, and municipalities have recommended many geoheritage sites, and sought their designation as geoparks. As of September 2017, China had over 300 provincial geoparks, 204 national, and 35 global geoparks (Ng, 2017). This is still by far the largest number of geoparks designated anywhere. They are members of the Asian Pacific Geoparks Network. One of the earliest was the Luochuan Loess National Geopark (LLNG) situated in central east China, approved in 2002. As a loess geopark it is still "unique in China and probably the world" (Dong et al., 2013: 4). It is geologically significant because China's loess is exceptionally thick compared with Europe. Quaternary loess is important in modern climate change studies. However, it research sites present some challenging geoconservation issues (Vasiljevic et al., 2014). Perusal of the literature on Chinese geoparks indicates a seemingly different emphasis in their establishment and management compared with Europe, a major focus on socioeconomic benefits (Zhao and Zhao, 2003) rather than geoconservation and promotion.

2.7 CLOSING THOUGHTS

Nearly 150 UNESCO global geoparks in over 40 countries and the many geoparks, with their outcomes focused on economic and social regeneration, will continue to develop the trends in interpretative provision developed during the 1970s industrial heritage boom that had similar desired outcomes. However, almost a decade after the suggestion that "Looking to the future and the development and promotion of Geoparks, greater and better use of the internet to illustrate and promote geosites can be expected; virtual field trips will become a feature of interpretative provision…" (Hose, 2010a: 55) this has not been fulfilled. Despites the advantages of electronic media's economy and easy revision, there is still an (over?) emphasis on hard-copy publications and on-site panels—probably because they are considered income generators; at least some geoparks have leaflets available as PDF files that can be downloaded from their websites. Despite the now widespread ownership of personal technology (such as smart-phones and tablet PCs), they are underutilized with little investment in dedicated applications to encourage and inform geosite visits.

It has been recognized in China and elsewhere that "The rise of geoparks not only potentially promoted regional economic development from tourism and created a new field of employment but also has attracted the attention of Chinese scholars" (Dong et al., 2013: 3). Impartial studies are needed that look into the costs and benefits of investment in geoparks.

Likewise, studies are also needed into the interactions of geoparks and their visitors in terms of the visitor experience and their environmental impact. Meanwhile, geoparks undoubtedly will become the places most visited by casual geotourists. This will pose a major sustainability issue, not just environmental but also the sociocultural impact, for geoparks. However, examination of numerous geopark and world heritage site proposals and management plans reveal little explicit mention of geoconservation (Hose, 2011), so emplacing geoconservation measures will continue to be an important issue to be addressed by geoparks. Consideration of the rise and fall in popularity of past scenic tourism destinations would be a worthwhile investment for geopark stakeholders.

Geotourism's emergence and development, which ultimately led to the creation of geoparks, required a fundamental shift in the way in which landscapes were perceived and then exploited for tourism; in particular, the recognition by travelers and tourists that wild landscapes were worthy places to visit was essential. The motivations of curiosity and esthetic value appeared before scientific value for travelers, although there was some later overlap with some artists and poets. Artists' (and later, photographers') visualizations and travelers' writings, and their impact on landscape recognition and promotion, continue to influence modern geotourism provision. The tourist guidebook is as popular as ever. Although today's geotourists continue the historic preference for perceived "wild" or "natural" landscapes, rather than the "controlled" agricultural landscapes and the "brutal" locations of heavy industry and mining, modern geotourism partly seeks to make the latter more acceptable places for geotourist visits. However, most, and certainly casual, geotourists are not adventurous in their choice of localities. Improvements during the 18th and 19th centuries in the physical and intellectual access, which encouraged elite tourists to visit previously virtually inaccessible locations (such as the Lake District and the Scottish Highlands and Islands), have 20th and 21st century parallels in niche tourism developments such as geotourism. These indicate how readily these elite and niche tourism destinations can become part of the mass tourism and packaged offering over time. The case studies indicate that the challenges related to increased public awareness of and access to wild areas, commonly considered as new issues by many of today's geotourism stakeholders, are akin to those that emerged almost 200 years ago; to rewrite the great Scottish geologist James Hutton's (1726–1797) famous uniformitarian principle, "the past is really the key to understanding and developing the present" and future geotourism provision.

KEYWORDS

- geo-interpretation
- geoparks
- geosites
- geoconservation
- geoheritage
- geotourism
- niche tourism

REFERENCES

Acott, T. G.; La Trobe, H. L.; Howard, S. H. An Evaluation of Deep Ecotourism and Shallow Ecotourism. *J. Sustainable Tour.* **1998,** *6* (2), 238–253.

Adams, A. *Parmelian Prints of the High Sierras*; Grabhorn Press: San Francisco, 1927.

Adams, A. *Yosemite and the Range of Light*; Little Brown and Company: Boston, 1979.

Adams, A. *Our National Parks*; Stillmam, A. G., Turnage, W. A., Eds.; Little Brown and Company: Boston, 1992.

Ager, D. V. Where Have All the Sections Gone? *Geol. Today* **1989,** *2* (3), 85–86.

Aldridge, D. *Guide to Countryside Interpretation, Part One: Principle of Countryside Interpretation and Interpretive Planning*; HMSO/Countryside Commission for Scotland/ Countryside Commission: London, 1975.

Alfrey, J.; Putnam, T. *The Industrial Heritage: Managing Resources and Uses*; Routledge: London, 1992.

Allbutt, M.; Moseley, J.; Rayner, C.; Toghill, P., Eds.; Itinerary 5: The Standard Ludlovian Section of Mortimer Forest. *Geologists' Association Guide No. 27: The Geology of South Shropshire*, 3rd Ed.; The Geologists' Association: London, 2002.

Allen, D. E. *The Naturalist in Britain: A Social History*; Allen Lane/Penguin Books: London, 1978.

Anon. *Guidelines for Collecting Fossils on the Isle of Wight*. Newport, Isle of Wight County Council, Newport, nd.

Anon. The Dudley Geochrom Where Past, Present and Future Meet. *Leisure Manager* **1990,** *8* (9), 12–13.

Anon. *County Durham Geological Conservation Strategy*; Durham, Durham County Council: Durham, 1994.
Anon. GCSE Geology Under Threat., *UKRIGS Newslett.* **2007,** *5* (2), 1.
Appleton, J. *The Experience of Countryside* (rev edn.); Wiley: New York, 1996.
Argles, T.; Minocha, S.; Burden, D. Virtual field Teaching Has Evolved: Benefits of a 3D Gaming Environment. *Geol. Today* **2015,** *31* (6), 222–226.
Arpat, E. İnsan ayağı izi fosilleri; yitirilen bir doğal anıt., *Yeryuvarı İnsan* **1976,** *1* (2), 45–49 [in Turkish].
Arpat, E.; Guner, Y. Goktaşı cukuru mu, cokme cukuru mu? *Yeryuvarı İnsan* **1976,** *1*, 12–13 [in Turkish].
Bademli, R. R. *National Action Plan for Environment: Protection of Natural, Historical and Cultural Sites and Matters*; State Planning Organisation: Ankara, 1997 [in Turkish].
Bailey, E. *Geological Survey of Great Britain*; Thomas Murby and Company: London, 1952.
Baird, J. C. Naked Rock and the Fear of Exposure. In *Geological and Landscape Conservation*; O'halloran, D., Green, C., Harley, M., Stanley, M., Knill, J., Eds.; The Geological Society: Bath, 1994; pp 335–336.
Baldwin, A.; Alderson, D. M. A Remarkable Survivor: A 19th Century Geological Trail in Rochdale, England. *Geol. Curator* **1996,** *6* (6), 227–231.
Barrow, G. Environmental Interpretation and Conservation in Britain. In *Conservation in Progress*; Goldsmith, F. B., Warren, A. Eds., Wiley: Chichester, 1993; pp 271–279.
Barton, M. E.; Delair, J. B. Oatlands Park Grotto and Its Ammonite Fossils. *Geol. Curator* **1982,** *3* (6), 375–387.
Bastion, R. De. The Private Sector—Threat or Opportunity? In *Geological and Landscape Conservation*; O'Halloran, D., Green, C., Harley, M., Stanley, M., Knill, J., Eds.; The Geological Society: Bath, 1994; pp 391–395.
Bedell, R. *The Anatomy of Nature: Geology and American Landscape Painting, 1825–1875*; Princeton University Press: Princeton, 2001.
Behrensmeyer, A. *Learning from Fossils: The Role of Museums in Understanding and Preserving Our Palaleontological Heritage.* Paper presented at "Partners in Paleontology: The Fourth Conference on Fossil Resources," October 1994, Colorado Springs, Colorado, USA.
Benghiat, A. *Interim Prospectus for a Peak District GeoPark Feasibility Study*; East Midlands Geological Society: Nottingham, 2015.
Bentivenga, M.; Geremia, F. *Geoheritage: Protecting and Sharing.* Papers presented at the 7th International Symposium ProGEO on the Conservation of the Geological Heritage, Bari, Italy, 24th–28th September 2012 (special issue). *Geoheritage* **2015,** *7* (1).
Black, G. P. Geological Conservation: A Review of Past Problems and Future Promise. In *Special Papers in Palaeontology No. 40—the Use and Conservation of Palaeontological Sites*; Crowther, P. R., Wimbledon, W. A., Eds.; Palaeontological Society: London, 1988, pp 105–111.
Black, J. *The British Abroad: The Grand Tour in the Eighteenth Century*; Stroud, Alan Sutton: Stroud, 1992.
Boo, E. *Ecotourism: The Potential and Pitfalls* (2 Vols); World Wildlife Fund: Washington, DC, 1990.
Boniface, P.; Fowler, P. J. *Heritage and Tourism in "The Global Village"*; Routledge: London, 1993.
Boon, G. Buried Treasure—Sheffield's Lost Fossil Forest Laid to Rest (Again). *Earth Heritage* **2004,** *22*, 8–9.

Boswell, J. *The Journal of a Tour to the Hebrides, with Samuel Johnson, LL.D, Containing Some Poetical Pieces by Dr. Johnson, Relative to the Tour, and Never before Published: A Series of his Conversation, Literary Anecdotes, and Opinions of Men and Books*; Charles Dilly: London, 1786.

Bray, W. *Tour Through Some of the Midland Counties, into Derbyshire and Yorkshire by William Bray, F.A.S. Performed in 1777*; Paraphrased in Volume 2, Mavor, W. (1798–1800) *The British Tourists; or Traveler's Pocket Companion, through England, Wales, Scotland, and Ireland. Comprehending the Most Celebrated Tours in the British Islands* (6 vols). London, 1783.

Bridgland, D. R. The Conservation of Quaternary Geology in Relation to the Sand and Gravel Extraction Industry. In *Geological and Landscape Conservation*; O'Halloran, D., Green, C., Harley, M., Stanley, M., Knill, J., Eds.; The Geological Society: Bath, 1994; pp 87–91.

Bridgland, D. R. Geoconservation of Quaternary Sites and Interests. *Proc. Geol. Assoc.* **2013**, *124*, 612–624.

Buckland, W. Anniversary Address to the Geological Society of London. *Proc. Geol. Soc. Lond.* **1840**, *III*, 231.

Buckley, R. Environmental Inputs and Outputs in Ecotourism: Geotourism with a Positive Triple Bottom Line? *J. Ecotour.* **2003**, *2* (1), 76–82.

Bunce, M. *The Countryside Ideal: Anglo-American Images of Landscape*; London, Routledge: London, 1994.

Burek, C. V. The Role of the Voluntary Sector in the Evolving Geoconservation Movement. In *The History of Geoconservation*; Burek, C. V., Prosser, C. D., Eds.; The Geological Society: London, 2008; pp 61–89.

Burek, C. V.; Prosser, C. D., Eds.; *The History of Geoconservation*; The Geological Society: London, 2008.

Burek, C. Geovolunteers: The Unpaid Force that Fuels Britain's Geoconservation Programme. *Earth Heritage* **2011**, *35*, 8–9.

Burek, C. V. The Role of LGAPs (Local Geodiversity Action Plans) and Welsh RIGS as Local Drivers for Geoconservation within Geotourism in Wales. *Geoheritage* **2012**, *4* (1–2), 45–63.

Burek, C. V. The Role of Local Societies in Early Modern Geotourism, a Case Study of the Chester Society of Natural Science and the Woolhope Naturalists' Field Club. In *Appreciating Physical Landscapes: Three Hundred Years of Geotourism*; Hose, T. A., Ed.; Special Publications, 417; The Geological Society: London, 2016; pp 95–116.

Burek, C. V.; France, D. E. NEWRIGS Uses a Steam Train and Town Geological Trail to Raise Public Awareness in Llangollen, North Wales. *Geoscientist* **1998**, *8* (9), 8–10.

Burek, C. V.; Potter, J. *Local Geodiversity Action Plans—Setting the Context for Geological Conservation. English Nature Research Reports, No. 560.* English Nature: Peterborough, 2006.

Farsani, N. T.; Coelho, C.; Costa, C. Geotourism and Geoparks as Novel Strategies for Socio-Economic Development in Rural Areas. *Int. J. Tour. Res.* **2011**, *13*, 68–81.

Calder, N. *The Restless Earth*; BBC Books: London, 1972.

Calder, N. *The Weather Machine*; BBC Books: London, 1974.

Calder, N. *The Comet is Coming!* Viking Press: New York, 1980.

Casey, J. N.; Stephenson, T. Putting Geology into Tourism—Some Tips and Practical Experiences. *Geological Society of Australia, 13th Australian Geological Convention, Canberra,* February 1996, Abstracts 41, 70.

Chen, A.; Li, W. Status Quo and Prospects of Tourism Earth-science. *Earth* **1985**, 4–16.

Chen, A.; Lu, Y.; Ng, Y. C. Y. *The Principles of Geotourism*; Springer and Science Press: Heidelberg and Beijing, 2015.

Clarke, G. Geology and the Public at the Natural History Museum. *Geol. Today* **1991,** *7* (6), 217–220.

Cleal, C. J.; Thomas, B. A. Wadsley Fossil Forest. In *Geological Conservation Review Volume 12: Palaeozoic Palaeobotany of Great Britain*; Chapman and Hall: London, 1995a; pp 208–210.

Cleal, C. J.; Thomas, B. A. Victoria Park. In *Geological Conservation Review Series No. 12: Palaeozoic Palaeobotany of Great Britain*; Chapman and Hall: London, 1995b; pp 188–191.

Cleal, C. J.; Thomas, B. A. *Geological Conservation Review Series No. 11: British Upper Carboniferous Stratigraphy*; Chapman and Hall: London, 1996.

Cleal, C. J.; Thomas, B. A.; Bevins, R. E.; Wimbledon, W. A. P. GEOSITES—An International Geoconservation Initiative. *Geol. Today* **1989,** *15* (2), 64–68.

Cohen, E. "Alternative Tourism"—A Critique. *Tour. Recreat. Res.* **1987,** *12* (2), 13–18.

Cotton, C. *The Wonders of the Peake*; London, 1681.

Cowham, J. H. *The School Journey: A Means of Teaching Geography, Physiography and Elementary Science*; Westminster School Book Depot: London, 1900.

Culshaw, M. G.; Bell, F. G.; Cripps, J. C.; O'Hara, M. Aspects of Geology in Planning. In *Planning and Engineering Geology—Geological Society Engineering Geology Special Publication No. 4-(Proceedings of the Twenty-Second Annual Conference of the Engineering Group of the Geological Society, Plymouth, September, 1986)*; Culshaw, M. G., Bell, F. G., Cripps, J. C., O'Hara M, M., Eds.; The Geological Society: London, 1987; pp 1–38.

Curtis, G. W. *Lotus-Eating: A Summer Book*; Hurst and Company: New York, 1852.

Dagnall, P. A New Era for the Earth Galleries. *Environ. Interpret.* **1995,** *10* (2), 4–5.

Davies, J.; Humpage, A.; Ramsay, T. Forest Fawr: A Welsh First. *Earth Heritage* **2006,** *26*, 10–11.

Defoe, D. *A Tour Through England and Wales Divided into Circuits* (2 vols); Dent, J. M.; London, nd.

Dingwall, P.; Weighell, T.; Badman, T. *Geological World Heritage: A Global Framework. A Contribution to the Global Theme Study of World Heritage Natural Sites, Protected Area Programme*; IUCN: Gland, 2005.

Dolan, B. *Ladies of the Grand Tour*; Harper Collins: London, 2001.

Dong, H.; et al. Geoconservation and Geotourism in Luochuan Loess National Geopark, China. *Quarter. Inter.* **2013,** *XXX*, 1–12.

Doughty, P. S. *The State and Status of Geology in UK Museums: Miscellaneous Paper No. 13*; The Geological Society: London, 1981.

Doughty, P. S. How Things Began: The Origins of Geological Conservation. In *The History of Geoconservation: Special Publication No 300*; Burek, C. V., Prosser, C. D., Eds.; The Geological Society: London, 2008; pp 7–16.

Dowling, R. K.; Newsome, D., Eds.; *Geotourism*; Elsevier: London, 2006.

Dowling, R. K. Geotourism's Global Growth. *Geoheritage* **2011,** *3*, 1–13.

Doyle, P. The Lessons of Crystal Palace. *Geol. Today* **1993,** *9* (3), 107–109.

Doyle, P. *Crystal Palace Park Geological Time Trail: Paxton's Heritage Trail Number Two*; London Borough of Bromley: Croydon, 1995.

Doyle, P.; Robinson, J. E. The Victorian Geological Illustrations of Crystal Palace Park. *Proc. Geol. Assoc.* **1993,** *104* (4), 181–194.

Duff, K. L.; McKirdy, A. P.; Harley, M. J., Eds.; *New Sites for Old: A Student's Guide to the Geology of the East MENDIPS*; Nature Conservancy Council: Peterborough, 1985.

Dunning, F. W. The Story of the Earth exhibition at the Geological Museum, London. *Museum* **1974**, *26* (2), 99–100.

Dunning, F. W. The Story of the Earth in the Geological Museum: A View From the Inside. *Geology* **1975**, *6*, 12–16.

Durham, Durham County Council, 1994. *County Durham Geological Conservation Strategy*; Durham County Council: Durham, 1994.

Eadington, W. R.; Redman, M. Economics and Tourism. *Ann. Tour. Res.* **1991**, *18* (1), 41–56.

Edmonds, R. P. H. The Potential of Visitor Centres: A Case History. In *English Nature Research Reports No.176—Earth Heritage Site Interpretation in England: A Review of Principle Techniques with Case Studies*; Page, K. N., Keene, P., Edmonds, R. P. H., Hose, T. A., Eds.; English Nature: Peterborough, 1996; pp 16–23.

Ellis, N. V., Eds.; *An Introduction to the Geological Conservation Review*; Joint Nature Conservation Commission: Peterborough, 1996.

Englefield, H. C. *A Description of the Principal Picturesque Beauties, Antiquities and Geological Phenomena of the Isle of Wight. With Additional Observations on the Strata of the Island, and Their Continuation in the Adjacent Parts of Dorsetshire by Thomas Webster*; T. Webster: London, 1816.

English Geodiversity Forum. *Geodiversity Charter for England*. English Geodiversity Forum, 2014 (accessed from http://www.englishgeodiversityforum.org/).

Fairbrother, N. *New Lives, New Landscapes*; Architectural: London, 1970.

Farsani, N. T.; Coelho, C. O. A.; de Costa, C. M. M.; de Neto de Carvalho, C., Eds.; *Geoparks and Geotourism: New Approaches to Sustainability for the 21st Century*; Brown Walker Press: Boca Raton, 2012.

Fiennes, C. *The Journeys of Celia Fiennes*; Morris, C., Ed.; Cresset Press: London, 1949.

Fisher, J. A. The Changing Nature of Earth Science Fieldwork Within the UK School Curriculum and the Implications for Conservation Policy and Site Development. In *Geological and Landscape Conservation*; O'Halloran, D., Green, C., Harley, M., Stanley, M., Knill, J. J., Eds.; The Geological Society: Bath, 1994; pp 477–481.

Foss, M. *On Tour: The British Traveller in Europe*; Michael O'Mara Books Ltd.: London, 1989.

Frey, M.-L.; Schafer, K.; Buchel, G.; Patzak, M. Geoparks—A Regional European and Global Policy. In *Geotourism*; Dowling, R. K., Newsome, D., Eds.; Elsevier: London, 2008; pp 95–117.

Gilpin, W. *Observations, Relative Chiefly to Picturesque Beauty, Made in the Year 1772, on Several Parts of England: Particularly the Mountains and Lakes of Cumberland, and Westmorland* (2 vols); London, 1786.

Gonsalves, P. S. Alternative Tourism—The Evolution of a Concept and Establishment of a Network. *Tour. Recreat. Res.* **1987**, *12* (2), 9–12.

Gray, J. M. *Geodiversity: Valuing and Conserving Abiotic Nature*; John Wiley & Sons Ltd., Chichester, 2004.

Gray, J. M. Geodiversity: The Origin and Evolution of a Paradigm. In *The History of Geoconservation: Special Publication No. 300*; Burek, C. V., Prosser, C. D., Eds.; The Geological Society: London, 2008; pp 31–36.

Grayson, A. Radio Geology. *Geol. Today* **1988a**, *4* (2), 62–63.

Grayson, A. *Rock Solid: Britain's Most Ancient Heritage*; Natural History Museum: London, 1988b.

Green, C. P. Excursions in the Past: A Review of the Field Meeting Reports in the First One Hundred Volumes of the Proceedings. *Proc. Geol. Assoc.* **1989,** *100* (1), 17–29.

Green, W. *The Tourist's New Guide, Containing a Description of the Lakes, Mountains, and Scenery in Cumberland, Westmorland, and Lancashire, with Some Account of Their Bordering Towns and Villages. Being the Result of Observations Made During a Residence of Eighteen Years in Ambleside and Keswick* (2 vols); Kendal, 1819.

Grub, A.; Albrechts, C. Earth Science Conservation in Germany. *Earth Sci. Conserv.* **1992,** *31*, 16–19.

Haines, T. *Walking with Dinosaurs: A Natural History*; BBC Worldwide: London, 1999.

Haines, T. *Walking with Beasts: A Prehistoric Safari*; BBC Worldwide: London, 2001.

Hall, C. M. Special Interest Travel: A Prime Force in the Expansion of Tourism? In *Geography in Action*; Welch, R., Ed.; University of Otago: Dunedin, 1989; pp 81–89.

Hall, C. M.; Weiler, B. What's Special about Special Interest Tourism? In *Special Interest Tourism*; Weiler, B.; Hall, C. M., Eds.; Belhaven: London, 1992; pp 1–14.

Hallaci, H.; Serjani, A. Database of the Geological Sites of Regional Importance in Albania: Selection and Completion with Database of Geological Sites of Regional Importance in Our Country—A Project of Geological Survey of Albania. In *Geological Heritage in South-Eastern Europe: Book of Abstracts*; Hlad, B., Herlec, U., Eds.; from the 12th Regional Conference on Geoconservation and ProGEO Working Group 1 Annual Meeting, Ljubljana, Slovenia, 5–9 September 2007, Ljubljana, Agencija RS za okolje; pp 19–20.

Hamilton, B. M. The Changing Place of Geology in Science Education in England. *J. Geol. Educ.* **1976,** *24*, 105–110.

Harley, M. J.; Robinson, E. RIGS—A Local Earth-Science Conservation Initiative. *Geol. Today* **1991,** *7* (2), 47–50.

Harley, M. J. Involving a Wider Public in Conserving Their Geological Heritage: A Major Challenge and Recipe for Success. In *Geoscience Education and Training in Schools and Universities, for Industry and Public Awareness*; Balkema: Rotterdam, 1996; pp 725–729.

Hawkins, B. W. On Visual Education as Applied to Geology Illustrated by Diagrams and Models of the Geological Restorations at the Crystal Palace. *J. Soc. Arts* and republished as a leaflet by James Tennant, 149 Strand: London, 1854.

Hawley, M. Conservation from the Chalkface. *Earth Heritage* **1994,** *2*, 26–27.

Hawley, D. Urban Geology and the National Curriculum. In *Geology On Your Doorstep*; Bennet, M. R., Doyle, P., Larwood, J. G., Prosser, C. D., Eds.; The Geological Society: London, 1996; pp 155–162.

Hawley, D. RIGS for Education. In *Proceedings of the First UK Rigs Conference*; Oliver, P. G., Ed.; Worcester, Herefordshire and Worcestershire RIGS Group: Worcester, 1998; pp. 65–78.

Henry, J. Pope's Grotto. *Newslett. Hist. Geol. Group Geol. Soc.* **2017,** *59*, 13.

Hewison, R. *The Heritage Industry: Britain in a Climate of Decline*; Methuen: London, 1987.

Heywood, P. Truth and Beauty in Landscape—Trends in Landscape and Leisure. *Landscape Australia* **1990,** *12* (1), 43–47.

Hibbert, C. *The Grand Tour*; Weidenfeld and Nicolson: London, 1969.

Himus, G. W. The Geologists' Association and its Field Meetings. *Proc. Geol. Assoc.* **1954,** *65* (1), 1–10.

Hindley, G. *Tourists, Travelers and Pilgrims*; London, Hutchinson: London, 1983.

Hobbes, T. *De Mirabilis Pecci: Being the Wonders of the Peak in Darby-shire, Commonly Called the Devil's Arse Peak. In English and Latine. The Latine Written by Thomas Hobbes of Malmsbury. The English by a Person of Quality*; William Crook: London, 1678.

Hooke, J., Eds.; *Coastal Defence and Earth Science Conservation*; The Geological Society: London, 1988.

Holmes, T. V.; Sherborn, C. D., Eds.; *Geologists' Association: A Record of Excursions Made Between 1860 and 1890*; Stanford: London, 1891.

Hose, T. A. Is It Any Fossicking Good? Or Behind the Signs—A Critique of Current Geotourism Interpretative Media. Unpublished keynote paper to Irish Geotourism Conference (DoENI/Geological SurveyNI/GeoConservation Commission), Ulster Museum, Belfast, 1988a.

Hose T. A. Selling Coastal Geology to Visitors. In *Coastal Defence and Earth Science Conservation*; Hooke, J., Ed.; The Geological Society: London, 1988b; pp 178–195.

Hose, T. A. *Interpreting Geology at Hunstanton Cliffs SSSI Norfolk: A Summative Evaluation (1994)*; The Buckinghamshire College: High Wycombe, 1994a.

Hose, T. A. *Rockhounds Welcome!—Assessing the United Kingdom Client Base for Site-Specific Geological Interpretation*. Unpublished Paper to the Visitor Studies Association Annual Conference, 26–30 July 1994, Raleigh, North Carolina, USA, 1994b.

Hose, T. A. Telling the Story of Stone—Assessing the Client Base. In *Geological and Landscape Conservation*; O'Halloran, D., Green, C., Harley, M., Stanley, M., Knill, S., Eds.; The Geological Society: Bath, 1994c; pp 451–457.

Hose, T. A. Selling the Story of Britain's Stone. *Environ. Interpret.* **1995a**, *10* (2), 16–17.

Hose, T. A. Evaluating Interpretation at Hunstanton. *Earth Heritage* **1995b**, *4*, 20.

Hose, T. A. Geotourism—Selling the Earth to Europe. In *Engineering Geology and the Environment*; Marinos, P. G., Koukis, G. C., Tsiamamaos, G. C., Stournass, G. C., Eds.; Balkema: Rotterdam, 1997; pp 2955–2960.

Hose, T. A. European Geotourism—Geological Interpretation and Geoconservation Promotion for Tourists. In *Geological Heritage: Its Conservation and Management*; Barretino, D., Wimbledon, W. P., Gallego, E., Eds.; Instituto Tecnologico Geominero de Espana: Madrid; 2000; pp 127–146.

Hose, T. A. Geotourism—Appreciating the Deep Time of Landscapes. In *Niche Tourism: Contemporary Issues, Trends and Cases*; Novelli, M., Ed.; Elsevier: Oxford, 2005; pp 27–37.

Hose, T. A. Leading the Field: A Contextual Analysis of the Field Excursion and the Fieldguide in England. In *Critical Issues in Leisure and Tourism Education: Current Trends and Developments in Pedagogy and Research*; Wickens, E., Hose, T.A., Humberstone, B., Eds.; Leisure and Tourism Education Research Centre: Buckinghamshire Chilterns University College, High Wycombe, 2006a; pp 115–132.

Hose, T. A. Geotourism and Interpretation. In *Geotourism*; Dowling, R. K.; Newsome, D., Eds.; Elsevier/Heineman: Oxford, 2006b; pp 221–241.

Hose, T. A. The Genesis of Geotourism and Its Management Implications. In *Abstracts Volume, 4th International Conference, GEOTOUR 2008, Geotourism and Mining Heritage*; AGH University of Science and Technology: Krakow, 2008a; pp 24–25.

Hose, T. A. Towards a History of Geotourism: Definitions, Antecedents and the Future. In *The History of Geoconservation: Special Publications, 300*; Burek, C. V., Prosser, C., Eds.; The Geological Society: London, 2008b; pp 37–60.

Hose, T. A. Geotourism in Almeria Province, Southeast Spain. *Tour Interdisciplinary J* **2008c**, *55* (3), 259–276.

Hose, T. A. Volcanic Geotourism in West Coast Scotland. In *Volcano and Geothermal Tourism: Sustainable Geo-Resources for Leisure and Recreation*; Erfurt-Cooper, P., Cooper, M., Eds.; Earthscan: London, 2010a; pp 259–271.

Hose, T. A. The Significance of Esthetic Landscape Appreciation to Modern Geotourism Provision. In *Geotourism: The Tourism of Geology and Landscapes*; Newsome, D., Dowling, R. K.; Eds.; Oxford, Goodfellow: London, 2010b; pp 13–25.

Hose, T. A. The English Origins of Geotourism (As a Vehicle for Geoconservation) and Their Relevance to Current Studies. *Acta Geogr. Slovenica* **2011**, *51*, 343–360.

Hose, T. A. 3G's for Modern Geotourism. *Geoheritage* **2012a**, *4* (1–2), 7–24.

Hose, T. A., Eds.; "Geotourism and Geoconservation" (special issue). *Geoheritage* **2012b**, *4* (1–2).

Hose, T. A., Eds.; *Geoheritage and Geotourism: A European Perspective*; Boydell and Brewer: Woodbridge, 2016a.

Hose, T. A. *Appreciating Physical Landscapes: Three Hundred Years of Geotourism*; The Geological Society: London, 2016b.

Hose, T. A. A Wheel in Edwardian Bedfordshire: A 1905 Geologists' Association Cycling Excursion Revisited and Contextualised. *Proc. Geol. Assoc.* **2018**, *129* (in press).

Household, I.; Sharples, C. Geodiversity in the Wilderness: A Brief History of Geoconservation in Tasmania. In *The History of Geoconservation: Special Publication No. 300*; Burek, C. V., Prosser, C. D., Eds.; The Geological Society: London, 2008; pp 257–272.

Hudson, B. J. *Complete Guide to the Lakes*; Hudson: Kendal, 1842.

Hudson, B. J. Waterfalls and the Romantic Traveler. In *Appreciating Physical Landscapes: Three Hundred Years of Geotourism, Special Publications, 417*; Hose, T. A., Ed.; The Geological Society: London, 2016; pp 41–57.

Hunter, M., Eds.; *Preserving the Past: The Rise of Heritage in Modern Britain*; Alan Sutton: Stroud, 1996.

Hutchinson, W. *An Excursion to the Lakes, in Westmoreland and Cumberland, August 1773*; London, 1774.

International Congress of Geotourism. *Arouca Declaration*, 2011, Available from: https://dl.dropboxusercontent.com/u/36358978/News/Declaration_Arouca_%5BEN%5D.pdf (accessed September 2, 2015).

James, L. *The Middle Class: A History*; Little Brown: London, 2006.

Jarzembowski, E. Writihlington Geological Nature Reserve. *Proc. Geol. Assoc.* **1989**, *100* (92), 219–234.

Jenkins, J. M. Fossickers and Rockhounds in Northern New South Wales. In *Special Interest Tourism*; Weiler, B., Hall, C. M., Eds.; Belhaven, London, 1992; pp 129–140.

Johnson, S. *Journey to the Western Islands of Scotland*; Strahan, W., Cadell, T.; London, 1775.

Johnston, D.; Sharpe, T. Visitor Behaviour at *The Evolution of Wales* Exhibition, National Museum and Gallery, Cardiff, Wales. *Geol Curator* **1997**, *6* (7), 255–266.

Joyce, E. B. Geological Heritage of Australia: Selecting the Best for Geosites and World Heritage and Telling the Story for Geotourism and Geoparks. *AESC 2006 Extended Abstract*; 2006b.

Joyce, B. Geotourism, Geosites and Geoparks: Working Together in Australia. Special Report, *The Australian Geologist* **2007**, *144*, 26–29.

Joyce, E. B. Australia's Geoheritage: History of Study. A New Inventory of Geosites and Applications to Geotourism and Geoparks. *Geoheritage* **2010**, *2* (1), 39–56.

Kazancı, N. Geological Background and Three Vulnerable Geosites of the Kızılcahamam Camlıdere Geopark Project in Ankara, Turkey. *Geoheritage* **2012**, *4* (4), 249–261.

Kelly, J. Cardiff Arms Its Park. *Museums J* **1994**, *94* (3), 25–26.

Ketin, İ. Turkiye'de onemli jeolojik aflormanların korunması. *Turkiye Jeoloji Kurumu Bulteni* **1970**, *13* (2), 90–93 [in Turkish].

King, J. H. Earth Science in the National Curriculum of England and Wales. *J. Geol. Educ.* **1993**, *41*, 318.

King, C.; Jones, B. The A Level Geology Challenge. *Teach. Earth Sci.* **2006**, *31* (1), 29–31.

King, C.; Orion, N.; Thompson, D. Earth Science in Britain and on the World Stage. *School Sci. Rev.* **1995**, *77*, 121–124.

Knell, S. J. The End for the Frodingham Ironstone? *Geol. Today* **1990**, *6*, 125–128.

Knell, S. No Stone Unturned. *Museums Journal* **1993**, *93* (3), 20.

Knell, S.; Taylor, M. A. Museums on the Rocks. *Museums J.* **1991**, *91* (1), 23–25.

Knell, S. J. *The Culture of English Geology 1815–1851: A Science Revealed Through Its Collecting*; Aldershot: Ashgate, 2000.

Komoo, I.; Deas, K. M. *Geotourism: An Effective Approach Towards Conservation of Geological Heritage*; unpublished paper. Malaysia, Symposium of Indonesia-Malaysia Culture, Badung, 1993.

Komoo, I. 1997. Conservation Geology: A Case for the Ecotourism Industry of Malaysia. In *Engineering Geology and the Environment*; Marinos, P. G., Koukis, G. C., Tsiambaos, G. C., Stournas, G. C., Eds.; Balkema: Rotterdam; pp 2969–2973.

Kotler, P.; Haider, D. H.; Rein, I. *Marketing Places: Attracting Investment, Industry, and Tourism to Cities, States, and Nations*; The Free Press: New York, 1993.

Krieg, W. Progress in the Management for Conservation of Geotopes in Europe. **Geol. Balcan.** **1996**, *26* (1), 13–14.

Krippendorf, J. Viewpoint: The New Tourist—Turning Point for Leisure and Travel. *Tour. Manage.* **1986**, *7* (2), 131–135.

Krippendorf, J. Ecological Approach to Tourism Marketing. *Tour. Manage.* **1987**, *8* (2), 174–176.

Kutay, K. The New Ethic in Adventure Travel. *Buzzworm: Environ. J.* **1989**, *1* (4), 31–36.

Lamb, S.; Singleton, D. *Earth Story: The Shaping Of Our World*; BBC Books: London, 1998.

Larwood, J.; Prosser, C. Geotourism, Conservation and Tourism. *Geol. Balacan.* **1998**, *28* (3–4), 97–100.

Larwood, J. G.; Markham, D. *Roads and Geological Conservation: A Discussion Document*; English Nature: Peterborough, 1995.

Last, J.; Brown, E. J.; Bridgland, D. R.; Harding, P. Quaternary Geoconservation and Palaeolithic Heritage Protection in the 21st Century: Developing a Collaborative Approach. *Proc. Geol. Assoc.* **2013**, *124* (4), 625–637.

Lawson, J. D. New Exposures on Forestry Roads near Ludlow. *Geologic. J.* **1973**, *8* (2), 279–284.

Lawson, J. D. *Mortimer Forest Geological Trail*; Nature Conservancy Council: London, 1977.

Lawson, J. D.; White, D. E. The Ludlow Series in the Ludlow area. In *A Global Standard for the Silurian System: Geological Series No. 10*; Holland, C. H., Bassett, M. G., Eds.; National Museum of Wales: Cardiff, 1989; pp 73–90.

Leafe, R. Conserving Our Coastal Heritage–A Conflict Resolved. In *Coastal Defence and Earth Science Conservation*; Hooke, J., Ed.; London, The Geological Society: London, 1998; pp 10–19.

Lincolnshire Geodiversity Group. *Local Geodiversity Action Plan for the Historic County of Lincolnshire (Lincolnshire, North Lincolnshire & North East Lincolnshire)*; Lincolnshire Biodiversity Partnership: Horncastle, 2010.

Lynch, J.; Barrett, L. *Walking with Cavemen: Eye-to-Wye With Your Ancestors*; BBC Worldwide: London, 2003.

Lyons, P. C.; Robertson, E. C.; Milton, L. C. Wroe Wolfe's Geology Course on Radio-Station WGBH (Boston) in 1954. *J. Geol. Educ.* **1993**, *41*, 170.

McCarthy, S.; Gilbert, M. *The Crystal Palace Dinosaurs: The Story of the World's First Prehistoric Sculptures*; Crystal Palace Foundation: London, 1994.

MacFadyen, C. The Vandalizing Effects of Irresponsible Core Sampling: A Call for a New Code of Conduct. *Geol. Today* **2010**, *26*, 146–151.

Mackay, C. *The Scenery and Poetry of the English Lakes: A Summer Ramble*; London, 1846.

Maini, J. S.; Carlisle, A. *Conservation in Canada: A Conspectus—Publication No. 1340*; Department of the Environment/Canadian Forestry Service: Ottawa, 1974.

Mandler, P. 1996. Nationalising the Country House. In *Preserving the Past: The Rise of Heritage in Modern Britain*; Hunter, M., Ed.; Alan Sutton: Stroud, 1996; pp 99–114.

Mantell, G. A. *Geological Excursion Round the Isle of Wight and Along the Adjacent Coast of Dorsetshire*; Henry G. Bohn: London, 1847.

Mantell, G. A. *Geological Excursion Round the Isle of Wight and Along the Adjacent Coast of Dorsetshire*, 2nd Ed.; Henry G. Bohn: London, 1851.

Mantell, G. A. *Geological Excursion Round the Isle of Wight and Along the Adjacent Coast of Dorsetshire*, 3rd Ed.; Henry G. Bohn: London, 1854.

Martini, G. The Protection of Geological Heritage and Economic Development: The Saga of the Digne Ammonite Slab in Japan. In *Geological and Landscape Conservation*; O'Halloran, D., Green, C., Harley, M., Stanley, M., Knill, J., Eds.; The Geological Society: Bath, 1992; pp 383–386.

Martini, G. Geological Heritage and Geotourism. In *Geological Heritage: Its Conservation and Management*; Barretino, D.; Wimbledon, W. P., Gallego, E., Eds.; Instituto Tecnologico Geominero de Espana: Madrid, 2000; pp 147–156.

Mayer, W. Geology and Tourism. *Geological Society of Australia, 13th Australian Geological Convention, Canberra, February 1996, Abstracts 41*. 278.

McKirdy, A. R. Protective Works and Geological Conservation. In *Planning and Engineering Geology—Geological Society Engineering Geology Special Publication No. 4 (Proceedings of the Twenty-Second Annual Conference of The Engineering Group of the Geological Society, Plymouth, September, 1986*; Culshaw, M. G., Bell, F. G., Cripps, J. C., O'Hara, M., Eds.; The Geological Society: London, 1987; pp 81–85.

McKirdy, A. P. *Earth Science Conservation in Great Britain—a Strategy: Appendices—A Handbook of Earth Science Conservation Techniques*; Nature Conservancy Council: Peterborough, 1991.

MGMR. *Circular on Establishing Geological Natural Reserves (GNR) (Proposed)*; Ministry of Geology and Mineral Resources of China: Beijing, 1987 [in Chinese].

MGMR. *Regulations on the Protection and Management of Geosites*; Ministry of Geology and Mineral Resources of China: Beijing, 1995 [in Chinese].

Millar, S. Heritage Management for Heritage Tourism. In *Managing Tourism*; Medlick, S., Ed.; Butterworth-Heinemann Ltd., Oxford, 1991; pp 115–121.

Moncton, H. W.; Herries, R. S., Eds.; *Geology in the Field: The Jubilee Volume of the Geologists' Association, 1858–1908*; Stanford: London, 1909.

Morritt, J. B. S. *The Letters of John B.S. Morritt of Rokeby Descriptive of Journeys in Europe and Asia Minor in the Years 1794–1796*; Marinden, G. E., Eds.; John Murray: London, 1914.

Mullen, R.; Munson, J. *"The Smell of the Continent": The British Discover Europe*; Macmillan: London, 2009.

Murray, H. *A Companion and Useful Guide to the Beauties of Scotland, to the Lakes of Westmorland, Cumberland, and Lancashire; and to the Curiosities in the District of Craven in the West Riding of Yorkshire. To Which Is Added, a More Particular Description of Scotland, Especially That Part of It, Called the Highlands*. London, 1799.

Myers, K. *The Catskills: Painters, Writers, and Tourists in the Mountains, 1820–1895*; Hudson River Museum: Yonkers, 1987.

Newsome, D.; Moore, S. A.; Dowling, R. K. *Natural Area Tourism: Ecology, Impacts and Management*; Channel View Publications: Cleveden, 2002.

Newsome, D.; Dowling, R. K., Eds.; *Geotourism: The Tourism of Geology and Landscape*; Goodfellow: Oxford, 2010.

Ng, Y. Economic Impacts of Geotourism and Geoparks in China; Keynote Address: Global Eco Asia-Pacific Conference, Adelaide, Australia, 2017, Nov. 27–29. Available at: https://2017.globaleco.com.au/perch/resources/Gallery/dr-young-ng.pdf.

North Pennines AONB Partnership. *North Pennines Area of Outstanding Natural Beauty and European Geopark: A Geodiversity Audit (revised March 2010)*. North Pennines AONB Partnership: Stanhope, 2010.

Novelli, M. (Ed.). *Niche Tourism: Contemporary Issues, Trends and Cases*; Elsevier: Oxford, 2005.

O'Halloran, D., Green, C., Harley, M. Stanley, M. and Knill, J., Eds.; *Geological and Landscape Conservation*; The Geological Society: Bath, 1994.

O'Halloran, M. Dinosaurs for Tourism: Picket Wire Canyon and the Rocky Mountain Paleotological Tourism Initiative, Colorado, USA. In *Practicing Responsible Tourism: International Case Studies in Tourism Planning, Policy, and Development*; Harrison, L. C., Husbands, W., Eds.; Wiley: New York, 1996; pp 495–508.

Oldroyd, D. R. *Earth, Water, Ice and Fire, Two Hundred Years of Geological Research in the English Lake District: Geological Society Memoir No. 25*; The Geological Society: London, 2002.

Ongur, T. Doğal anıtlarin korunmasında yasal dayanaklar. *Yeryuvarı ve İnsan* **1976**, *1* (4), 17–23 [in Turkish].

Orion, N.; King, C.; Krockover, G. H.; Adams, P. E. The Development and Status of Earth Science Education: A Comparison of Three Case Studies from Israel, England and Wales and the United States of America (Part I). *Sci. Educ. Int.* **1999a**, *10* (2), 13–23.

Orion, N.; King, C.; Krockover, G. H.; Adams, P. E. The Development and Status of Earth Science Education: A Comparison of Three Case Studies from Israel, England and Wales and the United States of America (Part II). *Sci. Educ. Int.* **1999b**, *10* (3), 19–27.

Otley, J. *A Concise Description of the English Lakes, and Adjacent Mountains, with General Directions to Tourists; and Observations on the Mineralogy and Geology of the District*; L. Otley: Keswick, 1823.

Ousby, I. *The Englishman's England: Taste, Travel and the Rise of Tourism*; Cambridge University Press: Cambridge, 1990.

Oxenham, J. R. *Reclaiming Derelict Land*; Faber and Faber: London, 1966.

Oxfordshire Geology Trust. *Draft Local Geodiversity Action Plan for North Wessex Downs AONB*; Oxfordshire Geology Trust: Faringdon, 2007.

Page, K. N. England's Earth Heritage Resource—An Asset for Everyone. In *Coastal Defence and Earth Science Conservation*; Hooke., J., Ed.; The Geological Society: London, 1998; pp 196–209.

Patzak, M.; Eder, W. "UNESCO GEOPARK". A New Programme – A New UNESCO label. *Geol. Balcan.* **1998**, *28*, 33–35.

Phillips, A. Interpreting the Countryside and the Natural Environment. In *Heritage Interpretation Volume 1—The Natural and Built Environment*; Uzzell, D. L., Ed.; Belhaven Press: London, 1989; pp 121–131.

Plog, S. C. Why Destination Areas Rise and Fall in Popularity. *The Cornell Hotel and Restaurant Administration Quarterly* **1974**, *15*, 55–58.

Prosser, C. Webster's Claypit SSSI—going, Going, Gone But Not Forgotten. *Earth Heritage* **2002**, *19*, 12.

Pullin, R. *Eugene von Guerard. Nature Revealed*; National Gallery of Victoria: Melbourne, 2011.

Pullin, R. The Artist as Geotourist: Eugene von Guerard and the Seminal Sites of Early Volcanic Research in Europe and Australia. In *Appreciating Physical Landscapes: Three Hundred Years of Geotourism, Special Publications, 417*; Hose, T. A., Ed.; The Geological Society: London, 2016; pp 59–69.

Rea, J. The Charmouth Public Inquiry: The Society's Position. *Geol. Soc. Newslett.* **1992**, *11* (5), 4–9.

Read, S. E. A Prime Force in the Expansion of Tourism in the Next Decade: Special Interest Travel. In *Tourism Marketing and Management Issues*; Hawkins, D. E., Shafer, E. L., Rovelstad, J. M., Eds.; George Washington University Press: Washington, DC, 1980; pp 193–202.

Reid, C. G. R. "The Time Trail": A Lesson Earned? *Geol. Today* **1994**, *10* (2), 68–69.

Reynard, E.; Coratza, P.; Giusti, C. "Geomorphosites and Geotourism" (Special Issue) *Geoheritage* **2011**, *3* (3), 2011.

Richards, L. Snailbeach Lead Mine—Reclamation and Conservation. *Earth Sci. Conserv.* **1992**, *31*, 8–12.

Richter, L. K. The Search for Appropriate Tourism. *Tour. Recreat. Res.* **1987**, *12* (2), 5–7.

Richter, L. K. *The Politics of Tourism in Asia*; University of Hawaii Press: Honolulu, 1989.

Robinson, J. E. *A Code of Conduct for Rock Coring*; Geologists" Association/Conservation Committee of the Geological Society of London: London, 1989.

Robinson, E. Arran 1999—A watershed? *Geol. Today* **1990**, *6* (3), 74–75.

Robinson, E. Who Would Buy a Coal Tip? *Earth Sci. Conserv.* **1993a**, *32*, 20–21.

Robinson, E. The European Community of Stone, Frome, Somerset. *Geol. Today* 1993b, *9* (1), 22–25.

Robinson, E. Geology Reserve: Wren's Nest Revised. *Geol. Today* **1996a**, *12* (6), 211–213.

Robinson, E. Earth Galleries Open: Impression 1. *Geol. Today* **1996b**, *12* (4), 128–130.

Robinson, E. Tourism in Geological Landscapes. *Geology* **1998**, *14*, 151–153.

Robinson, H. *Chester Society of Natural Science, Literature and Art—The First Hundred Years 1871–1971*; Chester Society of Natural Science, Literature and Art: Chester, 1971.

Robinson, E.; McCall, G. J. H. Geoscience Education in the Urban Setting. In *Urban Geoscience*; McCall, G. J. H., De Mulder, E. F. G., Marker, B. R., Eds.; Balkema: Rotterdam, 1996; pp 235–251.

Röhling, H.-G.; Schmidt-Thomé, M. Geoscience for the Public: Geotopes and National GeoParks in Germany. *Episodes* **2004**, *27* (4), 279–283.

Rolfe, W. D. I. Recent Pillage of Classic Localities by Foreign Collectors. *Inf. Cir. Geol. Physiogr. Sect., NCC* **1977**, *13*, 2–4.

Rolfe, W. D. I. Early Life on Land—The East Kirton Discoveries. *Earth Sci. Conserv.* **1988**, *5*, 22–28.

Rolfe, W. D. I.; Milner, A. C.; Hay, F. G. The Price of Fossils. In *Special Papers in Palaeontology, 40; The Use and Conservation of Palaeontological Sites*; Crowther, P. R., Wimbledon, W. A., Eds.; The Palaeontological Society: London, 1988; pp 139–171.

Ross, M. *Planning and the Heritage: Policy and Procedures*; Spon: London, 1991.

Santucci, V. L., Eds.; "Geodiversity and Geoconservation" (Special Issue). *George Wright Forum* **2005,** *22* (3).

Schools Council Geology Curriculum Review Group. *Schools Council Working Paper 58: Geology in the School Curriculum;* Evans Methuen: London, 1977.

Serjani, A. Geological Sites of the External Zones of Albania. *Geol. Balcan.* **1996,** *26* (2), 11–14.

Serjani, A.; Cara, F. List of Geological Sites of Albania. *Geol. Balcan.* **1996,** *26* (1), 57–60.

Serjani, A.; Neziraj, A.; Hallaci, H. Ten Years of Geological Heritage in SE Europe. In *Proceedings of the ProGEO WG-1 Sub-Regional Meeting and Field Trip;* Anon., Ed.; ProGEO: Tirana, 2005.

Sharpe, T.; Howe, S.; Howells, C. Gallery Review: Setting the Standard? The Earth Galleries at the Natural History Museum, London. *Geol. Curator* **1998,** *6* (10), 395–403.

Sharples, C. *A Methodology for the Identification of Significant Landforms and Geological Sites for Geoconservation Purposes;* Forestry Commission: Tasmania, 1993.

Sharples, C. *Concepts and Principles of Geoconservation,* Tasmanian Parks and Wildlife Service, 2002. http://dpipwe.tas.gov.au/Documents/geoconservation.pdf (accessed December 16, 2015).

Sillitoe, A. *Leading the Blind: A Century of Guidebook Travel 1815–1911;* Macmillan: London, 1995.

Slomka, T.; Kicinska-Swiderska, A. Geotourism—the Basic Concepts. *Geoturystyka* **2004,** *1,* 5–7.

Smith, P. J. Earth Galleries Open: Impression 2. *Geol. Today* **1996,** *12* (4), 130–132.

Sorby, H. C. On the Remains of a Fossil Forest in the Coal-Measures at Wadesley, near Sheffield, *Q. J. Geol. Soc.* **1875,** *31,* 458–459.

Spiteri, A. Malta: A Model for the Conservation of Limestone Regions. In *Geological and Landscape Conservation;* O'Halloran, D., Green, C., Harley, M., Stanley, M., Knill, J., Eds.; The Geological Society: Bath, 1994; pp 205–208.

Stanley, M. F. The National Scheme for Geological Site Documentation. In *Earth Science Conservation in Europe: Proceedings from the Third Meeting of the European Working Group of Earth Science Conservation;* Erikstad, L., Ed.; Norsk Institutt For Naturforskning: Oslo, 1993; pp 17–22.

Stow, D. A. V.; McCall, G. J. H., Eds.; *Geoscience Education and Training in Schools and Universities for Industry and Public Awareness;* Balkema: Rotterdam, 1996.

Stueve, A. M.; Cock, S. D.; Drew, D. *The Geotourism Study: Phase 1 Executive Summary,* 2002. www.tia.org/pubs/geotourismphasefinal.pdf.

Taylor, M. A. "Fine fossils for Sale": The Professional Collector and the Museum. *Geol. Curator* **1987,** *5* (2), 55–64.

Taylor, M. A. Palaeontological Site Conservation and the Professional Collector. In *Special Papers in Palaeontology;* Crowther, P. R., Wimbledon, W. P., Eds.; The Palaeontological Society: London, 1988; pp 123–134.

Taylor, M. A. The Local Geologist 1: Exporting Your Heritage? *Geol. Today* **1992,** *7* (1), 32–36.

Thomas, B. A.; Warren, L. M. Geological Conservation in the 19th and Early 20th Centuries. In *The History of Geoconservation: Special Publication No. 300;* Burek, C. V., Prosser, C. D., Eds.; The Geological Society: London, 2008; pp 17–30.

Thomas, I. A.; Prentice, J. E. A Consensus Approach at the National Stone Centre, UK. In *Geological and Landscape Conservation;* O'Halloran, D., Green, C., Harley, M., Stanley, M., Knill, J., Eds.; The Geological Society: Bath, 1994; pp 423–427.

Thomas, I. A.; Hughes, K. Reconstructing Ancient Environments. *Teach. Earth Sci.* **1993**, *18*, 17–19.

Tilden, F. *Interpreting Our Heritage*; The University of North Carolina Press: Chapel Hill, 1957.

Toghill, P. Geological Conservation in Shropshire. *J. Geol. Soc. Lond.* **1972**, *128*, 513–515.

Toghill, P. Involving Landowners, Local Societies and Statutory Bodies in Shropshire's Geological Conservation. In *Geological and Landscape Conservation*; Knill, S., O'Halloran, D., Green, C., Harley, M., Stanley, M., Eds.; The Geological Society: Bath, 1994; pp 463–466.

Trench, R. *Travelers in Britain: Three Centuries of Discovery*; Arum Press, London, 1990.

Tresise, G. 1966. *The Earth Before Man: A Guide to the Geology Gallery, Liverpool Museum*; Liverpool Museum: Liverpool, 1966.

Tresise, G. The Story of the Earth at the Geological Museum. *Museums J.* **1973**, *73* (2), 71–72.

Tresise, G. The Liverpool Geological Association 1880–1910. *North West Geol.* **2013**, *18*, 26–34.

Trimble, S. *Lasting Light—125 Years of Grand Canyon Photography*; Northland Publishing: Flagstaff, 2006.

UNESCO. *UNESCO Geoparks Programme Feasibility Study*; UNESCO: Paris, 2000.

UNWTO. *The Role of Recreation. Management in the Development of Active Holidays and Special Interest Tourism and the Consequent Enrichment of the Holiday Experience*; UNWTO Publications: Madrid, 1985.

UNWTO. *What Tourism Managers Need to Know. A Practical Guide for the Development and Application of Indicators of Sustainable Tourism*; UNWTO Publications: Madrid, 1997.

Urry, J. *The Tourist Gaze; Leisure and Travel in Contemporary Societies*; Sage: London, 1990.

Vasiljevic, D. A.; Markovic, S. B.; Hose, T. A.; Ding, Z.; Guo, Z.; Liu, X.; Smalley, I.; Lukic, T.; Vujicic, M. D. Loess Palaeosol Sequences in China and Europe: Common Values and Geoconservation Issues. *Catena* 2014, *117*, 108–118.

Von Heger, F. Ferdinand von Hochstetter. In *Mitteilungen der Kaiserlich-Koniglichen Geographischen Gesellschaft*; Communications of the Imperial-Royal Geographical Society, 1884; p 27.

Walter, H. M.; Hart, D. The Story of the Earth. *Geology* **1975**, *5*, 92–97.

Ward, R. *A Guide to the Peak of Derbyshire, Contains a Concise Account of Buxton, Matlock, and Castleton and Other Remarkable Places and Objects Chiefly in the Northerly Parts of that Very Interesting County*; William Ward: Birmingham, 1827. [available as a 1977 reprint, E. J. Morten, Didsbury].

Warner, R. *A Tour through the Northern Counties of England, and the Borders of Scotland*; G. and J. Robinson: London, 1802.

Weiler, B.; Hall, C. M., Eds.; *Special Interest Tourism*; Belhaven: London, 1992.

West, T. *A Guide to the Lakes: Dedicated to the Lovers of Landscape Studies, and to All Who Have Visited, or Intend to Visit the Lakes in Cumberland, Westmorland and Lancashire*; B. Law: London, 1778.

Whyte, I. William Wordsworth's Guide to the Lakes and the Geographical Tradition. *Area* **2000**, *32* (1), 101–106.

Wilkinson, J. *Select Views in Cumberland, Westmoreland and Lancashire by the Rev. Joseph Wilkinson, Rector of East and West Wretham in the County of Norfolk, and Chaplain to the Marquis of Huntley*; R. Ackerman: London, 1810.

Wilkes, G. A. *A Dictionary of Australian Colloquialisms*; Sydney University Press: Sydney, 1978.

Wimbledon, W. A. P. Geosites—A New Conservation Initiative. *Episodes* **1996**, *19* (3), 87–88.

Wimbledon, W. A. P.; et al. A First Attempt at a Geosites Framework for Europe—An IUGS Initiative to Support Recognition of World Heritage and European Geodiversity. *Geol. Balcan.* **1998,** *28* (3–4), 5–32.

Wimbledon, W. A. P.; Smith-Meyer, S. *Geoheritage in Europe and its Conservation*; ProGEO: Oslo, 2012.

Wood, R. M. *On the Rocks: A Geology of Britain*; BBC Books: London, 1978.

Wood, S. P. The Bearsden Project or Quarrying for Fossils on a Housing Estate. *Geol. Curator* **1983,** *3* (7), 423–434.

Woodward, H. B. *The History of the Geological Society of London*; The Geological Society: London, 1907.

Wordsworth, W. *A Guide Through the District of the Lakes of Northern England with a Description of the Scenery, & c. for the use of Tourists and Residents*; 5th ed.; Longman, Hurst, Rees, Orme and Brown: London, 1820.

Wordsworth, W. *The River Duddon, A Series of Sonnets; Vaudracour & Julia: and Other Poems. To Which is Annexed, A Topographical Description of the Country of the Lakes, in the North of England*; Hudson and Nicholson: Kendal, 1835.

Wyatt, J. *Wordsworth and the Geologists*; Cambridge University Press: Cambridge, 1995.

Young, A. *The Farmer's Tour through the East of England Being the Register of a Journey through Various Counties of this Kingdom, to Enquire into the State of Agriculture, & c. Containing I. The Particular Methods of Cultivating the Soil. II. The Conduct of Live Stock and the Modern System of Breeding. III. The State of the Population, the Poor, Labour, Provisions, & c. IV. The Rental Value of the Soil, and its Division into Farms, with Various Circumstances Attending their Size and State. V. The Minutes of above Five Hundred Original Experiments, Communicated by Several of the Nobility, Gentry, & c. With Other Subjects That Tend to Explain the Present State of English Husbandry*; W. Strahaan and W. Nicoll: London, 1771.

Young, A. *Arthur Young's Tour in Ireland (1776–1779) [edited with introduction and notes by Arthur Wollaston Hutton, with a bibliography by John P. Anderson of the British Museum]*; George Bell and Sons: London, 1892.

Young, A. *Travels During the Years 1787, 1788, and 1789, Undertaken More Particularly With a View of Ascertaining the Cultivation, Wealth, Resources, and National Prosperity of the Kingdom of France*. Bury St. Edmunds, 1792.

Young, J. Note on the Section of Strata in the Gillmore Quarry and Boulder Clay on the Site of the New University Building. *Tran. Geol. Soc. Glasgow* **1868,** *III*, 298.

Zhao, X.; Zhao, T. The Socio-Economic Benefits of Establishing National Geoparks in China. *Episodes* **2003,** *26* (4), 302–309.

Zouros, N. The European Geopark Network: Geological Heritage Protection and Local Development—A Tool for Geotourism Development in Europe. In *4th European Geoparks Meeting—Proceedings Volume*; Fassoulas, C., Skoula, Z., Pattakos, D., Eds., 2006; pp 15–24.

Zouros, N. European Geoparks: New Challenges and Innovative Tools towards Earth Heritage Management and Sustainable Local Development. In *Proceedings of the 12th European Geoparks Conference*; Aloia, A., Calcaterra, D., Cuomo, A., De Vita, A., Guida, D., Eds., National Park of Cilento, Vallo di Diano and Alburni Geopark, 2013; pp v–ix.

CHAPTER 3

Urban Geotourism in Poland

KRZYSZTOF GAIDZIK

Department of Fundamental Geology, Faculty of Earth Sciences, University of Silesia, Będzińska 60, 41-200 Sosnowiec, Poland

E-mail: krzysztof.gaidzik@us.edu.pl

ABSTRACT

Urban geotourism combines sightseeing tours of human monuments, architectonic structures, even cultural events along with exploring the natural history of Earth, and admiring nature's wonders because every town has a story to tell us that is somehow related to geology. Hence, countries like Poland, with its long history of facing upheavals that have resulted in an abundance of old cities and towns with architectonic monuments from different periods, rich cultural heritage, and mining tradition, together with high geodiversity, presenting an especially high potential for development of urban geotourism in this scenic country. This chapter presents the current state of urban geotourism in Poland based on a few large, medium- and small-sized towns, selected for this study. This specific form of tourism has been gaining acceptance and reputation in Poland for the past several years. However, its development should be considered rather spontaneous and, in many cases, not exactly well planned or well managed. Moreover, the geotouristic and geoeducational potential of many cities and towns still remains unexplored. However, this is not the case for large cities like Kraków, Kielce, and Wrocław, as well as for many small towns in Poland and postindustrial regions of Lower and Upper Silesia, where the combination of urban tourism with geoeducation and admiration for natural wonders was recognized by local governments as a great idea for promoting tourism in the town, leading to the increased inflow of tourists and therefore adding to the city's coffers. Furthermore, the extraordinary prospect of educating people in earth sciences in the city, which could not be more difficult than natural surroundings, is simply intriguing.

3.1 INTRODUCTION

Geology and earth sciences allow us to study and learn more about the immense, intriguing, and compelling history of our planet. Indeed, every rock has a story to tell, every structure gives us an opportunity to wonder about the natural process that produced it, every fossil lets us meditate about the animals once inhabiting our planet, and every mineral gives us the perception of the beauty created by nature. However, this prospect is usually limited to scientists and people passionate about earth sciences.

Geotourism serves as a way to spike interest in a nongeologist about geology and earth sciences and to give them a chance to understand and appreciate natural processes and structures. This form of tourist activity can be defined as "the provision of interpretive and service facilities to enable tourists to acquire knowledge and understanding of the geology and geomorphology of a site (including its contribution to the development of the Earth sciences) beyond the level of mere esthetic appreciation" (Hose, 1995, 2008; Hose and Vasiljević, 2012), or a form of tourism "based upon visits to geological objects (geosites) and recognition of geological processes integrated with esthetic experiences gained by a contact with a geosite" (Słomka and Kicińska-Świderska, 2004). Geotourism is a form of sustainable tourism that specifically focuses on geology and landscape (e.g., Słomka and Kicińska-Świderska, 2004; Newsome and Dowling, 2010; Dowling, 2015), geoconservation and an understanding of geological heritage—geoeducation (Hose, 1995, 2008). Thus, it is essentially related with exploration of natural geosites in the field. Nevertheless, as pointed out by Charsley (1996) every city has a story to tell that is probably related to geology. Thus, one can explore the history of Earth, study the natural processes, and admire the natural wonders adorning a city, simply by combining sightseeing of human monuments, architectonic structures, and cultural events with geotourism and geoeducation (e.g., Zagożdżon and Zagożdżon, 2016). This specific style of geotourism is known as urban geotourism which is increasingly gaining popularity and is corroborated by world-wide examples, such as Prague (Březinová et al., 1996) and Brno (Kubalíková et al., 2017) in Czech Republic, Vienna in Austria (Seemann and Summesberger, 1998), Kraków in Poland (Rajchel, 2004), Berlin in Germany (Schroeder, 2006), Lisbon in Portugal (Rodrigues et al., 2011; da Silva, 2019), Mexico City in Mexico (Palacio-Prieto, 2015), São Paulo City in Brazil (Del Lama et al., 2015), Paris in France (De Wever et al., 2017), Torino in Italy (Gambino et al., 2019), and so on.

Geosites in the cities include wide range of objects such as architectonic structures with vast history and interesting buildings or decorative stones (e.g., Březinová et al., 1996; Seemann and Summesberger, 1998; Bromowicz and Magiera, 2003, 2015; Rajchel, 2004; Schroeder, 2006; Rodrigues et al., 2011; Del Lama et al., 2015; da Silva, 2019), underground cities (e.g., De Wever et al., 2017), parks and gardens (e.g., Palacio-Prieto, 2015; Portal and Kerguillec, 2018), decorative arts, sculptures and structures of the city, even pavements (e.g., Březinová et al., 1996; Seemann and Summesberger, 1998; Bromowicz and Magiera, 2003, 2015; Rajchel, 2004), mines, quarries, and remnants of ore exploitation (e.g., Nita and Myga-Piątek, 2010; Zagożdżon and Zgożdżon, 2016; De Wever et al., 2017; Gaidzik and Chmielewska, 2020), landforms (e.g., Palacio-Prieto, 2015), interesting outcrops (e.g., Nita and Myga-Piątek, 2010; Palacio-Prieto, 2015), museum collections (Rajchel, 2004; Słomka et al., 2006; Pieńkowski, 2011; De Wever et al., 2017), cultural events, costumes and practices related to earth processes or structures, and so on. Also, recent advances in mobile technology contribute to the rapid development of specific forms of geotourism, offering a variety of urban geotourism tools, such as mobile applications aiding geotourism (Pica et al., 2018). This study presents the current state of the rapidly and rather spontaneously burgeoning urban geotourism in Poland, based on selected set of large-, medium-, and small-sized towns, with particular focus on the predominant type of urban geotourism in each of these locations. The high potential for urban geotourism in Poland is corroborated by its high geodiversity thanks to the location on three main European megaunits of different ages (Fig. 3.1), together with long and difficult history that the country has faced resulting in an abundance of old cities and towns with historical monuments, architectural wonders, and rich mining tradition. However, even though a significant development in geotourism can be observed in the past few years, the geotouristic and geoeducational potential of these cities have not been explored sufficiently, yet.

3.2 GEOTOURISTIC POTENTIAL OF POLAND

The geological structure of Poland is very interesting. Its location in Central Europe seated at the triple junction that comprises three principal European mega units of different ages: Precambrian East European Platform, Palaeozoic West European Platform, and Mesozoic and Cenozoic Carpathian Orogen (Fig. 3.1; Aleksandrowski, 2017). Its placement results in a high geodiversity and makes it a perfect geotouristic destination. Numerous processes throughout the time from Precambrian to this day have shaped the surface of Poland and created numerous absorbing and fascinating geological objects. Many of them show both outstanding esthetic values (e.g., monumental or unusual forms, scale of structures, beauty of landscape, etc.) and the enchanting geological history related to tectonic, volcanic, magmatic, eolithic, fluvial, erosional, and/or catastrophic processes.

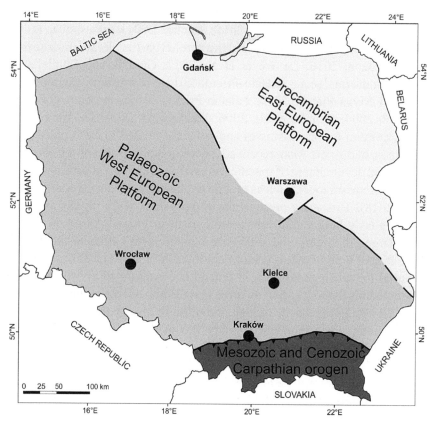

FIGURE 3.1 Tectonic provinces and main cities of Poland (after Aleksandrowski, 2017).

Natural beauty of these geosites will definitely draw attention of all visitors, irrespective of their level of understanding of geological processes. Wide geodiversity and high geotouristic potential of Poland are corroborated by the existence of three official national geoparks (Muskau Arch Geopark, St. Anne Hill Geopark, and KarkonoszeGeopark) of which only Muskau

Arch Geopark is included in the Global UNESCO Geoparks Network.[1] Apart from these geoparks there are some other initiatives that aspire to be included in the geoparks network, together with various catalogues of the most important, interesting or fascinating geosites of Poland or specific regions that could be part of the network, too (e.g., Słomka et al., 2006, 2012; Stachowiak et al., 2013). Many of those are presented in the digital form, such as the Catalogue of Geotouristic Sites of Poland from the Ministry of Environment of Poland,[2] and the Catalogue of Geotourist Sites in nature reserves and monuments,[3] the Central Catalogue of Geosites in Poland published by the Polish Geological Institute—National Research Institute,[4] or the local catalogue of geosites in the Silesian Voivodeship.[5]

These include mainly natural and anthropogenic outcrops of various types of rock, geomorphosites presenting interesting geomorphological features and abandoned mines/quarries (Słomka et al., 2006, 2011, 2012; Stachowiak et al., 2013).

3.3 URBAN GEOTOURISM IN POLAND

Despite the high geotouristic potential of Poland, long history and abundance of historical monuments, architectural wonders, and so on, so far, not too much attention has been given to urban geotourism. For example, among the 100 geosites included in the Catalogue of Geotouristic Sites of Poland published by the Ministry of the Environment of Poland (Słomka et al., 2006) only two can be related with urban geotourism, that is, the Petrified Wood from Siedliska and The Dragon's Cave in the Wawel Hill. The first geosite includes the occurrence of Tertiary petrified wood fragments considered as one of the natural peculiarities of Eastern Poland. The specimens can be seen in the museum in Siedliska, which has collected about 500 petrified woods varying from palm-sized ones to blocks weighing about 500 kg that happens to be the main attraction of the town and region. Actually, the entire tourist traffic in this town can be attributed to these petrified woods. The second geosite is a very popular site among tourists visiting Kraków. The Dragon's Cave is located in the Wawel Hill, under the Royal Castle, in the center of

---
[1] http://www.unesco.org/new/en/natural-sciences/environment/earth-sciences/
[2] https://archiwum.mos.gov.pl/kategoria/2398_katalog_obiektow_geoturystycznych_w_polsce
[3] http://www.kgos.agh.edu.pl/index.php?action=projekty&subaction=25.9.140.921(Retrieved: Jan. 01, 2020)
[4] http://geoportal.pgi.gov.pl/portal/page/portal/geostanowiska (Retrieved: Jan. 01, 2020)
[5] http://www.geosilesia.us.edu.pl/295,geostanowiska_wojewodztwa_slaskiego.html

Kraków. It is one of the best-known caves in Poland due to its localization and well-known legend about the dragon who lived under the Wawel Hill. The breathing fire dragon sculpture standing at the entrance to the cave is an additional tourist attraction (Firlet, 1996; Słomka et al., 2006). Urban geotourism in Poland is still a developing branch of geotourism and requires much more attention from scientists, governments, management, and tourists themselves. In recent years, a number of studies on geotouristic attractions of the largest Polish cities have been carried out. However, these focus mainly on Kraków (e.g., Rajchel, 2004, 2008), Wrocław (e.g., Zagożdżon and Śpiewak, 2011; Zagożdżon, 2012; Zagożdżon and Zagożdżon, 2015a–d, 2016), Kielce (e.g., Nita and Myga-Piątek, 2010), and postindustrial geotourism in the Upper Silesia region (e.g., Gaidzik and Chmielewska, 2020). Other cities, such as Warszawa, the capital of Poland, Gdańsk, or Poznań, even though opulent with geotouristic sites, have not been well studied and/or well promoted, yet.

3.3.1 URBAN GEOTOURISM IN LARGE CITIES

3.3.1.1 KRAKÓW

Kraków, the former capital of Poland, saw more than 12 million visitors in the year 2016 (Borkowski, 2016) and is one of the most important tourist destinations in Central Europe. It is the second-largest city in Poland located in the southern part of the country, situated on the banks of the Vistula River in the Lesser Poland (Małopolska) region (Fig. 3.2). It is also one of the oldest cities in Poland that dates back to the seventh century, whereas archaeological findings provide evidence that Wawel Hill (central part of the city with the Royal Castel) was settled as far back as the early Stone Age (Rajchel, 2004). The geotouristic potential of Kraków is very well studied and presented in a comprehensive book by Jacek Rajchel, published in 2004: *Kamienny Kraków. Spojrzenie geologa* (Stony Kraków. Geological point of view), together with several articles on specific issues, such as cephalopods in the architecture of Kraków (Kin and Rajchel, 2008), Dragon's Cave in Wawel Hill (Firlet, 1996), decorative stone in the Zygmunt Chapel in the Royal Castle (Bromowicz and Magiera, 2003, 2015), traditional pavements in Kraków (Rajchel, 2009a), travertine in architecture (Rajchel, 2009b), interesting imported decorative stones (Górny, 2009), remnants of mine industry (Dmytrowski and Kicińska, 2011; Sermet and Rolka, 2012), and apotropaical stony sculptures in Kraków architecture (Rajchel, 2014). The city has been impacted by innumerable changes over long time periods

Urban Geotourism in Poland

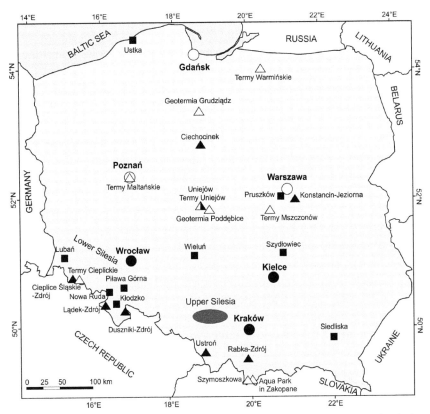

FIGURE 3.2 Location of selected urban geotourism sites in Poland; black circles—large cities with geotouristic provisions, white circles—large cities with poor geotouristic infrastructures, black squares—small towns known as the urban geotourism sites, gray ellipse—region promoting the post-industrial urban geotourism, black triangles—spa resorts, and white triangles—thermal centers (after Dryglas and Różycki, 2016).

leading to characteristics such as incomparable color and "atmosphere" that are related to high variability of building rock materials and decorative stones representing igneous, sedimentary, and metamorphic rocks. The selection of building rock materials always depends on political and territorial conditions; thus, historical monuments were built predominantly with accessible materials, that is, local Polish materials and from neighboring countries, for example, Hungarian marbles (Rajchel, 2004, 2008). Among the most important local materials are

1. Silesian-Cracow Monocline: white, Upper Jurassic limestone (Majer-Durman, 2012), yellowish Triassic diplopora dolomite, black Devonian

limestone from Dębnik, red-violet Permian porphyry, pinkish Lower Carboniferous "marble" from Paczółtowice, and Holocene travertine
2. The Carpathian Foredeep: white, Miocene limestone from Pińczów and Miocene alabaster
3. Carpathian orogen: brownish and greenish, Cretaceous and Tertiary sandstone, Miocene and a site from the Pieniny Mountains and Carboniferous granite from the Tatra Mountains
4. Świętokrzyskie Mountains: Devonian and Jurassic limestone and calcareous conglomerate, rusty-colored, yellow and white, Triassic and Jurassic sandstone
5. The Ukrainian shield (in the present day, outside the Polish borders): Precambrian plutonic and volcanic rocks and Devonian sandstone
6. Sudetes and the Fore-Sudetic monocline: Precambrian marble, Paleozoic granite and syenite, Permian and Cretaceous sandstone (Rajchel, 2004). In today's date, the mosaic of the city is produced by the combination of these traditional, historical building materials, and decorative stones and a wide spectrum of stone materials from all over the world appended just recently. The combination of various rock types from all over the world together with the rich and enchanting history of the city makes Kraków an excellent choice for urban geotourism and geoeducation purposes (Rajchel, 2008). The vast majority of the most important historical monuments of Kraków [e.g., Old Town and Kazimierz—the former Jewish Quarter, St. Mary's Basilica (Fig. 3.3a), Wawel Cathedral (Fig. 3.3b), Royal Castle on the Wawel Hill together with Dragon's Cave, St. Francis' Basilica, Main Market Square, Underground Town, St. Joseph's Church, Collegium Maius, St. Anne Collegiate, etc.] give tourists a chance to admire beauty and variability of world-wide decorative rocks. However, apart from historical and esthetic values these objects, if properly presented can be an indispensable laboratory for earth sciences education. Geotourist can obtain, learn, and admire the mineral composition; texture and structure; be it sedimentary, volcanic, or magmatic structures; tectonic structures; signs of weathering (that took place both in the deposit itself and after architectonic application); fossils, and so on. Apart from information on natural processes and structures, visitors to those places can also learn about the methods of stone quarrying and history and/or legends related to the same. The proposition of geotouristic paths and routes around Kraków can be found in Rajchel (2004). However, not only architecture of buildings, building stone materials or decorative stones can be of interest for

Urban Geotourism in Poland 101

FIGURE 3.3 Main attractions of Kraków. (a) St. Mary's Basilica on the main square and (b) Wawel Cathedral in the Royal Castle on the Wawel Hill.

urban geotourists. In Kraków, even materials used for paving streets are interesting and can be studied and appreciated by visitors. For example, main streets of Kraków: Floriańska Street, Grodzka Street and Wiślna Street were paved with blocks of the Strzegom granite (Lower Silesia region) more than 150 years ago (Rajchel, 2004, 2008, 2009a).

3.3.1.2 WROCŁAW

Wrocław in southwestern Poland (Fig. 3.2), like Kraków, is a city with a very long and complicated history. It is well known as "The Venice of the North" because of the highly developed network of Odra channels (the second longest river in Poland). In spite of being suitable, this nickname is also confusing, since it commonly refers to Saint Petersburg in Russia. Wrocław is characterized by a colorful and rich architecture based on a wide spectrum of rock types sourced from different parts of the world, with a predominance of local stones. This is due to the specific location of the city in the center of Lower Silesia, that is, stonemason capital of Poland. Many of the decorative or building stones used in various Polish cities and towns, and also abroad, originated from this region. For example, the traditional pavement in Kraków is made of granite from Strzegom (Rajchel, 2004, 2009a), a town located in Lower Silesia, about 50 km southwest of Wrocław. The same stone is widely used in Wrocław as pavement material and also as building, decorative and sculpture stone. Also, other rock types excavated in Lower Silesia can be found here. These include primarily other granite types (such as Strzelin granite, Karkonosze granite, etc.), marble (mainly from Sławniowice), syenite, sandstone and many others (Lorenc and Mazurek, 2010; Zagożdżon and Zagożdżon, 2015c, 2016). Apart from local material, stones imported from different parts of the world are also common in the architecture of Wrocław (Zagożdżon and Zagożdżon, 2015c).

The abundance of various types of rocks used as pavement material, building and decorative stones, or material for artworks (e.g., sculptures) together with its interesting history and beautiful landscapes make Wrocław a perfect destination for urban geotourism (Lorenc and Mazurek, 2010; Zagożdżon and Śpiewak, 2011; Zagożdżon and Zagożdżon, 2015a,b, 2016), as well as a great location for geoeducation (Zagożdżon and Zagożdżon, 2015c,d). Zagożdżon and Śpiewak (2011) presented a proposition of two geotouristic routes in the northern part of Wrocław (between Bridges and Ostrów Tumski) that include 29 geosites and more than 50 different types of igneous, sedimentary, and metamorphic rocks, together with variations in mineralogy, colors, characters, and a wide variety of structures and textures, such as schlieren bands, enclaves, xenoliths and veins (e.g., pegmatites, aplites, quartz veins) in igneous rocks related to the evolution of magma chambers and the formation of magmatic intrusions; variation of sedimentary structures and fossils that indicate changes in the paleoenvironment and paleogeography; brittle and ductile tectonic structures, especially in granite from Strzegom, as well as marble and gneiss that are evidences of tectonic forces acting in the lithosphere (Zagożdżon and Śpiewak, 2011; Zagożdżon and Zagożdżon, 2015a). Exogenous features can be observed in selected sites along the proposed routes as well. These features are predominantly related to weathering, caused by water, pollution, wind, acid rain, and so on. Some of the selected sites also show the effects of much more extraordinary destructive factors, such as artillery fire from the Second World War (Zagożdżon and Śpiewak, 2011; Zagożdżon, 2012; Zagożdżon and Zagożdżon, 2015c), making them perfect sites to combine education related with geological processes and human history. Moreover, the proposed routes give the prospect of presenting to a wide audience the techniques of exploitation, procurement, and processing of the building and decorative stones, as well as its application in art (historical and contemporary sculpture). Most important geosites with high esthetic, historical and cultural values, present in proposed routes include (Zagożdżon and Śpiewak, 2011): the route "Between Bridges" and the route "OstrówTumski." For the route "Between Bridges":

1. Mieszczańskie Bridges, Institute of Geological Sciences (University of Wrocław)
2. Bem's Square
3. Herbarium of University of Wrocław
4. Natural History Museum
5. St. Michael the Archangel Church
6. Stanisław-Tołpa Park
7. Faculty of Architecture (Wrocław University of Technology)
8. Scientific and Educational Center (Wrocław University of Environmental and Life Sciences)
9. Promenade in the Grunwald square, Grunwald Bridge and group of fountains

And for the route "OstrówTumski":

1. Statue of St. John of Nepomuk
2. Tenement houses on the Katedralna street
3. Statue of Blessed Virgin Mary with Child, John the Baptist Cathedral
4. Square at Assembly Hall of Pontifical Faculty of Theology
5. St. Giles church
6. Hotel of John Paul II
7. Holy Cross Church
8. Saint Martin Church
9. Mill Bridge

3.3.1.3 KIELCE

Kielce, the capital of Świętokrzyskie Voivodeship, on the route between Kraków and Warszawa (Fig. 3.2) is one of the largest cities in Poland that is particularly, closely related with geology and ore explorations. The exploitation of Flintstone in this area started early in the Neolithic times (~6000 BCE), and the remnants of those "prehistoric underground mines" can still be seen and visited in Krzemionki Opatowskie Museum.[6] Kielce is the only city in Europe with such diverse and protected geological formations (Nita and Myga-Piątek, 2010), and an abundance of outcrops of Palaeozoic, Triassic, and Cenozoic formations (Filonowicz, 1980). It is popularly known as the largest open-air geological museum (Nita and Myga-Piątek, 2010) with five nature reserves (Biesak-Białogon, Kadzielnia, Karczówka, Ślichowice, and Wietrznia) and one geopark (Kielce Geopark) within the city. Recently, the Kielce Geopark was included in a bigger project called "Geoland of the Holy Cross Mountains Geopark" (http://geopark.pl) that is applying to become a part of the Global UNESCO Geoparks network.[7] This new initiative covers five communities belonging to Kielce County: Kielce, Chęciny, Morawica, Sitkówka-Nowiny, and Piekoszów. However, because the main aim of this paper is urban geotourism, only the Kielce city and its geopark will be described here in detail.

Kielce Geopark located in the city comprises an active network of geotourist and educational facilities, situated within the geological reserves of Wietrznia, Kadzielnia, and Ślichowice, and it also includes the underground tourist route in Kadzielnia that runs through the interconnected caves of "Odkrywców" (*Explorers*), "Prochownia" (*Gunpowder Store*), and "Szczelin" (*Crevice*).[8]

---

[6]http://krzemionki.pl/en/
[7]http://geopark-kielce.pl/geoland-of-the-holy-cross-mountains-geopark/
[8]http://geopark-kielce.pl/en/

Located in the center of the city, the Devonian limestone quarry known as Kadzielnia, with about 20 caves, is of particular interest. Apart from its unique location, this geosite offers broad geoeducational potential presenting abundance of tectonic, sedimentological and karst structures together with fossils, for example, corals (Nita and Myga-Piątek, 2010). Apart from geotouristic attractions, the site is also adapted to the needs of contemporary artistic performances and festivals, due to modernized Amphitheatre Kadzielnia built here.[9] The Amphitheatre serves also as an excellent viewpoint to the Świętokrzyskie Mountains and the city of Kielce. The Wietrznia Nature Reserve covers several inactive mining pits indicating historic Devonian limestone and dolomite excavations. Nowadays, it serves mainly as Creative Work Centre "Wietrznia," that is, the Institution of Art and Culture of the City Hall of Kielce, created for the needs of sculptors and sculptural development in the area. The Jan Czarnocki Rock Reserve, Ślichowice, is an urban geosite with the highest geoeducational potential in all of Poland featuring textbook examples of tectonic folds in the upper-Devonian limestone (Fig. 3.4). As such, it was under legal protection as early as the interwar period.[10] Apart from quarries and caves integrated in a geopark, Kielce city also offers architectural specimens with interesting buildings adorned with decorative stones, such as the Palace of Kraków Bishops, the Basilica of the Assumption of the Blessed Virgin Mary (Barcicki, 2014), and the church and monastery complex on the Karczówka Hill (Jędrychowski, 2010, 2014; Złonkiewicz and Fijałkowska-Mader, 2018). What is particularly interesting is the last urban geosite that bears testimony to the geodiversity of local buildings and decorative stones used across centuries. These include decorative Bolechowice limestones (Upper Devonian), Zelejowa "rose-like" calcite (Upper Permian-Lower Triassic), Kunów and Szydłowiec sandstones (Lower Jurassic), Doły Biskupie and Wąchock sandstones (Lower Triassic), Sudetic marbles, and so on (Jędrychowski, 2010, 2014; Złonkiewicz and Fijałkowska-Mader, 2018).

3.3.1.4 OTHER BIG CITIES (WARSZAWA, POZNAŃ, AND GDAŃSK)

Other large cities of Poland, such as Warszawa, Poznań, and Gdańsk (Fig. 3.2), are rather poorly studied and described in the context of geodiversity and geotouristic potential they hold. This is rather odd, because Warszawa and Gdańsk, apart from Kraków, are one of the most visited cities in Poland,

---
[9]http://geopark-kielce.pl/en/
[10]http://geopark-kielce.pl/en/

with an abundance of historical monuments and interesting architectural structures. Indeed Warsaw, the capital of Poland and the largest city, with about 20 million visitors in 2016,[11] is the most visited city in Poland. Moreover, the possibility of using the petrographic and mineralogical variety of building and decorative stones in Warszawa for geoeducational and touristic purposes was already proposed in the "Geological Guidebook around Warszawa and the surrounding area: with a geological map" by Jan Lewiński et al. (1927). Nowadays, the only geotouristic possibilities that these cities offer with the proposition of geotoursitic routes and urban geosites is limited to sparse papers published in specialized stonemasons' journals (e.g., Walendowski, 2010) or geological conference materials (e.g., Tołkanowicz, 2008) which are almost inaccessible to tourists. An exception here is a paper published online on the website of the Polish Geological Institute—National Research Institute presenting the geoeducational and geotouristic potential of Warsaw subway (Tołkanowicz, 2009), and recent paper on geodiversity of Poznań city (Zwoliński et al., 2017). According to the authors, urban geosites in Poznań are represented mainly by Scandinavian erratics in the Millennium Park and landforms produced by glaciations (i.e., the Morasko Hill).

FIGURE 3.4 Tectonic folds in the upper-Devonian limestone in the Jan Czarnocki Rock Reserve Ślichowice in Kielce.

---

[11] http://wot.waw.pl/wp-content/uploads/2016/08/Turystyka_w_Warszawie_Raport_2016.pdf

3.3.2 URBAN GEOTOURISM IN SMALL TOWNS

Urban geotourism related to the small towns is more common in Poland than geotouristic routes, well-prepared geosites or geotouristic guide books for big cities. Tourist traffic in those places is usually concentrated and focused only on one specific geotouristic attraction (Zagożdżon and Zagożdżon, 2016), such as location of interesting fossils, minerals, specific and rare types of rocks, architecture of a town related to one specific stone material, monuments made of the decorative stone typical to the region's sculpture, infrastructure, and/or architectonic remnants of mine activity, and so on. A selected few small and medium-sized towns in Poland with their main attractions related to earth sciences are presented in Figure 3.2 (yellow circles) and in Table 3.1. These are mainly towns situated in southern Poland. The beauty of those places is principally due to their architectural structures, which showcase different types of buildings and decorative stones, for example, Wieluń (Śpiewak, 2014), Kłodzko (Zagożdżon and Zagożdżon, 2015e), Lubań (Myśliwiec and Wojciechowski, 2000), and Piława Górna (Gil, 2015), or stone exploitation sites within towns, along with monuments made of local material, for example, Szydłowiec (Kowalski and Urban, 2004), Lubań (Myśliwiec and Wojciechowski, 2000), Nowa Ruda (Borzęcki and Marek, 2013), Tarnowskie Góry (see Gaidzik and Chmielewska, 2020). Some of them gained the popularity based on a very specific and characteristic geosite, such as the Tertiary petrified wood fragments in Siedliska, mentioned earlier (Słomka et al., 2006) or Piława Górna, which promotes itself as a place of the 1879 Gnadenfrei meteorite fall (Gil, 2015). The geotouristic potential of other locations, like Pruszków, remains unknown to authorities and inhabitants (Górska-Zabielska and Zabielski, 2017). Another specific type of urban geotourism in small-sized towns mainly in southern Poland is spa-geotourism (Chowaniec and Zuber, 2008) and thermal geotourism (Dryglas and Różycki, 2016) (Fig. 3.2). Spa geotourism can be considered popular, especially in mountain towns in southern Poland, whereas thermal geotourism is still rather marginal, in spite the fact that there is a high consumer interest—most tourists staying in spa resorts indicate that they find the thermal pools the most desirable attraction (Dryglas and Różycki, 2016). This is also corroborated by a relatively small number of health and spa resorts in Poland using geothermal waters for therapeutic purposes (9 out of 45). The Polish tradition of using geothermal waters in health resorts can be traced back to the 12th century (Dryglas and Hadzik, 2016). The most popular spa resorts located in towns

TABLE 3.1 Main Geotouristic Attractions of Selected Small Towns in Poland (Ordered Alphabetically) (for Locations, See Fig. 3.1).

No.	Town	Voivodeship	Latitude	Longitude	Main attraction
1	Kłodzko	Dolnośląskie	50°26′16″N	16°39′10″E	Variety of building and decorative stone mainly from the Lower Silesia region, particularly in Town Hall, Museum of the Kłodzko Land, and the Church of Our Lady of the Rosary on Franciszkańska Square (Zagożdżon and Zagożdżon, 2015e)
2	Lubań	Dolnośląskie	51°07′05″N	15°17′21″E	Basalt exploitation sites and monuments (Myśliwiec and Wojciechowski, 2000)
3	Nowa Ruda	Dolnośląskie	50°34′45″N	16°30′05″E	Nowa Ruda postmining dump (minerals, fossils, esthetic values), together with underground tourist trail (Borzęcki and Marek, 2013)
4	Piława Górna	Dolnośląskie	50°40′54″N	16°45′17″E	Quarries, architecture monuments, place of 1879 Gnadenfrei meteorite fall (Gil, 2015)
5	Siedliska	Lubelskie	50°16′02″N	23°33′39″E	Tertiary petrified wood fragments (Słomka et al., 2006)
6	Wieluń	Łódzkie	51°13′21″N	18°34′26″E	More than 50 different types of building and decorative, both local and foreign material (Śpiewak, 2014)
7	Szydłowiec	Mazowieckie	51°13′26″N	20°51′26″E	Sandstone exploitation (Kowalski and Urban, 2004)
8	Pruszków	Mazowieckie	52°09′56″N	20°48′20″E	Museum of Ancient Mazovian Metallurgy, erratics, old glacial relief, Pliocene clays, till, bog iron, and water (Górska-Zabielska and Zabielski, 2017)
9	Ustka	Pomorskie	54°34′43″N	16°52′09″E	Petrographic variety of beach clasts and cliff coast (Zagożdżon, 2014)

include: CiepliceŚląskie-Zdrój, Lądek-Zdrój, Duszniki-Zdrój, Ciechocinek, Konstancin-Jeziorna, Ustroń, Rabka-Zdrój, and Uniejów. Apart from those, nine urban thermal centers are present on the map of Poland: Geotermia Grudziądz, Aqua Park in Zakopane, geothermal bathing pool Szymoszkowa in Zakopane, Termy Mszczonów, Termy Uniejów, Geotermia Poddębice, Termy Maltańskie in Poznań, Termy Cieplickie, and Termy Warmińskie in Lidzbark Warmiński (Fig. 3.2) (Dryglas and Różycki, 2016). Thus, even though this branch is improving every year, the supply of thermal tourism services does not yet match the demand in terms of the number and construction time of thermal facilities in statutory spa resorts (Dryglas and Hadzik, 2016).

3.3.3 POSTINDUSTRIAL URBAN GEOTOURISM—SILESIA REGION

Postindustrial geotourism is a specific form of urban geotourism intensively developed and popularized in southern Poland, that is, Dolnośląskie (Lower Silesia) and Śląskie (Silesia) Voivodeships (Fig. 3.2). The abundance of postmining towns, architectural style, infrastructure, and mining culture from the region's ore and bituminous coal exploitations makes the Silesian region the perfect geotouristic destination. Apart from the above the additional advantage of the Silesian region as a popular geotouristic destination is its vicinity to majority of geosites located in the medium-sized towns of Upper Silesia, for example, Gliwice, Zabrze, Ruda Śląska, Katowice, Sosnowiec, Świętochłowice, Chorzów, Piekary Śląskie, Siemianowice Śląskie, Dąbrowa Górnicza, Mysłowice, and Lower Silesia, for example, Wałbrzych, NowaRuda, ZłotyStok, Kletno, and so on. Such situation generates propitious conditions for urban geotourism development.

Postindustrial urban geotourism is particularly popular in Upper Silesia (Silesian Voivodeship). Among multiplicity of remnants of hundreds of years of history of mining in this area, the most important are (1) Historic Silver Mine and Black Trout Adit in Tarnowskie Góry, (2) Queen Louise Coal Mine and Main Key Hereditary Adit in Zabrze, and (3) Guido Coal Mine in Zabrze (Gaidzik and Chmielewska, 2020), the latter one was recently included in the UNESCO World Heritage List (http://whc.unesco.org/en/list/1539).

There is also the so-called Industrial Monuments Route of the Silesian Voivodeship (www.zabytkitechniki.pl/en-US) that includes 42 museums, heritage parks, inhabited workers' settlements and still running workplaces; many of them related with ore or bituminous coal exploitation and processing, for example, Chorzów: "President" Shaft; Czerwionka-Leszczyny: Familoki; Częstochowa: Museum of Iron Ore Mining, Dąbrowa Górnicza: Drill Mine of the "Sztygarka" Town Museum; Katowice: Wilson Shaft Gallery, Giszowiec Settlement, Nikiszowiec Settlement, Museum of Zinc Metallurgy; Ruda Śląska: Ficinus Workers' Settlement; Rybnik: "Ignacy" Historic Mine; Świętochłowice: coal mine, KWK Polska; Zabrze: Museum of Coal Mining, "Queen Louise" Adit, Maciej Shaft, "Guido" Coal Mine. However, extremely popular objects included in the Industrial Monuments Route are not the only geotouristic attractions of this region. Most of the towns from this region also have a very long history that results in interesting architectural structures, such as churches, town halls, monuments, and sculptures in Gliwice (Labus, 2005), Bytom or Katowice, made of local and predominantly imported building materials and decorative stones. However, these are usually not sufficiently described and/or promoted. For example, the geotouristic potential of the largest cathedral in Poland, that is, Arch cathedral of Christ the King in Katowice, with geoeducational examples of tectonic structures (mainly en échelon, tension gashes, pull-apart basins, brittle and ductile shear zone, and many others) in dolomite from Imielin that covers the inner part of the Arch cathedral, is not studied, or described, neither promoted (Fig. 3.5). And there are a lot of other buildings, objects, and monuments with equal or possibly even higher geotouristic potential that are still waiting to be discovered and promoted.

3.4 CONCLUSIONS

High geodiversity of Poland and its location on the three main European megaunits of different ages and thus profusion of fascinating geological objects, together with its long and difficult history resulting in an abundance of old cities and towns with historical monuments, architectural wonders, and so on, makes it an ideal place for urban geotourism development. This specific form of tourism has been gaining acceptance and reputation in Poland for the last several years. Nevertheless, in general, the geotouristic and geoeducational potential of many cities remains unexplored. This is especially true for large cities. For example, there is no guidebook or at least scientific description of the geotouristic potential, proposition of geotouristic routes or geosites of the capital of Poland and its largest city, Warszawa. The same situation goes for Gdańsk or Poznań. Worldwide examples, and those of Kraków, Kielce, Wrocław, and many small towns in Poland, along with postindustrial urban geotourism in Upper Silesia show that the combination of urban tourism with geoeducation and admiration of natural wonders can be a great idea for promotion of tourism in a town (acting as a "business card") leading to the growth of tourist traffic and therefore augmenting the city's revenue. Furthermore, the extraordinary prospect of educating people in earth sciences in a city, which could not be more different than the lap of mother nature, is simply intriguing.

FIGURE 3.5 Archcathedral of Christ the King in Katowice (a), with tension gashes forming two conjugate sets, pull-apart basins, and shear zone in dolomite from Imielin that covers the inner part of the archcathedral (b and c).

ACKNOWLEDGMENTS

I would like to thank Ewa Welc for the discussion and valuable suggestions and comments that helped to improve the final version of the manuscript. Also, I would like to thank Proma Nautiyal for English editing of the manuscript.

KEYWORDS

- urban geotourism
- Poland
- geosite
- architecture
- geoeducation

REFERENCES

Aleksandrowski, P. Prowincje Tektoniczne Polski. In *Atlas geologiczny Polski*; Nawrocki, J., Becker, A., Eds.; Państwowy Instytut Geologiczny: Warszawa, 2017.

Barcicki, M. Uwarunkowania Rozwoju Geoturystyki Miejskiej na Przykładzie Kielc. *Logistyka* **2014**, *3*, 326–339.

Borkowski, K. Prezentacja badań ruchu turystycznego w roku 2016 w Krakowie. In *XXVIII Forum Turystyki—podsumowania sezonu turystycznego w Krakowie*: Kraków, 2016.

Březinová, D.; Bukovanská, M.; Dudková, I.; Rybařik, V. *Praha kamenná*. Narodni muzeum: Praha, 1996; 287 pp.

Bromowicz, J.; Magiera, J. Materiał kamienny wnętrza Kaplicy Zygmuntowskiej. In *Kamień architektoniczny i dekoracyjny*; Bromowicz, J., Ed.; 23–24.09.2003, AGH: Kraków, 2003; pp 5–12.

Bromowicz J.; Magiera J. *Kamienie wczesnośredniowiecznych budowli Krakowa. Ich pochodzenie na tle geologii miasta*. Wydawnictwa AGH: Kraków, 2012.

Charsley, T. J. Urban Geology: Mapping it Out. In *Geology on Your Doorstep*; Bennett, M. R., Doyle, P., Larwood, J. G., Prosser, C. D., Eds.; The Geological Society, Bath, 1996; pp 11–18.

Chowaniec, J.; Zuber, A. Touristic Geoattractions of Polish Spas. *Przegląd Geol* **2008**, *56*, 706–711.

Chylińska, D.; Kołodziejczyk, K. Geotourism in an Urban Space? *Open Geosci*. **2018**, *10* (1), 297–310.

da Silva, C. M. Urban Geodiversity and Decorative Arts: the Curious Case of the "Rudist Tiles" of Lisbon (Portugal). *Geoheritage* **2019**, *11*, 151–163.

De Wever, P.; Baudin, F.; Pereira, D.; Cornée, A.; Egoroff, G.; Page, K. The Importance of Geosites and Heritage Stones in Cities—A Review. *Geoheritage* **2017**, *9* (4), 561–575.

Del Lama, E. A.; Bacci, D. D. L. C.; Martins, L.; Garcia, M. D. G. M.; Dehira, L. K. Urban Geotourism and the Old Centre of São Paulo City, Brazil. *Geoheritage* **2015**, *7* (2), 147–164.

Dmytrowski, P.; Kicińska, A. Waloryzacja geoturystyczna obiektów przyrody nieożywionej i jej znaczenie w perspektywierozwoju geoparków. *Probl Ekol Krajobr* **2011**, *29*, 11–20.

Dowling, R. K. Geotourism. In *Encyclopedia of Tourism*; Jafari J., Xiao H., Eds.; Springer: Berlin, 2015; pp 389–391.

Dryglas, D.; Hadzik, A. The Development of the Thermal Tourism Market in Poland. *Geotourism/Geoturystyka* **2016**, *46–47*, 27–42.

Dryglas D.; Różycki P. European Spa Resorts in the Perception of Non-commercial and Commercial Patients and Tourists: The Case Study of Poland. *e-Re Tour Res* **2016**, *13* (1–2), 382–400.

Filonowicz, P. Mapa podstawowa 1:50 000 ark. Kielce. In *Mapa geologiczna Polski 200 000*. Wydawnictwo Instytutu Geologicznego: Warszawa, 1980.

Firlet, E. M. *Smocza Jama na Wawelu. Historia, Legenda, Smoki*. Universitas: Kraków, 1996.

Gaidzik, G.; Chmielewska, M. Post-Mining Objects as Geotourist Attractions: Upper Silesian Coal Basin (Poland). In *The Geotourism Industry in the 21st Century*; Sadry, B. N., Ed.; The Apple Academic Publishers Inc., 2020; pp 505–524.

Gambino, F.; Borghi, A.; d'Atri, A.; Gallo, L. M.; Ghiraldi, L.; Giardino, M.; Martire, L.; Palomba, M.; Perotti, L.; Macadam, J. TOURinSTONES: A Free Mobile Application for Promoting Geological Heritage in the City of Torino (NW Italy). *Geoheritage* **2019**, *11*, 3–17.

Gil, G. *Inwentaryzacja stanowisk na potrzeby opracowania Miejskiej Trasy Geoturystycznej w Piławie Górnej*. Oprac. Inst. Nauk Geol. Uniw. Wroc.: Wrocław, 2015. http://www.pilawagorna.pl/asp/pliki/201601/inwentaryzacja_pg2015.pdf (accessed January 29, 2018).

Górny, Z. Wybrane przykłady kamienia naturalnego z Włoch i Niemiec zastosowane w obiektach architektonicznych Krakowa—krótka wycieczka geologiczna. *Geoturystyka* **2009**, *1–2* (16–17), 61–70.

Górska-Zabielska, M.; Zabielski, R. Potential Values of Urban Geotourism Development in a Small Polish Town (Pruszków, Central Mazovia, Poland). *Quaest. Geogr.* **2017**, *36* (3), 75–86.

Hose, T. A. Selling the Story of Britain's Stone. *Environ. Interpret.* **1995**, *10* (2), 16–17.

Hose, T. A. Towards a History of Geotourism: Definitions, Antecedents and the Future. The History of Geoconservation. *Geol. Soc. London Spec. Publ.* **2008**, *300* (1), 37–60.

Hose, T. A.; Vasiljević, D. A. Defining the Nature and Purpose of Modern Geotourism with Particular Reference to the United Kingdom and South-East Europe. *Geoheritage* **2012**, *4* (1–2), 25–43.

Jędrychowski, J. *Karczówka, góra mocy i kruszców*. Georaj: Kielce, 2010.

Jędrychowski, J. *Kamień w architekturze regionu świętokrzyskiego*. Georaj: Kielce, 2014.

Kin, A.; Rajchel, J. Skamieniałości głowonogów w architekturze Krakowa. In *Abstrakty Pierwszego Polskiego Kongresu Geologicznego*, 26–28 czerwca 2008, Wyd. PTG: Kraków, 2008, p. 53.

Kowalski, W.; Urban, J. *Szydłowiec—miasto na kamieniu*. Wyd. Urz. Miejski w Szydłowcu: Szydłowiec, 2004.

Kubalíková, L.; Kirchner, K.; Bajer, A. Secondary Geodiversity and Its Potential for Urban Geotourism: A Case Study from Brno City, Czech Republic. *Quaest. Geograph.* **2017**, *36* (3), 63–73.

Labus, M. Geoturystyka miejska na przykładzie Gliwic. *Zeszyty Naukowe Politechnik Śląskiej, Seria: Górnictwo* **2005**, *269*, 221–230.

Lewiński, J.; Łuniewski, A.; Małkowski, A.; Samsonowicz, J. *Przewodnik geologiczny po Warszawie i okolicy*. Wydawnictwo Oddziału Warszawskiej Komisji Fizjograficznej Polskiej Akademii Umiejętności: Warszawa, 1927, 178 pp.

Lorenc, M. W.; Mazurek, S. Wybrane, nowe propozycje atrakcji geoturystycznych z Dolnego Śląska. *Geoturystyka* **2010**, *3–4*, 3–18.

Majer-Durman, A. The occurrence of Upper Jurassic limestones in Cracow area and examples of their usage in the city architecture. *Geotourism/Geoturystyka* **2012**, *28–29*, 13–22.

Myśliwiec, B.; Wojciechowski, K. *Szlakiem wygasłych wulkanów—ścieżka dydaktyczna pieszo-rowerowa* (folder). Wyd. Urz. Miasta Lubań: Lubań, 2000.

Newsome, D.; Dowling, R., Eds.; *Geotourism: The Tourism of Geology and Landscape*; Goodfellow Publishers Limited: Oxford, UK, 2010.

Nita, J.; Myga-Piątek, U. Georóżnorodność i geoturystyka w terenach poeksploatacyjnych na przykładzie regionu chęcińsko-kieleckiego. *Geoturystyka* **2010**, *22–23*, 51–58.

Palacio-Prieto, J. L. Geoheritage within Cities: Urban Geosites in Mexico City. *Geoheritage* **2015**, *7* (4), 365–373.

Pica, A.; Reynard, E.; Grangier, L.; Kaiser, C.; Ghiraldi, L.; Perotti, L.; Del Monte, M. GeoGuides, Urban Geotourism Offer Powered by Mobile Application Technology. *Geoheritage* **2018**, *10* (2), 311–326.

Pieńkowski, G. Geologiczne muzea i parki tematyczne Dźwignią edukacji, rozwoju i biznesu. *Przegląd Geol.* **2011**, *59*, 323–328.

Portal, C.; Kerguillec, R. The Shape of a City: Geomorphological Landscapes, Abiotic Urban Environment, and Geoheritage in the Western World: The Example of Parks and Gardens. *Geoheritage* **2018**, *10*, 67–78.

Rajchel, J. *Kamienny Kraków. Spojrzenie geologa*. Uczelniane Wydawnictwa Naukowo-Dydaktyczne AGH: Kraków, 2004.

Rajchel, J. The Stony Cracow: Geological Valors of Its Architecture. *Przegląd Geol.* **2008**, *56*, 653–662.

Rajchel, J. Tradycyjne bruki w krajobrazie Krakowa. *Geologia/Akademia Górniczo-Hutnicza im. Stanisława Staszica w Krakowie* **2009a**, *35* (1), 41–55.

Rajchel, J. Martwice wapienne w architekturze Krakowa. *Geologia/Akademia Górniczo-Hutnicza im. Stanisława Staszica w Krakowie* **2009b**, *35* (2/1), 313–322.

Rajchel, J. Kamienne apotropaiczne rzeźby w architekturze Krakowa. *Przegląd Geol.* **2014**, *62* (3), 156–162.

Rodrigues, M. L.; Machado, C. R.; Freire, E. Geotourism Routes in Urban Areas: A Preliminary Approach to the Lisbon Geoheritage Survey. *GeoJ. Tour. Geosites* **2011**, *8* (2), 281–294.

Schroeder, J. H. Naturwerksteine in Architektur und Bau-geschichte von Berlin. *Führer zur Geologie von Berlin und Branden-burg* **2006**, 6.

Seemann, R.; Summesberger, H. *WienerSteinwan-derwege. Die Geologie der Großstadt*. Verlag Christian Brandstätter: Wien, 1998.

Sermet, E.; Rolka, G. *Walory geoturystyczne kamieniołomów na krakowskim Zakrzówku—możliwości zagospodarowania*; Abstrakty II. Pols. Kongr. Geol.: Warszawa, 2012; p 76.

Słomka, T., Ed. *Katalog obiektów geoturystycznych w obrębie pomników i rezerwatów przyrody nieożywionej (The Catalogue of Geotourist Sites in Nature Reserves and Monuments)*. AGH Akademia Górniczo-Hutnicza. Wydział Geologii Geofizyki i Ochrony Środowiska. Katedra Geologii Ogólnej i Geoturystyki: Kraków, 2012. http://www.kgos.agh.edu.pl/index.php?action=projekty&subaction=25.9.140.921

Słomka, T.; Kicińska-Świderska, A. Geoturystyka–podstawowe pojęcia. *Geoturystyka* **2004**, *1* (1), 5–8.

Słomka, T.; Kicińska-Świderska, A.; Doktor, M.; Joniec, A. *Katalog obiektów geoturystycznych w Polsce*. Projekt finansowany przez Narodowy Fundusz Ochrony Środowiska i Gospodarki Wodnej: Kraków, 2006.

Śpiewak, A. Building Stone Used in Architectural Objects in the Town of Wieluń. *Geotourism/Geoturystyka* **2014**, *38–39*, 19–28.

Stachowiak, A.; Cwojdziński, S.; Ihnatowicz, A.; Pacuła, J.; Mrázová, Š.; Skácelová, D.; Otava, J.; Pecina, V.; Rejchert, M.; Skácelová, Z.; Večeřa, J. *Geostrada Sudecka–Przewodnik Geologiczno-Turystyczny*. PIG–PIB, ČGS: Warszawa–Praga, 2013.

Tołkanowicz, E. Kamień w obiektach architektonicznych Warszawy—spojrzenie geologa. In: *Abstrakty Pierwszego Polskiego Kongresu Geologicznego*, 26–28 czerwca 2008. Wyd. PTG: Kraków, 2008; p. 123.

Tołkanowicz, E. *Miejska geologia—metro warszawskie*, 2009. https://www.pgi.gov.pl/dane-geologiczne/133-kopalnia-wiedzy-nowe/skau-i-mineray/1414-miejska-geologia-metro-warszawskie.html (accessed January 28, 2017).

Walendowski, H. Poznaj Poznań. *Nowy Kamieniarz* **2010**, *50* (7), 78–83.

Zagożdżon, P. P. Blizny wojny w kamieniu ryte—śladydziałań wojennych w wybranych obiektach kamiennych na terenieWrocławia. *Pr. Nauk. Inst. Górn. PWroc.*, *135, Studia i Mater* **2012**, *42*, 147–162.

Zagożdżon, P. P. Geoturystyka w Ustce—urlopowe reminiscencje. *Pryzmat—Wiadomości Polityczne Wrocławia*, **2014**. http://www.pryzmat.pwr.edu.pl/wiadomosci/894 (accessed January 26, 2018).

Zagożdżon, P. P.; Śpiewak, A. Kamień w architekturzea geoturystyka miejska—przykłady z terenu Wrocławia. *Górn. i Geol. XVI. Pr. Nauk. Inst. Górn. PWroc., 133, Studia I Materiały* **2011**, *40*, 123–143.

Zagożdżon, P. P.; Zagożdżon, K. D. Zjawiska i struktury geologiczne w architekturze—krótki przewodnik geoturystyczny po Wrocławiu. In *Abstrakty IV Ogólnopolskiej Konferencji Naukowej Złoża kopalin—aktualne problemy prac poszukiwawczych, badawczych i dokumentacyjnych*, 15–17 kwietnia 2015; Wyd. PIG-PIB: Warszawa, 2015a; pp 128–129.

Zagożdżon, P. P.; Zagożdżon, K. D. Sieć miejskich tras geoturystycznych w Kłodzku jako nowe narzędzie w zakresie edukacji i popularyzacji geologii. In *Abstrakty IV Ogólnopolskiej Konferencji Naukowej Złoża kopalin—aktualne problemy prac poszukiwawczych, badawczych i dokumentacyjnych*, 15–17 kwietnia 2015; Wyd. PIG-PIB: Warszawa, 2015b; pp 130–131.

Zagożdżon, P. P.; Zagożdżon, K. D. Kamienne elementy architektury miejskiej jako geologiczne zaplecze edukacyjne—przykłady wrocławskie. *Przegląd Geol* **2015c**, *63*, 150–154.

Zagożdżon, P. P.; Zagożdżon, K. D. Możliwości edukacji geologii na podstawie wrocławskiego kamienia architektonicznego. *Przegląd Geol* **2015d**, *63*, 284–288.

Zagożdżon, P. P.; Zagożdżon, K. D. Nowe oblicze staregoKłodzka—sieć tras geoturystycznych jako nowatorska metodapromocji miasta. *Zeszyty Muzeum Ziemi Kłodzkiej* **2015e**, *13*, 138–147.

Zagożdżon, P. P.; Zagożdżon, K. D. Wybrane aspekty geoturystyki w Polsce—obiekty podziemne i geoturystyka miejska. *Przegląd Geol*. **2016**, *64*, 739–750.

Złonkiewicz, Z.; Fijałkowska-Mader, A. Building Stones in Architecture of the Church and Monastery Complex on the Karczówka Hill in Kielce (Holy Cross Mts., South-Central Poland). *Przegląd Geol*. **2018**, *66*, 421–435.

Zwoliński, Z.; Hildebrandt-Radke, I.; Mazurek, M.; Makohonienko, M. Existing and Proposed Urban Geosites Values Resulting from Geodiversity of Poznań City. *Quest. Geogr.* **2017**, *36* (3), 125–149.

INTERNET SOURCES

https://archiwum.mos.gov.pl/kategoria/2398_katalog_obiektow_geoturystycznych_w_polsce—Catalogue of Geotouristic Sites of Poland from the Ministry of the Environment of Poland (accessed December 28, 2017).

http://geopark-kielce.pl/en/—Official Website of the Kielce Geopark (accessed January 29, 2018).

http://geopark.pl—official website of the Geoland of the Holy Cross Mountains Geopark (accessed October 4, 2018).

http://geopark-kielce.pl/geoland-of-the-holy-cross-mountains-geopark/—Application of the Geoland of the Holy Cross Mountains Geopark for the UNESCO Global Geoparks Network (accessed October 4, 2018).

http://geoportal.pgi.gov.pl/portal/page/portal/geostanowiska—Central Catalogue of Geosites in Poland Made by Polish Geological Institute—National Research Institute (accessed December 28, 2017).

http://www.geosilesia.us.edu.pl/295, geostanowiska_wojewodztwa_slaskiego.html—Local Catalogue of Geosites in Silesian Voivodeship (accessed January 19, 2018).

http://krzemionki.pl/en/—Official Website of the KrzemionkiOpatowskie Museum (accessed January 29, 2018).

http://whc.unesco.org/en/list/1539—TarnowskieGóry Lead–Silver–Zinc Mine and Its Underground Water Management in the UNESCO World Heritage List (accessed January 19, 2018).

http://wot.waw.pl/wp-content/uploads/2016/08/Turystyka_w_Warszawie_Raport_2016.pdf—Turystyka w Warszawie. Raport 2016 (accessed January 19, 2018).

www.zabytkitechniki.pl/en-US—Official Website of the Industrial Monuments Route of the Silesian Voivodeship (accessed January 19, 2018).

CHAPTER 4

Mining Heritage as Geotourism Attractions in Brazil

ANTONIO LICCARDO[1*], VIRGINIO MANTESSO-NETO[2], and MARCOS ANTONIO LEITE DO NASCIMENTO[3]

[1]Geoscience Department, State University of Ponta Grossa, Ponta Grossa, PR, Brazil

[2]Member of State of São Paulo Council on Geological Monuments, São Paulo, SP, Brazil

[3]Geology Department, Federal University of Rio Grande do Norte, Natal, RN, Brazil

*Corresponding author. E-mail: aliccardo@uepg.br

ABSTRACT

Mining in Brazil has a much more recent history than in the Old World but has had great importance at various times over the last three centuries. Because of its role in the international scenario or because of the importance of mineral extraction in the development of the country, some mines are considered cultural heritage at regional, national, and international levels. Six cases chosen throughout Brazil's large territory, representing different periods of our history are here presented and discussed also in function of the existing tourism. These sites, and their respective main mineral resource, are Ouro Preto (state of Minas Gerais—gold), Diamantina (state of Minas Gerais—diamonds), Chapada Diamantina (state of Bahia—carbonados), Ametista do Sul (state of Rio Grande do Sul—amethyst), Pedro II (state of Piauí—noble opal), Currais Novos (state of Rio Grande do Norte—scheelite/tungsten), and Itu (state of São Paulo—dimension stone). The offer in these places of different degrees of geoscientific information adequately presented to the general public proposes the development of geotourism as an aggregating factor in cultural, scientific, and educational tourism. The expectation is that this strategy will contribute to the protection of the mining heritage and the geoconservation of the sites.

4.1 INTRODUCTION

Brazil was discovered by the Portuguese in 1500 and, unlike Spanish territory in the so-called New World, did not initially show much potential for *mining*. In the first 200 years, the economy of this colony was linked to the vegetal extractivism and to the cultivation of sugar cane, with very few occurrences of alluvial gold in the southeast. Although the Spanish conquistadors found abundant gold, silver, and emeralds in the Inca, Aztec, and Maia territories from the moment of their arrival in America, the Portuguese found cultures that had not developed the skills of mining and metallurgy, which made it difficult to discover and use the existing, if any, mineral deposits.

In 1695, the enormous gold deposits of Minas Gerais were found, which began a new economic cycle and a new history that changed the course not only of the colony but also of Portugal and of Europe itself throughout the 18th century. An almost century-long period of extraordinary richness changed reality and culture in Brazil and Portugal, with worldwide sociopolitical consequences and with repercussions still present in the culture of both countries.

The volume and importance of these three centuries of mining in Portuguese America, nowadays Brazil, have left deep marks in the economy and in the social structure of the country. With its territory of continental dimensions, Brazil has many important deposits of a large number of minerals and ores, characterizing even a kind of territorial vocation for mining activity. In the first cycle of mineral exploitation, essentially coinciding with the 18th century, gold and precious stones were, by far, the most important products, and the remaining vestiges of that activity are currently a consistent and fascinating cultural heritage.

With the worldwide demand for new minerals after the Industrial Revolution, mining in Brazil has diversified its production of metallic ores, minerals and industrial rocks and colored gems. In terms of participation in the world market, Brazil detains 92% of the niobium production, 20% of iron ore, 22% of tantalite, 19% of manganese, 11% of aluminum and asbestos, and 19% of graphite, among others (Barreto, 2001). According to the National Department of Mineral Production (DNPM, 2015), the country's mineral resources are meaningful and diversified, currently covering production of 72 mineral substances, being 23 metallic, 45 nonmetallic, and 4 energetic. There are 1820 mining operations and 13,250 mining licenses. According to the Brazilian Institute of Gems and Metals (IBGM, 2010), no less than a third of the world's gemstones production is Brazilian and, in terms of gold production, in 2008, production reached 12th place in the worldwide ranking.

Authors such as Prieto (1976) argue that the mining activity in America was a creator of cities and nations, causing rapid occupation and rapid development, compared to other human activities, and that knowledge of the territory was fostered by the search for precious minerals. For this author, who analyzed in depth the history of mining in Latin America, the exploited gold was transmuted into *imponderable essences*, such as the very formation of the peoples and the concept of nationalities in Hispanic and Portuguese America.

Among these imponderable essences, the mining culture, deeply rooted in many parts of Brazil, has a strong potential to foster tourism (geological, cultural, or heritage) including, but not limited to, visits to the old (and in some cases still active) extraction sites. Tourist experience in mining areas involves socializing with local communities, visits to quarries and/or to underground galleries, interpretive museums, learning and enjoyment with landscapes and special sensations, only possible in loco—such as the unforgettable experience of tourists who remain in absolute darkness for some minutes and then see the glow of an infinite number of luminescent scheelite grains that dot the walls of the scheelite mine in Currais Novos (Rio Grande do Norte). A study of already consolidated cases of mining heritage linked to tourism is presented and discussed in different places in Brazil (Fig. 4.1). Gold, diamonds, and imperial topaz represent the mining of the 18th and 19th centuries in Minas Gerais and Bahia. Colored gemstones such as amethyst in Rio Grande do Sul and opal in Piauí were produced only from the 20th century on and represent the continuity of extraction of these fascinating materials. Also from the 20th century, specific resources such as scheelite (tungsten) or varvite (basically a dimension stone, but also used for the manufacture of rough-style furniture, barbecue stands, decoration items, etc.) are symbolic of the evolution of mining and the different demands of today's society.

Considering these factors, the following localities were selected for the geotourism analysis in areas of mining heritage in Brazil: Ouro Preto and Diamantina (state of Minas Gerais); Chapada Diamantina (state of Bahia); Ametista do Sul (state of Rio Grande do Sul); Pedro II (state of Piauí); Currais Novos (state of Rio Grande do Norte); and Itu (state of São Paulo). These sites present, in addition to the importance of their cultural landscapes and exceptional geological content, a thematic tourism already consolidated for years, with different degrees of maturity and structuring.

FIGURE 4.1 Map of Brazil indicating the areas mentioned in this text where mining heritage represents a meaningful asset to local tourism. Chapada Diamantina means "Diamond Plateau" (a large area), whereas the other five names indicate cities.

Tourism in Brazil is still strongly characterized by the sun/beach binomial, privileging the coastal areas in terms of infrastructure. On the other hand, the presence of geological elements in the main places of tourist attraction is remarkable. Analyzing the 2010 edition of the periodical publication of the Brazilian Ministry of Tourism, entitled "Roteiros do Brasil. 94 Belos Motivos Para Viajar Pelo Brasil: Tudo o que você precisa saber para curtir férias inequecíveis" (Itineraries in Brazil. 94 Beautiful Motives for Traveling in Brazil: Everything you need to know to enjoy unforgettable holidays), Mantesso-Neto et al. (2012) showed that of the motives (in practice, places) listed, 26.1% have an element of geodiversity as the main, or sole, attraction; 23.4% have a strong connection with geology; 23.1% have a moderate connection; and only

27.4% of the motives have no connection with geological sites or features. In other words, even without an explicit mention or explanation of this presence, elements of geodiversity participate in 72.6% of the "94 Beautiful Motives for Traveling Through Brazil."

Two of best-known Brazilian world-class tourist attractions are geological features: Pão de Açucar, or Sugar Loaf, a 560-my-old granitic-gneissic complex with the characteristic sugarloaf shape, and Cataratas do Iguaçu, or Iguazu Falls, a majestic binational amphitheater of about 275 falls carved by the Paraná River (which in that stretch makes the international Brazil-Argentine border) in the 130-my-old, 1500-m thick Paraná basalt plateau. Both are UNESCO's World Heritage properties.

In this context of geological masterpieces, geotourism tunes in with cultural tourism in that it seeks to value information and knowledge as attractions anchored to places of remarkable value of cultural or natural heritage. The development of geotourism, especially in the historical mining sites, brings an important argument for cultural tourism, which presents itself as an alternative of social sustainability for areas of the interior of Brazil, as well as a factor of strengthening of the identity of these communities forged by mining.

4.2 MINING HERITAGE AND GEOTOURISM

The concept of mining heritage includes all vestiges of mining activities of the past (recent or old), to which a social group attributes historical, social or cultural values (Puche Riart et al., 1994). Sánchez (2011) further indicates that the constituent elements of the mining heritage can be part of the historical, archaeological, industrial, cultural, ethnographic, and even geological heritage—so it can be the object of study of archeology, technology, economic history, and social history. Until the 1960s, the notion of mineral heritage referred to the remains of mineral extraction in the preindustrial or protoindustrial periods, that is, prior to the First Industrial Revolution and was almost exclusively restricted to the field of archeology. Starting from the period of the creation of the first machines in the mining industry, industrial archeology was developed as a new field of history, opening the possibility of including mining activities of the contemporary age. With this, mining heritage came to be considered a subdivision of the industrial heritage (Guiollard, 2005).

At the regional and local levels, contemporary mining heritage means the physical memory of a past and its recovery, and the possibility of reestablishing the bonds that existed between the community and the mineral resources of the territory (Marchán and Sánchez, 2013).

Pérez de Perceval Verde (2010) and Marchán and Sánchez (2013) indicate a heterogeneous set of elements susceptible of constituting a heritage legacy of mining, which can be classified as immobile elements (such as large installations), mobile elements (such as objects that have been used in production throughout its history), documents, immaterial heritage, and the cultural landscape. The valorization of this legacy occurs, usually with scientific, didactic, or touristic objectives, by the creation of museums and cultural spaces, visitation to the inactive work areas, the cultural landscape, the natural spaces with traces of mining, and so on. The qualification and appreciation of the traces of the exploitation as components of a mining heritage usually carry a heavy load of subjectivity, since it is not easy to individualize the criteria for this appreciation. For Pearson and McGowan (2000), assigning heritage value to a particular mining site may be the result of widespread opinion, a specialized group (regional or local) or an independent researcher.

Traditionally, little research has been done on the touristic potential of mines, quarries, and old mining facilities, perhaps because mineral extraction areas are peripheral as economically viable tourist circuits (Edwards and Llurdés i Coit, 1996). However, more recently, a new appreciation of beauty in mining landscapes has been developed, based on its historical and cultural significance (Cole, 2004). For Frew (2008), these sites have facilities for tourism management, as in other types of manufacturing processes, such as visits to wineries in France or the manufacture of cheese in the Netherlands. A demand for new cultural contexts and constant learning has been a decisive factor for the quality tourism experience. Thus, an overlapping of geotourism in areas of mining heritage enriches the theoretical support for the optimization of cultural tourism. Geotourism presents a series of possible definitions, such as that of Frey (1998), where the term refers to a new occupational and business sector in which the main scope is the transfer and communication of geoscientific knowledge to the general public. For some authors, the prefix *geo* is related to the concepts of human geography, which is concerned with analyzing the interaction with local communities. However, for most definitions, geological aspects play a fundamental role in the tourism proposal, especially today when they can be correlated to UNESCO's geopark projects. The first definition of geotourism was published in England (Hose, 1995), as a proposal to facilitate the understanding and provide facilities of services so that tourists acquire knowledge of the geology and geomorphology of a site, going beyond mere spectators of esthetic beauty. Pretes (2002) highlighted mining sites as of particular relevance to geotourism, as in these places the impact of geology on people and communities is clear. Ancient mining facilities offer a touristic interest as they present the contrast between natural landscape and environmental degradation or industrial landscape.

According to Dowling and Newsome (2006), geotourism is a segment linked to tourism in natural areas that emphasizes the geological aspects (rocks, minerals, fossils, geoforms, water and, mainly, geological processes). For Gouthro and Palmer (2011), tourism in industrial heritage is already a niche well established within the heritage tourism industry. It is necessary to consider the evolution of the concept of geotourism together with the evolution of the different segments of the tourist trade, especially cultural or heritage tourism, for an effective implementation of actions.

Since 2002, Brazil has adopted a policy focused on the expansion and qualitative development of tourism in municipalities. Among the phenomena that have occurred in this sector in recent years, segmentation was the most symptomatic. The concepts and knowledge about tourism planning presented a strong evolution and the visitor himself today shows a more complex and demanding profile. The tourism of sun and beach, to which the image of Brazil has always been associated, still exists and is consolidated, but other fronts have arisen and are presenting themselves as development alternatives. Ecotourism, rural tourism, and cultural tourism are among the most important segments. Experiencing the local culture or nature and getting to know the identity of the municipalities visited reveal a type of visitor more conscious and willing to obtain information and new experiences associated with leisure. Geotourism encompasses this philosophy, that is, it provides the tourist, in his leisure time, easy access to the geological information, products, and attractions available in that area.

The selective gathering of technical information and its "translation" into accessible language in the form of panels, pamphlets, and graphic material is part of the essence of cultural tourism. Living in natural environments is the essence of ecotourism. Geotourism proposes that technical information be made available on natural or altered environments, among them deposits and landscapes of old mining or sites of geological heritage. Geotourism, therefore, incorporates characteristics of cultural tourism into ecotourism and is practiced, in most cases, in rural areas. It's worth mentioning that specific activities of geotourism can be, and effectively are, put to practice in urban areas, but those cases are beyond the scope of this paper.

There is a challenge in innovating and/or interpreting mining areas by pointing out how much cultural and natural values are interacting. Landscape analysis is a means of identifying different layers of historical meanings. Cultural landscapes should be associated with stories and human experiences to enrich the tourism context, as pointed out by Reeves and Mcconville (2011). The geological history of places can be one of the layers of interpretation in mining heritage environments.

4.2.1 GOLD AND GEMSTONES IN MINAS GERAIS

The 18th century became known as the mining cycle in Brazil due to the discovery of gold and diamond in Minas Gerais. The main urban centers responsible for this activity were Ouro Preto—gold was discovered in this region in 1695—and Diamantina, with the discovery of diamond in 1714.

Virtually, all the jewels of the European nobility of this period were built with raw material from Brazil. Tons of gold were extracted in the surface mines and in the first underground galleries in Ouro Preto, with peculiar mining techniques and with an abundance of wealth that was reflected in the architecture, lifestyle, art, culture and other aspects of daily life, some of which are still noticeable today. In fact, due basically to the Methuen Treaty of 1703, though indirectly, gold produced in this region "effectively funded Britain's industrial revolution" as "gold flowed steadily out of Portugal to Britain" (Boxer, 1962).

Ouro Preto is also the region of the sole occurrences of imperial topaz in the world. This special variety of topaz was discovered in 1772 and its existence was commented on by European travelers–naturalists who passed through the region, such as John Mawe (1812). One of the production areas receives frequent tourist visits, which contribute to the support of the local community. Although very limited, the production of imperial topaz is the basis of an important jewelers' trade in Ouro Preto, with more than 20 stores selling original design jewelry, which reflects the local identity (AJOP, 2009).

Soapstone is another mineral resource of Ouro Preto with strong touristic appeal. Technically known as steatite, this is a low-degree metamorphic rock that has a high content of talc as one of its main components. The presence of this mineral makes it a soft rock, suitable for sculptures of great plastic beauty. In the city, there are many works of Antonio Francisco Lisboa (known as Aleijadinho, 1730–1814), considered the greatest sculptor in Brazilian history. The soapstone plays an important role in the architectural composition of the city, as it is often used in association with other rocks, such as quartzite and shale (Pereira et al., 2007). Its extraction zone receives some visitation, without any kind of infrastructure, but generating an intense production of handicrafts.

Due to its historical, social, and cultural importance, as well as to its natural and architectural beauty (Fig. 4.2), Ouro Preto owes a good portion of its recent development to tourism. Available data show that in 2004, this activity accounted for 10.4% of the gross local product and 21.8% of the tax revenue. About 11% of the region's active population was directly or indirectly related to tourism in 2002 (Flecha et al., 2011). Ouro Preto has a rich and varied natural environment that surrounds the whole municipality, with waterfalls, prospectors' trails and native forests, partly protected by State Parks, partially explored by ecotourism. Paradoxically, this ecosystem contrasts with the anthropic action that has been occurring in the region for centuries, but which has resulted in historical richness and a cultural link with mining, the oldest and most active economic activity in the region (Flecha et al., 2011).

FIGURE 4.2 View of the historic center of Ouro Preto. The city is a World Heritage Site because of its preserved architecture from the 18th and 19th centuries, which originally reflected the wealth of gold mining (photo: A. Liccardo).

The city was classified as a World Heritage Site by UNESCO in 1980 and among its attractions for mining tourism are: traces of excavations and structures to wash the ore in open air; old hand-excavated underground galleries; Passagem Gold Mine (organized structure of gold extraction in the 19th century); scientific, thematic, and historical museums; 13 churches in Baroque style, many with large portions of their inside walls lined with gold produced in the 18th century; active mining of imperial topaz open to visitation; trade in precious stones and handicrafts in soapstone. The region has unlimited potential for the development of tourism, especially in specific segments such as mineral tourism, associated or not with the ecotourism already existing in the surroundings. In fact, in Ouro Preto, tourism, history, geology, and mining are inextricably connected and makeup one big reality.

Diamantina, about 380 km (by road) north of Ouro Preto, has also been a UNESCO World Heritage Site since 1999. For a period of more than 150 years, between the exhaustion of the Indian mines and the discovery of diamonds in South Africa, or from the early 19th century to the 1860s, Diamantina was practically the only source of diamonds in the world. This material had participation even in the relations of Portugal with Napoleon, when diamonds were used in attempts to buy peace with France. It was in Diamantina that, for the first time, a diamond encrusted in unaltered rock was discovered—a pre-Cambrian conglomerate (placer), which was thought to be the parent rock. This fact, among others, attests the exceptionality of regional geological heritage and its importance in the history of geology and mining.

The city features baroque architecture characteristic of the 18th century and well preserved. The region presents an exceptional geodiversity that has been receiving international scientific visitation for decades, as well as an interesting relation between the cultural and natural landscapes. Among its attractions for mining, heritage tourism is active and inactive prospective mines, diamond museum, mineral trade, and geosites recognized worldwide as didactic/scientific for geology.

It is interesting to point out how the local names reflect the mineral richness of that region of the country: the cities mentioned are called Diamantina (something like "the village of diamonds"), Ouro Preto ("black gold," due to the dark impurities that often accompanied the ore), and there are other city names of similar origin; nowadays the state is, and formerly the province was, named Minas Gerais ("general mines").

4.2.2 DIAMONDS AND CARBONADOS IN BAHIA

In the 19th century, the heavily exploited diamond deposits of Bahia supplanted Minas Gerais in production. The cities of Lençóis, Andaraí, Igatu, and Mucugê—then great centers of production, nowadays, quaint old-looking villages—are now renowned touristic centers. The region known as Chapada Diamantina (Diamond Plateau) is the heart of one of the most beautiful National Parks in Brazil and its geological context is similar to that of Diamantina (MG), an elongated geomorphological mega-structure known as Serra do Espinhaço.

Mining Heritage as Geotourism Attractions in Brazil

The Chapada Diamantina National Park is one of the main tourist attractions of Brazil due to its geological landscape, geomorphology, exceptional rivers and waterfalls, diamond mining history, fauna, and flora. Mining has now been banned in favor of the conservation unit, but its culture has been preserved in its many, varied aspects (Fig. 4.3).

FIGURE 4.3 Ruins related to the mining history turned into an open-air museum in Igatu, former diamond-mining center in Chapada Diamantina, Bahia (photo: A. Liccardo).

Officially, diamonds were discovered in Mucugê in 1844 by prospectors looking for gold in the then-unknown lands of the North. The large quantity (with good quality) found in alluvial deposits and in conglomerates throughout the Chapada soon caused the birth of several villages that prospered due to mining. The production in 1850 and 1860 reached its peak, with more than 70,000 carats/year (Cornejo and Bartorelli, 2010). Lençóis, a village founded in 1845, came to have a large cosmopolitan population which payed for imported fashion, style, and novelties from Europe, and a rich architecture, partially maintained until today. According to Giudice and Souza (2012), the richness of diamonds became the center of such a strong international attraction that a French vice-consulate was installed in Lençóis. However, according to the author, this office functioned more than anything else like a facilitator for the purchase of the stones by European jewelers from the local prospectors (garimpeiros). After a golden phase of approximately 25 years, diamond mining's decline started in 1871. Some initial attempts to mechanize mining were done in the first half of the 20th century (Catharino, 1986) and in the 1980s, mechanized mining was reintroduced in the region. Installed in the river beds, inside and outside the National Park; these mines were definitively closed in 1996. Even after 150 years of mineral extraction in the Chapada Diamantina, there is some rudimentary artisanal mining done only with the use of hand tools.

In the first two decades of the 20th century, there was an outbreak of carbonado mining. The carbonado is a black, opaque, polycrystalline variety of diamond, and a very efficient abrasive, superior even to the diamond itself. The first carbonado had been found in that region in 1841, but it was the continuous unfolding of the Industrial Revolution that turned it into a highly sought after product, with its value reaching up to 6 times that of the diamond. As the mines of the Chapada Diamantina were, at that time, the only known source of carbonados (Funch, 1997), this gave a renewed impulse to mining activities in the region. A known historical fact is that the carbonates found in Bahia were applied in the manufacture of cutting and drilling machinery used in the construction of the Panama Canal.

Carbonados have never been found in primary diamond mines, only in alluvial deposits in Brazil and, more recently, in the Central African Republic. Recent theories suggest a possible extraterrestrial origin for this material (Haggerty, 2017). Some miners still find carbonados in the Chapada Diamantina and sell them informally in Lençóis, Andaraí, and Mucugê.

The cities of the Chapada Diamantina have typical 19th-century architecture, a result of the plentiful economy of the mining era. Very rich in terms of geodiversity, the Chapada receives intensive tourism (both domestic and international) of adventure and nature, largely related to geoforms, waterfalls, rivers, rock formations, and mining traces. Among its attractions for tourism in mining heritage are active and inactive prospective gold mines, excavations and constructions characteristic of diamond mining, ruins of mining villages, tourist museums of diamond and mining, mineral trade, old diamond cutting workshops, geosites recognized worldwide as didactic/scientific for geology. According to Giudice and Souza (2012), tourism is the most promising of the activities developed in the cities of Chapada Diamantina, mainly Lençóis, Mucugê, and Andaraí, which have good hotels and service structure. Tourism has the potential to become the main sustainable economic base, as there is a rich and varied heritage that can be continuously explored. Remains of mining activities include technological attractions that, along with natural attractions, have come to be seen under a new perspective, adding value to conventional tourism.

4.2.3 AMETHYST IN RIO GRANDE DO SUL

In southern Brazil, in Rio Grande do Sul, the largest amethyst deposits on the planet are located in geodes formed in the thick basalt layers resulting from the extensive volcanism that occurred 135 million years ago during the separation of continents (Africa and South America). Around the several mines installed, a village of miners was formed, which became the present municipality of Ametista do Sul, whose existence is due solely to the activities of mineral exploitation.

Amethyst mining has been occurring in this region since the 19th century, with knowledge brought by German immigrants. In this region, visible inside horizontal tunnels open by miners in the fresh basalt, geodes of metrical dimensions, that reach a few hundred kilos, are often found. Associated with the amethyst are other minerals such as agate, hyaline quartz, and rose quartz, calcite (in a great variety of shapes and colors), gypsite (selenite variety), zeolites, and barite. These minerals generally make up beautiful mineralogical aggregates, in samples that are disputed by collectors and coveted by the world's great museums (Liccardo and Juchem, 2008).

Since the beginning of the 2000s, the city has been investing in tourist development as an economic alternative for the moments of crisis in the mineral sector. Thus, Ametista Park was installed, a tourist complex that includes old mining galleries, museums, stores, and other attractions (Fig. 4.4). This park/museum has special characteristics that showcase the potential of integration of economic, cultural, scientific, and environmental aspects. The project was carried out by taking advantage of two mining galleries with more than 20 years of exploitation, in which can be observed, still embedded in the basalt, several geodes with interesting minerals of commercial value that were not extracted, allowing to the tourists an on-site analysis, and a visual enjoyment of a unique kind.

The underground galleries converge to a large room with objects and machinery previously used in mining activities from where the visitor can move to another tunnel, which gives access to a mining front where miners are in real activity. There are galleries that are now deposits of wine bottles (which can be bought at the local store). The park also has an infrastructure composed of exhibition halls of geodes and mineral samples, and a well-organized structure of commerce and services. The landscaped surroundings include a viewpoint, from which tourists can see several mining areas that are active on the slopes of the neighboring mountains.

FIGURE 4.4 Ametista Park, built in an old mining area with visits to the galleries. Inside the galleries, large geodes can still be seen enclosed in the thick basalt layers (photo: A. Liccardo).

The museum's collection of samples is really outstanding and of great mineralogical and strategic importance, as it allows special samples to remain not only in national territory but adjacent to their original location. A 2.5-t amethyst geode with high gemological quality is considered the largest ever removed and is the main sample on display. The aim of this private institution (the entrepreneur is a local stone producer and merchant) is to promote the development of the region in the touristic and educational aspects, as well as to effectively contribute to the diversification of the economy. To create a new attraction related to the mineral tourism, the city's main church, Saint Gabriel, received, in 2004, a refurbishing of its interior, in which some walls were lined with real amethyst crystals, as well as fonts and other pieces of furniture which were made from large geodes collected from the local mines. It is the only church in the world clad with this precious stone and certainly a great tourist attraction for Brazil. A themed hotel (Hotel das Pedras—literally, The Rock Hotel) was built in 2008, also with some walls covered with amethyst and citrine crystals (yellow variety resulting from the thermal treatment of the amethyst) and offers all the rooms decorated with different gemological materials. Even the smallest details are themed: soap bars have the typical shape of an amethyst "tooth" crystal, toothpicks have a plastic purple "mini-crystal" at one end, and so on.

Currently, there are more than one hundred licensed mines for amethyst and agate production and most of them have guided visits offered by the region's miners' cooperative. Due to its volcanic soil and a strong Italian colonization, the region has become in recent years a wine producer. Abandoned galleries have been transformed into wineries and there are programmed wine tastings inside the mines that are examples of great cultural appreciation of the *terroir*. The amethyst brand is present in many aspects, associated with handicrafts and other local productions. Tourism in Ametista do Sul today is a social reality and the municipality receives visitors from all over the world due to geological exceptionality and a living experience of underground mining. Over the last few years, with this tourism-oriented planning, many mineral and gemstones shops have emerged, as well as small stonecutting workshops and artisanal jewelry makers. Infrastructure and services have also shown great development in the last decade. Among its attractions for tourism in mining heritage, Ametista do Sul offers visits to active and inactive mines, physical contact with the ore inside the mining galleries, first-hand knowledge of mining processes and equipment, exceptional mineralogical museum, trade-in minerals and gemological handicrafts, geosites recognized worldwide as didactic/scientific for geology.

4.2.4 OPALS IN PIAUÍ

The main noble opal deposits in the world are in Australia and Brazil. The only Brazilian deposits of this gem are found in the northeast portion of the country, in the state of Piauí. An old mining town, Pedro II is now a city in the vicinity of these mines that began their exploration in the 1940s.

There are about 30 mines, some active, others inactive, and the biggest and most important one is the so-called Boi Morto (Dead Ox). Until some years ago, the mining was done in small galleries, in very precarious conditions, with many risks for the miners.

The time of greatest production was when the Empresa de Minérios Brasil Norte-Nordeste—EMIBRA (North–Northeast Brazil Minerals Company) operated the Boi Morto mine and reached, between 1960 and 1976, more than 80 employees. During this period, there were about 30 mining sites occupied on different occasions. Firms operated larger deposits, the primary ones, while secondary deposits were exploited mainly by prospectors (Milanez and Puppim, 2009). From the 1980s, companies began to leave Pedro II and that included the closing of EMIBRA. At that time, there were 22 abandoned, three paralyzed and only three active mines. The Boi Morto mine was then informally occupied by miners who began working in the tailings of the company, sometimes trying their luck in some galleries.

In the following decade, some movements of public and private institutions began to try to strengthen the opal production chain in Pedro II. The main strategy at this time was the training of goldsmiths and jewelers, adding value to their products and increasing the share of the income that remained in the city (Milanez and Puppim, 2009). The result of this mobilization was the creation of the Arranjo Produtivo Local (APL, Local Productive Arrangement) project in 2005 (Milanez and Puppim, 2009). An APL is a sort of cooperative plan involving a number of institutions aiming at improving the efficiency of the local production. To make the project viable, a diagnosis was made which identified that the opal production chain in the region had a low level of technological and economic efficiency at all stages. It was then proposed: formalization of activities related to opal extraction, improving working conditions and reducing the environmental impacts of mining, value added in opal processing, and management and business strengthening of the opal chain, which included support for the formation of associations and cooperatives, training in business management, brand building, negotiation of specific lines of credit, and development of mineral tourism project (FINEP, 2005). Today, more than 2000 people are involved in the opal production chain of Pedro II, taking into consideration direct and indirect jobs. All the work is done jointly with about 150 cooperated miners in that region (Milanez and Puppim, 2009). To strengthen the APL, the Association of Jewelers and Lapidaries of Pedro II and the Cooperative of Miners of Pedro II were created giving a new dimension to their activities.

There are projects of standardization of the areas of mining and rational development of tourism. One of the actions proposed is the creation of the Opal Museum in an attempt to keep part of the mobile assets (rare opal samples) in the national territory and accessible to the public (Carvalho and Liccardo, 2011). In line with this development, a geological and mining heritage survey was carried out to develop geotourism, a more inclusive type of tourism that contemplates environmental and social valorization, generating income and occupation in the region. The main criterion was the use of the unique geological content and the deep history of the mining as a tool of economic structuring by the tourism and as a consequence its preservation and valorization by the population (Carvalho and Liccardo, 2011; Carvalho, 2015).

Thus, 14 points of geotouristic interest were raised, including old opal mines (Boi Morto and Mamoeiro), relevant points of geomorphology (Mirador do Gritador, "Screamer's Belvedere" and Cachoeira do Salto Liso, "Smooth Falls"), information about the water supply and aquifers, as well as rocks used in buildings and exposed in the historical center. Also included were private properties that could in the future become tourist attractions or for research purposes, since they cover geological, geomorphological, faunal, and floristic aspects of relevance in the same area (Carvalho and Liccardo, 2011). The initial strategy used was to improve the potential of established natural tourism sites by adding scientific information in a didactic and attractive way, in harmony with the cultural, environmental, economic, and social aspects, both for visitors and for the local community. In addition, new points based exclusively on scientific content, such as diabase dikes (Carvalho, 2015) are proposed. Most of the tourism that occurs in this place is linked to the production chain of this gem, with production and sale of jewelry or commercialization of crude and polished opals to merchants (mostly foreign), which made the municipality better known internationally than in the country itself. Mineral tourism presents itself as a popular attraction throughout the state, but the infrastructure is still primitive.

Handicrafts and local products are the typical tourist goods marketed. At this point, the role of design in the creation of objects, the development of new stone cuts and jewelry products plays a key role, because this design often depends on local identity and adds strength to the opals as tourist attractions. An example of this was the program implemented in Pedro II, with the development of a local design for the use of the opal locally mined (Liccardo and Chodur, 2009). The jewelry in Pedro II today presents national-level quality, with its own design identity and a thriving trade in the city, with several jewelry and stonecutting shops (see Fig. 4.5) and workshops. The tourist experience in Pedro II includes visits to

FIGURE 4.5 Various shops and stores working with stonecutting and jewelry opened in Pedro II, where tourists can purchase a variety of original products produced with opals from the neighboring mines (photo: A. Liccardo).

all the production processes, from the mining to the final jewel. Important archaeological, geological, and geomorphological sites are present in the region, including the Sete Cidades National Park, an important tourist attraction.

4.2.5 SCHEELITE IN RIO GRANDE DO NORTE

The state of Rio Grande do Norte, in northeastern Brazil, hosts the main occurrences of scheelite (tungsten ore) in the country. They are represented by stratiform deposits associated to calcissilicatic rocks, resulting from contact metamorphism between granite and carbonate rocks (marbles). According to Dantas (2008), the discovery of the scheelite and then the creation of the Brejuí mine have leveraged the local economy in the 1940s. Tungsten was a much-needed component for steel production during World War II, so Currais Novos became a very active labor pole. Based on the revenue provided by the high production of scheelite, the tungsten ore, the owner of the Brejuí mine, Tomaz Salustino, gave free course to his entrepreneurship and created several important buildings, such as the airstrip (that later gave origin to today's airport), the Tungstênio Hotel, the movie theater, the football (soccer) field, a radio station, the miners' village, a club, sports courts, schools, a laboratory and the Church of Saint Theresa "...a real city, all around the mine." There are other mines in the region (Barra Verde, Boca de Laje, Bodó), but Brejuí is by far the most important one; in fact, it is considered South America's largest scheelite producer. The mine opened in 1943 and had its apogee along the remaining years of World War II. After the end of the war, it slowly declined, and the fall in prices in the international market led to its closing in 1996.

In 2000, seeking an economic alternative to the mine, its owners saw in tourism and education the possibility of using equipment, underground galleries and the existing museum as attractions and thus began their activities, opening it for visits by tourists and students. With this, Mina Brejuí became the largest theme park in Rio Grande do Norte, being visited daily by tourists and students from various parts of the Northeast, from the rest of Brazil and, ever increasingly, from abroad. In 2005, the extraction activities were resumed; so, tourism and mining began to live along in perfect harmony.

In the Brejuí Mine, Touristic Complex visitors can appreciate the historical and cultural content of the mine, which involves visiting underground galleries, tailings hills (described in the tour as "dunes"), Memorial Tomaz Salustino, Mário Moacyr Oporto Mineral Museum, the cave of Santa Bárbara, and the church of Santa Tereza D'Ávila.

One of the most interesting aspects of the visit to the theme park is the visit to some of the tunnels through which scheelite was extracted. The mine has about 60 km of underground tunnels, of which 300 m have been carefully adapted for tourist visitation. Visits are conducted by a specially trained tour guide; they start by a presentation about safety (in the mining activities and in the tour) and an explanation about the methods of working in subterranean galleries; elements of this underground network, such as the chimney where the ore passed from one level to another and a spout that served to forward the ore production are shown and explained. The visitor is then taken to a large room, when there is a pause of a few minutes, all lights are put off, and the visitors are asked to remain silent.

One of the recommendations of modern tourism techniques is to allow tourists to live new experiences, including sensorial ones. At this point, then, the basic idea is to have the visitor experience the physical sensation of total darkness and total silence, which are conditions that people can hardly ever experience nowadays. Suddenly, in that already unusual context, "at a given moment ultraviolet lamps are turned on, and an infinite number of scheelite grains scattered throughout the rock walls exhibit their fluorescence ... and visitors stare in awe!" Needless to say, that is the high point of the visit, which is wrapped up in another hall, with the presentation of engineering techniques, such as the columns to support the ceiling and chimneys for ventilation, totaling about 20–30 min of walking. From what was briefly explained above, it is quite clear that the Brejuí Mine (Fig. 4.6) has become a milestone in its region, particularly for the local community, where it has caused important changes in work conditions, living processes, and even personal relationships. The extraction of tungsten has become a cultural symbol, capable of generating the attribution of names of ores in heritage sites (buildings, squares), streets, businesses, festivals, and even religious missionary sectors. Geotourism, which has been taking place since the year 2000 in Currais Novos, has promoted awareness of the importance and value of regional geodiversity with evident socioeconomic repercussions. A map of geological interest was published in 2009 to integrate the site with other points of geotouristic characteristics in the region, mainly related to gemstones (Liccardo and Nascimento, 2008).

FIGURE 4.6 Brejuí mine: underground gallery, reserved for visitors' tours. Machines and equipment used for scheelite extraction, started in 1943 and still active in other parts of the mining complex are shown (photo: A. Liccardo).

4.2.6 VARVITO IN SÃO PAULO

The Varvito Park is located in the city of Itu, in the southeastern portion of Brazil, in the state of São Paulo, about 100 km from the capital (also named São Paulo) and 50 km from Campinas (second largest city in the state). This geological park, under the municipality's administration, was installed in 1995 in an old dimension stone quarry, which, in recent times, was in operation from the 1940s until the 1980s. Historically, there are records of the quarry being in operation at least since the second decade of the 19th century. Most likely, though, extraction started long before. The village was founded in 1610, became a regional commercial hub in the early/middle 18th century, and a center of cane sugar production and export—therefore a rich city—in the middle 1700s. Varvite plaques of various sizes (in some cases reaching more than 1 × 2 m), have been extensively used for paving the streets, the squares and the sidewalks, creating a certain visual/cultural identity to the city. Commercially, those plaques are sometimes referred to as ardósia, the Portuguese word for "slate," even though it is a different rock.

Varvite is the name used to describe a finely stratified sedimentary rock, formed by sediments deposited in an environment with glacial influence. It is typically formed by repetitive succession of pairs of horizontal layers of clay-silt (black-gray), or clay-fine sand (light gray), arranged in horizontal strata. This package of sedimentary rocks shows a very didactic evidence of glaciation that occurred about 280 million years ago, when that region was part of Gondwana. Other outcrops of varvite occur in various places of the country, but this is the most important one, due to a series of positive factors: its size, the fact that is nicely exposed (Fig. 4.7), the number of articles that have been written about it, and, last but not the least, particularly for its use as a geotouristic site, the fact that it is located in an urban area. In practical terms, one of its remarkable characteristics is its easiness of access, since there is a bus stop of a regular urban line about 30 m away from its entrance gate.

FIGURE 4.7 Old varvito (type of dimension stone) quarry, transformed into a geological/cultural park in Itu. Direct, hands-on contact with a rock formed about 280 million years ago, and panels, which bring the adequate amount of information in easy-to-understand language, are among the reasons for its success (photo: A. Liccardo).

The varvite has been carefully studied by many authors (e.g., Almeida, 1948; Rocha-Campos and Varvito de Itu, 2002) since 1946, but the site itself and its geological context have faced many threats of destruction by mining. Fortunately, in the last decades, it has been recognized and registered as a geological and mining heritage due to its importance (Carneiro, 2016).

Since its installation, the Varvito Park has received more than 500 thousand visitors, among tourists, students, and researchers. Carneiro (2016) analyzed the last five years of visitation, which present a frequency of 60,000 visitors/year. The site has good infrastructure and excellent planning in the dissemination of scientific content, such as an open-air museum. Points of special scientific and/or tourist interest are identified and receive attractive names, such as Eco Square, Yesteryears' Waterfall, Fossil Lake, Pebble Trail, Benthonic Trail, Gondwana Amphitheater, Teardrop Grotto, Boulders Forest, and Jurassic Lake. There is also a space for temporary exhibitions and lectures. Monitors accompany schools and groups, and the main topics are geological time, climate change, sedimentary and glacial processes, mining techniques, and heritage appreciation. Carneiro (2016) considered in his didactic–scientific analysis that the Varvito Park is a successful case of visitation and that the geological context and the information made available in the place are effective strategies for its operation. The already mentioned convenient logistics, the various universities that situated in various cities within a radius of 200 km and the continuous presence of news about the park in the media in Itu and the rest of the state are probably the factors that allowed the development of geotourism, for several decades, in this ancient area of mining.

4.3 CHALLENGES TO DEVELOPMENT OF GEOTOURISM IN MINING HERITAGE IN BRAZIL

The appreciation of the geological heritage associated with old mining areas opens up the possibility of offering a new layer of knowledge about historical and socioeconomic contents. Mining heritage attractions can be very successful if they are presented as part of a broader tourism plan (Edwards and Llurdés i Colt, 1996). The interpretation is an opportunity to provide visitors with experiences of cultural insights (Carr, 2004) and can increase the value of the attraction, allowing the visitor to learn informally and enjoy the place visited (Light, 1995).

Six different cases have been exposed in this text, presenting very different situations in terms of history, socioeconomic contexts, and enormous regional differences in a country of continental dimensions. However, there is a factor in common in all these cases, which is the presence of mining heritage functioning as the main attraction in tourism. The simple fact of being in the old mining environment may be the main motivating factor for the visitation of these areas, but the offer of geoscientific information seems to make a difference in the quality of the tourist experience. Within the complexity of tourism analysis, it is possible to introduce other factors that may interfere with the efficiency of the use of a geotouristic approach in mining areas that are already consolidated as tourist attractions. A comparison of some key attributes in support of tourism activities shows that the quality of access to sites, for example, has a major impact on the development of tourism. While Ouro Preto and Itu are close to airports and highways with excellent access, other sites such as Ametista do Sul and Pedro II represent logistical challenges which in some cases include even long secondary roads. In relation to the facilities implemented, the quality of lodging, restaurants, etc., must be taken into account. In this aspect, Diamantina and Chapada Diamantina have been the focus of conventional tourism investments, such as cultural tourism in the first case and ecotourism in the second. Ouro Preto is the most consolidated of all of our six cases and certainly offers the best infrastructure. In Itu, the urban location of the site and the proximity to other medium-sized cities guarantee these facilities. In relation to the tourist publicity in the media or even within the tourist trade, Ouro Preto and Chapada Diamantina stand out, in part for the maturity of the tourist process in their territories. Currais Novos, even with greater restrictions on other attributes, stands out for an intense campaign of regional and even national disclosure. The prospect of creating a geopark (the Geopark Seridó) in this region contributes effectively to this tourist promotion and is based entirely on geotourism. As for the inclusion of these cases in regional tourism programs, only Ouro Preto and Chapada Diamantina are consolidated as classic circuits. Itu and Currais Novos have characteristics more related to scientific or educational tourism and, even regionally, do not yet participate in integrated circuits.

    This comparison points out that the greater development of geotourism within the conventional tourism which already exists in these different places could also contribute to the development of the other attributes. It would seem that geotourism would be easier to implement in areas where cultural tourism already exists as a motivating factor for the visit. However, cultural interpretation can be significantly improved in all cases with the provision of geoscientific information, which would result in social and educational developments and gains.

The availability of road maps depends upon adequate technical surveys, whether of a mineralogical or geological nature, or related to tourism criteria, including logistics and infrastructure. Some itineraries are already being implemented, such as the Gems and Jewels Route in Rio Grande do Sul or the Seridó Geopark Roadmap, in Rio Grande do Norte. On the other hand, scientific itineraries have been used for decades by researchers, professors, and university students, as well as specialized groups of foreigners who participate in scientific events in Brazil. In several of these events, suggestions have been published for itineraries that, although specifically aimed at the participants of these congresses, are coveted by presenting excellent logistic surveys, such as distances between points of interest, and nearby main centers for lodging and food. Examples of these itineraries were published by Mantovani et al. (2000), which allowed the visit to the main gem deposits in Rio Grande do Sul, and by César-Mendes and Gandini (2000) and César-Mendes et al. (2004), which pointed out the main gemstones deposits in Minas Gerais. Also, Liccardo (2007), Liccardo and Juchem (2008) and Liccardo and Nascimento (2008) published specialized itineraries in various regions of Brazil, which were used by foreign scientists during technical excursions or even by specialized tourism.

As for Pedro II, Carvalho (2015) deeply analyzed its tourism development projects and states that, despite the advances made by the government program (APL) for the production of opal and its proposal of a geotouristic itinerary, the municipality, as a whole, still needs a comprehensive, adequate urban and touristic plan. Despite its present deficiencies, it has enough attractions to foster economic growth based on the opal production chain.

4.4 CONCLUSIONS

The basic idea of geotourism is to add scientific and educational knowledge to natural heritage in a pleasant and understandable way, valuing it and enabling a sustainable touristic visits to take place (Nascimento et al., 2008). In this sense, appreciation of cultural diversity and conservation of natural resources, as well as their esthetic value and geographic characteristics, all put together in a visit causing minimal impacts, are the basis for this tourism niche. Geotourism in areas of mineral heritage proposes not only simple visitation, but a cultural and educational gain, as well as results that improve the quality of life of the populations.

Tourism in areas of mining heritage has already been functioning in Brazil for many decades but in an empirical and often spontaneous way.

Groups of amateur mineralogists, scientists, collectors, jewelers, and traders have been visiting the main mines for a long time and patronizing a strong, even if limited, local trade, an effect that is particularly welcomed in areas of artisanal mineral extraction. In the areas of mining of rare gem and minerals in Brazil, a movement of people around this niche can be characterized as tourism or geotourism, since it requires the development of hotel infrastructure, food, and transportation, among others, and geoscientific information plays a fundamental role.

This type of action can help the development of local economies, generate new sources of income and jobs, or contribute to the awareness of visitors and the population about the environment and its relationship with the human being. Ideally, revenues generated by tourism (or specifically geotourism) should be essentially, or even exclusively, invested in the development of infrastructures and the maintenance of local attractions, generating important social evolution. It is expected that this process will ultimately lead to natural heritage protection and geoconservation, through sustainable development and awareness of the natural processes that govern the planet.

KEYWORDS

- **geotourism**
- **mining heritage**
- **mining geotourism**
- **Brazil**
- **South America mining history**

REFERENCES

AJOP. *Roteiro Joias e Gemas de Ouro Preto*. Folheto de divulgação da Associação de Joalheiros de Ouro Preto, 2009.

Almeida, F. F. M. A "Roche Moutonnée" de Salto, Estado de São Paulo. São Paulo, *Bol. Geol. Metal.* **5,** *112–123*, 1948.

Barreto, M. L. *Mineração e Desenvolvimento Sustentável: Desafios Para o Brasil*. Centro de Tecnologia Mineral. Mineração e Desenvolvimento Sustentável: Desafios Para o Brasil. CETEM: Rio de Janeiro; MCT, 2001, 216 p.

Boxer, C. *The Golden Age of Brazil 1659–1750*. Berkeley, 1962.

Carneiro, C. D. R. Glaciação Antiga no Brasil: Parques Geológicos do Varvito e da Rocha Moutonnée nos Municípios de Itu e Salto, SP. *Terræ Didat.* **2016,** *12* (3), 209–219.

Carr, A. Mountain Places, Cultural Spaces: The Interpretation of Culturally Significant Landscapes. *J. Sustain. Tour.* **2004,** *12* (5), 432–459.

Carvalho, C. A.; Liccardo, A. Patrimônio Geológico-Mineiro e Turismo em Pedro II, Piauí. In: *Proceedings of I Simposio Brasileiro de Patrimônio Geológico e II Congresso Latino-Americano e do Caribe sobre Iniciativas em Geoturismo*, Rio de Janeiro, 2011.

Carvalho, C. A. O Papel do APL da Opala de Pedro II, Piauí, Na Estruturação do Turismo Mineral do Município. Master's Dissertation, Escola de Artes, Ciências e Humanidades, University of São Paulo: São Paulo, 2015.

Catharino, J. M. *Garimpo—garimpeiro—garimpagem: Chapada Diamantina, Bahia, Vol. 5.* Philobiblion, 1986.

César-Mendes, J.; Gandini, A. L. Guide to the Major Colored Gemstone Deposits in the Vicinity of Belo Horizonte, MG, Brazil. Post-Congress Field Trip. In *31st International Geological Congress*, Rio de Janeiro, Brazil, August 6–17, 2000. Field trip Aft 20, 2000; 58 p.

César-Mendes, J. C.; Liccardo, A.; Duarte, L. C.; Gandini, A. L.; Karfunkel, J.; Addad, J. E.; Roeser, H. *Guide to Brazilian Gemstones Deposits. Field Trip Guide*. In *8th International Congress on Applied Mineralogy, 2004, Águas de Lindóia. A Field Trip Guide to Selected Brazilian Mineral Provinces*; Neumman, R., Ed.; São Paulo: International Council on Applied Mineralogy, 2004, pp. 67–87.

Cole, D. Exploring the Sustainability of Mining Heritage Tourism. *J. Sustain. Tour.* **2004,** *12* (6), 480–494.

Cornejo, C.; Bartorelli, A. *Minerals & Precious Stones of Brazil*; Solaris Cultural Productions, 2010, 704 p.

Dantas, G. C. *Mina Brejuí: A Maior Produtora de Scheelita do Brasil.* RN/UNP/BCNC: Natal, 2008.

DNPM. *Sumário Mineral*; Departamento Nacional de Produção Mineral: Brasília, 2015; 135 p.

Dowling, R. K.; Newsome, D., Eds.; *Geotourism.* Elsevier/Heineman: Oxford, 2006.

Edwards, J. A.; Llurdés i Coit, J. C. Mines and Quarries, Industrial Heritage Tourism. *Ann. Tour. Res.* **1996,** *23* (2), 341–363.

FINEP. Convênio Ref. 3686/04. Financiadora de Estudos e Projetos: Rio de Janeiro, 2005.

Flecha, A. C.; Lohmann, G. M.; Knupp, M. E. C. G.; Liccardo, A. Mining Tourism in Ouro Preto, Brazil: Opportunities and Challenges. In *Mining Heritage and Tourism: A Global Synthesis*; Conlin, M. V.; Jolliffe, L., Eds.; Routledge, Abingdon, UK, New York, 2011; pp 194–202.

Frew, E. A. Industrial Tourism Theory and Implemented Strategies. *Advances in Culture, Tourism and Hospitality Research*; Emerald Group Publishing Limited, 2008; pp 27–42.

Frey, M. L. Geologie Geotourismus–Umweltbildung: Themen and Atigkeitsbereiche im Spannungsfeld Okonomie and Nachhaltige Entwicklung. *Programme and Summary of the Meeting Contributions, Technical University Berlin*, 1998.

Funch, R. *Um guia para o visitante a Chapada Diamantina: o circuito do diamante: a Parque Nacional da Chapada Diamantina, Lençóis, Palmeiras, Mucugê, Andaraí.* Secretaria da Cultura e Turismo do Estado da Bahia, 1997.

Giudice, D. S.; Souza, R. M. *Geodiversidade e lógicas territoriais na chapada diamantina.* CBPM: Salvador, Bahia, 2012; 126 p.

Gouthro, M. B.; Palmer, C. Pilgrimage in Heritage Tourism. Finding Meaning and Identity in the Industrial Past. In *Mining Heritage and Tourism: A Global Synthesis*; Routledge Advances in Tourism. Routledge: Oxon, New York, pp. 33–43.

Guiollard, P. C. *Conservation et valorisation du patrimoine minier contemporaine. Mines de charbon, d'or et d'uranium en France métropolitaine.* Edition Pierre-Christian Guiollard. Z A Le Cherbois: France, 2005; 117 p.

Haggerty, S. E. Carbonado Diamond: A Review of Properties and Origin. *Gems Gemol.* **2017**, *53* (2), 168–179.
Hose, T. A. Selling the story of Britain's Stone. *Environ. Interpret.* **1995**, *10* (2),16–17.
IBGM. *O setor de gemas e joias no Brasil.* Instituto Brasileiro de Gemas e Metais. Brasília, 2010. Disponível em: http://www.infojoia.com.br/pdf/banco/setor_grandes_numeros_2009-20100816-124710.pdf.
Liccardo A.; Chodur N. L. Turismo Mineral no Brasil—Gemologia e Geoturismo. In *Anais do I Seminário sobre Design e Gemologia de Pedras, Gemas e Joias do Rio Grande do Sul*, 2009. Disponível em: http://usuarios.upf.br/~ctpedras/sdgem/artigos/Art20_Liccardo_FINAL.pdf.
Liccardo, A.; Juchem, P. L. Geoturismo—roteiro de turismo mineral na região Sul. In *XLIV Congresso Brasileiro de Geologia*, 2008. Sociedade Brasileira de Geologia: Curitiba, PR, 2008; p 393.
Liccardo, A.; Nascimento, M. L. Geoturismo—Roteiro de Turismo Mineral no Nordeste. In *XLIV Congresso Brasileiro de Geologia*, 2008. Sociedade Brasileira de Geologia: Curitiba, PR, 2008; vol. 1, p 394.
Liccardo, A. Turismo Mineral em Minas Gerais, Brasil. *Rev. Glob. Tour.* **2007**, *3*, 1–17.
Light, D. Heritage as Informal Education. In *Heritage, Tourism and Society*; Herbert, D. T., Ed.; Mansell: London, 1995; pp 117–145.
Mantesso-Neto V.; Mansur K. L.; Ruchkys U.; Nascimento M. A. L. O Que Há de Geológico nos Atrativos Turísticos Convencionais no Brasil. *Anuário do Instituto de Geociências — UFRJ* **2012**, *1*, 49–57.
Mantovani, M. S. M.; Wildner, W.; Juchem, P. L. Paraná Basin Magmatism, Stratigraphy and Mineralization (Southern Brazil). In *31st International Geological Congress, 2000, Rio de Janeiro. Pre-Congress Field Trip Guide - Bft 01*; Sociedade Brasileira de Geologia: São Paulo, 2000; pp 1–63.
Marchán, C.; Sánchez, A. Consideraciones sobre el patrimonio minero desde la perspectiva de un servicio geológico nacional. *Bol. Paran. Geoci.* **2013**, *70*, 77–86.
Mawe, J. *Travels in the Interior of Brazil: Particularly in the Gold and Diamond Districts of that Country, by Authority of the Prince Regent of Portugal: Including a Voyage to the Rio de Le Plata and a Historical Sketch of the Revolution of Buenos Ayres.* Illustrated with Engravings (No. 2). Longman, Hurst, Rees, Orme, and Brown, 1812.
Milanez, B.; Puppim, J. A. Opalas de Pedro II: o APL como remediação da grande mina. In *Recursos Minerais e Sustentabilidade Territorial*; Fernandes, F. R. C., Enriquez, M. A., Alamino, R. C. J., Eds.; CETEM/MCTI: Rio de Janeiro, 2011; Vol. 2, pp 69–88.
Nascimento, M. A. L.; Ruchkys, U. A.; Mantesso-Neto, V. *Geodiversidade, Geoconservação e Geoturismo: Trinômio Importante Para a Proteção do Patrimônio Geológico;* SBGEO, 2008; 82 p.
Pearson, M.; Mcgowan, B. *Mining Heritage Places. Assessment Manual.* Australian Council of National Trusts and Australian Heritage Commission: Camberra, 2000; 212 p. http://www.environment.gov.au/heritage.
Pereira, C. A.; Liccardo, A.; Silva, F. G. *A Arte da Cantaria*; ComArte: Belo Horizonte, 2007; 120 p.
Pérez de Perceval Verde, M. A. Patrimonio Minero: Un Variopinto y Problemático Mundo de Vestigios. *ÁREAS. Revista Internacional de Ciencias Sociales. El Patrimonio Industrial, el Legado Material de la História Económica*; Ediciones de la Universidad de Murcia, Fundación CajaMurcia: Murcia, 2010; Vol 29, pp 51–59.

Pretes, M. Touring Mines and Mining Tourists. *Ann. Tour. Res.* **2002**, *29* (2), 439–456.

Prieto, C. *La minería em el Nuevo Mundo*. Ediciones de La Revista de Occidente: Madrid, 1976; 350 p.

Puche Riart, O.; García Cortés, Á.; Mata Perelló, J. M. Conservación del patrimonio histórico minerometalúrgico español. In *Proceedings of IX Congreso Internacional de Minería y Metalurgia*, Tomo 5, León, España, 1994; pp 433–448.

Reeves, K.; Mcconville, C. Cultural Landscape and Goldfield Heritage: Towards a land management framework for the historic south-west pacific gold mining landscapes. *Landsc. Res.* **2011**, *36* (2), 191–207.

Rocha-Campos A. C. Varvito de Itu, S. P. Registro clássico da glaciação neopaleozóica. In *Sítios Geológicos e Paleontológicos do Brasil*; Schobbenhaus, C., Campos, D. A., de Queiroz, E. T., Winge, M., Berbert-Born, M. L. C., Eds.; DNPM BRASIL. Depto. Nac. Prod. Mineral (DNPM)/Serv. Geol. Brasil (CPRM). Com. Bras. Sítios Geol. e Paleobiol. (SIGEP): Brasília, 2002; p. 147–154.

Sánchez, R. A. Estudio del patrimonio minero de Extremadura. In *Una visión multidisciplinar del patrimonio geológico y minero. Cuadernos del Museo Geominero*; Florido, P., Rábano, I., Eds.; Instituto Geológico y Minero de España: Madrid, 2011; Vol 12, pp 3–30.

PART II
Geoassessments: Geoheritage Assessments for Geotourism

CHAPTER 5

Geomorphosites: Esthetic Landscape Features or Earth History Heritage?

EMMANUEL REYNARD

Institute of Geography and Sustainability and Interdisciplinary Centre for Mountain Research, University of Lausanne, CH–1015 Lausanne, Switzerland

E-mail: emmanuel.reynard@unil.ch

ABSTRACT

Geomorphosites are landforms that are considered as heritage sites. They may have several values, one being their esthetic character, which may explain why, in many countries, they are recognized as natural monuments. The esthetic character of geomorphosites often plays a key role in developing geotourism aimed at enhancing the geographical character of a place, as geomorphosites are landscape elements that attract tourism. On the other hand, their esthetic character may mask their interest for the reconstruction of the history of the Earth, which is at a core of a more history oriented geotourism aimed at explaining the Earth's history and processes. This chapter discusses the issue of the esthetic dimension of geomorphosites, and its importance for geotourism, and illustrates it with two examples: glacier tourism and geocultural sites.

5.1 INTRODUCTION

Geomorphosites (or geomorphological sites) are landforms considered as heritage sites (*in-situ* geoheritage) (Panizza, 2001; Reynard, 2009a; Reynard and Coratza, 2013; Coratza and Hobléa, 2018). They constitute one of the multiple categories of geodiversity sites or geosites (see Brilha, 2016 for a discussion on the difference between the two types). As is the case of other geosites, the fact they are considered as heritage sites is the result of a social process ('heritage making') that consists in attributing a certain number of values to landforms recognized by more or less large sections of society. In this sense, considering a landform as a geomorphosite is the result of a cultural process (Reynard and Giusti, 2018) that transforms them from simple landscape elements to valued parts of the territory that can be exploited and promoted by society as territorial resources.

The question of the values attributed to geoheritage sites has been largely discussed (Panizza and Piacente, 1993; Reynard, 2005; Brilha, 2016), as have the various methods developed to assess the qualities and importance of geosites or geomorphosites (e.g., Brilha, 2002; Coratza and Giusti, 2005; Serrano and González-Trueba, 2005; Bruschi and Cendrero, 2005; Pereira et al., 2007; Reynard et al., 2007; Zouros, 2007; Bruschi and Cendrero, 2009; Lima et al., 2010; Kavčič and Peljhan, 2010; Bruschi et al., 2011; Fassoulas et al., 2011; Pellitero et al., 2011; Coratza et al., 2012; Rocha et al., 2014; Reynard et al., 2016; Sellier, 2016; Brilha, 2016; Costa Mucivuna et al., 2019).

Among the different values attributed to geosites, that is, their interest for society, one can distinguish intrinsic value and use values. Intrinsic value is related to the quality of the sites, independently of their potential or effective use by society, whereas use values are related to the "exploitation" of the site characteristics in different fields (tourism and education, for instance) (e.g., Serrano and González-Trueba, 2005; Bruschi and Cendrero, 2005; Pereira et al., 2007). Sometimes the intrinsic value is divided into a central (scientific) value and several additional values (Reynard, 2005), that is, the ecological, cultural or esthetic interest of the site.

Geotourism is a socioeconomic activity that exploits this intrinsic value to varying extents. In this sense, it activates several use values of geosites. It will be recalled that geotourism developed in two parallel ways in the last two decades (see e.g., Hose, 2000, 2012; Dowling and Newsome, 2006; Megerle, 2008; Newsome and Dowling, 2010, 2018; Hose and Vasiljević, 2012; Hose, 2016a, b): one, influenced by the tradition of geographical travels, promoted in particular by the journal *National Geographic*, considers that geotourism is a form of tourism aimed at the discovery of the geographic character of a region, that is, that which makes it different from another territory (Stueve et al., 2002), whereas the other, promoted mainly by Earth scientists, considers geotourism in a more restrictive way, as a form of tourism aimed at the discovery of the geology of a region. The Arouca Declaration, adopted in 2011, tries to reconcile the two approaches, considering that geotourism is a form of "tourism which sustains and enhances the identity of a territory, taking into consideration its geology, environment, culture, esthetics, heritage, and the well-being of its residents. Geological tourism is one of the multiple components of geotourism" (Arouca Declaration, 2011). According to this declaration, geotourism should focus "not only on the environment and geological heritage, but also on cultural, historical, and scenic value." However, the debate is not closed and some authors (e.g., Hose, 2016b: 6–7) consider that the Declaration has added confusion rather than solved the problem.

In this chapter, our aim is not to compare these two approaches but to discuss the place of landforms and geomorphosites in the field of geotourism, and in particular, to address the issue of the esthetic value of geomorphosites in the context of geoheritage promotion.

5.2 THE ESTHETIC VALUE OF GEOMORPHOSITES

Compared to other types of geosites, geomorphological sites have three specific characteristics (Reynard, 2005, 2009a; Coratza and Brilha, 2018): the dynamic dimension, the imbrication of scales, and the esthetic dimension. More than other categories of geosites, geomorphosites enable observation of active Earth processes. But not all geomorphosites are active; therefore, Pelfini and Bollati (2014) recently proposed three categories: passive sites, active sites, and passive evolving sites, e.g., landforms that were formed in morphoclimatic conditions that no longer occur, but that are evolving under new morphoclimatic processes. A moraine deposited during the last glaciation, which is now eroded by gullying processes is an example of a passive evolving site. The scale issue is related to the fact that the size of geomorphosites may range from small, one-off sites (e.g., sinkhole, erratic boulders) to large landscapes (e.g., a glacier forefield, valley, dune field, etc.) (Reynard, 2005, 2009a; Coratza and Hobléa, 2018). In this chapter, we focus on the third dimension, that is, the esthetic one, already partly discussed in previous publications (Reynard, 2009b; Reynard and Giusti, 2018) but addressed here in the context of geotourism.

The question of the esthetics relates to the landscape value of geomorphosites (Reynard, 2005; Reynard and Giusti, 2018). Along with geological, hydrological, and biological features, landforms are one of the key elements composing the biophysical part of the so-called objective landscape, that is, the association of physical, biological, and anthropic elements that form all the landscapes in the world. As the landscape concept is not restricted to this objective component but also implies a relationship with observers, one should also consider the subjective part of landscapes, which entails a process of perception by observers or users (Reynard, 2005). It is this process of perception that confers certain values to landforms, one of them being the esthetic dimension, which can evolve over time and according to the cultural context or to the individual. Indeed, the perception of the "beauty" of landforms is not easy to address and varies greatly from one person to another. The case of waterfalls in the Swiss Alps is typical. In the 18th–19th centuries, when tourism was emerging, waterfalls that linked hanging glacial valleys with the main valleys were greatly appreciated by travelers and were among the most visited sites in the Alps. Today, this interest has considerably declined under the effect of two factors: on one hand, the flow of waterfalls has decreased due to the construction of hydroelectric facilities in the upstream part of the catchments; on the other hand, most of the valley bottoms have become corridors for travelers in a hurry to reach high altitude tourist sites. The waterfalls, which were tourist highlights two or three centuries ago, have therefore lost most of their prestige (Gauchon, 2002).

Nevertheless, as discussed by Reynard and Giusti (2018), there are many cases in which the esthetic dimension of landforms is greatly appreciated and used by the society. Exemplary are the numerous UNESCO World Heritage sites classified according to criterion vii ("to contain superlative natural phenomena or areas of exceptional natural beauty and esthetic importance") that have been formed by geomorphological processes (Migoń, 2009, 2018). Exceptional or peculiar landforms may also constitute a "visiting card" for places of worldwide renowned (Reynard et al., 2017), for example, the sugarloaf of Rio de Janeiro, the mesa landscape of Cape Town or the karstic landscape of the Ha Long Bay in Vietnam. In hundreds of cases, the esthetic dimension of relief forms plays a central role, even if these reliefs are not on a heritage list. This is particularly the case in certain morphoclimatic contexts such as deserts, volcanoes, coasts or mountain ranges, which often offer spectacular views.

Tourism has fully exploited the esthetic and picturesque character of landforms, in particular in the Alps (Hose, 2010, 2011, 2016; Reynard et al., 2011; Hobléa et al., 2017), where the first forms of tourism emerged three centuries ago, and in North American national parks, in which tourism combining wilderness and spectacular landscapes was developed (Héritier, 2003). Hose (2016) demonstrates that the roots of geotourism must be sought in what he calls the European landscape tourism, in particular the Grand Tour (Towner, 1985), especially during the Romantic period, from the end of the 18th century, when interest in wild nature and the sublime became increasingly important for travelers. The development of scenic tourism in northern England (Cope, 2016; Henry and Hose, 2016) and Scotland (Gordon and Baker,

2016) also played a key role in the beginnings of geotourism. As an example, Portal (2014) studied the heritage trajectory of the Killarney National Park in Ireland; she demonstrated the importance of the picturesque character of the landscape, as well as the pictorial and literary expression of some landforms, in particular lakes and artificial landforms such as copper mines. The famous weathered blocks of the Fontainebleau Forest, near Paris, popularized by the impressionist painters of the School of Barbizon, are an example of a geomorphological landscape promoted for its picturesque and esthetic character (Peulvast et al., 2014).

Two problems emerge from the dominance of the esthetic character of geomorphosites. On the one hand, it can condemn less spectacular sites, ordinary landscapes, to oblivion in the eyes of some authors (Portal, 2013). Indeed, there is a risk of retaining only impressive sites in geosite inventories, with less beautiful landforms—scree, rock glaciers, pediments, for example—running the risk of being forgotten in favor of the most esthetic landforms. On the other hand, the esthetic character of geomorphosites may reduce the interest of these sites to their landscape value alone to the detriment of their interest as witnesses to the history of the Earth. This non-visibility of some landforms was studied by Cayla et al. (2012), who demonstrated that perception filters may interpose between the landforms and the observers, and may dissemble the real and fundamental interest of the site. The scientific interest of picturesque and long-known sites often remains masked by the spectacular character of the site. The challenge in developing this type of site is to go beyond mere appearances and to reveal the deep nature of the sites, the stages of their evolution (their morphogenesis) and their interest for the reconstruction of Earth's history.

5.3 THE CONCEPT OF NATURAL MONUMENTS

Sites of high esthetic value have often been described as "natural monuments" or "monuments of nature" (Conwentz, 1909). This concept seems to have been invented by Alexander von Humboldt *(Naturdenkmal)* as early as 1814 (Walter, 2004). Conwentz (1909) stressed the analogy between cultural and natural monuments. The term monument is generally applied to objects that are the result of human activity but can also concern nature: "Just as the ornamented stone obelisk is a monument of art *(Kunstdenkmal)*, and as the rude stone, erected by man in former ages to the memory of the dead, is a prehistoric monument, so too the erratic block, transported in past time by natural forces, constitutes a natural monument" (Conwentz, 1909: 2).

At the end of the 19th century, various movements emerged with the aim of protecting natural monuments in reaction to industrial, energy and urban development, and the concept was introduced in several nature conservation policies as a conservation category for the protection of single, generally rare, and impressive objects, both biological (e.g., trees) or geological (e.g., erratic blocks; Fig. 5.1A). Their esthetic character was often emphasized. In 1899, the Swedish Parliament created a Waterfall Committee for the registration and charting of all the public waterfalls with the objective of preserving a few falls considered as natural monuments from industrial exploitation (Conwentz, 1909). The National Trust for Places of Historical Interest or Natural Beauty was created in England, Wales and Ireland in 1895. In France, the Protection of Natural Sites and Monuments of Artistic Character Act was adopted in 1906 and commissions were created for sites and natural monuments, while in Italy, the League for the Protection of Natural Monuments was created in 1913 (Guichard-Anguis and Héritier, 2008).

The concept was also disseminated by associations such as Touring or Alpine clubs (Guichard-Anguis and Héritier, 2008), created for the development of tourism. For example, in France, the Alpine Club and the Touring Club actively participated in the commissions for sites and natural monuments established by the 1906 law. Their aim was to protect the sites as a base for the development of tourism (Gauchon, 2002).

In the United States, the Antiquities Act was adopted in 1906. It gave the President the authority, by presidential proclamation, to create *national monuments* on federal lands to protect significant cultural or natural features. National monuments include both cultural and natural monuments, and are divided into four general classes: prehistoric, historic, geologic, and biologic. Examples of geosites classified as national monuments are the Devils Tower (monolith made of igneous rock columns), the Glacier Bay in Alaska, including a number of tidewater glaciers and the Petrified Forest of Arizona.

In the former Habsburg Empire, a "Law regarding the protection of natural and historic monuments" was adopted in 1908 (Geacu et al., 2012). It influenced legislation in several Central European countries, including Romania where the first law for the protection of natural monuments was passed in 1930, and a Romanian Academy Commission for the Monuments of Nature still exists today. Currently, 230 protected sites are considered as natural monuments (Fig. 5.1B,C) (Geacu et al., 2012). In Germany, according to the Nature and Landscape Protection Act (1976), natural monuments are a protected category; sites are protected according criteria such as rarity, unicity, and beauty.

In Brazil, according to the National System of Units of Conservation (Federal Law 9.985, 18 July 2000) natural monuments are "rare natural sites, both singular or of great scenic beauty." In Georgia, natural monuments are "relatively small areas of national importance, represented by ecosystems of rare, unique, and highly esthetic features, specific geographical and hydrological formations, and individual samples of plants or fossils of living organisms."[1] Switzerland has an inventory of landscapes and natural monuments of national importance, which cover 19% of the national territory; they are "elements of landscape protection in which the focus is on the overall appearance of the landscape."[2] Typical examples of natural monuments in Switzerland are erratic blocks or erosional landforms (Fig. 5.1D).

Finally, since 1978, "Natural Monument or Feature" constitutes Category III of the IUCN Protected Areas Management Categories (Dudley, 2008). It can be "a landform, sea mount, submarine cavern, geological feature such as a cave or even a living feature such as an ancient grove. They are generally quite small protected areas and often have high visitor value." "It is perhaps the most heavily influenced of all the categories by human perceptions of what is of value in a landscape or seascape rather than by any more quantitative assessments of value." Natural monuments have a lower level of protection than strict natural reserves (Category Ia), wilderness areas (Category Ib) and national parks (Category II).

All these examples on different continents and in different countries, are evidence for the worldwide dissemination of the natural monument concept.

To come back to geomorphosites, there is the risk of seeing only the monumental, that is, the esthetic character of sites classified as natural monuments, and of forgetting their interest for the knowledge and reconstruction of Earth's history. The task of geotourism promoters should be to recall the importance of this aspect. In the following section, we give two examples to illustrate this dichotomy.

5.4 EXAMPLES

5.4.1 GLACIER TOURISM

In some regions (Antarctica, Scandinavia, Canada, the European and New Zealand Alps, the Himalayas), glaciers and glacier landforms are

---
[1] http://apa.gov.ge/en/protected-areas/Naturalmonument, accessed 20.12.2019
[2] http://www.enhk.admin.ch/en/topics/landscapes-and-natural-monuments/, accessed 20.12.2019

major tourist attractions. For example, along with mountains, lakes and pastures, glaciers are part of the image of the Swiss landscapes. Glaciers are also at the heart of the attractiveness of national parks in western Canada (Héritier, 2003).

FIGURE 5.1 **A.** *Pierre à Dzo* erratic boulder, Monthey, Switzerland. This is an example of natural monument. The block was transported by the Rhone glacier and is therefore a testimony of the Earth's climate history. The block also has several inscriptions reminding us of scientists involved in the development of the theory of glaciation in the first part of the 19th century and is therefore of great cultural importance as a testimony to the history of the Earth sciences (Photo: E. Reynard, 2011). **B.** The Sphynx, Bucegi Mountains, Romania. This landform shaped by weathering of sandstones has been recognized as a natural monument of geomorphic origin by the Romanian government (Photo: E. Reynard, 2017). **C.** The Pâclele Mari mud volcanoes, Buzau valley, Romania. This landform was recognized as a natural monument in 1924 and is also a natural reserve of botanical and geological interest (Photo: E. Reynard, 2015). **D.** The Euseigne Pyramids, Swiss Alps. These landforms have been shaped by gullying processes in glacial sediments and the site has been part of the Swiss inventory of landscapes and natural monuments since 1977 (Photo: E. Reynard, 2017).

Glaciers were one of the main features that attracted tourists to the Alps at the turn of the 18th and 19th centuries. Together with waterfalls and rocks, they were among the elements of nature that gave the romantic travelers the sense of the sublime that, as romantic tourism developed, replaced the former fear of mountains (Reichler, 2002). The Chamonix glaciers or the Rhone glacier, in Switzerland, were popular destinations for Grand Tour travelers. If we consider that these were the first generation of geotourists (Hose, 2016), we can conclude that glaciers have played a central role in geotourism since its beginnings. Above all, it was their spectacular character that attracted tourists, but the role of climate was not ignored. Indeed, in the first half of the 19th century, in the framework of the controversy concerning the origin of erratic blocks, the glacier theory developed and was widely disseminated by Louis Agassiz (1837). Glaciers were recognized as relics of the cold periods in the Earth's history and their deposits and the traces they left in the landscape (moraines, erratic boulders, glacial striae, and *roches moutonnées*) were considered as clues to past climates.

Interest in glaciers did not decrease over time, and persisted until recently. The Jungfrau-Aletsch area is one of the key tourist regions in Switzerland, which became particularly popular with Asian tourists after it was put on the UNESCO World Heritage list in 2001. The glaciers in the Mont-Blanc massif (France, Italy) are the core attraction in the region. In some mountain tourist resorts, glaciers are also used for summer skiing (e.g., Zermatt or Saas Fee in Switzerland, Hintertux in Austria, Whistler in Canada or Les Deux Alpes in France)[3].

Glacier tourism is now threatened by climate warming (Koenig and Abbegg, 1997; Steiger et al., 2019). Several summer skiing glacier are as that existed in the 1980s are now closed, including Les Diablerets and Mont-Fort in Switzerland. Another problem is that glaciers become darker as they are covered with increasing quantities of debris (Fig. 5.2A) falling from the glacier walls as a result of melting permafrost. They thus lose their landscape appeal for summer tourism. Glacier melting also causes significant geomorphological transformations along the proglacial margins. On the other hand, new forms that are attractive for tourism, may develop. This is the case of proglacial lakes, like the one that recently appeared at the front of the Rhone glacier in Switzerland (Fig. 5.2B). The front of the glacier, which was located downstream from a glacial lock for several decades, recently retreated upstream and a lake formed in the umbilicus cleared of ice and blocked by the lock. However, these lakes can be dangerous, especially if they are fed by ice falls. Thus, at the Miage glacier, in the Mont Blanc massif (Italy), an ice collapse in the proglacial lake caused several injuries in summer 2009.[4]

---

[3]https://www.telegraph.co.uk/travel/ski/articles/Thebest-glaciers-for-summer-skiing/, accessed 20.12.2019
[4]https://www.youtube.com/watch?v=P2WsNI1fn-M, accessed 20.12.2019

The paths leading to glacier tongues are also strongly impacted by geomorphological processes induced by global warming. These trails may be cut by landslides or debris flows and the safety of hikers is often threatened by rock falls. The Forni Glacier in the Italian Alps has been studied for years by researchers at the University of Milan. The glacier has receded 2 kilometers since 1895. However, as has been the case of most Alpine glaciers, the retreat has not been continuous. After a major retreat between 1920 and 1970, the glacier then advanced 300 meters in 10 years; since 1982, it has retreated an average of 27 meters per year (Diolaiuti and Smiraglia, 2010). The recession induces gravitational paraglacial processes that impact mountain trails. The current acceleration of climate warming also reduces flow-related dynamics and increases the number of collapse structures on the glacier itself (Azzoni et al., 2017). The release of large areas previously occupied by glaciers is creating new geomorphosites, including proglacial lakes, retreating moraines, and the release of new *roches moutonnées*. What is more, the variability of the distribution of supraglacial debris cover alters ablation processes and results in the formation of epiglacial morphologies: cryoconites, dirt cones and glacier tables thus appear at the glacier surface, which can increase the attractiveness of the area, as is also the case of newly formed ice caves (Diolaiuti and Smiraglia, 2010).

A new form of tourism is now emerging as a result of climate change: last-chance tourism, meaning visiting destinations that could disappear due to climate warming. Glaciers are one of these destinations, and some countries including Iceland are enjoying a boom related to this new form of tourism.

One could imagine that this new interest in glaciers represents an opportunity to disseminate information on the history of climate, cryosphere or morphogenesis of mountain or polar regions to a wider public. However, this does not appear to be the case, as shown by several studies. In the case of the *Mer de Glace* glacier in Chamonix (France), Pralong (2006) pointed out that, although the site is of high scientific value as a highly didactic example of a valley glacier, (it was at the heart of the development of glaciology in the 19th century by James David Forbes, 1843) and is striking evidence of global warming, its development by actors of the tourism sector is mainly based on esthetic criteria and experience of the ice, by offering the opportunity to actually go inside the glacier through a gallery dug in the ice. Incidentally this experience is possible in many other glaciers, in particular, the Swiss Titlis, Jungfrau and Rhone glaciers. In Iceland, Burfoot (2017) investigated several tourist destinations, to see if and how ecosystem services are communicated to tourists. One of the trips analyzed was a boat tour of *Vatnajökull* glacial lagoon. The author reported that very little information on environmental issues was communicated by the tour guide, and scientific concepts were not discussed in detail.

These examples show that the interest in using glaciers and glacier landforms as support for the understanding of Earth history is often overshadowed by their spectacular characteristics and their purely esthetic dimension. However, the situation at the Mer de Glace has changed in recent years, with glaciologists present on site throughout the summer season to provide scientific explanations.

As evidence for the active processes and changes related to climate warming, glaciers may be important places for environmental education on the dynamics of Earth's system and Earth's history (Reynard and Coratza, 2016). In mountain and high-latitude areas, there is great potential for the development of climate related geotourism as already achieved on the Tsanfleuron Glacier in the Swiss Alps. The rapid retreat of the glacier located on limestone bedrock revealed spectacular new glaciokarstic landscapes (Fig. 5.2C) that has increased the region's tourist appeal. Several interpretive products have been prepared to capitalize on the spectacular effect of glacial retreat: a geotourist map (Martin and Reynard, 2009), whose content is focused on two main messages—the hydrological and geomorphological characteristics of karst environments and the fast glacier regression, as well as a map and panels in the landscape showing the rapid glacier retreat since the end of the Little Ice Age (1860). The educational panels (Fig. 5.3) have been set up at different points at the site, close to existing infrastructures (mountain hut, public transport stop, ski lift). These panels help visitors to understand the formation of the site and its evolution, the current transformations induced by climate change, and the interactions between glacio-karst phenomena and human activities.

5.4.2 CULTURAL GEOMORPHOSITES AND GEOTOURISM

Geocultural sites are geosites that interact closely with elements of cultural heritage (Panizza and Piacente, 2003; Reynard and Giusti, 2018). This is the case of hills of various origins (cuesta fronts, glacial locks, and volcanic necks) on which military or religious buildings have been constructed. This is also the case for geo-archaeological sites. These are archaeological remains in specific geological contexts, and sometimes particular landforms or geological structures combined with forms of intangible cultural heritage, e.g., legends linked to processes such as earthquakes or glacier movements, toponymy, and specific religious traditions. In all these cases, the value of the site, and hence

158 The Geotourism Industry in the 21st Century

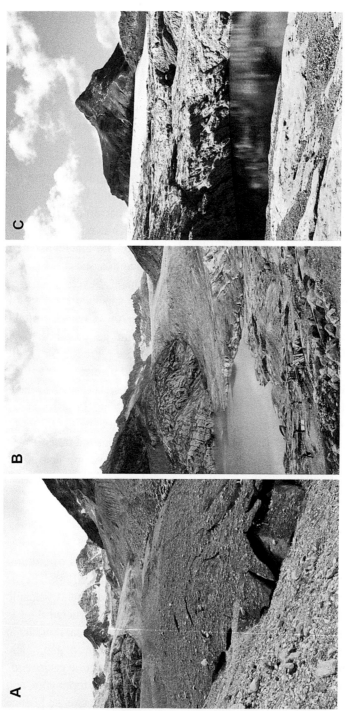

FIGURE 5.2 Three examples of glacier landscapes in the Swiss Alps and their geotourism potential. **A.** Due to climate change, the debris cover on the Moiry glacial tongue has increased, which reduces the esthetic attraction of the glacier (Photo: E. Reynard, 2017). **B.** The recession of the Rhone glacier has led to the formation of a proglacial lake, which is a new tourist attraction in the area (Photo: E. Reynard, 2017). **C.** The fast retreating Tsanfleuron glacier has revealed spectacular new glacio-karstic landscapes. Several interpretive products have been prepared to explain the landscape changes caused by the glacier recession (Photo: E. Reynard, 2014).

Geomorphosites: Esthetic Landscape Features

FIGURE 5.3 Interpretive panels (and geotourists) in front of a mountain hut on the glaciokarstic Tsanfleuron geomorphosite (Photo: E. Reynard, 2017).

its possible exploitation for tourism, derives both from the cultural importance and from the geological value of the site. The Mont-Saint-Michel in France (Bonnot-Courtois et al., 2013) and the Venice lagoon in Italy (Bondesan, 2017) are good examples of geocultural sites. Because of their spectacular character, landforms are often associated with cultural heritage. They can therefore provide a base for the development of a particular mode of geotourism one could term "geocultural tourism." The target audience for this type of tourism is visitors interested in different forms of culture, including scientific culture and knowledge of natural environments. We illustrate this point with a recent study carried out in southeastern Tunisia with the aim of conducting an inventory of regional geoheritage for the creation of a geopark (Reynard et al., 2018). The study area corresponds to the structural plateau of Matmata-Dahar which is formed by a system of superposed cuestas, inclined towards the west (*Grand Erg Occidental*). The cuesta fronts face east and dominate the Jeffara coastal plain. The backslopes have been dissected by valleys and depressions, partially filled with Pleistocene sediments, particularly fluvial conglomerates and aeolian silts (loess) (Ben Fraj, 2012). The climate is classified as arid, mean annual rainfall is less than 200 mm (183.4 mm/a in Medenine during the period 1961–1990) with more than 50% interannual variability; potential evapotranspiration (calculated using the Penman formula) is around 1500 mm in Medenine, and the mean annual air temperature is around 20°C, with the mean monthly temperature reaching 36°C in August in Medenine.

The cultural material heritage is made up of three main elements (Ben Ouezdou, 2001): the *ksour*, the *jessour* hydraulic infrastructure, and the cave dwellings. The three types of infrastructure were built for protection against enemies (*ksour*), aridity (*jessour*), and heat (troglodytes). The *ksour* are fortified granaries set upon high points of the relief (cuesta fronts, witness buttes, and outliers, rocky spurs). They had both an economic function (storage of grain in family granaries called *ghorfas*), a social function (as a meeting place in the inner courtyard of the fortified complex) and a defense function (a place of retreat in the case of attack) (Fig. 5.4A). The *jessour* (Fig. 5.4B) are small agricultural hydraulic installations. They consist of a series of dikes (called *tabia*), transverse to the water flow in the ravines and wadis and stepped on the longitudinal profile. At the rear of the dike is a relatively flat surface for retaining water and sediments, on which orchards (olive trees in particular) are cultivated, associated with annual crops (beans, lentils, and vegetables) when rainfall conditions allow. Each *jesr* is connected to the *jesr* downstream by a weir that allows part of the water to flow downstream. This sophisticated system has made it possible to cultivate olive trees well beyond their natural ecological limits.

The links between cave dwellings and geomorphology were recently studied by Boukhchim et al. (2018), who demonstrated that the two main types of troglodyte habitats—lateral and vertical-lateral—depend to a great extent on the geomorphological configuration. In areas where loess accumulated, troglodytic habitats are based on a vertical model (Fig. 5.4C): an inner courtyard is first dug in the loess, then underground rooms are installed laterally around the courtyard. Seen from above, the landscape of these troglodytic habitats is lunar in appearance, the wind accumulations being dotted with 'craters' corresponding to the courtyards. On the edges of cuestas, the difference in hardness between sedimentary layers was exploited to install a horizontal troglodyte cave system. The caves were dug in the marl layers, and were protected by the upper limestone layer that formed the roof of the rooms (Fig. 5.4D). The backslope of the lower cuesta level was often used to install outdoor infrastructure, such as livestock enclosures.

The geosites were inventoried with a view to creating a regional geopark (Reynard et al., 2018). Of the 29 geosites selected, half are of major geomorphological value, and the others being mainly sedimentological and paleontological sites. Three sites are primarily of landscape value. An inventory including a description of access conditions, a scientific description of the site, as well as an analysis of threats, and proposed protection measures was prepared for each site.

FIGURE 5.4 Examples of geocultural sites on the Matmata-Dahar plateau, Southeast Tunisia. **A.** Ksar Hallouf (Photo: E. Reynard, 2017). **B.** Series of *Jessour* along a wadi (Photo: E. Reynard, 2015). **C.** Vertical cave dwellings and crater landscape in aeolian silts in Haddej (Photo: E. Reynard, 2015). **D.** Horizontal cave dwellings in a cuesta landscape in Guermessa (Photo: E. Reynard, 2015). These examples provide the basis for the promotion of cultural geotourism in the planned geopark.

In the second step, the potential of each geosite for geotourism was analyzed. The sites were classified according to different forms of tourism:

- nine sites have potential for geological tourism; these are essentially paleontological or sedimentological geosites and the tourism activities that could be developed mainly target specialists (Earth science enthusiasts, researchers, and students);
- eighteen sites lend themselves to the development of geocultural tourism, that is, the opportunity to discover the strong interactions between the way of life of local societies and their geomorphological and climatic environment, often in interaction with more strictly geological dimensions;
- two sites have potential for the development of ecotourism.

5.5 CONCLUSION

Because of their spectacular character, geomorphosites are an important territorial resource for tourism. Indeed, in many cases, the original tourist offers comprised landscapes, which themselves are strongly influenced by landforms. However, for the development of geotourism, the esthetic character of landforms can be double-edged. Indeed, if we consider geotourism as a form of discovery of the particular geographical characteristics of places, the so-called sense of place, then the spectacular component of certain landscapes may be an attractive tourism product. Similarly, if the objective is to develop more specific geological tourism, via geotourism viewed as the discovery of the geological history of places, and it is often by explaining the geomorphology of landscapes that one can discuss the more complex issues of the history of the Earth and the processes associated with it.

At the same time, however, there is a serious risk that only particularly esthetic and spectacular forms will be used by the tourist sector, solely for their character as "natural monuments." This implies a double risk for geotourism: on the one hand, it limits the interest of geomorphosites to their simple status as picturesque or spectacular features of the landscape; on the other hand, this selective approach to the landscape would have the corollary of making visitors forget the so-called ordinary landscapes, which nevertheless represent the vast majority of the landscapes visited. In both cases, there is the risk that the proposed geotourism offer does not allow the discovery of the historical character of landscapes and more generally of the importance of taking long periods into account to understand human history.

The geocultural approach, which consists in establishing links between the different forms of cultural heritage, both tangible and intangible, and the geological and geomorphological components of the landscape, appear to be one way of going beyond a purely esthetic tourist enhancement of the geomorphological heritage and of offering visitors an integrated cultural discovery of the regions they visit, one that combines the geological heritage, the biotic heritage, and the cultural heritage.

KEYWORDS

- **geomorphological heritage**
- **geomorphosites**
- **geocultural sites**
- **esthetics**
- **natural monuments**

REFERENCES

Agassiz, L. *Etudes sur les glaciers*; Jent et Gasmann: Neuchâtel, 1840 (in French).

Arouca Declaration on Geotourism, November 12, 2011. http://www.europeangeoparks.org/?p=223 (accessed Dec 20, 2019)

Azzoni, R. S.; Fugazza, D.; Zennaro, M.; Zucali, M.; D'Agata, C.; Maragno, D.; Cernuschi, M.; Smiraglia, C.; Diolaiuti G. A. Recent Structural Evolution of Forni Glacier Tongue (Ortles-Cevedale Group, Central Italian Alps). *J. Maps* **2017**, *13* (2), 870–878.

Ben Fraj, T. Proposition d'un schéma chronostratigraphique des héritages quaternaires continentaux de la Jeffara septentrionale et de la partie nord-orientale du plateau de Dahar-Matmata (sud-est tunisien). *Quaternaire* [Online] **2012**, *23* (2), http://quaternaire.revues.org/6242 (accessed Dec 20, 2019) (in French).

Ben Ouezdou, H. *Découvrir la Tunisie du Sud. De Matmata à Tataouine: Ksour, Jessour et Troglodytes*. Tunis, 2001.

Bondesan, A. Geomorphological Processes and Landscape Evolution of the Lagoon of Venice. In *Landscapes and Landforms of Italy*; Soldati, M., Marchetti, M., Eds.; Springer International Publishing: Cham, 2017, pp 181–191.

Bonnot-Courtois, C.; Walter-Simonnet, A. V.; Baltzer, A. The Mont-Saint-Michel Bay: An Exceptional Megatidal Environment Influenced by Natural Evolution and Man-Made Modifications. In *Landscapes and Landforms of France* Fort, M., André M. F., Eds.; Springer: Dordrecht, 2013, pp 41–51.

Boukhchim, N.; Ben Fraj, T.; Reynard, E. Lateral and "Vertico-Lateral" Cave Dwellings in Haddej and Guermessa. Characteristic Geocultural Heritage of Southeast Tunisia. *Geoheritage* **2018**, *10* (4), 575–590.

Brilha, J. Geoconservation and Protected Areas. *Environ. Conserv.* **2002**, *29* (3), 273–276.

Brilha, J. Inventory and Quantitative Assessment of Geosites and Geodiversity Sites: A Review. *Geoheritage* **2016**, *8* (2), 119–134.

Bruschi, V. M.; Cendrero, A. Geosite Evaluation. Can We Measure Intangible Values? *Il Quaternario* **2005**, *18* (1), 293–306.

Bruschi, V. M.; Cendrero, A. Direct and Parametric Methods for the Assessment of Geosites and Geomorphosites. In *Geomorphosites*; Reynard E., Coratza P., Regolini-Bissig G., Eds.; Pfeil: München, 2009; pp 73–88.

Bruschi, V. M.; Cendrero, A.; Albertos, J. A. C. A Statistical Approach to the Validation and Optimisation of Geoheritage Assessment Procedures. *Geoheritage* **2011**, *3* (3), 131–149.

Burfoot, C. Tourism as a Tool for Communicating Complex Environmental Issues: Applying the Ecosystem Services Framework to Nature-Based Tourism Activities Across Iceland. Master Dissertation in Environmental Sciences, University of Södertörn, Sweeden, 2017.

Cayla, N.; Hoblea, F.; Biot, V.; Delamette, M.; Guyomard, A. From the Invisibility of Geomorphosites to their Geoheritage Revelation. *Geocarrefour* **2012**, *87*, 171–186 (in French).

Conwentz, H. *The Care of Natural Monuments with a Special Reference to Great Britain and Germany*; Cambridge University Press: Cambridge, 1909.

Cope, M. A. Three Centuries of Open Access to the Caves in Stoney Middleton Dale Site of Special Scientific Interest, Derbyshire. In *Geoheritage and Geotourism: A European Perspective*; Hose, T. A., Ed.; Geological Society: London, 2016, 157–170.

Coratza, P.; Giusti, C. Methodological Proposal for the Assessment of the Scientific Quality of Geomorphosites. *Il Quaternario* **2005**, *18* (1), 307–313.

Coratza, P.; Hobléa, F. The Specificities of Geomorphological Heritage. In *Geoheritage: Assessment, Protection, and Management*; Reynard, E., Brilha J., Eds.; Elsevier: Amsterdam, 2018, pp 87–106.

Coratza, P.; Galve, J. P.; Soldati, M.; Tonelli, C. Recognition and Assessment of Sinkholes as Geosites: Lessons from the Island of Gozo (Malta). *Quaest. Geogr.* **2012**, *31* (1), 25–35.

Costa Mucivuna, V; Reynard, E.; da Glória Motta Garcia, M. Geomorphosites Assessment Methods: Comparative Analysis and Typology. Geoheritage **2019**, https://doi.org/10.1007/s12371-019-00394-x.

Diolaiuti, G. A.; Smiraglia, C. Changing Glaciers in a Changing Climate: How Vanishing Geomorphosites have been Driving Deep Changes in Mountain Landscapes and Environments. *Géomorphol.: Relief, Processus, Environ.* **2010**, *16*, 131–152.

Dowling, R. K.; Newsome, D. *Geotourism*, Eds.; Elsevier/Heineman: Oxford, 2006.

Dudley, N. *Guidelines for Applying Protected Area Management Categories*; IUCN: Gland, 2008.

Fassoulas, C.; Mouriki, D.; Dimitriou-Nikolakis, P.; Iliopoulos, G. Quantitative Assessment of Geotopes as an Effective Tool for Geoheritage Management. *Geoheritage* **2012**, *4* (3), 177–193.

Forbes, J. D. *Travels through the Alps of Savoy and Other Parts of the Pennine Chain, with Observations on the Phenomena of Glaciers*; Black: Edinburgh, 1843.

Gauchon, C. Les sites naturels classés entre 1906 et 1930 dans les Alpes du Nord : entre tourisme et protection, bilan et actualité. *Rev. Geogr. Alpine* **2002**, *90*, 15–31 (in French).

Geacu, S.; Dumitrascu, M.; Maxim, J. The Evolution of the Natural Protected Areas Network in Romania. *Rom. J. Geogr.* **2012**, *56* (1), 33–41.

Gordon, J. E.; Baker M. Appreciating Geology and the Physical Landscape in Scotland: From Tourism of Awe to Experiential Re-Engagement. In *Geoheritage and Geotourism: A European Perspective*; Hose, T. A., Ed.; Geological Society: London, 2016; pp 25–40.

Guichard-Anguis, S.; Héritier, S. *Le patrimoine naturel entre culture et ressource. Géographie et cultures* **2008**, *66*, 1–138 (in French).
Henry, C. J.; Hose, T. A. The Contribution of Maps to Appreciating Physical Landscape: Examples from Derbyshire's Peak District. In *Geoheritage and Geotourism: A European Perspective*; Hose, T. A., Ed.; Geological Society: London, 2016; pp 131–156.
Héritier, S. Tourisme et activités récréatives dans les parcs nationaux des montagnes de l'Ouest canadien: impacts et enjeux spatiaux (Parcs nationaux Banff, Jasper, Yoho, Kootenay, Lacs Waterton, Mount Revelstoke et des Glaciers). *Ann. Géol.* **2003**, *62* (9), 23–46 (in French).
Hobléa, F.; Cayla, N.; Giusti, C.; Peyrache-Gadeau, V.; Poiraud, A.; Reynard, E. Les géopatrimoines des Alpes occidentales: émergence d'une ressource territoriale. *Ann. Géol.* **2017**, *7* (17), 566–597 (in French).
Hose, T. A. European Geotourism—Geological Interpretation and Geoconservation Promotion for Tourists. In *Geological Heritage: its Conservation and Management*; Barretino, D.; Wimbledon, W. A. P.; Gallego, E., Eds.; Sociedad Geologica de España, Instituto TechnologicoGeoMinero de España; ProGEO: Madrid, 2000; pp 127–146.
Hose, T. A. The Significance of Esthetic Landscape Appreciation to Modern Geotourism Provision. In *Geotourism: The Tourism of Geology and Landscape*; Newsome, D., Dowling, R. K., Eds.; Goodfellow: Oxford, 2010, pp 13–25.
Hose, T. A. The English Origins of Geotourism (as a Vehicle for Geoconservation) and their Relevance to Current Studies. *Acta geogr. Slov.* **2011**, *51* (2), 343–360.
Hose, T. A. 3G's for Modern Geotourism. *Geoheritage* **2012**, *4*, 7–24.
Hose, T. A., Eds.; *Appreciating Physical Landscapes: Three Hundred Years of Geotourism*; The Geological Society Special Publication: London, 2016a.
Hose, T. A., Eds.; *Geoheritage and Geotourism: A European Perspective*; The Boydell Press: Woodbridge, 2016b.
Hose, T. A.; Vasiljević, D. A. Defining the Nature and Purpose of Modern Geotourism with Particular Reference to the United Kingdom and South-East Europe. *Geoheritage* **2012**, *4*, 25–43.
Kavčič, M.; Peljhan M. Geological Heritage as an Integral Part of Natural Heritage Conservation Through its Sustainable Use in the Idrija region (Slovenia). *Geoheritage* **2010**, *2* (3–4), 137–154.
Koenig, U.; Abegg, B. Impacts of Climate Change on Winter Tourism in the Swiss Alps. *J. Sust. Tourism* **1997**, *5* (1), 46–58.
Lima, F. F.; Brilha, J. B.; Salamuni, E. Inventorying Geological Heritage in Large Territories: a Methodological Proposal Applied to Brazil. *Geoheritage* **2010**, *2* (3-4), 91–99.
Martin, S.; Reynard, E. How can a Complex Geotourist Map be Made More Effective? Popularisation of the Tsanfleuron Heritage (Valais, Switzerland). In *6th European Congress on Regional Geoscientific Cartography and Information Systems, Munich, 9-12 June 2009, Proceedings*, 2009; vol. 2, pp 261–264.
Megerle, H. *Geotourismus. Innovative Ansätze zur touristischen Inwertsetzung und nachhaltigen Regionalentwicklung*. Kerrsting: Nürnberg, 2008 (in German).
Migoń, P. Geoheritage and the World Heritage List of UNESCO. In *Geomorphosites*; Reynard E.; Coratza P., Regolini-Bissig G., Eds.; Pfeil: München, 2009, pp 121–132.
Migoń, P. Geoheritage and World Heritage Sites. In *Geoheritage. Assessment, Protection, and Management*; Reynard, E., Brilha J., Eds.; Elsevier: Amsterdam, 2018; pp 237–249.
Newsome, D.; Dowling, R. K., Eds.; *Geotourism: The Tourism of Geology and Landscape*; Goodfellow: Oxford, 2010.

Newsome, D.; Dowling, R. K. Geoheritage and Geotourism. In *Geoheritage. Assessment, Protection, and Management*; Reynard, E., Brilha J., Eds.; Elsevier: Amsterdam, 2018; pp 305–321.

Panizza, M. Geomorphosites: Concepts, Methods and Example of Geomorphological Survey. *Chin. Sci. Bull.* **2001**, *46*, Suppl. Bd, 4–6.

Panizza, M.; Piacente, S. Geomorphological Assets Evaluation. *Zeitschr. Geomorph., N.F.* **1993**, Suppl. Bd *87*, 13–18.

Panizza, M.; Piacente, S. *Geomorfologia culturale*. Pitagora: Bologna, 2003 (in Italian).

Pelfini, M.; Bollati, I. Landforms and Geomorphosites Ongoing Changes: Concepts and Implications for Geoheritage. *Quaest. Geogr.* **2014**, *33* (1), 131–143.

Pellitero, R.; González-Amuchastegui, M. J.; Ruiz-Flaño, P.; Serrano, E. Geodiversity and Geomorphosite Assessment Applied to a Natural Protected Area: the Ebro and Rudron Gorges Natural Park (Spain). *Geoheritage* **2011**, *2*, 163–174.

Pereira, P.; Pereira, D.; Caetano Alves, M. I. Geomorphosite Assessment in Montesinho Natural Park (Portugal). *Geogr. Helv.* **2007**, *62*, 159–168.

Peulvast, J.-P.; Bétard, F.; Giusti, C. The Seine River from Ile-de-France to Normandy: Geomorphological and Cultural Landscapes of a Large Meandering Valley. In *Landscapes and landforms of France*; Fort, M., André M. F., Eds.; Springer: Dordrecht, 2013, pp 17–28.

Portal, C. Patrimonialiser la nature abiotique ordinaire. Réflexions à partir des Pays de la Loire (France). *L'espace géographique* **2013**, *42*, 213–226 (in French).

Portal, C. Appréhender le patrimoine géomorphologique. Approche géohistorique de la patrimonialité des reliefs par les documents d'archives. L'exemple du Parc National de Killarney (Kerry, Irlande). *Géomorphol.: Relief, Processus, Environ.* **2014**, *1*, 15–26 (in French).

Pralong, J.-P. Géotourisme et utilisation de sites naturels d'intérêt pour les sciences de la Terre: les régions de Crans-Montana-Sierre (Valais, Alpes suisses) et de Chamonix-Mont-Blanc (Haute-Savoie, Alpes françaises). PhD Dissertation, University of Lausanne, 2006 (in French).

Reichler, C. *La découverte des Alpes et la question du paysage*. Georg: Genève, 2002 (in French).

Reynard, E. Géomorphosites et paysages. *Géomorphologie: relief, processus, environnement* **2005**, *3*, 181–188.

Reynard, E. Geomorphosites: Definition and Characteristics. In *Geomorphosites*; Reynard E.; Coratza P., Regolini-Bissig G., Eds.; Pfeil: München, 2009a; pp 9–20.

Reynard, E. 2009b. Geomorphosites and Landscapes. In *Geomorphosites*; Reynard E.; Coratza P., Regolini-Bissig G., Eds.; Pfeil: München, 2009b; pp 21–34.

Reynard, E.; Coratza, P. Scientific Research on Geomorphosites. A Review of the Activities of the IAG Working Group on Geomorphosites over the Last Twelve Years. *Geogr. Fis. Din. Quat.* **2013**, *36*, 159–168.

Reynard, E.; Coratza, P. The Importance of Mountain Geomorphosites for Environmental Education. Examples from the Italian Dolomites and the Swiss Alps. *Acta geogr. Slov.* **2016**, *56* (2), 291–303.

Reynard, E.; Fontana, G.; Kozlik, L.; Scapozza, C. A Method for Assessing the Scientific and Additional Values of Geomorphosites. *Geogr. Helv.* **2007**, *62*, 148–158.

Reynard, E.; Giusti, C. The Landscape and the Cultural Value of Geoheritage. In *Geoheritage. Assessment, Protection, and Management*; Reynard, E., Brilha, J., Eds.; Elsevier: Amsterdam, The Netherlands, 2018; pp 147–165.

Reynard, E.; Hoblea, F.; Cayla, N.; Gauchon, C. Iconic Sites for Alpine Geology and Geomorphology. Rediscovering Heritage? *J. Alpine Res.* [Online] **2011,** *99* (2) http://rga.revues.org/1435 (accessed Dec. 20, 2019).

Reynard, E.; Perret, A.; Bussard, J.; Grangier, L.; Martin, S. Integrated Approach for the Inventory and Management of Geomorphological Heritage at the Regional Scale. *Geoheritage* **2016,** *8,* 43–60.

Reynard, E.; Pica, A.; Coratza, P. Urban Geomorphological Heritage. An Overview. *Quaest. Geogr.* **2017,** *36* (3), 7–20.

Reynard, E.; Ben Ouezdou, H.; Ouaja, M.; Ben Fraj, T.; Abichou, H.; Clivaz, M.; Ghram Messedi, A. Géoparc du Dahar. Feuille de route pour la création d'un géoparc UNESCO dans le Sud-est tunisien; Swisscontact: Tunis, 2018 (in French).

Rocha, J.; Brilha, J.; Henriques, M. H. Assessment of the Geological Heritage of Cape Mondego Natural Monument (Central Portugal). *Proc. Geol. Ass.* **2014,** *125,* 107–113.

Sellier, D. A Deductive Method for the Selection of Geomorphosites. Application to Mont Ventoux, Provence, France. *Geoheritage* **2016,** *8* (1), 15–29.

Serrano, E.; González-Trueba, J.J. Assessment of Geomorphosites in Natural Protected Areas: the Picos de Europa National Park (Spain). *Géomorphol.: Relief, Processus, Environ.* **2005,** *3,* 197–208.

Steiger, R.; Scott, D.; Abegg, B.; Pons, M.; Aall, C. A Critical Review of Climate Change Risk for Ski Tourism. *Curr. Issues Tour.* **2019,** *22* (11), 1343–1379.

Stueve, A. M.; Cook, S. D.; Drew, D. The Geotourism Study: Phase 1 Executive Summary. National Geographic: Washington, 2002.

Towner, J. The Grand Tour. A Key Phase in the History of Tourism. *Ann. Tour. Res.* **1985,** *12,* 297–333.

Walter, F. *Les figures paysagères de la nation: territoire et paysage en Europe (XVIe-XXe siècle).* Ed. EHESS: Paris (in French).

Zouros, N. Geomorphosite Assessment and Management in Protected Areas of Greece. Case Study of the Lesvos Island Coastal Geomorphosites. *Geogr. Helv.* **2007,** *62,* 169–180.

CHAPTER 6

Geoheritage and Geotourism in Albania

AFAT SERJANI

ProGEO-Albania, Tirana, Albania

E-mail: afatserjani@gmail.com

ABSTRACT

This chapter introduces the first geotourism book in Albania. Common data on the conceptual basis for geodiversity and geotourism are given in this chapter. In detail, they are presented the main chapters of the first book of the series: "Geoheritage and Geotourism in Albania." As the main achievements, we outline criteria for the selection of important Earth science sites and the classification and categorization of geological sites. Geotourism in Albania is described and illustrated by four major geological tours. As a conclusion, I have written about global geotourism as an important trend of the 21st century. Geotourism is considered as the most important of all kinds of tourism.

6.1 INTRODUCTION

Preparation of a new book on geotourism was, for me, an inspiration and a nice surprise, because of some important reasons:

First, I have worked as a geologist for about 50 years and most of my research has been done in geological prospecting and also promotion. Geotourism is impossible without geology and geoheritage. Geology is the main base for developing geotourism. Geodiversity can be studied mainly in the field.

Second, since 1995, I have been involved in ProGEO (The European Association for the Conservation of the Geological Heritage), and have participated in every meeting and symposium. I have visited a lot of outstanding geological sites and geoparks in different countries.

Third, geotourism uses geodiversity and geoheritage for sustainable development, and so it is going to be main economic trend in many countries of our planet. At the same time, geotourism embraces cultural heritage and values as well.

Developing research on geotourism we should never forget the protection of geoheritage for the coming generations.

6.1.1 ABOUT THE "GEODIVERSITY" CONCEPT

Before publishing the book "Geoheritage and Geotourism in Albania" we have used and presented the "Geotourism" concept at some symposiums, workshops, and congresses. In the Third ProGEO Symposium in Madrid, for the first time, this concept was used in the oral presentation: Geosciences Significance and Tourist Values of some "UNESCO/IUGS Geosites and Geoparks in Albania" (Serjani et al., 1999). "Geotours in Albania" were presented in Dublin, Ireland (Serjani et al., 2002), and "Geotourist values of the Albanian Alps" in Pirot, Serbia (Serjani et al., 2003).

I think it is interesting to remember the publication of Beck and Cable (1998), some years before on: *"Interpretation for the 21st century: Fifteen Guiding Principles for Interpreting Nature and Culture"* (USA, Sagamore Publishing); this publication, I think, it is just leading to the geotourism development in the 21st century.

The geodiversity concept includes geological–geomorphologic aspects of the natural heritage. Geological sites represent rare, unique, and pattern unrepeatable phenomena, reflecting in separated intervals of the time of the history of the Earth's crust. Geological sites in most cases, testify to the history of humankind since ancient times (Turner, 2009). Names of geological sites sometimes are related to historical events and figures, preserved in legends, or are linked with religious belief, and cult objects.

6.1.2 GEOTOURISM CONCEPT

Geotourism includes all scientific, didactic, exploration, and tourist activities concentrated in geological sites and geomorphologic landscapes. According to Sadry (2009), all selected geodiversities considered as potential geosites, will be changed to a real geosite (means as a geotourism site/destination) after the tourism infrastructures are provided and geoconservation measures are taken.

Just after publishing the book *"Geoheritage and Geotourism in Albania"* (2003), I saw the word "Geotourism" at the IGC in Firenze, Italy. On the desk of the Iranian delegation, I received a CD: "Geotourism-Kazemi: Geological Phenomena in Iran," including two clips, 15 min each, by Alireza Amrikazemi, Geological Survey of Iran.

One year later, Thomas Hose pointed out: Also an interpretative strategy, such as geotourism, should generate the public pressure required for the promotion and protection of the "geoheritage" (Hose, 2005).

Later, others published about geotourism: Dowling and Newsome (2006); Nekouie-Sadry: *"Fundamentals of Geotourism: With Special Emphasis on Iran,"* (2009); Asrat, Demissie, Mogessie: *"Geotourism in Ethiopia"* (2009); Raukas, Bauert, Willman, Puurmann, Ratas: *"Geotourism Highlights of The Saaremaa and Hiiumaa Islands, GeoGuide Baltoscandia, Tallinn"* (2009); Dowling and Newsome: *"Global Geotourism Perspectives"* (2010); Newsome and Dowling: *"Geotourism: The Tourism of Geology and Landscape"* (2010); Garofano: *"Geotourism. The geological Attractions of Italy for Tourists"* (2010); Erfurt-Cooper and Cooper: *"Volcano and Geothermal Tourism: Sustainable Geo-Resources for Leisure and Recreation"* (2010); Farsani, Coelho, Costa, Neto de Carvalho: *"Geoparks & Geotourism: new approaches to sustainability for the 21st century"* (2012); But before all else, Barettino, Wimbledon and Gallego published the seminal work: *"Geological Heritage: Its Conservation and Management"* (Madrid (Spain), 2000) include contributions on geotourism from European countries as the fruit of the aforementioned Third ProGEO Symposium that was held in Madrid in 1999.

Urban and Gagol, in 2008, published: *"Geological Heritage of the OE ewiêtokrzyskie (Holy Cross) Mountains,"* Central Poland, where they pointed: "The geoheritage of the OE wiêtokrzyskie Mt. should be properly used for public profit and advantage of Earth-science education, which is severely threatened in Europe (van Loon, 2008). Apart from scientific value, the necessary conditions for geotourism are accessibility and illustration (Alexandrowicz et al., 1992)."

Wimbledon and Smith-Mayer published the book: *"Geoheritage in Europe and its Conservation"* (Oslo, September 2012), where are included contributions on geoheritage and geological sites from 37 European countries. In the preface of this book, authors pointed out: "The only record of the history of our planet lies in the rocks beneath our feet. Rocks and the landscape are the memory of the Earth. Here, and only here, is it possible to trace the processes, changes, and upheavals which have formed our planet over thousands of millions of years. The more recent part of this record, of course, includes the evolution of life, including man. The record preserved in the rocks and landscape is unique, and much of it is surprisingly fragile. Today, it is threatened more than ever. What is lost can never be recovered and, therefore, there is an urgent need to understand and protect what remains of this our common heritage."

More recently, there are two 2016 books edited by Thomas A. Hose (2016a, 2016b) *"Geoheritage and Geotourism: a European Perspective,"* has contributors: Kevin Crawford, Peter Davis, John E. Gordon, Thomas A. Hose, Jonathan G. Larwood, Slobodan B. Markovic, Martin Munt, Emmanuel Reynard, Nemanja Tomic, Djordjije A. Vasiljevic, Margaret Wood, and Volker Wrede. The other 2016 work edited by Hose is *"Appreciating Physical Landscapes: Three Hundred Years of Geotourism,"* published by The Geological Society of London. It was the outcome of the "Appreciating Physical Landscapes: Geotourism" conference, convened and organized by Hose at the Geological Society on 23rd October 2012. The conference assembled an international cast of some 15 speakers from Australia, South Africa, North America and Europe.

The most recent reference books include *"Geoheritage: Assessment, Protection and Management"* (Reynard and Brilha 2018) and *"A Handbook of Geotourism"* (Dowling and Newsome, 2018), each of them has been written by at least 45 authors and experts around the world. In writing the present reference book, *"The Geotourism Industry in the 21st Century,"* 35 geotourism experts around the world have contributed. Added to these, conferences such as the one on September 2017 in Krakow, "A geology and mining region of Poland," organized by the Faculty of Geology, Geophysics, and Environmental Protection at AGH University of Science and Technology was an international conference on Geotourism, Mining Tourism, Sustainable Development, and Environmental Protection. The annual international GEOTOUR conference in Krakow, Poland and actually as a series of conferences held in all Eastern European Countries since 2005 (e.g., recently in 2019, held in Czech Republic), has been dedicated to all aspects of geotourism.

Three global geotourism conferences, including the inaugural one, have been convened by Ross Dowling and his partners in Australia, Malaysia and Oman during 2008, 2010 and 2011, respectively. The number of these conferences on geology, geomorphology and tourism focusing on geotourism and relevant subjects such as geological heritage, geomorphological heritage, mining heritage and also geoparks are dramatically increasing (the forthcoming 36th International Geological Congress in Delhi, India, with a sub-theme including geotourism, which will be held on November 2020 is another example). In fact, there are plenty of these conferences which will not be discussed in this chapter anymore. Here I like to remember why geotourism must be considered the most important of all kinds of tourism. Geotourism is based on geology of the planets and on geological construction of the region where we are living. In many geological sites, geological outcrops can be seen deep planetary processes, and visitor can be imaging the history and transformation of the Earth millions of years before. Life and commonly biodiversity are developed on the land, soil, and seawater. Geological sites and geomorphologic landscapes are important for education of new generations and can be used for new researches and for exploration of natural processes.

All kinds of tourism are based on the geodiversity. Geotourism is strongly connected to land geodiversity and geoheritage. The geodiversity concept includes geological–geomorphological aspects of natural heritage. Every geosite, esthetic landscape, astonishing geological outcrop has unrepeatable beauty. It is needed to discover, to hear, and to feel by every human being. Geological sites represent rare, unique, and pattern unrepeatable phenomena of the history of the Earth's crust. Geodiversity and geoheritage are used for development of the tourist industry. All kinds of human activities on the land, such as: Geological sites, geomorphologic landscapes, geoparks, esthetic geology, mining, and anthropological sites belong to geotourism. Geodiversity and geological heritage in all cases are connected with different kinds of cultural heritage.

Studies on geodiversity can be on different scales: Global or planetary, continental, regional, national, and local. Geotourism activities can be for study, exploration, didactic, or education, curiosity, amusement, rest, or curative character.

6.2 THE BOOK "GEOLOGICAL HERITAGE AND GEOTOURISM IN ALBANIA"

This book was compiled and published in framework of the UNESCO Program. It is published in both Albanian and English versions (Fig. 6.1).

In the introduction, there are treated concepts of geological heritage and geological sites. A short overview on Geological Heritage in Europe is written.

Europe, the old continent, the cradle of human culture, has done the main first step in Geological Heritage Conservation. Europe, the native land of geological sciences, where the most part of geological phenomena, which testify about the history of our planet, was discovered for the first time, has done the conclusion of discovering during the 20th century and determined the strategy of knowledge, inventory, management, and

FIGURE 6.1 Front cover of the book: Geological Heritage and Geotourism in Albania published in 2003 in Albanian and English versions.
Source: Serjani et al. (2003).

using of geological sites. The first discovering in geomonumental aspect of nature and European geology were done since the beginning of the European Renaissance by geologists Nicolas Steno, Agostino Scilla, etc. (Hose, 1998).

Concerning Albanian nature, the first evaluation belongs to the famous Renaissance figure Sami Frasheri, who since 1888 wrote: "Albania is one of the best regions of Balkan, with high mountains, wide fields, sea sides, bays, and a lot of harbors, with nice rivers and lagoons."

Below in the book are given data about the beginning of natural heritage and geoheritage in European countries and in Albania, about organization structures and management in some European countries and in Albania.

6.3 THE CLASSIFICATION OF GEOLOGICAL SITES

In detail is described the classification and categorization of geological sites of Albania. Geological sites are classified and selected according to kinds, types, genetically, and rocky belonging, depending on geological phenomena that influenced their formation. On the other hand, geological sites are categorized according to their scientific, didactic, and esthetic importance.

The first publications on the criteria and methods of classification of natural heritage were done a long time before (Gonggrijp and Bockschoten, 1981; Gonggrijp, 1992), whereas the criteria for classification of geological sites are published later by Wimbledon et al. (1996). To the questions of criteria and methods for classification of natural monuments, there are devoted some special works of A. V. Lapo and M. Vdovets in Russia. The criteria and methods for classification of geological sites of Albania were presented in ProGEO symposium in Tallinn, Estonia (Serjani, 1997). The criteria for selection of geological sites are presented in Table 6.1.

The definition and evaluation of the different outcrops, mineral deposits, and geological phenomena as geological sites are based on some principles (methods) of classification as below:

- Level of knowledge based on the scale of geological survey of the region.
- Complex evaluation of geological site. In most cases, geological sites are result of some geological processes.

TABLE 6.1 Criteria for Selection of the Important Earth Science Sites.

Group Gea-The Netherland (1976)	After G. P. Gonggrijp and G. J. Boekschoten (1981)	After G. P. Gonggrijp (1992)	After W. Krieg (1996) (more important criteria for selection geotopes)	Criteria used for selection of geosites in Albania
1. Rarity	1. Rarity	1. Rarity	1. Rarity	1. Rarity
2. Soundness	2. Condition	2. Present conditions	2. Danger by planned constructions	2. Representativity
3. Representativity	3. Scientific importance	3. Representativity	3. Conspicuousness	3. Diversity
4. Educational and scientific value	4. Educational importance	4. Diversity scientific and educational importance	4. Attractivity	4. Scientific values
5. Diversity	5. Representativity	**Additional criteria**	5. Ability for conservation	5. Didactic importance
		5. Size		6. Esthetic values
		6. Clarity		7. Accessibility
		7. Accessibility		8. Clarity of representatives
		8. Vulnerability		9. Irreplacebility
		9. Irreplaceability		10. Vulnerability
				11. Size
				12. Actuality or Soundness
				13. Danger
				14. Ability for conservation

- Scientific values of geological sites. Here we mean the interest for new studies and scientific values of each geosite to the geology of our country, to geology of Balkan Peninsula, Alpine Chain and wider, in framework of Eastern Mediterranean Chain: Dinarides-Albanides-Hellenides-Taurides.
- Method of definition and evidence of the specific features.
- Natural esthetic view.
- Unique features in framework of tectonic zone in Albania and wider.
- Original features in framework of magmatic massifs and rocks, carbonate shallow water platforms and pelagic basins.
- Typical features or those of classical-didactical character, especially in case of sedimentary geology and stratigraphical sections.
- Special features in a tectonically structural aspect.
- Practical nowadays actuality, climate-curative and recreation values.
- Comparison method with similar geosites in neighboring countries and geological regional structures.

Geological sites of Albania in this book are classified into 11 groups as follows:

(1) Stratigraphical sites; (2) paleoenvironmental sites; (3) paleobotanic sites; (4) magmatic, sedimentary, and metamorphic processes; (5) geological sites of metallogenic-economic character; (6) tectonically structural geosites; (7) geological sites of oceanic and continental dimensions; (8) relationship between tectonic plates; (9) submarine with transgressions and breaks in sedimentations, during different geological periods; (10) geological sites of complex values, where are included geomorphologic, geological, hydrogeology sites; (11) historical sites, some outcrops, and section of continuously discussions and controversies.

The classification of geological sites that advanced a few years later (Serjani et al., 2008) is presented in Table 6.2.

6.3.1 THE CATEGORIZATION OF GEOLOGICAL SITES

For the first time, Wimbledon (1990) has done the categorization of natural monuments of the United Kingdom as follows: The best sites, unique sites, the first sites, and the pattern sites.

In second ProGEO Symposium, in Rome, was treated the idea of the World list of geological sites (Wimbledon et al., 1996), based on the GILGES Project, proposing local, national, and international categories. In Albania,

TABLE 6.2 Classification of the Geological Sites.

After Anon (1995)	After Zogorchev and Trankov (1996)	After Wimbledon et al. (1998)	After Serjani (2014)
1. Paleontologic	1. Mineral outcrops (mineralogic, metallic, and nonmetallic deposits)	1. Stratigraphic	1. Stratigraphic
2. Geomorphologic	2. Water springs	2. Magmatic, metamorphic, sedimentary, petrol., text., and struct., events, provinces	2. Paleoenvironmental
3. Paleoenvironmental	3. Rocks (petrology)	3. Mineralogic–economic (metallogenic–economic)	3. Paleobotanic
4. Magmatic, metamorphic, sedimentary	4. Fossils	4. Structural (tectonic–structural)	4. Magmatic, metamorphic, and sedimentary rocky complexes
5. Stratigraphic	5. Stratigraphic sections	5. Geomorphologic, erosion and relief formation processes	5. Metallogenic–economic
6. Mineralogic	6. Geostructural features	6. Astroproblems, craters	6. Tectonical–structural
7. Structural	7. Complex geosites	7. Geological processes of continental and oceanic sizes relations between tectonic plates	7. Geomorphologic
8. Economic	8. Old quarries and mines	8. Historic for development of geological sciences	8. Astro problems
9. Historic	9. Unique objects and geosites	9. Paleoenvironmental	9. Continental–oceanic sizes geosites
10. Relationship of tectonic plates	10. Geomorphologic features	10. Paleobiologic	9. Relationship of tectonic plates
11. Astroproblems (Astro phenomena)	11. Coal and Fuel deposits		10. Submarine, underwater processes
12. Continental–oceanic of global sizes	12. Geological information		11. Historic geosites
13. Under waters			12. Geoarchaeology and geomythology

we have proposed the following categories of geological sites (Serjani and Heba, 1995): Geological sites of international, national, regional, and local importance.

In the 5th chapter of this book, there are described shortly and illustrated with pictures, regional geological sites of Albania. Albania is the country of the special geological interest, representing the most suitable territory for the observation of the most important geological phenomena, which represent geological sites of regional (international) importance. The definition of the regional geological sites of Albania is done in geological context of Balkan Peninsula, south-eastern Europe, and Mediterranean region. Some unique geological sites, and the first findings, are determined much wider, in the European context.

The general geological features of the geology of Albania are conditioned by its geographical position in Alpine, Mediterranean Belt, and by the evolution during past geological periods, together with neighboring countries, with which it has shared the continuation of tectogenesis, structures, paleogeography of sedimentary sequences, magmatic and metamorphic processes. In geological aspect, Albania is part of the folded "Mediterranean Alpine Chain."

The oldest deposits in Albania are metamorphic rocks of Ordovician–Silurian age in the Korabi (Pelagonian) zone. The region during these old periods it was yet part of the Gondwanan continent. Permian sequences that coincide in time with the division of Pangea by the Hercynian Appalachian System, outcrop in the Korabi and Albanian Alps zones. Traces of Hercynian Tectogenesis during late Paleozoic are found in the Korabi and Gashi zones.

During Late Triassic and Lower Jurassic, Albanian territory together with neighboring countries were located just at the division contact of Pangea, between two orogenic belts. During that period, the thick, limestone sequences in so-called Tethys Ocean were laid down, forming the basis for the modern-day Mediterranean.

During these geological periods, appeared the Cimmerian Orogeny, with the Mirdita Ophiolite, which continued up to the Early Cretaceous. The Mirdita ophiolite is of classical stratification. Mantel harzburgite–dunite of the Eastern Belt is rich in chromite ore, where it is prospected and explorated the Bulqiza chromite deposit, a unique one concerning the morphology of a folded ore body. In the framework of the above geological evolution, in the Mirdita ophiolite zone there are determined some regional geosites such as: Gzhiq-Shenpal-Shenmeri of Mirdita oceanic spreading with pillow lavas, Gjegjan-Kalimash-Runa petrologic sequences of ultrabasic and basic rocks, one of the best sections of ophiolites amongst the Mediterranean and Alpine chains from the Pyrenees to Pamirs and Himalayas (Serjani, 1967); Bregu Bibes as a typically paragenetical association of PGE and Fe-Ni-Cu-Cr, Kaçinar-Munella-Domgjon sheeted dyke complex, which represents rare phenomena of worldwide importance, Derveni eclogites, which can be compared with similar outcrops in Algeria and Marocco, Gziqi-Shenpal site, a very good outcrop of radiolarite beds of oceanic crust, which represents the top (cover) of ophiolite activity.

Alpine tectogenesis, beginning since the Early Neogene, is represented and documented widely in the Inner and External Albanides, in all the territory of Albania. Molasse of the Pre- Adriatic Depression and in inner depressions forming big coal-bearing basins are a rare example all over the Mediterranean. There are a lot of full Molasse structures and sections, representing stratotypes in regional context, with good preservation of fossil beds. Evaporate domes in the Peshkopi region and salt diapirs of the Ionian zone, as rare geological phenomena that were formed during that period as well.

The red outcrops of "Terra Rossa," of Pliocene–Quaternary lake sediments in the Ionian zone, are the best all over the Mediterranean countries. Deep seismogenic, transform faults, and thrusts of tectonic units (described some times as "tectonic nappes") of Albania are of regional, Mediterranean scale. The Dinarides-Albanides-Helenides belt was compressed between the Balkanides east and the Apulian microplate west. The foundation of African foreland outcrops just in Llogara, where is the contact between the African Plate (Adria Microplate) and Orogen, whereas in central and north Albania there are documented some regional overthrust tectonics of inner structures east on external structures west.

Late and new tectonics have played the main role in uplifting and relief formation of Albania. Albanian relief was formed since the Middle-Upper Miocene and continued intensively during the Pliocene and Quaternary. There are distinguished four main geomorphologic cycles:

1. During the early (young) cycle in Albanian Alps, Eastern-Central and Southern mountainous geographical units, as result of glacial, erosion, structural-erosion, and karst processes, the high mountainous chains, ridges, horsts, and depressions were formed, presenting severe, but specific, esthetic geological sites and landscapes.
2. The areas of wide river valleys between mountainous chains belonging to the middle stage or to the maturity stage. River-erosion processes have caused the formation of deep narrow canyons, gorges,

and esthetic valleys, which served as the best places for prehistoric human settlements.
3. The areas in late maturity stage, where belong hilly regions of Pre-Adriatic depression and some wide river valleys with predomination of the erosion–denudation processes forming smooth landscapes.
4. The western field areas of Adriatic coast and lagoons belonging to the last, old geomorphologic stage. Along with Adriatic and Ionian Sea coasts, there are placed Patoku, Karavasta, Narta, Orikumi, and Butrinti lagoons, which represent important ecosystems all over the Mediterranean.

6.4 GEOTOURISM IN ALBANIA

At the beginning, I wrote about Albanian nature and its values, about geoscientific features of geosites. The trend of developing of geotourism is done in detail. Through nature and culture, the documentation of the Earth's history in that territory of the Planet and the documentation of the History of civilization of Albanian people is being done.

Amongst the protection and preservation of picturesque landscapes of esthetic, scientific, teaching values and their usage for study, is the problem of using in general and especially geological sites for complex purposes of tourist developing in our country. The humans as natural being educate into the nature breast. The ancient traditions of people for religious pilgrimage, for gathering, for celebration of natural dates, for organization of collective excursions and picnics in the past were the main activities of recreation and educational character. The organization of such activities serves the aim of making people more sensitive to nature, to the environment, and to the problems linked with them for understanding their relations to the environment and how to be conscientious for active participation in the improvement of environment and protection of geological sites, to understand that the actual ecological crisis is a problem not only of specialists but for all human society in cities, towns, and villages.

The formative education of pupils and students is linked with knowledge of our nature and its numerous geological sites. Teaching, practices, excursions, collecting, examinations of nature must be strongly supported by state organizations, by foreign funds, and by private Albanian companies and firms.

Being a Mediterranean country, with many sunny days, with beautiful nature, with numerous geosites, with native beaches and mountainous climate-curative spots, with natural parks and geoparks, with lakes and cold water springs next to ancient towns, to the archaeological and architectural places, and to the historical monuments, Albania has the guaranteed promises for development of a tourist industry in general and especially of geotourism. Naturally, for development of tourism and geotourism, there is need to solve many problems concerning legislation, infrastructure, management, claim, and publications. As concrete examples must serve the cases of Shkodra, Ohri, and Prespa interborder lakes, proclaimed strict natural parks in Albania and in neighboring countries. Central organs must establish the rules of management and usage. The private initiatives for protection and usage of geoparks and geosites must be supported. The claim of geoparks and natural tourist areas must continue through TV, leaflets, guidebooks in different languages. The main recommended geotours and exploration areas in Albania are as follows (Fig. 6.2):

1. Southern Geotour: Tirana-Berati-Gjirokastra-Saranda. Along with this geotour, there are located hydrocarbon-bearing sedimentary basins, with underwater breaks in sedimentation and transgressions. During this geotour, can be visited UNESCO heritage in Albania: Berati and Gjrokastra old, museum cities and Butrinti archeological site.
2. Northern geotour: Tirana-Kruja-Lezhe-Shkoder-Bajram Curri-Valbona geopark. This geotour crosses sedimentary rocks of Pre-Adriatic depression, Kruja, Krasta-Cukali, and Albanian Alps tectonic zones. Kruja historic town and castle that attained its great fame between 1443 and 1468 when the National hero, George Kastrioti (1405–1468), also known as Skenderbeg, made Kruja his seat. The most interesting and beautiful in this itinerary are Komani structural-tectonical geosite and Valbona geopark at the foots of Jezerca mountain, the highest peak of Albanian Alps.
3. Northeastern Geotour (Mirdita ophiolite): Tirana-Rubik-Burrel-Bulqize-Peshkopi. This geotour is for examination of Mirdita ophiolites, especially Bulqiza ultrabasic massif, where it is prospected and explorated Bulqiza chromite deposit, the unique chromite deposit in Europe, concerning their folded ore body of large sizes.
4. Southeastern geotour: Tirana-Elbasan-Librazhd-Pogradec-Korce. It is a complex itinerary of magmatic and sedimentary complexes of nice landscapes and geosites of geomorphologic character.

Geoheritage and Geotourism in Albania 183

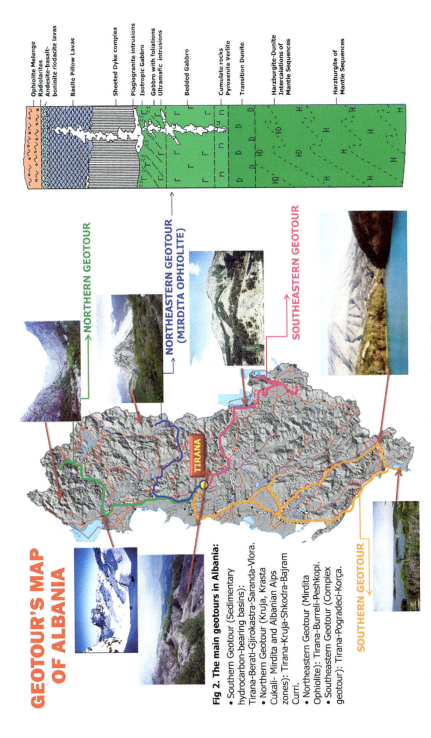

FIGURE 6.2 The main recommended geotours and exploration areas in Albania.
Source: Serjani (2004).

At last, there are recommended, for exploration, some areas and geotours. The purpose of exploration is glacial, karst, and erosion processes in Albanian Alps, Korabi highland, Kurveleshi Plateau. For exploration of river canyons, in upper stream of Drini, Mati, Shkumbini, and Vjosa rivers. For exploration of the seaside lagoons and magmatic geosites in Mirdita ophiolites.

The questions of geotourism, geotours, exploration groups, their organization, camping, and adventures are treated in detail in the Monograph: "Upper Kurveleshi Highland a Museum of Albanian Nature" published in Tirana, by Mediaprint (Serjani, 2009).

6.5 CONCLUSION: ABOUT THE GEOHERITAGE AND GEOTOURISM TRENDS IN THE 21ST CENTURY

About the concepts of geoheritage, geosite, geotourism it is written at the beginning of this century, especially in European countries, in Australia and in North America. Due to the very speedy communication and traveling, many companies in every country are developing enjoyment and rest tourism in the best places of the World. Most people in every country are interested nowadays about our planet and continents. Thus, it is time to put an innovative approach of concepts of geoheritage and geotourism and toward nature conservation. Some global publications do this. Geological sites and geomorphologic landscapes can be grouped according to their values in global, continental, regional, national, and local ones. According to this grouping, it is better to do the organization of geotourist activities, associations, groups, geotrips, and geotourist theory and practice.

During the last 30 years, there are discovered and classified geological sites of different genetic origins. The question is: How many types of geosites are unexplored?

Geomorphologic geosites and landforms constitute the largest group, maybe in every country, and they are used for tourism and geotourism, because they, first of all, reflect esthetic values. But their geological aspect in most cases is not known or not discovered. In the publication by Reynard et al. (2016), it is pointed out that "currently research on geomorphosites is developing in the mythological issues and on specific geomorphologic context of mountain environments."

Hydrogeological geosites, karst springs, and basins commonly are presented and described, but not fully explained. Every landscape has its geological background. For example the beautiful beaches used widely for tourism and summer rest have their geological genesis such as: Panalu'u Black Beach formed by black volcanic sand, Papakolea Green Beach in Gway Island, formed by chrysotile crystals, White Nyams Beach in New South Wales, Australia, formed by the most white sands in the Planet, Violet Hills around Pfeifer Beach in Big Sur Vin California rich in granite minerals, or Red Beach Kaihalulu by red sands from Kauki volcano. Geological aspect or background is missed even to the outstanding geological phenomena of the Planet and geological sites in different countries.

Interesting it is the Big Book of Polly Manguel: Bounty books (2013) for the development of global tourism. The author described the most beautiful natural parks and wild places of the world and invited people to visit them. Using it for geotourism, it would need to add and describe geological aspect of (geo) parks.

Global geotourism in our world can be based on the main geological, planetary phenomena that caused the formation of outstanding geosites. As the main geological nets and lines of the planet, as an alternative for global geotourism we can recommend: Tectonic Plates and their contacts, ophiolite belts in the world, metallogenic provinces, sedimentary stratigraphy, circle geological structures, such as "Eye of Sahara," geotectonic relief formation movements, which caused landscapes and geomorphologic sites, etc.

Many tourist activities have developed since the beginning of this century, even from one continent to the other, whereas tourist activities inside countries are going to be commonplace. All activities done in nature and natural monuments belong to geotourism. A lot of geosciences activities, such as international congresses, conferences, and symposiums, commonly accompanied by geotrips, are helping geotourist development. There is a potential for development of geotourism, almost in all countries.

Geological associations, such as ProGEO (The European Association for the conservation of Geological Heritage), and Geological Society of Australia are making many efforts with their studies and activities to intercalate tourism with geotourism. Recently, Ari Brozinski (Finland) has proposed the foundation of The International Association for Geotourism (http://www.aribro.com) with the following aim:

"Our aim is to promote geo and mining heritage sites as tourist attractions, to promote interdisciplinary scientific research and comprehensive protection of nature and mining heritage sites, to promote business activity in geotourism and to grant scholarships for students and young scientists active in the field of nature and mining heritage protection."

ACKNOWLEDGMENTS

First of all, I would like to express my thanks to Bahram Nekouie Sadry, who discovered my achievements in the field of geotourism and invited me to compile a chapter of this book. It was a pleasure for me receiving, on 27 August 2017, the invitation by editor to contribute to the compilation of this part of the book. My contacts with Dr. Sadry began about 10 years ago, after the publication of my book "Geoheritage and Geotourism in Albania."

This message, for preparation of new book on geotourism, was for me, an inspiration and nice surprise, because of some important reasons that I mentioned in the chapter. Finally, thanks to Dr. Susan Turner (Brisbane) and Wimbledon Bill for help with English.

KEYWORDS

- Albania
- geodiversity
- geosite
- geotourism
- geoheritage

REFERENCES

Amrikazemi, A. *Geotourism-Kazemi*; IGC: Florence, Italy, 2004.
Asrat, A.; Demissie, M.; Mogessie, A. *GeoTourism in Ethiopia*; Shama Books: Ethiopia, 2009.
Barretino, D., Wimbledon, W. P., Gallego, E. *Geological Heritage: Its Conservation and Management*; Eds.; Instituto Tecnologico Geominero de Espana: Madrid, 2000.
Beck, L.; Cable, T. T. *Interpretation for the 21st Century: Fifteen Guiding Principles for Interpreting Nature and Culture*; Sagamore Publishing: USA, 1998; p 242.
Dowling, R. K.; Newsome, D. Ecotourism's Issues and Challenges. In *Geotourism*; Dowling, R., Newsome, D., Eds.; Elsevier/Heinemann Publishers: Oxford, UK, 2006; pp 242–254.
Dowling, R. K.; Newsome, D., Eds.; *Handbook of Geotourism*; Edward Elgar Publishing: Cheltenham, Gloucestershire, 2018.
Erfurt-cooper, P.; Cooper, M. *Volcano and Geothermal Tourism: Sustainable Geo-Resources for Leisure and Recreation*; 2010.
Farsani, N. T.; Coelho, C.; Costa, C.; Neto de Carvalho, C., Eds.; *Geoparks and Geotourism: New Approaches to Sustainability for the 21st Century*; Brown Walker Press: Boca Raton, 2012.

Gonggrijp, G. P.; Bockschoten, G. J. Earth Science Conservation: No Science Without Conservation. *Geol. en Mijnbow* **1981**, *60* (3), Wiggers issue, 433–445.

Gonggrijp G. Earth Science Conservation in Europe. Present Activities and Recommended Procedures. In *Planing the Use of the Earth Surface; Lecture Notes In Earth Sciences, No. 42*. Springer Verlag: Berlin, 1992; pp 371–392.

Garofano M. Geotourism. *The Geological Attractions of Italy for Tourists;* 2010; p 179.

Hose, T. A. European *"Geotourism"*. Third International ProGEO Symposium, Madrid, 2000.

Hose, T. A. Geotourism – Appreciating the Deep Time of Landscapes. In *Niche Tourism: Contemporary Issues, Trends and Cases -27-38;* Marina, N., Ed.; Oxford: Burlington, 2005.

Hose, T. A. Geotourism and Interpretation. In *Geotourism*; Dowling, R. K., Newsome, D., Eds.; Elsevier Butterworth-Heinemann: Oxford, Burlington, 2006, 2006; pp 221–241.

Hose, T. A., Eds.; *Geoheritage and Geotourism: A European Perspective*; The Boydell Press: Woodbridge, 2016a.

Hose, T. A., Eds.; *Appreciating Physical Landscapes: Three Hundred Years of Geotourism*; The Geological Society Special Publication: London, 2016b.

Manguel, P. Bounty Books, a Division of Octopus Publishing Group Limited Endeavour House, 189 Shaftesbury Avenue, London, WC2H 8JY, 2013.

Moisiu, L.; et al. Listing of Geological Sites of Albania. In The 5th International ProGEO Symposium, Rab Island, Croatia, Proceedings, Zagreb, 2008.

Newsome, D.; Dowling, R. K. *Geotourism: The Tourism of Geology and Landscape. Global Geotourism Perspectives*; Good fellow Publishers: Oxford, 2010.

Raukas, A.; Bauert, H.; Willman, S.; Puurmann, E.; Ratas, U. Geotourism Highlights of the Saaremaa and Hiiumaa Islands; NGO GEO-Guide Baltoscandia: Tallinn, 2009.

Reynard, E.; Coratza, P.; Hobles, F. Current Research on Geomorphosites. *Geoheritage J* **2016**, *8*, 1–3. DOI 10.1007/s12371-016-0174-3.

Reynard, E.; Coratza, P.; Hobles, F. *Current Research on Geomorphosites*; 2016.

Reynard, E.; Brilha, J., Eds.; *Geoheritage: Assessment, Protection and Management*; Elsevier Inc., Amsterdam, 2018.

Sadry, B. N. *Fundamentals of Geotourism: With Special Emphasis on Iran*; SAMT Publishers: Tehran, 2009 (English summary available online at: http://physio-geo.revues.org/4873?file=1 ; Retrieved: Jan. 01, 2020) (in Persian).

Sadry, B. N., Ed.; *The Geotourism Industry in the 21st Century: The Origin, Principles, and Futuristic Approach*; The Apple Academic Publishers Inc., 2020.

Serjani, A. Stratification of Kukesi Ultrabasic Massif and His Perspective for High Quality Chromite Ores. *Bul. U. T. Seria e Shk. Nat.* Nr. 4, Tirane, 1967; pp 60–71.

Serjani, A.; Heba, G. Natural Monuments in Albania. *Studime gjeografike* Nr. 8, Tirana, 1996.

Serjani, A.; Neziraj, A.; Jozja, N. *Criteria and Methods used for Classification and Selection of the Geological Sites of Albania*, ProGEO Meeting, Tallinn, Lahema National, Park Estonia, June 2–4, 1997, Proceedings, 1997, pp 58–67.

Serjani, A.; Neziraj, A.; Jozja, N. *Preliminary Classification of the Geological Sites of Albania. Contribution to the 8th Congress of the GSG*, Patra, Proceedings of Congress Volume XXXII, Nr. 1, 1998, pp 33–40.

Serjani, A. et al. Geosience Significance and Tourist Values of some "UNESCO/IUGS Geosites and Geoparks" in Albania. Third ProGEO Intern. Symposium, Madrid, Spain, Session 2, 1999; pp 94–101.

Serjani A.; et al. Geological Heritage Conservation and Geotourism in Albania. The Book in Albanian and English versions. *Botim i Shtypshkronjes "Marin Barleti,"* Tirana, 2003.

Serjani, A.; Avxhi, A.; Neziraj, A. Geotourist Albania, The Fifth International Symposium on Eastern Mediterranean Geology, Thessaloniki, Greece. Chatzipetros, A. A., Pavlides, S. B., Eds.; 2004, Vol. 1, pp 419–422.

Serjani A., *Monograph: Upper Kurveleshi Highland a Museum of Albanian Nature*. Media Print: Tirana, 2009.

Turner, S. Saving Planet Earth: What we can learn from Geoparks. In *East Asiaian Geoparks-Vision, Problems and Prospects*. Taiwan East Asia International Geopark Conference, abstract's volume, Lin, J.-C., Ed.; Taipei, 2009, pp 119–121.

Urban, J.; Jerzy G. *Geological Heritage of the OE ewiêtokrzyskie (Holy Cross) Mountains*; Przeglad Geologiczny: Central Poland, 2008; Vol. 56, nr. 8/1, pp 618–628.

Wimbledon, W. A. P.; Anderson, S.; Cleal, C. J.; Cowie, J. W.; Erikstad, L.; Gonggrijp, G. P.; Johansson, C. E.; Karis, L. O.; Suominen, V. Geological World Heritage GEOSITES, a Global Comparative Site Inventory to Enable Prioritisation for Conservation, ProGEO Symposium, 1996, Roma.

Wimbledon W. A. P.; et al. First Attempt at a Geosites Framework For EUROPE-an IUGS Initiative to Support Recognition of World Heritage and European Geodiversity. ProGEO'98 Meeting, Belogradchik, Bulgaria (Published in Geologica Balcaniaca, 28.3-4), 1998; pp 5–32.

Wimbledon, W. A. P.; Smith-Mayer, S., Eds.; *Geoheritage in Europe and Its Conservation*; ProGEO: Oslo, 2012; p 405.

CHAPTER 7

Establishing an Appropriate Methodology for the Management of Geological Heritage for Geotourism Development in the Azores UNESCO Global Geopark

EVA ALMEIDA LIMA[1*] and MARISA MACHADO[2]

[1]*Department of Geosciences, University of the Azores, Apartado 1422, 9501-801 Ponta Delgada, Portugal*

[2]*Azores UNESCO Global Geopark, Centro Empresas da Horta, Rua do Pasteleiro s/n; 9900-069 Horta, Portugal*

*Corresponding author. E-mail: eva.mc.lima@uac.pt

ABSTRACT

The Azores archipelago is distinguished by its volcanic origin, and has a very rich and remarkable geodiversity that includes a variety of landscapes filled with innumerable craters, volcanic lakes, cliffs and calderas, fumaroles, hot springs, volcanic caves, fault scarps, and marine fossil deposits. The recognition of the value and relevance of its geological heritage resulted in the integration of the archipelago in the Global and European Geoparks Networks and in the UNESCO Global Geoparks program. The volcanic landscape is the main focus of interest, and development of geotourism in the archipelago presents a wide range of possibilities for sustainable use, where several activities can be performed and associated tourist products developed with the creation of the Geopark. The main challenge to this new form of tourism is to keep a balance between the enjoyment in the geolandscapes, geological heritage, and geoconservation.

7.1 INTRODUCTION: THE AZORES ARCHIPELAGO

The Azores archipelago, located in the North Atlantic Ocean, is a Portuguese autonomous region composed of nine islands, several islets, and the surrounding seafloor, of volcanic origin.

It is characterized by the small size of the islands (between 17 and 745 sq km), by its dispersion (distributed along approximately 600 km) and for its distance from the European and American continents (1815 km from Portugal mainland and 2625 km from Canada) being a geostrategic bridge between these two continents.

The islands are divided into three groups: The Western Group with Flores and Corvo islands; the Central Group with Terceira, Graciosa, São Jorge, Pico, and Faial islands; and the Eastern Group formed by São Miguel and Santa Maria islands. All the islands are inhabited with a population of 245,283 inhabitants (SREA, 2016) and totaling 2324 sq. km of land surface, being administratively formed by 19 municipalities and 156 parishes.

The Azores geodiversity presents elements closely linked to the dynamics of Planet Earth, in particular, the volcanism and geotectonics of this Atlantic region, with the archipelago as a natural laboratory of volcanic geodiversity. Its morphology is characterized by different types of volcanoes, hydrothermal fields, volcanic ridges, volcanic lakes, black sand beaches and volcanic caves, among others (Lima, 2009).

7.2 THE AZORES UNESCO GLOBAL GEOPARK

The international relevance of the Azorean geodiversity, the high number and quality of its geosites, and the undoubted importance of its geological heritage (Brilha et al., 2005; Lima, 2007; Nunes et al., 2011; Nunes and Lima, 2013), together with a rich biological and cultural heritage, supported a major effort of the Azores Government to implement geoconservation and environmental education policies, which resulted in the creation of the Azores Geopark (Lima et al., 2010; Nunes et al., 2011). The Azores Geopark was the first archipelagic global geopark, being integrated in the European Geoparks Network and Global Geoparks Network in March 2013, and in November 2015 in the UNESCO Global Geoparks program. It is based on a network of 121 geosites spread across the nine islands and the surrounding seafloor:

i) that ensures representativeness of the geodiversity that characterizes the Azorean territory,

ii) that reflects its geological and eruptive histories of about 10 million years,
iii) with common conservation and promotion strategies, and
iv) based on a decentralized management structure with support in all the islands. Among those there are six geosites of international relevance: the Mid-Atlantic Ridge (located between Faial and Flores islands) and associated deep-sea hydrothermal fields, the Furnas volcano caldera (São Miguel island), the Pico mountain volcano (Pico island), the caldera and Furna do Enxofre volcanic cave (Graciosa island), the Capelinhos volcano (Faial island), and the Algar do Carvão volcanic pit (Terceira island) (Nunes et al., 2011).

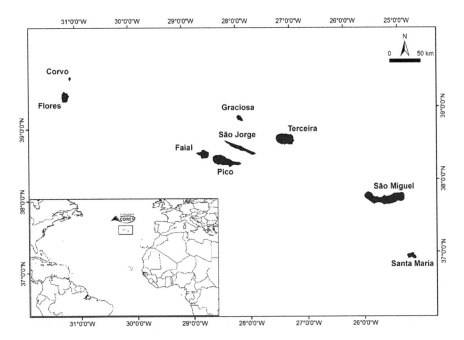

FIGURE 7.1 Setting of the archipelago of the Azores.

7.2.1 AZOREAN CULTURE MAN AND VOLCANOES

Since the settlement, in the XV century, the Azorean people have a strong connection with "their" volcanoes, living through several volcanic eruptions and experiencing earthquakes.

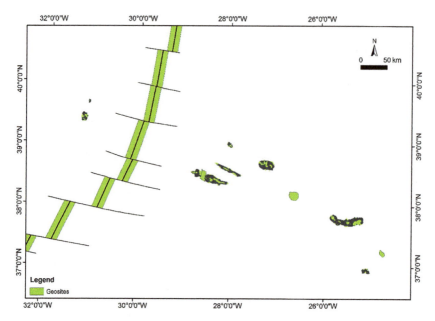

FIGURE 7.2 The Azores UNESCO Global Geopark and its 121 geosites.

Given the strong catholic faith of the inhabitants and lack of scientific knowledge about these catastrophic natural events, the Azoreans created religious events in order to "calm down the wrath of God," such as processions and pilgrimages. In the architecture, it is possible to see old manor houses, monasteries, churches, and fortresses built with the rocks existing on each island. This relationship is also imprinted on the stone walls that divide some rural terrains, the enjoyment of thermal baths, the gastronomy (namely the famous Azorean dish "Cozido das Furnas," cooked on the geothermal soil), or different types of mineral and thermal waters with therapeutic properties commonly used by the population, and also in the literature and music (e.g., the famous poem of José Ferreira "ilhas de Bruma/ Islands of Mystery" (Viveiros et al., 2012).

7.3 GEOCONSERVATION STRATEGY OF THE AZORES UNESCO GLOBAL GEOPARK

Since 2009, a geoconservation strategy has been implemented in the archipelago, which is based on a working methodology that systematizes the tasks for conservation of the geological heritage (Brilha, 2005). So the regional strategy is based on the following steps (Lima et al., 2014):

- Inventory and characterization: 121 geosites were characterized and analyzed. In this process, during the analysis and characterization of the geosites, 93 geosites were distinguished for tourist use; but in a new review carried out in 2017, nowadays there are 105 geosites for tourist use. These geosites are well equipped with infrastructure to support tourism and leisure activities; some also offer nature and outdoor activities, which allows visitors to take advantage of their morphological, ecological, and landscape characteristics;
- Quantification of the value or relevance: based on the methodology from Lima (2007);
- Classification: most of the geosites that were justified to have protection were already protected areas; however, during the revision of the regional protected area were classified and reclassified some geosites;
- Conservation: a geoconservation strategy should give concrete and practical answers to a preliminary assessment on the threats that may relate to the geosites—maintenance actions in the geosites;
- Valorization and promotion: public actions of information and interpretation about the value of the geosites and the geological heritage in general;
- Monitoring: verification and analysis of the evolution of the geosites conservation, to ensure the maintenance of its value and relevance. Since 2014, the monitoring of terrestrial geosites is done every month, and among other parameters, are analyzing the number of visitors, the activities they develop, whether they respect the rules for visiting, the conditions of the place, the state of conservation of the geological characteristics of interest.

There is currently a scientific project in progress, called TURGEO, "Definition of carrying capacity for touristic use of geosites: a tool for the sustainability and tourism valuing of the natural resources of the Azores." The project also aims to improve the process of decision making in the management of geosites.

7.4 GEOTOURISM IN THE AZORES UNESCO GLOBAL GEOPARK

The Azorean landscapes are promoted from the very beginning, through tourism campaigns at national and international levels. However, since the creation of the Azores Geopark, there has been a spread of information and promotion, so the Azoreans can recognize the value and importance of their geological heritage often used in the daily activities, being very important to the development of geotourism (Lima et al., 2013; Lima and Machado, 2018).

Tourism is one of the sectors with a high development potential in the Azores. Traditionally, the visitors of this region are looking for the volcanic landscapes and the surrounding sea for mere contemplation, enjoyment, or for the practice of different activities in the unique natural environment that the Azores offer. The nature tourism is the product defined as strategic for the Azores within the "National Strategic Plan for Tourism," is a major one of the main tourist products of the Region and is the main component of the Azorean tourist industry (Nune et al., 2011; Lima et al., 2013). Geotourism is defined as priority touristic product, being the only considered anchor for the development in the nine islands in the recent Strategic and Marketing Plan of the Azores Tourism (IPDT, 2016). The tourism developed in the Azores is tourism of experiences. The volcanic landscape presents a wide range of possibilities for sustainable use, where several activities can be performed and associated tourist products are developed (Machado et al., 2013; Lima et al., 2013; Lima and Machado, 2018). There are several activities and infrastructures available in the archipelago (Viveiros et al., 2012):

1. Walking trails, around 100 walks classified with different types of difficulty levels;
2. Volcanic caves open to the public;
3. Coastal geosite with Bathing zones, such as natural swimming pools on the volcanic rocks, hydrothermal zones, and volcanic craters;
4. Belvederes with outstanding views;
5. Boat trips to observe the coastline and the islets, complementing the whale watching or big game fishing;
6. Hydrothermal experiences related with health and wellness tourism, taking advantage of baths in thermal waters of recognized therapeutic properties, the intake of carbonated and mineral waters, and the use of mud as Peloids, used traditionally since the 17th century;
7. Interpretative centers and museums dedicated to environmental or cultural themes;
8. Panoramic places to nature photography;
9. Ruins that are testimonies of the occurrence of geological phenomena, that can be visited, such as volcanic eruptions and earthquakes;
10. Outdoor activities like speleology, climbing, canoeing, coasteering, snorkeling, diving, geocaching, among many others;
11. Several inclusive activities and experiences, adapted and accessible to all the people according to their conditions and needs.

Establishing an Appropriate Methodology

FIGURE 7.3   Visit to the interpretative centers.

FIGURE 7.4   Boat trips.

FIGURE 7.5 Walking trails.

FIGURE 7.6 Visit to belvederes.

7.5 THE AZORES GEOPARK STRATEGY FOR GEOTOURISM AND SUSTAINABLE DEVELOPMENT

In the last decade, tourism has proved to be an economic activity with great potential in the archipelago, which is strongly focused on providing emotions and experiences to visitors mostly taking advantage of the volcanic landscape, the main promotional icon of the archipelago, which has an undeniable geotouristic potential that can be better exploited with the creation of the Azores Geopark, so the Azores Tourism Promotion Board, the association responsible for the promoting the Azores as a tourist destination and for qualification of the tourism offered in the Region, defined the Azores Geopark as responsible for establishing the regional strategy for the development of geotourism (Nunes et al., 2014; Lima and Machado, 2018). The geotourism strategy in the Azores UNESCO Global Geopark is supported on the exploitation, maximization, and organization of the existing services and tourist infrastructures, taking profit of the available resources and enhancing joint synergies. For this purpose, partnerships with several stakeholders have been set up, with tourism companies, with interpretation and visitors centers, with the Azores Promotion Board and among others (Nunes et al., 2011). These partnerships are a relevant tool on the management of geotourism in the region and the implementation of policies of conservation that ensure promotion of the geotourism as a tourism product of outstanding quality and reputation (Nunes et al., 2014). In the recent years, the geopark has developed several activities and support materials, with the aim of promotion and development of geotourism (Viveiros et al., 2012; Lima et al., 2013; Machado et al., 2013; Machado et al., 2015):

1. "Geosites Charts" per island, with areas and information about each geosite and also about some support facilities in the island;
2. "Azores geosites leaflet," with a simple map, identification, and photos of the geosites of the archipelago;
3. the "Geotourism interactive brochure of the Azores" available at www.visitazores.com, that include suggestions of places to visit, leisure, and interpretation activities, and other useful information, to support travelers and tourists that visit the islands, but also to assist tourism companies to promote their products and services;
4. "Thematic Routes," including Volcanic Caves Route to "discover the subterranean world of the islands," through the volcanic caves open

to the public; the Belvederes Route to "discover, by car, the Azores volcanic landscapes"; the Walking Trails to "discover, by foot, the Azores geosites"; the Science and Interpretation Centers Route to "learn and interpret the Azorean volcanic phenomena," valuing the science, interpretation, and visitors centers; the Thermal Route to "discover and enjoy the power of the Azores volcanism"; the Urban Routes to "discover the geodiversity present in my village/town"; and the Litoral Routes "to discover the coastal landscapes";

5. "Geotourism courses" for the touristic companies' staff (training and refresher courses);
6. "Geotourism workshops" targeted to students of tourism technical courses and degrees;
7. the "Azores Geopark Passport" to promote the thematic routes and traveling between the islands and the partners of the geopark;
8. development of original geo-products like the "bomb biscuit" developed by a local pastry-maker or baker, the "Queijo do Morro" (a cheese produced nearby the geosite Morro do CasteloBranco, in Faial island), and the "Queijo do Vale" (also a cheese produced in the geositeFurnas Volcano caldera, in São Miguel island);
9. wood posts at sites of special geological interest (see Fig. 7.10);
10. interpretative geological panels on the terrestrial geosites of international relevance;
11. "photo-panoramic panels" in iconic Azorean geo-landscapes;
12. information panels in some geosites;
13. participation in regional Tourism plans, "Strategic and Marketing Plan of the Azores Tourism";
14. and participation in the touristic fairs and national and international events.

7.5.1 NEW TARGETS FROM THE AZORES UGG FOR THE DEVELOPMENT OF GEOTOURISM: GEOCACHING

Geocaching is an adventure game, a treasure hunting that promotes the interaction with nature and local culture, through the search for caches or even earthcaches/landscapes. This game has a strong environmental component, which seeks to promote respect for the environment.

The Geocaching, in the Azores UGG territory is assumed as a "fun" tool to explore the sites; it is also assumed as a geotouristic and nature tourism activity with huge potential. This activity allows the geocachers

Establishing an Appropriate Methodology 199

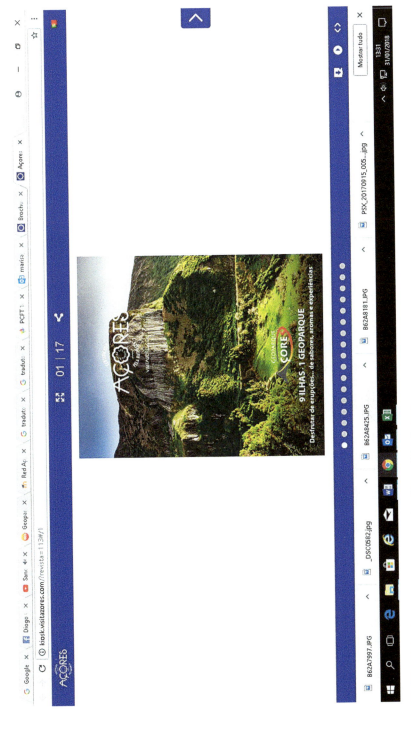

FIGURE 7.7 Geotourism interactive brochure of the Azores.

to explore volcanoes, calderas, lakes, volcanic caves, fault scarps, and many other geolandscaps and cultural landscapes, as well as to experience the traditions that persist today in the daily life of the Azorean people. It is important to mention that through practice of geocaching, the geocachers can practice other geotouristic activities to find the caches, such as walking trails, visit the viewpoints, dive in the Azorean sea, visit museums, discover historical monuments, practice canyoning or climbing, among many others.

FIGURE 7.8 Geotourism courses.

The Azores UGG is partnership of the "GeoTour Azores" a project of the Regional Government, with geocaches for all the islands, the main goal is to make people discover all the archipelago surrounded by the Azorean nature and culture.

The Azores geopark is creating its own geocaches, to promote the geodiversity of the region promoting also geoconservation of the geosites. These adventures are so enriched by the geodiversity and by the emotions that are provided in the Azores discovery.

Establishing an Appropriate Methodology 201

FIGURE 7.9   Geoproducts "Queijo do Morro."

7.5.2 NEW TARGETS FROM THE AZORES UGG FOR THE DEVELOPMENT OF THE GEOTOURISM: GEOPARK ACCESSIBLE TO ALL

The Azores geopark was aimed at creating several activities and materials that can be adapted to any person, depending on their needs, a very general term which embraces persons with disabilities, old people, and persons that are generally excluded from many activities in society (Marisa et al., 2016).

FIGURE 7.10 Info wood post at site of geological interest.

FIGURE 7.11 Interpretative geological panels on the terrestrial geosites of international relevance.

FIGURE 7.12 Participation in the touristic fairs.

This program include: (1) the study and evaluation of the needs of communities in the Azores Geopark; (2) the creation of adaptive experimental activities, such as providing information in different formats to address the unique needs of each visitor (e.g., Braille and audio reading materials); (3) creation of activities that develop all the senses and emotions of the participants; and (4) creation of activities that promote the inclusion of people in society (Marisa et al., 2016).

To facilitate the creation of this program, the geopark has been establishing partnerships with several regional social inclusive organizations, which provides support for former drug addicts and their families, association for persons with visually impairments, and with an inclusive cultural tourism agency, and also realized some experimental inclusive, such as interpretive visits to several geosites, some practical activities with volcanoes to stimulate the senses of the participants, and virtual visit 360° to some geosites to handicapped users.

7.6 FINAL REMARKS

The promotion and development of geotourism is a key issue and the first goal of the Azores Geopark, based on the rich Azorean geodiversity; the management of this geodiversity results from systematic conservation work of the geopark, which also contributes to the development of geotourism and also to sustainable tourism, promoting local economy and minimizing socio-cultural and environmental impacts of tourism.

Thus, integration of the existing services and infrastructures with new interpretative services and products, allow the implementation of a high-quality geotourism in the archipelago, in close relationship to other domains of Nature Tourism.

As a consequence of all these works, the region has been recognized with several national and international prizes and rewards. In 2010, the Sete Cidades Lake and the Volcanic Landscape of Pico Island were considered two of the "7 Natural Wonders of Portugal"; in 2011, the Azores were selected as the second best islands in the world for Sustainable Tourism by National Geographic Traveler; in 2015, 2016, and 2017, the region was distinguished as one of the "100 most sustainable tourism destinations in the world" by Green Destinations Organization.

Come to the Azores and meet the volcanoes, the lakes, the tea plantations, watch whales, dive in the blue ocean, walk around the islands and enjoy the stew of Furnas, and the regional sweets and wines. These are only a few of the many proposals to enjoy in the archipelago.

ACKNOWLEDGMENTS

This work is a contribution to the doctoral research project "Definition of a methodology for the management of geological heritage. An application to the Azores archipelago (Portugal)," Ref. M3.1.2/F/033/2011, supported by the Science Regional Fund of the Azores Government, and co-financed by the European Social Fund through the EU Pro-Employment Program.

KEYWORDS

- geoconservation
- geotourism
- The Azores Global UNESCO Geopark
- the management of geological heritage
- geosite

REFERENCES

Brilha, J.; Andrade, C.; Azerêdo, A.; Barriga, F. J. A. S.; Cachão, M.; Couto, H.; Cunha, P. P.; Crispim, J. A.; Dantas, P.; Duarte, L. V.; Freitas, M. C.; Granja, M. H.; Henriques, M. H.; Henriques, P.; Lopes, L.; Madeira, J.; Matos, J. M. X.; Noronha, F.; Pais, J.; Piçarra, J.; Ramalho, M. M.; Relvas, J. M. R. S.; Ribeiro, A.; Santos, A.; Santos, V.; Terrinha P.; Definition of the Portuguese Frameworks with International Relevance as an Input for the European Geological Heritage Characterisation. *Episodes* **2005**, *28* (3), 177–186.

Lima E. A. Património Geológico dos Açores: Valorização de Locais com Interesse Geológico das Áreas Ambientais, Contributo para o Ordenamento do Território. Master Thesis on Land and Environmental Planning, Azores University, 2007.

Lima, E. A.; Machado, M. Geotourism in the Azores UNESCO Global Geopark, Portugal. In *Handbook of Geotourism*; Dowling, R., Newsome, D., Eds.; Edward Elgar Publishing: Gloucestershire, 2018; pp 320–328.

Lima, E. A.; Machado, M.; Nunes, J. C. Geotourism Development in the Azores Archipelago (Portugal), As an Environmental Awareness Tool. *Czech J. Tour.* **2013**, (3), pp 126–142.

Lima, E. A.; Nunes, J. C.; Costa, M. P.; Porteiro, A. O Geoturismo Como Instrumento de Valorização do Geoparque Açores. In *Geoturismo e Desenvolvimento Local*; Carvalho, C. N., Rodrigues, J., Jacinto, A., Eds.; Câmara Municipal de Idanha-a-Nova/Geoparque NaturTejo: Lda, 2009; pp 149–160.

Lima, E. A.; Nunes, J. C.; Costa, M. P.; Porteiro, A.; Azores Geopark: An Atlantic Geopark. Abstract "4th International UNESCO Conference on Geoparks," Langkawi, Malasia; April, 2009; p 97.

Machado, M; Lima, E. L.; Ponte, J.; Nunes, J. C. Contributions of the Azores Geopark to the Geotourism Development. Communication presented at the VIII Seminário de Recursos Geológicos, Ambientais e Ordenamento do Território, Trás-os-Montes University, Vila Real, Portugal, November 6, 2015.

Machado, M.; Viveiros, C.; Lima, A. F.; Lima, E. A. Contributions of the Azores Geopark to the Geotourism Development in the Archipelago. Communication presented at the 12th European Geoparks Conference, National Park of Cilento, Vallo di Diano e Alburni, Italy, September 4, 2013.

Machado, M.; Lima, E. A.; Costa, M. P. Geopark Accessible To All. Communication presented at the 14th Global Geoparks Conference, English Riviera Geopark, United Kingdom, September 2016.

Machado, M.; Lima, E.; Machado, L.; Silva, C. *Geocaching in the Azores Geopark: A Tool to Explore the Geological Heritage*. In Abstract book, 13th European Geoparks Conference; Katja Saari, Jarkko Saarinen, Mari Saastomoinen, Eds.; Rokua Geopark, Finlândia, 2015; pp 74, ISBN: 978-952-93-5939-4.

Nunes, J.; Analysis of Geosites with Touristic Use of the Azores UNESCO Global Geopark: A Contribution to Enhance Sustainable Tourism Policies, 14th European Geoparks Conference, Azores UNESCO Global Geopark, Abstract book, 2017; p 27.

Nunes, J. C.; Lima E. A. Ranking and Classifying the Azores Islands Geosites. *Rendiconti Online della Società Geologica Italiana*, Special Issue for the 12th European Geoparks Conference, Ascea, Italy, **2013**, (20), 192–196.

Nunes, J. C.; Lima, E. A.; Costa, M. P.; Porteiro, A. Azores Islands Volcanism and Volcanic Landscapes: Its Contribution to the Geotourism and the Azores Geopark Project. *Geosci. Online J.* **2010**, *18* (16), 4.

Nunes, J. C.; Lima, E. A.; Ponte, D.; Costa, M. P.; Castro, R. *Azores Geopark Application to European Geoparks Network*. Azores Geopark, Eds., Horta, Portugal, 2010.

Nunes, J. C.; Machado, M.; Lima, E. A.; Leandro, C.; Castro, R.; Toste, J. The Partnership Azores Geopark/Azores Tourism Promotion Board: Strengthening the Geotourism in the Azores Archipelago. The 6th International UNESCO Conference on Global Geoparks, Stonehammer Geopark, Saint John New Brunswick Canada, Abstract book, 64 and 65; p, 2014.

Viveiros, C.; Lima, E. A.; Nunes, J. C. Geotours and the Development of Rural Areas in the Azores islands'. Proceedings of the 11th European Geoparks Conference, Arouca Geopark, Portugal, Sep. 21, 2012; pp. 291–292.

CHAPTER 8

"Simply the Best": The Search for the World's Top Geotourism Destinations

MURRAY GRAY

*Honorary Professor of Geography, School of Geography,
Queen Mary University of London, Mile EndRoad, London E1 4NS,
United Kingdom*

E-mail: j.m.gray@qmul.ac.uk

ABSTRACT

Several attempts have been made over recent years to list the world's top tourism, nature or geological sites. In this chapter an attempt is made to identify the world's top geotourism destinations using the criteria of visual impact, site quality, educational potential, and reasonable tourist accessibility. An additional requirement is to ensure that the list presents sites that are both internally geodiverse and represents the world's geodiversity as far as this is possible in a limited assessment. The list should also attempt to include sites from all the continents. The proposal is that the world's top 10 geotourism destinations are:

1. Iguaçu/Iguazu Falls (Brazil/Argentina) or Victoria Falls (Zimbabwe/Zambia)
2. Grand Canyon National Park (USA)
3. Great Barrier Reef (Australia)
4. South China Karst to Ha Long Bay (China/Vietnam)
5. Yellowstone National Park (USA) or Hawaii Volcanoes National Park (USA)
6. Uluru (Australia)
7. Central Swiss Alps (Switzerland)
8. Cal Orcko Parque Cretácico (Bolivia)
9. The "Golden Circle" Route (Iceland)
10. Table Mountain, Cape Town (South Africa), or Rio de Janeiro (Brazil)

This proposed list of sites is intended to be the start rather than end of a process that may eventually lead to an internationally endorsed list of top geotourism destinations that can then be used to raise the public profile of geology, geotourism, and geoheritage.

8.1 INTRODUCTION

"...better than all the rest" (Tina Turner, 1989)

A popular current occupation is to compile lists of the public's favorite things, whether it's the "100 best pop songs ever," the "50 best movies of all time," or "favorite boy and girl names." But this is not an entirely recent trend. For example, the list of the Seven Wonders of the Ancient World was created many centuries ago, and according to *Wikipedia*, has been followed up by lists of the "Seven Wonders of the Modern World," "New Seven Wonders," "Seven Natural Wonders of the World," "New7Wonders of the World," "New7Wonders of Nature," "New7Wonders Cities," "Seven Wonders of the Underwater World," "Seven Wonders of the Industrial World" and "Seven Wonders of the Solar System," as well as "Seven Wonders" of individual countries. Tourism and geology have not been immune from this idea and the aims of this chapter are to outline some of these attempts, to propose a provisional best geotourism destinations list, and to suggest that although the process may be subjective and frivolous, it could still be useful in raising the profile of geology, geotourism, and geoheritage.

8.2 ATTEMPTS TO LIST THE TOP TOURISM/NATURE SITES

In 2002, the BBC produced a television program on "50 Places to See Before You Die" based on 20,000 viewer votes. The list includes many cultural sites including Singapore, Las Vegas, and the Great Wall of China, but it also includes natural sites. In fact the Grand Canyon (USA) topped the list and the Great Barrier Reef (Australia) came second. There were also no less than four waterfalls on the list—Angel Falls (Venezuela, at 47), Iguacu Falls (Brazil/Argentina, at 26), Victoria Falls (Zambia/Zimbabwe at 21), and Niagara Falls (USA, at 15). Other geotourism sites on the list included the Matterhorn (Switzerland, at 46), Iceland (at 44), Mount Everest (Nepal, at 30), Alaska (USA, at 28), Yosemite National Park (USA, at 23), Uluru (Australia, at 12), Lake Louise (Canada, at 11), and South Island, New

Zealand (at 4). The list illustrates many of the potential flaws in this type of project. First, the list reflects the travels, activities, preferences, and prejudices of those taking part, in this case British television viewers. Second, the scale of the sites ranges unhelpfully from whole countries (e.g., Iceland) to individual sites (e.g., Abu Simbel, Egypt). And third, how was the voting conducted and could it have been open to manipulation by vested interests?

Starting in 1997, CNN produced an unordered list of the "7 Natural Wonders of the World" as follows:

- Aurora Borealis (various countries)
- Grand Canyon (USA)
- Great Barrier Reef (Australia)
- Harbor at Rio de Janeiro (Brazil)
- Mount Everest (Nepal)
- Paricutin Volcano (Mexico)
- Victoria Falls (Zambia/Zimbabwe)

Apart from the Aurora Borealis, all of these can be considered as geotourism sites. Grand Canyon, Great Barrier Reef, Mount Everest, and Victoria Falls are repeated from the BBC List, which also included Rio de Janeiro as a city location. The Harbor at Rio de Janeiro in this list is presumably meant to reflect the mountainous and coastal setting of the city, but its inscription on the World Heritage List in 2012 was based only on its cultural status, including its artistic traditions (criteria v and vi). It is also worth commenting on the Great Barrier Reef as it is clearly an active ecological site but its massive structure has developed over thousands of years and mainly comprises fossil corals. The main surprise on the CNN list is Paricutin Volcano, 230 km west of Mexico City. First erupting in 1943, its cinder cone built to 425 m over a nine-year period and was the first eruption of this type whose life cycle could be studied by volcanologists. This raises the important point that sites important to geoscientists are not necessarily the most attractive to geotourists. In the case of Paricutin, its form and composition are largely unremarkable and it is rarely included in any other top site listing. It is not designated or protected and visitors are able to climb it and walk around the circular rim.

A challenge to these "7 Natural Wonders" was made between 2007 and 2011 when a list of the "New7Wonders of Nature" was compiled on the initiative of Canadian Bernard Weber and administered by his Swiss-based New7Wonders Foundation. The process included nomination, national qualification, expert evaluation, and 100 million votes cast from around the world for the shortlisted sites. The list, again unordered, has no overlap with the CNN promoted one:

- Amazon Rainforest and River (Brazil et al.)
- Halong Bay (Vietnam)
- Iguaçu/Iguazu National Parks (Brazil/Argentina)
- Jeju Island (South Korea)
- Komodo Island National Park (Indonesia)
- Puerto Princesa Subterranean River (The Philippines)
- Table Mountain National Park (South Africa)

However, the project has been the subject of criticism, including allegations of repetitive voting driven by national pride and the projected economic benefits of increased tourism. Certainly Jeju Island is the surprise on this list, while Komodo Island's fame is largely related to its giant lizards ("Komodo Dragons").

In response to this new list, CNN formed the organization Seven Natural Wonders, in 2008 with the aim of reinforcing their previously announced list of sites. The organization also produced lists of the Seven Natural Wonders for each continent and made the point that there would be no entry fees, sponsorship, marketing, and so on. associated with the project in contrast to the New7Wonders scheme. The winning wonders were to be determined by experts from around the world using "statistical and traditional significance, uniqueness, and pure splendour" (Wikipedia, accessed 4/11/17).

8.3　ATTEMPTS TO LIST THE TOP GEOLOGICAL SITES

In 2015, the US-based *Forbes Science* produced an unordered list of "13 Geologic Wonders of the Natural World" compiled by Dr Trevor Nace, a graduate of Duke University[1]:

- Tibetan Plateau (China)
- The Door to Hell (Turkmenistan)
- Fumaroles, Yellowstone, Wyoming (USA)
- Mount Roraima (Brazil/Guyana/Venezuela)
- Giants Causeway (Northern Ireland)

---
[1] https://www.forbes.com/sites/trevornace/2015/11/16/13-geologic-wonders-of-the-natural-world

- Kilauea Volcano, Hawaii (USA)
- The Blue Hole (Belize)
- Antelope Canyon, Arizona (USA)
- Cretaceous–Palaeogene (K/Pg) boundary (The Netherlands)
- Crystal Cave (Mexico)
- Travertine Deposits, Pamukkale (Turkey)
- Salar de Uyuni (Bolivia)
- Grand Canyon, Arizona (USA)

The most interesting inclusion on this list is the Cretaceous-Palaeogene boundary at Geulhem in the Netherlands, representing the largest worldwide extinction event 66 million years ago. Dr Dace remarks that at this site "you can put your finger on the remnants of three quarters of the world's species. It's an incredible lesson in geology and our earth's history in that so much can happen in a few centimeters of sediment." However, this is not the official GSSP boundary for this event, which is at Oued Djerfane in Tunisia. The Door to Hell in Turkmenistan is a burning crater initiated by natural gas drilling, followed by ground collapse and ignition of the escaping gases, so it does not really count as a natural site.

The UK-based Rough Guides produced an ordered list of "20 Geological Wonders of the World"[2] in 2016:

1. Danakil Depression (Ethiopia)
2. Cal Orcko (Bolivia)
3. Parc National de l'Ankarana (Madagascar)
4. Colca Canyon (Peru)
5. Underground River, Palawan (The Philippines)
6. Mingsha singing sand dunes (China)
7. Mount Roraima (Brazil/Guyana/Venezuela)
8. Knockan Crag (Scotland)
9. Thrihnukagigar Volcano (Iceland)
10. Kilauea Volcano, Hawaii (USA)
11. The Blue Hole (Belize)
12. Antelope Canyon, Arizona (USA)
13. Ha Long Bay (Vietnam)
14. Nitmiluk Gorge (Australia)
15. Brighstone Bay, Isle of Wight (England)
16. Sperrgibiet National Park (Namibia)

---

[2]https://www.roughguides.com/gallery/20-geological-wonders-of-the-world/

17. Pulpit Rock (Norway)
18. Gold Reef City, Johannesburg (South Africa)
19. Giant Crystal Caves (Mexico)
20. Meteor Crater, Arizona (USA)

This is a strangely inconsistent list, with five sites in common with the *Forbes Science* list, and some repeats from other lists (e.g., Ha Long Bay) but also some sites that are of only local interest (e.g., Nitmiluk Gorge in Australia). The list also raises another important issue because although it puts the Danakil Depression at the top its list, it also warns visitors to "Check… before travelling as this is a geologically and politically volatile area." Therefore not all sites included in some lists are necessarily easily or safely accessible by tourists.

In the United Kingdom for several years up to 2018, the *Complete University Guide* has included an ordered list of the "Top 10 Geological Wonders of the World"[3]:

1. Antelope Canyon, Arizona (USA)
2. Cave of Crystals (Mexico)
3. Great Blue Hole (Belize)
4. Grand Canyon, Arizona (USA)
5. Pamukkale (Turkey)
6. Mount Roraima (Brazil/Guyana/Venezuela)
7. Reed Flute Cave (China)
8. Door to Hell (Turkmenistan)
9. Chocolate Hills (The Philippines)
10. The Eye of the Sahara (Mauritania)

This list repeats several sites from other lists though it also contains three new ones at positions 7, 9, and 10.

There have also been attempts to list top sites for individual continents or countries. For example, in 2009, the *Smithsonian* magazine produced an ordered list of the "10 most spectacular geologic sites in the United States"[4]:

1. Grand Canyon, Arizona
2. Yellowstone National Park, Wyoming, Idaho, Montana

---

[3]https://www.thecompleteuniversityguide.co.uk/courses/geology/top-10-geological-wonders-of-the-world/
[4]https://www.smithsonianmag.com/science-nature/the-ten-most-spectacular-geologic-sites-38476122/?q=

3. Niagara Falls, New York
4. Meteor Crater, Arizona
5. Mount St Helens National Volcanic Monument, Washington
6. La Brea Tar Pits, California
7. San Andreas Fault at the Carrizo Plain, California
8. Mammoth Cave National Park, Kentucky
9. Ice Age Flood Trail, Washington, Oregon, Idaho
10. Lava Beds National Monument, California

Although the content and ordering could be debated, this looks like a very reasonable listing of the top US geotourism sites. Also in the United States, Albert Dickas has produced two books listing "101 American geosites you've gotta see" (2012) and "101 American fossil sites you've gotta see" (2018). The sites are listed alphabetically by state and the latter includes a site from every US state.

Anhaeusser et al. (2016) produced a book on *Africa's Top Geological Sites* that describes 44 sites mainly in eastern and southern Africa including Victoria Falls, Table Mountain, and the Danakil Depression. However, there is no indication of how the sites were selected, and there are certainly some notable omissions such as the World Heritage Site at Wadi Al-Hitan in Egypt famous for its whale fossils, the two GSSPs in Morocco, and the West Coast Fossil Park in South Africa's Western Cape. However, this last site is one of those featured in Whitfield's (2015) book *50 Must-See Geological Sites in South Africa,* which also features Table Mountain, The Vredfort Dome and the Cullinan Diamond Mine.

The Scottish Geodiversity Forum has prepared a list of the "51 best places to see Scotland's geology"[5]. It includes several sites famous for the historical development of ideas/concepts in geology, including Knockan Crag (thrusts), Siccar Point (unconformities), and Glen Roy (glacier-dammed lakes).

In 2014, the Geological Society of London invited nominations for "100 Great Geosites" in the United Kingdom and Ireland.[6] Over 400 sites were nominated and these were supplemented by sites suggested by an expert group. The public were then invited to vote for their favorite sites and 1500 votes were received. The 100 most popular sites were placed into 10 groups of sites, the categories being: landscape, industrial and economic, historical and scientific importance, educational, adventurous, human habitation, coastal, outcrops, folding and faulting, fire and ice. The

---
[5] www.scottishgeology.com
[6] https://www.geolsoc.org.uk/100geosites

selected sites were published and publicized in October 2014 and led to the largest ever media interest in a Geological Society of London topic with several UK newspapers publishing the story.

8.4 DISCUSSION AND PROPOSALS

This last point is important given the low profile of geoheritage and geotourism in most countries. So there is a case for the geoscience community to agree and publicize a list of what it considers to be the world's top geotourism destinations. Some will see this as a subjective and frivolous exercise given that no one has visited all potential sites and we all have our own favorites, but the need to raise the profile of geology, geotourism, and geoheritage might overcome such reservations and justify the exercise.

Of course, geoscientists have been assessing and selecting scientifically important sites for geoconservation purposes in many countries for decades, including the use of inventorying methods (e.g., Brilha, 2016). There have also been attempts to select scientific "Global Geosites" (Cleal et al., 2001), though this approach has effectively stalled except in some European countries (e.g., Garcia-Cortés et al., 2009 in Spain). Other scientifically important geosites are included on the UNESCO World Heritage List, or within the UNESCO Global Geoparks Network, or as the IUGS's Global Stratotype Sections and Points (GSSPs). But several of these scientifically important geosites have fairly limited geotourism potential. In fact, there have been few attempts by geoscientists to compile a list of the world's top geotourism sites, yet they are uniquely qualified to create such a list. Contrast that with the many surveys of top/best ecotourism sites that can be found on the Internet.

So what are my top ten geotourism sites? First, we need to know what constitutes a geotourism site. Hose (1995) originally defined geotourism as "The provision of interpretive and service facilities to enable tourists to acquire knowledge and understanding of the geology and geomorphology of a site...beyond the level of mere esthetic appreciation." Nowadays, tourists gain information through many, diverse, media sources, and most geosites promote geological/geomorphological understanding of their area through these sources, even if there is little interpretive provision on the ground. It is therefore difficult to know what level of geological knowledge and understanding that visitors to a geosite have, or whether they have merely an "esthetic appreciation." In these circumstances, we should not be overconcerned with the details of visitor geological understanding.

Second, and related to the above, there need to be some criteria for selecting the top geotourism sites and these ought to include visual impact, site quality, educational potential, and reasonable tourist accessibility/facilities. Two other factors are suggested as important. First, there should be a reasonably wide distribution of sites from across the globe, rather than many sites being from a single continent. Second, I'd also suggest that geodiversity (Gray, 2013) should be a factor, both within individual sites and for the list overall, for example, a list of 10 geodestinations should not include the four waterfall sites on the BBC list.

But it should include one, and my favorite is the Iguaçu/Iguazu Falls on the Brazil/Argentine border and lying within two National Parks. This is a spectacular site that includes 275 individual falls when at mid-level discharge and was inscribed onto the World Heritage List in 1984. The National Parks on both sides of the river/border are easily accessible and within the parks themselves accessibility to the falls is via several trails. There are even some raised walkways leading out over the water where falls can be observed from very close quarters. There are also helicopters and boat trips, while the Argentine side has a short train ride. Overall, therefore, it meets the visual impact, accessibility, and internal geodiversity criteria. Where it falls down is on educational value, since there is very little geological interpretation at the site itself though this is available on the Internet. Three geological interpretive panels were placed on the site by the Parana Geological Survey in the 2000s (Moreira, 2012) but strangely these have now been removed. The diversity of the falls is due to the width and shallowness of the Iguaçu River, the position of the falls on a tight meander bend and the horizontality of the lava flow forming the falls. The water on the inside (Brazil) and particularly on the outside of the meander (Argentina) misses the main falls at the "Devil's Throat" part of the channel and instead emerges as individual waterfalls over a distance of almost 3 km (Fig. 8.1). There are two main steps within the falls representing the two main lava flows in the area. It is not the highest fall in the world or the one with the highest discharge but it is the biggest waterfall *system* in the world. The main alternatives are the Niagara Falls (USA) and Victoria Falls (Zambia/Zimbabwe) both of which are also UNESCO World Heritage Sites. On seeing the Iguaçu/Iguazu Falls, US First Lady Eleanor Roosevelt is reputed to have said "poor Niagara," since the latter has only three main falls. However, the Victoria Falls are impressive and will be the preferred waterfall site by some assessors. The reason for placing waterfalls first on the list of geotourism sites is that they are dominated by water movement. It often seems as though geoheritage loses out to bio-conservation in public attitudes for the major reason that animals move whereas geosites are usually static. But at waterfalls water movement dominates the landscape and the public love it.

FIGURE 8.1 An example of waterfall diversity at Iguaçu/Iguazu Falls, Brazil/Argentina.

Second on my list is the Grand Canyon (USA). This is also an easily accessed site and within the National Park there is easy access along the canyon rims and into the canyon via several trails. There are educational talks by rangers, including geological ones and considerable geological interpretation in the visitor centers. As well as being a National Park it is also a World Heritage Site. It certainly presents a spectacular visual impact as a canyon up to 30 km wide, 1500 m deep and 450 km long. It has been formed over the last 6 million years by down cutting erosion by the Colorado River and its tributaries as the Colorado Plateau uplifted. If the canyon had been eroded into a single rock type it would still be an impressive feature, but a major attraction of the site is its internal stratigraphic and morphological diversity. The horizontal strata are of varying resistances resulting in a stepped vertical profile, while there is also longitudinal morphological diversity (see Fig. 8.2). It is one of the most visited tourism sites in the world, and meets all the selection criteria.

"Simply the Best": The Search for the World's Top Geotourism Destinations 217

FIGURE 8.2 Geotourists admiring the internal geodiversity of the Grand Canyon, the United States.

Third on my list is the Great Barrier Reef, and is another feature that demonstrates significant internal geodiversity since it is composed of nearly 3000 individual reefs and almost 900 islands stretching over a distance of 2300 km. It is believed that the reef started to grow 20,000 years ago when world sea level was 120 m lower during the Last Glacial Maximum. As sea levels rose, coral growth was able to keep pace and has created the massive structure we see today. The reef is a World Heritage Site and part is protected as a Marine Park, but these labels are not preventing coral bleaching brought about by warming ocean waters and other factors. Accessibility is good with several resorts on islands on the reef, boat trips (including glass-bottomed boats), and cruises, as well as helicopter flights. The Great Barrier Reef Foundation attempts to educate the public about the reef and fund raise to protect it. The only reservation about including this site on the list is that many visitors are more interested in the reef ecology rather than its geology/geomorphology.

There ought to be a karst site on the list and there is no shortage of potential candidates—for example, Mammoth Cave (USA), Carlsbad Caverns (USA), the Dinaric Karst in Europe, Ha Long Bay (Vietnam), South China Karst (China), Gunung Mulu National Park (Malaysia), and the Puerto Princesa Underground River (The Philippines). The poor accessibility of the last two sites must count against them even though the caves are spectacular. Mammoth Cave and Carlsbad Caverns are huge underground networks of passages and caves but only a limited number are open to the public. This leaves Ha Long Bay and the South China Karst, which in fact together form part of a karst transect stretching for a distance of over 1000 km from the 2000 m high Yunnan-Guizhou Plateau to the lowland plains and river valleys of the eastern Guangxi basin and then on to the drowned limestone tower karst landscapes of northeast Vietnam (Williams, 2015). These sites contain world-class examples of tower karst and cone karst as well as giant dolines, river gorges, cave systems, and the stone forest of Shilin. Together, these sites constitute the world's finest examples of humid tropical and subtropical karst. The South China Karst is a serial World Heritage Site, one part of which is the Guilin tower karst mainly accessible by boat on the River Li (Fig. 8.3). The transect is continued southwards into Vietnam to include the lowland tower karst landscape of Trang An and reaching the coast at Ha Long Bay, both of which are World Heritage Sites. Ha Long Bay represents an area of flooded tower karst with over 1500 islands and is popular with tourist boats and cruise ships. The individual parts of the transect are therefore mainly accessible by boat, and represent a spectacular diversity of tower karst landscapes and caves. However, local interpretation about the formation of these landscapes is sadly lacking.

FIGURE 8.3 Tourist boats on the River Li within the Guilin section of the South China Karst.

A volcanic site also merits being on the list, and again there is no shortage of potential candidates from which to choose, from the Mid-Atlantic Ridge sites of Iceland or the Azores, to the subduction zone sites of Japan or Indonesia, and to the crustal hot-spot sites of Hawaii or Yellowstone National Park. The last two of these are alternatives on my list (but see also Iceland's "Golden Circle" route below). Yellowstone merits inclusion firstly for historical geoheritage reasons—it became the world's first national park in 1872 primarily because of its spectacular geothermal phenomena—geysers, hot springs, mud pots, terraces, and steam vents. The *Yellowstone National Park Act* (1872) states that it was "set apart as a public park or pleasuring ground for the benefit and enjoyment of the people." And it has certainly achieved that aim, becoming one of the most visited geotouristic destinations in the world. Second, it is estimated that there are around 10,000 individual geothermal features in the park giving it an unrivalled diversity of features, including over 300 geysers, the most famous of which is "Old Faithful." This represents the densest concentration of geothermal features anywhere in the world. These features are the result of Yellowstone sitting on top of a volcanic hotspot away from any tectonic plate boundary. Many of the features are easily accessible from roads and trails, and the Park also includes the Falls on the Yellowstone River. The last major eruption of the "supervolcano" was around 640,000 years ago and its caldera measures a staggering 75 × 45 km (Smith and Siegel, 2000). Yellowstone was inscribed onto the UNESCO World Heritage List in 1978 and there is a good level of geological interpretation within the park. The main alternative to Yellowstone is the Hawaii Volcanoes National Park comprising two active volcanoes, Mauna Loa and Kilauea. Intense activity at the latter site in 2018 means that geotourism opportunities here may be spectacular but restricted.

Uluru (formerly Ayers Rock) in central Australia is a 350 m high, steep-sided, bare, arkosic sandstone hill, 9.4 km in circumference. It's vertical bedding and red color, which changes with the lighting conditions, gives it a dramatic appearance. It is a World Heritage Site and forms part of the Uluru-Kata Tjuta National Park. It is one of the most recognized physical features on the planet and receives hundreds of thousands of visitors each year. Uluru, its caves and other features are sacred to the local aboriginal people (Anangu) who have never been happy with visitors climbing the rock, and in 2017 the National Park Board voted unanimously to ban climbing from October 2019. It is still possible to walk the perimeter trail or take a helicopter flight around it.

A mountain site makes sense as does a glacial one and these two elements are combined spectacularly in the Central Swiss Alps. Accessibility is a key here as no other mountain range in the world is as accessible in winter and summer as the Swiss Alps given the large number of ski lifts, cable cars, railways, and funiculars. Of particular note is the Jungfrau railway that terminates at 3454 m at the Jungfraujoch Station, Europe's highest. From there, visitors have a stunning view of the Aletsch Glacier, Europe's largest at 23 km long and can also walk across the ice to the Mönchsjoch Hut and other locations. Not far away at Murren, visitors can take a cable car to the summit of the Schilthorn at 2970 m where, from the *Piz Gloria* revolving restaurant as featured in the James Bond film *On Her Majesty's Secret Service*, they can enjoy views of the three great mountain peaks—Jungfrau (4158 m), Mönch (4107 m), and Eiger (3967 m). The north face of the latter is a significant draw for experienced climbers. But an even higher revolving restaurant at 3457 m occurs at Mittelallalin, reached by cable car and underground funicular from Saas-Fee. From the top, visitors can explore the Ice Pavilion, the world's largest ice grotto. Zermatt, in the neighboring valley, gives spectacular views of the Matterhorn (4478 m) and Monte Rosa (4634 m) on the border with Italy, and the local Gornergrat rack railway gives good access to a number of valley glaciers whose Little Ice Age trim lines are exceedingly clear. In fact the whole area of the Central Swiss mountains and glaciers are a major tourist location for hikers, climbers, or those just wanting to enjoy the spectacular scenery.

A palaeontological site should be on the list and given the public's obsession with dinosaurs, it makes sense to include a dinosaur site. There are several excellent dinosaur exhibits in museums, including the Royal Tyrell Museum of Palaeontology in Alberta, Canada, Field Museum, Chicago, USA, Smithsonian Museum, Washington DC, USA, Museum für Naturkunde, Berlin, Germany, and National Dinosaur Museum, Canberra, Australia. Some Museums or Visitor Centers are built over outcrops with *in situ* dinosaur fossils, including the Zigong Dinosaur Museum in China and the Bone Quarry at Dinosaur National Monument, USA. But the most impressive *in situ* dinosaur site in the world must be the dinosaur footprint site at Cal Orcko near the city of Surce in southern Bolivia. First discovered by quarry workers in 1985, the quarry face now displays over 5000 footprints in 462 trails from 15 dinosaur species and is the largest dinosaur footprint site in the world. One of the most impressive trails is 347 m long and made by a baby *Tyrannosaurus rex* locally known as Johnny Walker. The site is known as Parque Cretácico and has been on Bolivia's World Heritage Tentative List since 2003. The footsteps were made in clay on a former shoreline, but they are very fragile and are subject to weathering. There is a museum on the site from which guided tours for visitors are run around lunchtime when the lighting conditions are at their best.

The so-called Golden Circle is a mainly geological tourist route in southwest Iceland (Gudmundsson, 2017). It includes the impressive tectonic site of Thingvellir where the North American Plate and European Plate are moving apart at c. 3 cm per year resulting in open fissures some of which have walking or canoe trails through them. Also on the Golden Circle route is the Strokkur Geyser, which erupts every 5–10 minutes to a typical height of 15–20 m but can be higher, the impressive Gullfoss waterfall and the Kerio volcanic crater. It is also possible to visit the local Nejavellir and Hellisheidarvirkjun geothermal power plants and a number of nearby geological excursions (Gudmondsson, 2017).

The last two alternative sites are closely related to the cities where they are found and therefore reflect the important and close association between human settlements and their physical settings. The first is Table Mountain

National Park (South Africa) and second is the mountain landscape and harbor at Rio de Janeiro (Brazil).

Table Mountain itself rises to over 1000 m and dominates the City of Cape Town (Fig. 8.4). It can be scaled by hiking trails or by cable car and is therefore easily accessible. The steep western face of Table Mountain is composed of subhorizontal, resistant, Cambrian sandstone beds. Subsidiary summits include Devil's Peak to the north, and Lion's Head and Signal Hill to the south, the latter being composed of deformed and metamorphosed shales. The underlying intrusion of the Cape granite is evident on the coastal rocks at Sea Point, a National Monument site, where the contact with thermally metamorphosed shales (hornfels) is well exposed (Whitfield, 2015; Compton, 2016). Table Mountain National Park extends southwards toward the Cape of Good Hope and Cape Point and is in three sections—Table Mountain, Silvermine, and Cape of Good Hope.

FIGURE 8.4 Table Mountain dominating the city of Cape Town, South Africa.

The Brazilian City of Rio de Janeiro is similarly dominated by its physical setting, in this case being built around and between a number of steep-sided bornhardts composed of augen gneiss and meta-sediments dating to c. 570 ma. They include Sugar Loaf and Corcovado, the latter with its huge statue of Christ the Redeemer. The coastal plain is formed of Quaternary sediments and the coastline itself is dominated by the sandy beaches of Copacabana, Ipanema, and Leblom. It has a huge natural harbor, mistaken for a river by early explorers in January, 1502 and hence the name of the city. The surrounding mountains rise steeply to over 1000 m. Sugar Loaf is accessed by two cable cars, while the summit of Corcovado can be reached by train or car/minibus. From these summits the beauty of the city's setting can be appreciated (Fig. 8.5). Inscribed as a cultural World Heritage Site in 2012, the citation refers to its "exceptionally dramatic landscape... of great beauty."

FIGURE 8.5 The view of Sugar Loaf Mountain and part of the Harbor from Corcovado, Rio de Janeiro, Brazil.

8.5 CONCLUSIONS

And so my proposed list of the world's top 10 geotourism destinations are:

1. Iguaçu/Iguazu Falls (Brazil/Argentina) or Victoria Falls (Zimbabwe/Zambia)

2. Grand Canyon National Park (USA)
3. Great Barrier Reef (Australia)
4. South China Karst to Ha Long Bay (China/Vietnam)
5. Yellowstone National Park (USA) or Hawaii Volcanoes National Park (USA)
6. Uluru (Australia)
7. Central Swiss Alps (Switzerland)
8. Cal Orcko Parque Cretácico (Bolivia)
9. The "Golden Circle" Route (Iceland)
10. Table Mountain, Cape Town (South Africa) or Rio de Janeiro (Brazil)

These 10 sites not only demonstrate the diversity of the physical world, but also are all internally geodiverse. There is also representation from 6 continents (South America, North America, Africa, Asia, Europe, and Australia—see Fig. 8.6). Only Antarctica is missing, but could have been included but for its limited accessibility/facilities. There is also a marine based site (Great Barrier Reef) while Hawaii Island is part of a shield volcano rising from the seabed.

These sites are not intended to be the final, definitive list. Rather, they are intended to be the start of a process that may eventually lead to a list of top geotourism sites endorsed by the geological community and can then be promoted to the public through the media. This endorsement might come through the IUGS's *International Commission on Geoheritage* and/or the IUCN's *Geoheritage Specialist Group*.

One issue to be decided is whether to have a list of 7, 10, 20, 50, 100 or some other number of sites. Why has the Scottish Geodiversity Forum listed "51 best places to see in Scotland" rather than 50? According to scottishgeology.com, they were aiming for a list of 50 sites "but our panel of experts just couldn't agree"! So, whatever number of sites are selected and whatever method is used, one point is very clear from the above discussion. Despite the title of this chapter selecting the world's top geotourism destinations will not be simple.

ACKNOWLEDGMENTS

I am very grateful to José Brilha, Jasmine Moreira, and Pauline Allen for their comments on a draft of this chapter, which significantly improved it. Ed Oliver of Queen Mary University of London kindly drew the map.

"Simply the Best": The Search for the World's Top Geotourism Destinations   225

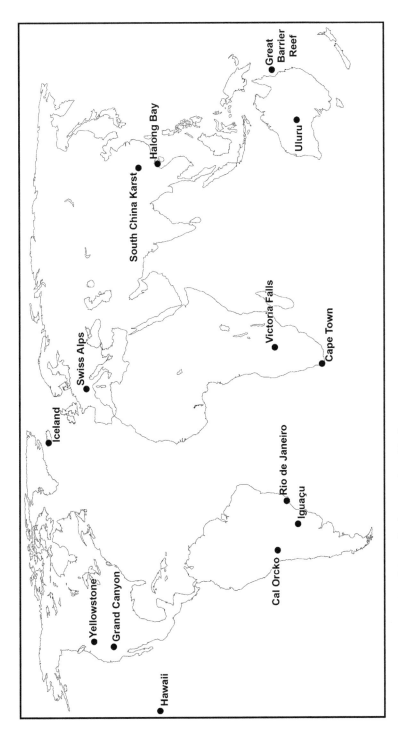

FIGURE 8.6 Distribution of the main sites mentioned in the text.

KEYWORDS

- geotourism
- nature
- tourism
- geodiversity
- geoheritage

REFERENCES

Anhaeusser, C. R.; Viljoen, M. J.; Viljoen, R. P.; Eds.; *Africa's Top Geological Sites*. Struik Nature: Cape Town, 2016.

Brilha, J. B. Inventory and Quantitative Assessment of Geosites and Geodiversity Sites: A Review. *Geoheritage* **2016**, *8*, 119–134.

Cleal, C.; Thomas, B.; Bevins, R.; Wimbledon, B. Deciding on a New World Order. *Earth Herit.* **2001**, *16*, 10–13.

Compton, J. Table Mountain and the Cape Peninsula. In *Africa's Top Geological Sites;* Anhaeusser, C. R., Viljoen, M. J., Viljoen, R. P., Eds.; Struik Nature: Cape Town, 2016; pp 111–115.

Dickas, A. B. *101 American Geo-Sites You've Gotta See*; Mountain Press Publishing Co.: Missoula, 2012; p 250.

Dickas, A. B. *101 American Fossil Sites You've Gotta See*; Mountain Press Publishing Co.: Missoula, 2018; p 249.

Garcia-Cortés, A., Ed.; *Spanish Geological Framework and Geosites: And Approach to Spanish Geological Heritage of International Significance*. Instituto Geologico y Minero de Espana: Madrid, 2016.

Gray, M. *Geodiversity: Valuing and Conserving Abiotic Nature*, 2nd ed.; Wiley-Blackwell: Chichester, 2013.

Gudmundsson, A. *The Glorious Geology of Iceland's Golden Circle*. Springer Nature: Switzerland, 2017; p 334.

Hose, T. A. Selling the Story of Britain's Stone. *Environ. Interp.* **1995**, *10*, 16–17.

Moreira, J. C. Interpretative Panels about the Geological Heritage—a Case Study at the Iguassu Falls National Park (Brazil).*Geoheritage* **2012**, *4*, 127–137.

Smith, R. B.; Siegel, L. J. *Windows into the Earth: The Geologic Story of Yellowstone and Grand Teton National Parks*. Oxford University Press: Oxford, 2000.

Whitfield, G. *50 Must-See Geological Sites in South Africa*. Struik Nature: Cape Town, 2015.

Williams, P. L. Lecture Delivered to the 1st International Conference on Geoheritage, Huanjiang, China, 5th June 2015.

PART III
Geointerpretation: Interpreting Geological and Mining Heritage

CHAPTER 9

Interpreting Mining: A Case Study of a Coal Mine Exhibit

TED T. CABLE

Park Management and Conservation, Kansas State University, Manhattan, KS66506, USA

E-mail: tcable@ksu.edu

ABSTRACT

This chapter presents multiple definitions of heritage interpretation and the use of well-established professional interpretive principles and practices at a popular coal mine exhibit in the Museum of Science and Industry in Chicago, Illinois (USA), the Western Hemisphere's largest science museum. It presents potential benefits of heritage interpretation to both the visiting public and to the sponsoring institutions.

9.1 INTRODUCTION

The chapter presents definitions and principles of heritage interpretation and discusses the application of interpretation approaches at the famous coal mine exhibit in the Museum of Science and Industry in Chicago, Illinois (USA). It presents potential benefits of using heritage interpretation in interpreting mining and promoting geotourism, and the author encourages the application of professional interpretation practices to interpret mining.

9.2 WHAT IS INTERPRETATION?

Many definitions of interpretation have been published in an attempt to capture this complex discipline. The most widely quoted definition came from Tilden (1977) in his classic book Interpreting Our Heritage: "*An educational activity which aims to reveal meanings and relationships through the use of original objects, by firsthand experience, and by illustrative media, rather than simply to communicate factual information.*" Sam Ham (2013) captures the current 21st century thinking about the term in his book "Interpretation- Making a Difference on Purpose" by stating that "*Interpretation is a mission-based approach to communication aimed at provoking in audiences the discovery of personal meaning and the forging of personal connections with things, places, people, and concepts.*" The National Association for Interpretation, the largest professional organization of interpreters, offers this succinct definition: "*Interpretation is a mission-based communication process that forges emotional and intellectual connections between the interests of the audience and meanings inherent in the resource.*"

In the context of geotourism, and specifically the subject of mining, interpretation translates (i.e., brings meaning to people about) natural or cultural heritage aspects of geologic resources. Just as a language interpreter brings meaning to an unknown foreign language, interpreters help people understand, connect to, and enjoy objects, resources, and landscapes in geoparks, mines, and museums that present unfamiliar geologic artifacts or themes.

These definitions indicate that interpretation offers more than instruction through facts. The interpreter must acquire facts, but then put them in a context to make their presentation enjoyable, meaningful, and relevant to the visitor. Interpretation connects people emotionally and intellectually with geologic resources, specific sites, events, and landscapes. Interpretation stimulates interest and observation. It helps people to develop skills, to read their landscape, and to relive their history. Ultimately, it helps them to further understand their home environment and the world around them (Beck et al., 2018).

9.3 PRINCIPLES OF INTERPRETATION

In the 1950s, Freeman Tilden published intertwined and inspiring "principles of interpretation." Tilden's (1977) principles have guided interpreters since then. These principles serve as guidelines for performance, for evaluation, and for training throughout the profession and across the globe. Tilden's six principles are as follows:

1. Any interpretation that does not somehow relate what is being displayed or described to something within the personality or experience of the visitor will be sterile.

2. Information, as such, is not interpretation. Interpretation is revelation based upon information. But they are entirely different things. However, all interpretation includes information.
3. Interpretation is an art, which combines many arts, whether the materials presented are scientific, historical or architectural. Any art is to some degree teachable.
4. The chief aim of interpretation is not instruction, but provocation.
5. Interpretation should aim to present a whole rather than a part and must address itself to the whole person rather than any phase.
6. Interpretation addressed to children should not be a dilution of the presentation to adults, but should follow a fundamentally different approach. To be at its best it will require a separate program.

Tilden also taught that love of the place and love for the visitors constituted the priceless ingredient—the overriding principle of interpretation (Tilden, 1977). Information, then, is merely the raw material that the interpreter shapes and presents in ways that entice, engage, interest, inspire, and clarify. Interpreters use facts and phenomena to communicate meaning and to develop deep understanding.

Building on Tilden, an expanded set of interpretive principles was developed (Beck and Cable, 2011). The first six principles are consistent with Tilden's since these principles are so well known among interpreters, although they have been updated.

1. To spark an interest, interpreters must relate the subject to the lives of the people in the audience.
2. The purpose of interpretation goes beyond providing information to reveal deeper meaning and truth.
3. The interpretive presentation—as a work of art—should be designed as a story that informs, entertains, and enlightens.
4. The purpose of the interpretive story is to inspire and provoke people to broaden their horizons.
5. Interpretation should present a complete theme or thesis and address the whole person.
6. Interpretation for children, teenagers, and seniors—when these comprise uniform groups—should follow fundamentally different approaches.
7. Every place has a history. Interpreters can bring the past alive to make the present more enjoyable and the future more meaningful.

8. Technology can reveal the world in exciting new ways. However, incorporating this technology into the interpretive program must be done with foresight and thoughtful care.
9. Interpreters must concern themselves with the quantity and quality (selection and accuracy) of the information presented. Focused, well-researched interpretation will be more powerful than a longer discourse.
10. Before applying the arts in interpretation, the interpreter must be familiar with basic communication techniques. Quality interpretation depends on the interpreter's knowledge and skills, which must be continually developed over time.
11. Interpretive writing should address what readers would like to know, with the authority of wisdom and its accompanying humility and care.
12. The overall interpretive program must be capable of attracting support—financial, volunteer, political, administrative—whatever support is needed for the program to flourish.
13. Interpretation should instill in people the ability, and the desire, to sense the beauty in their surroundings—to provide spiritual uplift and to encourage resource preservation.
14. Interpreters can promote optimal experiences through intentional and thoughtful program and facility design.
15. Passion is the essential ingredient for powerful and effective interpretation—passion for the resource and for those people who come to be inspired by it.

9.4 BENEFITS OF INTERPRETATION

As noted in the definitions and principles, successful interpreters communicate the significance of cultural and natural resources. They guide understanding, appreciation, and a sense of place. Importantly interpreters also guide visitors in how they can best experience and enjoy the resources. Interpreters give the gifts of knowledge, joy, beauty, among many others (Beck and Cable, 2011). They also can contribute to building family and community unity, strength, and health. In this way, interpreters contribute multiple significant benefits to society (Beck et al., 2018).

Interpretation also generates benefits for the organizations and agencies offering it. The use of interpretation as a management tool has a long and well-documented history. Theories and empirical studies support the many anecdotal success stories. Beck et al. (2018) provide a comprehensive summary of agency benefits including a reduction in vandalism and other depreciative behavior, increased visitor safety, and increased public support and cooperation for the agency or organization, increased visitor satisfaction, and even increased philanthropy (e.g., Powell and Ham, 2008).

9.5 INTERPRETATION OF MINING

Mine tourism is popular around the world. At most visited sites, interpretation is used to enhance the educational, recreational, inspirational, and community-building values of visits to mines and mining exhibits. The National Mining Association (nma.org) lists more than 20 actual mine sites that welcome visitors and offer mine tours in the United States and Canada. Add to these the many museums, visitor centers, and reconstructed and historic mining towns that also interpret mining heritage and it is obvious that mine-related tourism is a popular touristic activity. Mine-related tourism is also popular outside of North America. Table 9.1 lists some mine sites that welcome visitors and offer interpretive tours to the public. This list in not comprehensive but illustrates the geographic spread of mine tourism. As mentioned regarding North America, worldwide many museums, visitor centers, and historic sites interpret mining heritage.

9.6 CASE STUDY: THE MUSEUM OF SCIENCE AND INDUSTRY COAL MINE EXHIBIT

The Museum of Science and Industry in Chicago is the Western Hemisphere's largest science museum and one of the largest in the world. The museum building is the only remaining building constructed for the 1893 World's Columbian Exposition where it served as the Palace of Fine Arts. The museum opened in 1933 as part of the Century of Progress World's Fair.

The Museum of Science and Industry has been immensely popular and has served a vital function as an educational institution. More than 180 million people have visited the museum since it opened in 1933. In 2014, 1.4 million people visited including 344,000 children on school field trips. In that same year, the museum's website had 5.5 million visits (msichicago.org).

Today the museum is home to 35,000 artifacts and 400,000 square-feet of hands-on exhibits. Popular exhibits include a World War II German submarine, 5-story movie screen, 40-foot tornado, and a Boeing 727 passenger jet. However, the most iconic exhibit and one of the most-popular is the Old Ben No. 17 Coal Mine.

TABLE 9.1 A Selection of Some Excellent Mine Sites that Interpret Mining Heritage. This Partial List Illustrates the Widespread Popularity of Mine Tourism and the Potential Influence of Mine Interpretation on Audiences Worldwide.

Name	Location	Website
Coober Pedy	Australia	cooberpedy.sa.gov.au/tourism
Kalgoorlie Super Pit	Australia	superpit.com.au
Atlas Coal Mine National Historic Site	Canada	atlascoalmine.ab.ca
Springhill Miners' Museum	Canada	springhillminersmuseum.blogspot.com
Cape Breton Miners Museum	Canada	minersmuseum.com
BergwerkSchauinsland	Germany	schauinsland.de/engl/home
Zeche Zollverein	Germany	zollverein.de
Tytyri Mining Museum	Finland	tytyrielamyskaivos.fi
Park Timna	Israel	parktimna.co.il
Morning Star Mine—Zealandia	New Zealand	visitzealandia.com/About/History
Silvermines at Kongsberg	Norway	norsk-bergverksmuseum.no
RorosMining Town	Norway	roros.no
Cobalt MinesBlaafarvevaerket	Norway	blaa.no
Khewra Salt Mine	Pakistan	khewra-saltmines.blogspot.com
Wieliczka Salt Mine	Poland	wieliczka-saltmine.com
Turda Salt Mine	Romania	salinaturda.eu
Kimberley Mine	South Africa	thebighole.co.za
Falun Mine	Sweden	falugruva.se
Kiruna Mine	Sweden	kirunalapland.se
Poldark Mine	the United Kingdom	poldarkmine.org.uk
LLechwed Slate Caverns	the United Kingdom	llechwedd-slate-caverns.co.uk

TABLE 9.1 (Continued)

Name	Location	Website
Dolaucothi Gold Mines	the United Kingdom	nationaltrust.org.uk/dolaucothi-gold-mines
Big Pit National Coal Museum	the United Kingdom	museum.wales/bigpit
Silver Mountain Experience	the United Kingdom	silvermountainexperience.co.uk
Todmorden Moor Geology Heritage Trail	the United Kingdom	todmordenmoor.org.uk/trail
Black Diamond Mines Regional Preserve	the United States	ebparks.org/parks/black_diamond
Asarco Mineral Discovery Center	the United States	asarco.com
Copper Queen Mine Underground Tour	the United States	queenminetour.com
Lebanon Mine	the United States	georgetownlooprr.com
Old One Hundred Mine	the United States	minetour.com
Crystal Gold Mine	the United States	goldmine-idaho.com
Stratica: Kansas Underground Salt Museum	the United States	underkansas.org
Sterling Hill Mining Museum	the United States	sterlinghillminingmuseum.org
Sierra Silver Mine Underground Tour	the United States	silverminetour.org
Hill Annex Mine State Park	the United States	dnr.state.mn.us/state_parks
Iron World Discovery Center	the United States	mndiscoverycenter.com
Soudan Underground Mine State Park	the United States	dnr.state.mn.us/state_parks
Hibbing Taconite Mine Tours	the United States	ironrange.org
Missouri Mines State Historic Site	the United States	mostateparks.com
Exhibition Coal Mine	the United States	beckley.org/underground-tour

FIGURE 9.1 A visitor reads an interpretive sign as they enter the Museum of Science and Industry's Coal Mine exhibit. [Photo taken by Ted Cable (Author)]

The Old Ben No. 17 Coal Mine exhibit was the museum's first exhibit. For generations visitors have descended down a simulated mine shaft into an authentically reproduced working coal mine. Grandparents travel down with their grandchildren just as they were taken down into the mine by their parents and grandparents. Since the museum's beginning the Coal Mine has become the museum's most iconic and beloved exhibit (Pridemore, 1996). Otto Kreusser the Director who joined the museum in 1932 is mostly responsible for acquiring and installing this exhibit. When Kreusser learned about the Old Ben No. 17 Coal Mine in Johnson City, Illinois that had closed due to the Great Depression of the 1930s, he contacted the bankers who had reposed the equipment. They happily sold these materials to Kreusser at a bargain price of $10,000 USD. The construction of the exhibit used the head frame, hoists, and other equipment from this closed mine. To add to the authenticity of the experience, Kreusser worked with perfume manufactures to create an odor in the exhibit that simulated the damp musty smell of a real coal mine. The coal mine exhibit's first visitors were people attending the World Fair who had read about this amazing new experience in newspapers around the world.

Back in the 1930s and still today visitors enter a cage elevator which rattles and jerks as it moves downward a short distance. Because of a moving wall they feel as if they are going deep underground. When the elevator "arrives" in the mine many visitors believe they are deep underground. Once

Interpreting Mining: A Case Study of a Coal Mine Exhibit

inside the exhibit, visitors board a work train to experience a coal miner's "daily commute" as they seem to travel through the veins of coal. Visitors experience different rooms of the mine and see working examples of how extraction machinery has evolved from the pickaxe to longwall machines that can carve out enormous sections of shale. The illusion of travel underground and mine operation is so realistic that some visitors have asked if the sale of coal from the mine subsidizes the operation of the museum.

FIGURE 9.2 The entrance to Chicago's Museum of Science and Industry's historic Coal Mine exhibit. [Photo taken by Ted Cable (Author)]

FIGURE 9.3 Interpreter guide at Chicago's Museum of Science and Industry's Coal Mine provides an educational and entertaining tour to visitors. [Photo taken by Ted Cable (Author)]

Interpreting Mining: A Case Study of a Coal Mine Exhibit 239

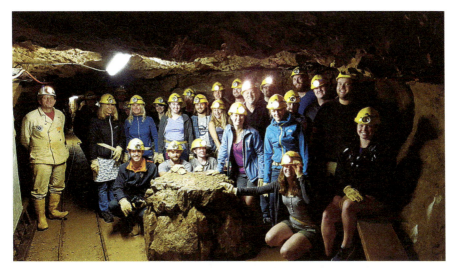

FIGURE 9.4 German and American college students experience a silver mine near Freiburg, Germany with their professors and an interpretive guide. [Photo taken by Ted Cable (Author)]

The museum is open 363 days a year and coal mine tours can start every 15 minutes with about 20 people on each tour. Over the past 5 years (2012–2017), on average 161,155 visitors experienced the coal mine exhibit annually with annual ticket sales ranging from 144,647 (2017) to 176,243 (2015). A tour guide interpreter dressed as a miner accompanies the groups of visitors down on the cage elevator and throughout the tour. Initially, these guides were actual retired coal miners. Today they are trained guides some of whom have a background performing professionally in Chicago's theater community.

9.7 THE ROLE OF INTERPRETATION AT THE COAL MINE EXHIBIT

In light of the aforementioned formal definitions and principles of interpretation it is clear that interpretation plays a key role in shaping the visitor experience at the Museum of Science and Industry's Coal Mine exhibit. A review of the history of the exhibit provides evidence of a long tradition of excellent interpretation at the exhibit and a recent (2017) visit by the author confirmed that visitors and the institution are both benefiting from outstanding interpretation.

Approaches consistent with Tilden's definition and principles of interpretation were established by early museum administrators. The museum's initial association with the World's Fair demonstrated to leadership the importance of balancing education and entertainment and going beyond merely communicating facts. This was particularly true when depending on public visitation and support of the institution. Curators emphasized original objects and hands on approaches consistent with Tilden rather than merely attaching labels and technical text to the artifacts. Julius Rosenwald the business and philanthropist who initially proposed and financed the museum anticipated and foreshadowed interpretive approaches when he promised a "push button museum" with "no hands off signs" (Pridemore, 1996).

FIGURE 9.5 Tourists visiting a German silver mine huddle together in tight passageways as they learn about mining from the interpretive guide. [Photo taken by Ted Cable (Author)]

One of the principles of interpretation is the importance of joyous enthusiasm and passion (see principle 15 above). This principle was applied by the retired miners who served as "demonstrators." According to Pridemore (1996) these "grizzled" men who knew so much about the mine and its equipment fascinated the city folk who came to the museum. Otto Kreusser's son Richard is quoted in Pridemore (1996: 56) as saying, "The miners knew their stuff. And they were wonderfully enthusiastic. They had no idea that so many people gave a damn about mining." It was a joy for these former miners to share their intimate knowledge and passion with the tourists and locals who visited the Coal Mine exhibit.

In 1940, Lennox Lohr became director of the museum with the goal to make the museum an exciting place. He stated, "I visualize the institution as a great national museum, a show, museum in which science and industry will live as dramatic things" (Pridemore, 1996: 65). In fact, he preferred the word "theater" to "museum" to describe the institution. His goal to relate to audiences in an entertaining and engaging fashion is clearly in the spirit of interpretive foundations and principles, particular the first and second principles of interpretation. Not only did Lohr seek to relate to audiences' primary interests but he aspired to going beyond merely providing information to provide personal meaning. His efforts at entertainment and engagement also took the form of storytelling as prescribed in Beck and Cable's third principle (Beck and Cable, 2011).

Under Lohr's leadership labels were written in simple English which is specified by both Tilden and Beck and Cable in their first principles. Moreover, Lohr addressed their second and fifth principles by ensuring that exhibits addressed the "meaning behind the machines" (e.g., how machines contributed to America's wealth, why mass production is not the cause of America's ills, or how the railway changed modern life). Their third principle of using the arts and storytelling was applied by strategically by making the museum a place to tell the story of corporate America. He wanted to make the museum "an embassy of ideas" (Pridemore, 1996: 68) and he used light historical drama on a national radio network to convey these ideas theatrically to the masses. Lohr addressed the fourth principle of provocation by using the museum to promote "the friendlier spirit and better understanding between capital and labor, producer and consumer." He further identified the theme of the museum (principle 5) as the "partnership of science and industry."

Interpretive techniques and philosophy at the museum were greatly advanced by a Curator of Physics named Lucy Nielsen. She devoted much effort to replace labels written by scientists "to impress and confuse people" with labels the average person could understand (Principle 1). She helped answer the question "So what?" by explaining why the visitor should care about these concepts of physics (Principle 2). In order to relate to audiences (Principle 1), a presenter must know their audience—who are the visitors and why they are attending. To that end, Lucy conducted formal visitor studies. She applied the sixth principle of interpretation by including children in these studies, taking young visitors seriously, and by extending the museum into schools to talk specifically about the coal mine and other exhibits. In applying the sixth principle of interpretation the goal was to make the museum place for children without being childish.

This institutional commitment to the principles of heritage interpretation continued into recent decades. The museum still understands that they offer informal education and therefore must find the right mix of entertainment and fun as well as learning. They tell their stories through the use of television and the Internet to inform, educate, and entertain those who cannot visit, including taking stories into school classrooms. Museum curators still emphasize interactive exhibits that relate to the personal lives and interests of their audience and they study their audience so that they can target those interests more effectively. In the 1990s, they began creating thematic zones in the museum recognizing that thematic interpretation (Beck and Cable, 2011) is more effective than merely presenting miscellaneous facts in an unorganized manner.

Currently, as prescribed by Beck and Cable's 10th principle of interpretation, the museum offers continual training on presentation techniques to their guides and docents. Twice a week before starting their shifts the staff is led in improvisational activities to keep their skills sharp. Teams of guides are formed and assigned to "coaches" who routinely observe presentations and offer training and feedback. In this way, the interpretive guides are always growing in their skills and abilities to communicate with the visitors.

9.8 PERSONAL INTERPRETATION AND COAL MINE EXPERIENCE

In addition to the institutional commitment to exhibits which apply the principles of interpretation, the museum's tour guides also skillfully demonstrate interpretive principles. On 30 October 2017, the author went to the museum and experienced the coal mine exhibit as a normal public visitor. During the visit, the application of Tilden's six principles by the interpretive guide was noted. The results are as follows:

Principle 1. Any interpretation that does not somehow relate what is being displayed or described to something within the personality or experience of the visitor will be sterile. Immediately at the beginning of the tour the guide learned where everybody was from and then personalized references throughout the tour to those locations. For example, some visitors were from Brazil and so periodically throughout the talk he referenced mining in Brazil to relate to these audience members. Additionally, he constantly made comparisons to things we all could relate to such as the blue flame on a gas stove or other such common relatable things.

Principle 2. Information, as such, is not interpretation. Interpretation is revelation based upon information. But they are entirely different things. However, all interpretation includes information. The guide made a special effort to go beyond mere facts and statistics to help us think about intangible meanings in coal mining. He challenged us to think about what life was like before electricity, what it must have been like to spend your day working in a mine, and the emotional impacts on families and communities.

Principle 3. Interpretation is an art, which combines many arts, whether the materials presented are scientific, historical or architectural. Any art is to some degree teachable. The tour guide was an experienced actor in the Chicago theater community. He was a consummate entertainer and storyteller as he also educated us about coal mines. His "underground humor" was outstanding and kept everybody laughing as they learned. This is an important interpretive technique because if audiences are laughing, they are listening. Humor helps hold the audience's attention and he used it artfully.

Principle 4. The chief aim of interpretation is not instruction, but provocation. The guide repeatedly provoked his audience to think of energy in a new way. He provoked us to think about where the energy comes from and the dedicated people who work under difficult conditions to provide that energy whenever we flip on a light switch. He encouraged visitors to not take coal-produced electricity and its many benefits for granted.

Principle 5. Interpretation should aim to present a whole rather than a part and must address itself to the whole person rather than any phase. This principle is really two principles in one. The first aspect of this principle is to present a whole theme—an important big idea or "take-home" message, rather than many miscellaneous unrelated facts. The guide clearly had an important theme that coal is an important part of improving our lives (even finding its way into our toothpaste!). He addressed the whole person by asking many open-ended questions which allowed for the participants to think about themselves and ask questions relevant to their own situations and interests.

Moreover, he told dramatic stories of how coalmining affected miners' health and the economic challenges that faced miners and their families. These stories touched the "whole" audience member by making both intellectual and personal emotional connections between miners and the visitors.

Principle 6. Interpretation addressed to children should not be a dilution of the presentation to adults but should follow a fundamentally different approach. The guide made special effort to talk to the children and ask them questions. He had them participate in touching or holding items and attempted to get them excited about science at their level. Some of his humor was directed specifically at the children to hold their attention. The parents appreciated the attention he showed the young people on the tour. Consistent with the Danish proverb, "You take a child by the hand, you take the mother by the heart" he connected with the adults on the tour through their children.

Overall the guide strongly applied each of Tilden's six principles of interpretation. Moreover, he applied Beck and Cable's 13th principle by revealing the beauty of the coal and the beauty of the courage and resource fullness of the coal miners as well as their 15th principle by demonstrating a passion for mining and for the audience that truly was inspiring.

9.9 CONCLUSION

Interpretation of mines and mining offers benefits to both the visiting public and to the sponsoring institutions. Because of the widespread popularity and public interest in mining interpreters at these sites have the opportunity to educate and entertain millions of visitors worldwide. In doing so, they make personal connections between people and the geologic resources that supply the world's energy and contribute to our quality of life. They can build support for the energy industry and the people who work in it. Interpreters at these sites can enhance the appreciation of energy sources and provoke conservation of those resources.

The iconic Coal Mine exhibit at Chicago's Museum of Science and Industry has been engaging and educating people for more than eight decades. This multisensory experience build around authentic equipment and entertaining guides allows people to gain a new perspective on energy and the ingenuity of those who have mined coal in the past and still bring this "black gold" to the surface. The Coal Mine exhibit still provokes curiosity among visitors and is still "a realistic and partially mysterious environment that continues to enchant" (Pridemore, 1996: 177).

The motto inscribed upon the Museum of Science and Industry's dome states: "Science discerns the laws of nature. Industry applies them to the needs of man." Interpreters at its Coal Mine exhibit, and at mining sites and museums around the world, tell the story of how humans benefit from this fruitful relationship between the science of geology and the industry of mining.

KEYWORDS

- **heritage interpretation**
- **coal mine exhibits**
- **Museum of Science and Industry**
- **mining geotourism**
- **science museums**
- **mining education**
- **mining interpretation**

REFERENCES

Beck, L.; Cable, T. *The gifts of Interpretation: Fifteen Guiding Principles for Interpreting Nature and Culture*, 3rd ed.; Sagamore: Urbana, 2011.

Beck, L.; Cable, T.; Knudson, D. *Interpreting Cultural and Natural Heritage: For a Better World*; Sagamore/Venture: Urbana, 2018.

Ham, S. *Interpretation: Making a Difference on Purpose*; Fulcrum Publishing: Golden, 2013.

Powell, R.; Ham, S. Can Ecotourism Interpretation Really Lead to Pro-Conservation Knowledge, Attitudes and Behavior? Evidence from the Galapagos Islands. *J. Sustain. Tour.* **2008**, *16*, 467–489.

Pridemore, J. *Inventive Genius: The History of the Museum of Science and Industry Chicago*; The Museum of Science and Industry: Chicago, 1996.

Tilden, F. *Interpreting our Heritage*, 3rd ed.; The University of North Carolina Press: Chapel Hill, 1977.

CHAPTER 10

Geotrails

THOMAS A. HOSE

Honorary Research Associate, School of Earth Sciences, University of Bristol, Wills Memorial Building, Queens Road, Clifton, Bristol, BS8 1RJ, UK

E-mail: gltah@bristol.ac.uk

ABSTRACT

This chapter provides a summary of the history, development, and nature of geotrails within the United Kingdom. Geotrails are probably the commonest and most geographically widespread form, in the United Kingdom and elsewhere, of modern geotourism provision. Whilst most postdate the 1970s, their origins can be found in late-19th century provision in England; this was unusual, both then and now, in being established in an urban area. Most of today's geotrails are in either rural or supposedly wild landscape locations. The chapter necessarily explores the different types of geo-interpretative media, together with a brief consideration of their efficacy, associated with geotrails. It also examines the nature and needs of those persons who access geotrails. Further, it suggests that and indicates some of the ways in which, new geotrails might better appeal and be accessible to wider and younger audiences than those usually targeted by geotrail providers.

10.1 AN INTRODUCTION TO GEOTRAILS AND THEIR USERS

Geotrails are one of the commonest and most geographically widespread types of modern geotourism provision, mainly postdating the 1970s, and have been recognized as such in the only published histories of UK (Hose, 2016a) and European (Hose, 2016b) geotourism. The latter account, particularly noted, especially for Italy, the association of geotrails with protected landscapes. In many ways, modern geotrails supplant, although the practice partly continues with the modern guided geological walks for tourists, the services of a personal guide in the field, something common for the wealthier geologists of the 19th century. The advantage of geotrails is one of cost and convenience to the user, although they can require considerable inputs in personnel, materials and time in their construction, design, and promotion; usually, this is undertaken by the local government and voluntary conservation sectors, rarely, except perhaps in the USA, by commercial organizations.

Two major types of users (Hose, 1996) are likely to interact with geotrail provision:

i) Casual/recreational: Amateur geologists and outdoor recreationalists, as individuals or in parties, who like to look for fossils and attractive minerals and rocks, ideally in an esthetically appealing landscape, whose expertise ranges from beginner to knowledgeable lifelong enthusiast, and for whom geoparks and preserved mine sites are specially intended, but who will also use geotrails.

ii) Dedicated/educational: High-school to postgraduate students and staff, together with researchers, undertaking geological inquiries part of geographical or environmental study programs and for whom much modern dedicated geotourism provision in the form of visitor centers and trails is particularly intended.

Geotrails are a significant element in geotourism provision. Geotourism has been defined in Europe as "The provision of interpretative and service facilities for geosites and geomorphosites and their encompassing topography, together with their associated in situ and ex situ artifacts, to constituency-build for their conservation by generating appreciation, learning and research by and for current and future generations" (Hose, 2012: 11). In Australasia, a similar approach has been adopted, where it is "…a form of natural area tourism that specifically focuses on geology and landscape. It promotes tourism to geosites and the conservation of geodiversity and an understanding of earth sciences through appreciation and learning. This is achieved through independent visits to geological features, use of geotrails and viewpoints, guided tours, geoactivities, and patronage of geosite visitor centers" (Dowling and Newsome, 2010: 232). Both definitions indicate the importance of geo-interpretation, with the latter specifically mentioning trails.

Indeed, geotourism is a geology-focused and visitor-centered development of environmental interpretation, "… the art of explaining the meaning and significance of sites visited by the public" (Badman, 1994: 429). Such an approach to environmental interpretation evolved from developments in the

USA in promoting sport-based wildlife recreation and can be simply defined as "...translating the technical language of a natural science or a related field in terms and ideas that people who are not scientists can readily understand. And it involves doing it in a way that is entertaining and interesting to these people" (Ham, 1992: 3). Tilden (1967) defined interpretation as "... an educational activity which aims to reveal meanings and relationships through the use of original objects, by first-hand experiences and by illustrative media, rather than simply to communicate factual information." In the UK this approach dates from the mid-1960s, following the establishment of the first temporary (during 'National Nature Week' in 1964) and permanent (in 1966 at the Forestry Commission's Grizedale Forest, in the Lake District National Park) nature trails. In the UK the emergence of heritage tourism in the mid-1980s was a positive agent in interpretative development which was then aimed at making its sites meaningful to visitors "...through stimulating and arousing their imagination and curiosity...as a means of enhancing the quality of a heritage site, and contributing to the satisfaction and enjoyment of visitors" (Light, 1995: 132). Environmental interpretation should not be confused with education and/or short-term knowledge acquisition. Although both involve information exchange, the former is based upon revelation (Tilden, 1977: 18). Overall, interpretation has three main functions:

1. Assisting visitors to appreciate a site's significance;
2. Aiding in a site's management;
3. Promoting understanding of the site's owner's/manager's/agency's policies.

The significance assigned to these three functions varies with the priorities of a site's managing and sponsoring agencies. The last is common with statutory agencies (such as Natural England in the UK) the larger business-like conservation charities (such as the National Trust in the UK) keen to demonstrate their public engagement and usefulness; hence, the plethora of logos, alongside those of numerous sponsors commonly adorning, or is that overwhelming, many trailside panels?

However, with specific regard to geotrails the most useful approach is that it is about "...explaining the significance of and encouraging an awareness and understanding of, the landscape and the physical forces which bring about its change and have led to the present appearance" (Countryside Commission, 1979). Two conceptual models for geointerpretative provision (Hose 1997, 2003) show the interrelationships between its various elements within overall geotourism provision, and also emphasizes the integrated geoarchaeological approach (Fig. 10.1) and how they can be categorized for users (Fig. 10.2). However, with the emergence of geocaching, there is some confusion over the use of the term geotrail (and geo-trail).

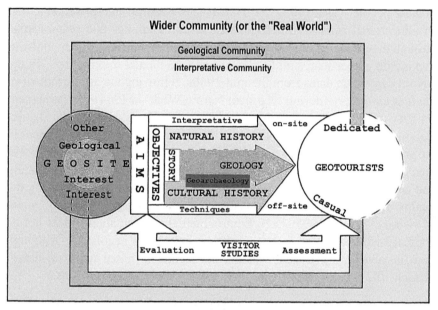

FIGURE 10.1 A conceptual model for geo-interpretative provision. This shows the interrelationships between its various elements within overall geotourism provision, and also emphasizes the integrated geoarchaeological approach.
Source: Hose, 2003.

Perhaps, the most useful working definition, as adopted by the author, of a geotrail is "A usually way-marked guided or self-guided route around a geological and/or geomorphological site on which points of interest are indicated and explained." The points of interest are usually marked by signposts, listening posts (which "dispenses" an audio-message either at the press of a button or on the approach of the visitor using a passive infra-red switch), interpretative (hopefully rather than purely informational) panels, or indicated on a printed route description or, better still, a map in a compact A5–A3 folded leaflet. Sophisticated trails can employ audio/audio–visual/touchscreen and even (although these are usually restricted to museums) electro-mechanical technology, geocaching technology, and the use of quick response (QR) labeling by which a smartphone/tablet computer after scanning the symbol downloads information for each specific location from a mobile phone signal or local area broadband network. Whilst geocaching, involves the use of satellite navigation devices or suitably equipped smartphones to locate positions on a preloaded route, QR approaches need access to a mobile phone or, and this rarely, a local area's wireless-broadband signal and this is much less widespread and reliable in many rural areas. Whilst the old audio-guides employed cassette or CD/DVD players, the new ones are solid-state (often an SD memory card) drive.

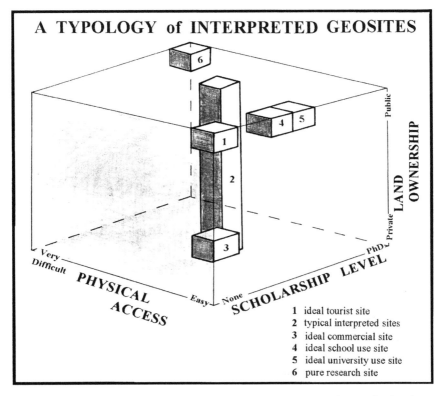

FIGURE 10.2 Geosite Cube Plot showing the relationship between three main site characteristics and how they define the nature of likely geosite use.
Source: Hose, 2003.

10.2 GEOTRAIL INTERPRETATIVE MEDIA

Geotrail interpretative provision for casual geotourists and recreationalists has traditionally centered around trailside panels, guide-posts, and leaflets, although in recent years there has been some shift toward smartphone and tablet downloadable files and applications (possibly utilizing quick response code, QR, shortcuts, and even geocached routes). Leaflets can vary in print quality from photocopied A4 paper (sometimes on colored paper) sheets to full-color gloss semi-board card stock. Bindings, if required vary from center-fold staples, ring/spiral binding to perfect (glued) binding. Or course, costs, and production, let alone revision and reprinting, times increase with the sophistication of the printing and binding methods. The chief advantages of leaflets, over trailside panels, are that they can be prepared in several language versions and can be updated fairly easily with relatively little capital investment; they can also, as the Lochaber Geopark in Scotland, generate income through sales and incorporated advertising creating income, although most continue to be free. Of course, their preparation and production costs can also be covered, as with trailside panels and guideposts, through advertising and sponsorship.

Trailside panels were commonly employed at most interpreted sites until the late-2000s when their high cost, with the local impact of the global economic downturn, became an issue. Most new trails employ leaflets and few or no marked locations on the ground. However, it is still useful to examine trailside panels because they are the most accessible medium for unplanned casual usage and they are commonly employed across the world in major tourist areas. Several examples are discussed within the text as they exemplify specific points and their location is broadly indicated on a sketch map (Fig. 10.3). What is surprising is the continued focus on permanent annual rather than seasonal panel content and illustrations; likewise, the continued use of robust but expensive ceramic, fiberglass, and enameled materials and carved/routed wood for panels. Despite the widespread availability of self-adhesive vinyl/polyester, there are few examples of temporary and seasonal panels employed at interpreted geosites. The author has previously employed laminated, (A3–A2) printed sheets for several events at interpreted geosites; his informal experiments have shown that these, when printed on high-quality paper with good-quality inkjet inks, will last through seasonal inclement weather, and in full sunlight, over at least a couple of years with no signs of degradation.

Panels can draw visitors' attention to a site and direct their usage within it. They can be examined throughout the day and year by users, and there is no cost to them for their use. They are even more interesting and improve trailside accessibility for the visually-impaired when they include tactile components, as at the Nerja cave geoarchaeological site in Spain's Costa del Sol at which one of the panels (Fig. 10.4) includes actual rock specimens but not the raised-edge illustrations and braille labeling of specially dedicated panels; they would then truly "…have the additional value of beneficially

Geotrails 253

raising the public profile of the implementing sponsor by demonstrating a commitment to environmental and heritage issues" (Page, 1994: 433) for almost all likely users.

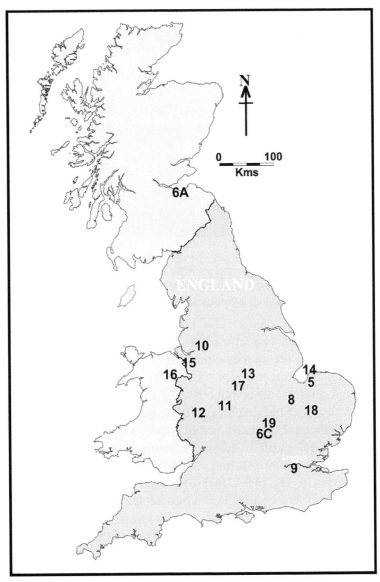

FIGURE 10.3 Location map for the illustrated UK trailside panels, which can be named from the figure number, mentioned in the text. Their essentially central England distribution reflects a combination of ideal geology and the nearness of urban areas (not shown) with potential users.

FIGURE 10.4 A panel at the Nerja cave geoarchaeological site in Spain's Costa del Sol that includes actual rock specimens. The short 2-km linear trail, within a gated and fenced enclosure, up to a viewpoint has several panels that provide accounts of the local geology, particularly in relation to the site's subterranean features and ecology.

Most dedicated geological trailside panels, just like leaflets and other interpretive media incorporate either "Traditional Geological" or "Casual User" themes. The former topics (favored by academic geologists, educationalists, and students) are commonly: (1) Lithological: Description of rock units and types; (2) Stratigraphical: The sequence of rocks and the implications; (3) Structural: Tectonic features and their implications; (4) Palaeontological/mineralogical: To admire, examine, or collect. The latter topics (favored by casual geotourists and recreationalists) are commonly: (1) The Unusual or unique: The oldest rock or fossil; (2) Economic: Finding and exploiting mineral resources; (3) Environmental: What life, the land, and climate were like in the remote past; (4) Historical: The lives and work of geologists; (5) Land-shaping processes: Action of the sea, rivers, ice, or volcanoes; and (6) Internal forces: Faulting and folding. However, the integrated rather than the dedicated approach is better for casual geotourists and best for recreationalists; for example, geoarchaeological rather than purely geological content. It is commonly not considered that the mindsets of those who "'consume" panels are different to those who commission or inform their content; likewise, their past experiences, knowledge, and understandings. This, if not adequately considered, can result in a mismatch between what is provided for trail users and what they actually need or would like; there are too many panels at an intellectual level beyond that of most viewers (Hose, 1997, 1998, 2005).

10.3 ATTRACTING AND HOLDING GEOTOURISTS AND RECREATIONALISTS

When a trailside panel is either ignored or examined in a cursory manner by trail users, it suggests a combination of poor design, uninteresting content, inappropriate location, and siting, and possibly also too much competition from adjacent environmental elements; probably other panels or viewpoints. Trailside panel siting, location, and on-site visibility are crucial to their effectiveness. They should be placed within safe well-illuminated (but avoiding the attachment of dedicated artificial light sources) areas and potential viewers should not be made to follow unnatural or difficult off-trail detours. Before the sites for trailside panels and signage are literally set in concrete it is always wise to conduct a walk-through and undertake observation studies in a potential trail route area to identify the paths taken by visitors and their loci—they often are not the official surfaced and signposted routes; the author's research has found that otherwise well-designed panels, such as that at Wolferton (Fig. 10.5) in Norfolk, are almost completely ineffectual because they are sited where few visitors are likely to walk or cycle past them. Trailside panels should not obstruct paths or create circulation bottlenecks. They should be in or very near, and obviously seen from, locations with many likely visitors.

Limited research has explored what attracts and distracts trailside panel viewers. It has examined the relationship between panel graphical/text elements and holding and attracting powers, together with communicative efficacy. "Holding power" is a measure of the amount of time users view a panel; it can be expressed as a range or a mean. It is not a measure of communicative success. A panel might have high holding power because its viewers might spend a lot of time examining it because they are either confused or are attempting to comprehend a difficult text and storyline—most casual outdoor recreationalists, unlike dedicated geotourists, as viewers are not that persistent! Further:

Attracting Power = Potential number of persons viewing panel

FIGURE 10.5 A panel placed atop an ancient landlocked Ice Age sea-cliff, facing seawards (A) at Wolferton in Norfolk; despite its excellent typography (B) and content it has comparatively little usage apart from local dog-walkers and is perhaps an excellent example of a panel that really could not be sited anywhere else, but perhaps no visitor survey was undertaken prior to the decision to present this otherwise excellent piece of interpretative provision.

10.3.1 ACTUAL NUMBER OF PERSONS VIEWING PANEL

It is a useful but crude method of analyzing a panel's efficacy. The main idea is to increase the expectation of, and decrease the anticipated effort to obtain that, reward through the use of competent typographic layout and content:

$$\text{Likelihood of Selection} = \text{Expectation of Reward}$$

10.3.2 EFFORT REQUIRED

Expectation of reward can be best be achieved by uncluttered or bold designs with limited text. Further, this is aided by utilizing several small panels with little content, rather than a single large one or even two packed with content. Such a multipanel approach enables several user groups to progress

Geotrails 257

sequentially from one information set to another, and also not to have rely upon too much short memory recall along the trail.

A trailside panel's banner or face, however, mounted (on posts, plinths, or walls), should be at a comfortable viewing height and distance. Panels should be of a size, type, construction materials, and siting appropriate to the host environment and the expected number of users. Plinths are probably best avoided because they are too easily subjected to misuse; for example, remarkably as portable barbeque stands (Fig. 10.6A, B), but usually as standpoints and seats (Fig. 10.6C). However, there are also those who see any upright surface as a means to express their "art"—seemingly termed graffiti, that usually be removed unless engraved with some hard stone (to hand or on the ring finger?), rather the vandalism of an expensive piece of interpretative provision!

FIGURE 10.6 Two examples of panel misuse. At Hutton's section (**A**), where the contact between volcanic and sedimentary rocks helped to begin, in Edinburgh, Scotland a panel on a low plinth has clearly been used as a portable barbeque stand (**B**). At Bradwell in Milton Keynes, this mainly architectural, although the origin of the locally-used Jurassic building stone is mentioned, mid-height plinth panel is an ideal seat (**C**), despite all the alternative nearby benches!

10.4 PERCEIVED BEST PRACTICE

Published purported best-practice, for example, Knudsen et al. (1999), Tilden (1967), Ververka (1994), although usually unsupported by any current empirical studies, observation studies and the author's own research studies

(Hose, 2003) indicate that trailside panels with high attracting and holding power, and apparently communicatively competent for informational exchange, are characteristically graphics-rich and text-poor, with much 'white space' (ideally in the ratio of approximately 2:1:1, respectively). The overall color scheme of a panel should harmonize with its specific location and be of appropriate materials/construction for the environmental setting. Panel banner or face format is an important design esthetic in attracting users; oblongs, in the ratio of 5:3 and 5:4, are more visually appealing than square or panoramic ones.

For communicative competence (not to mention cost!) graphic elements, such as drawings, maps, and photographs, should be used sparingly. Indeed, a few bold graphic elements are better than a multitude of elements packed into the space. Most drawings and all maps and plans should have a comparative graphic or bar (and never numeric) scale. Location maps and site plans, with inadequate map and visualization skills inherent in at least two-thirds of the population, require special care in their preparation; enlarged copies of officially sourced topographic maps (Fig. 10.7A) are unsuitable. "Bird's eye" or orthographic perspective view sketch maps and block diagrams (Fig. 10.7B) are probably the most readily understood by casual recreationalists.

Graphics that duplicate what the user can readily see unless they either add explanation or engage them in focused observation to show that which would normally be overlooked, should be avoided. Photo-realistic graphics can beneficially affect their attracting power but users have difficulty in recalling them, whereas simple or cartoon graphics are quite memorable. Graphics with a strong seasonal bias, as commonly found on many ecological panels at sites with geological interest, such as at Barnack Holes and Hills National Nature Reserve (NNR) near Peterborough, are best avoided; the depiction of deciduous vegetation is an especial problem. Sadly, whilst the various on-site panels at Barnack Holes mention something of the site's geological interest (Fig. 10.8), which includes the quarried Jurassic limestone's use in the building of medieval Ely and Peterborough cathedrals, there is not one dedicated to it.

The selection of font type, size, and embellishment to establish a text hierarchy is significant in attracting or deterring users. Ideally, the author's "rule of three," established from working with students on tourism design projects, should be applied to the production of panels that should have no more than

Geotrails 259

1. three different fonts, sizes, and embellishments;
2. three main blocks of text, body text hierarchies;
3. major concepts, themes, and facts.

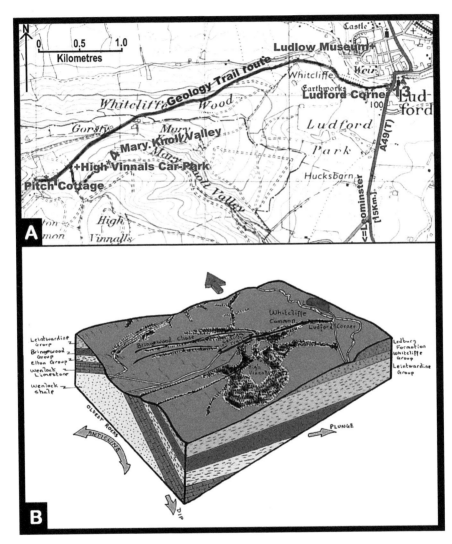

FIGURE 10.7 These two route overviews, for the Mortimer Forest Geology Trail, show the classic 1: 25,000 map (**A**) and the more user-friendly orthorhombic projection (**B**) employed for a block diagram. However, both have their drawbacks, with map requiring quite good map skill from the user and the block diagram being of little navigation use in the field; a true "bird's eye" map is the compromise.

In addition, the textual style and vocabulary should be suitable for a reading and comprehension age of 13 years or less, relatively easy to assess with modern word-processing software. Whilst this chapter does not attempt to either present an analysis of the text content of specific panels or to provide an account of the various tools to do such, it must briefly note the basics of panel text content. It should, by relating to their lives and experiences, provoke readers' interest or curiosity; the main approach should be to build site empathy rather than plain knowledge impartation. The text should also reveal a trail site's story as part of an overall theme, present its information at an appropriate level and finally fire readers' imaginations by imparting something of geology's fascination.

FIGURE 10.8 Barnack Holes and Hills NNR is protected for its biodiversity which is much dependent on its underlying Jurassic limestones disturbed by quarrying since medieval times. Most of the features highlighted are of rather short seasonal duration and interest—unlike the geology!

10.5 HISTORICAL AND MODERN GEOTRAILS IN THE UNITED KINGDOM

This brief account of geotrails focuses on the United Kingdom, partly, because it is the author's home domicile but also because the country rather led the way in the past in the development of geotrails with some of Europe's and the world's earliest such provision. Indeed, the world's first attempt to create a geotrail was associated with an innovative mid-19th-century geological theme park in south London. This consisted of accurately rendered geological sections, with 15 three-dimensional reconstructions of prehistoric animals and some plants, commissioned in 1852 for the Crystal Palace (Doyle, 1993, 1995; Doyle and Robinson, 1993; McCarthy and Gilbert, 1994). The iron-reinforced concrete and brick reconstructions included the three dinosaurs (Fig. 10.9), known from just a few bones at the time to the scientific world. Today, even with their inaccuracies, on first seeing them it is possible to understand why they caused a national sensation; unlike in Victorian times, today's journals and newspapers would not applaud their frightening realism! After a £4 m restoration, completed in 2007, they were designated by English Heritage as grade I architectural structures. An accompanying contemporary leaflet (Hawkins, 1854) guided visitors around the park.

FIGURE 10.9 One of the Crystal Palace Park dinosaur reconstructions; the stratigraphically correct plinth on which it stands is barely visible behind it due to the overgrowth of shrubs and herbaceous plants, unlike when it was first placed on the site.

The world's first standalone urban geology trail, at Rochdale (Baldwin and Alderson, 1996: 227) near Manchester, was established sometime before 1881 in a public cemetery by a local authority; the exact date it was constructed is uncertain but the eminent geologist William Boyd-Dawkins praised it in 1881 when he organized a field trip to the trail. When Rochdale Public Cemetery was laid out in 1855 its planner was assisted by two local geologists, so it is probable that the geotrail dates from the early 1860s. It has some 30 upstanding stone pillars, with each rock type named (Fig. 10.10). The start and finish stones are accompanied by a biblical quote; that for the last is "He made the Earth by His power. He hath established the world by his wisdom. The series of stones commencing here Boulder Stones and in the descending order terminating with Lava elucidates the arrangement of the Strata of the Earth's crust in the order they were formed by the creator. Speak to the Earth and it will teach thee." It is appropriate to the moral educational role attributed to natural science studies, such as geology, in the late 19th century. Supposedly, these would lead to a greater understanding and respect for the Creator's work. It is a short, at around 400 m, semicircular geotrail that starts inside the cemetery and finishes by the lodge at its entrance. Located within relatively controlled surroundings it would have been particularly useful to elementary school and Sunday School teachers for around an hour or so of instruction. No leaflet appears to have been published for the trail.

Both of these early UK trails survive to the present-day but the Rochdale geotrail required conservation work by 2010. By then, several of the pillars had been damaged or lost; they were repaired, cleaned, and replaced as necessary. However, until the second half of the 20th century little more geotrail innovation was achieved in the UK. The first new development was at the Wren's Nest (Prosser and Larwood, 2008) at Dudley within the Birmingham conurbation. Following its 1956 designation as the UK's first purely geological NNR it was recognized by the Nature Conservancy Council (the statutory conservation body) that it faced a major challenge in managing the site because "...there was a danger from over-collecting, and the approach changed to one of look-and-see rather than hammer-and-take" (Robinson, 1996a: 211). A proactive educational geoconservation approach was adopted at this redundant Silurian limestone quarry (Fig. 10.11A) with the development of a geotrail waymarked by metal posts (Fig. 10.11B) and explained in a traditional guide-book. Several later versions (Fig. 10.11C) of this were eventually supplemented by an on-site display (Robinson, 1996) in 1996. However, interpretative panels were not employed because of the high risk of vandalism on the site; it is one of the few such sites that the author and colleagues can recall being pelted with stones! This strictly geological interpretation began before the real big mid-20th-century boost in countryside interpretation with National Nature Week (see above) and the development of numerous local nature trails from the mid-1960s; perhaps not the first time geologists have been ahead of their biological cousins in developing innovative educational and interpretative projects!

FIGURE 10.10 One of the original stone plinths of the Rochdale urban geotrail, established in a public cemetery in the early 1850s. It is an Upper Carboniferous sandstone from Middle Hill at Whitworth near Manchester.

FIGURE 10.11 The precipitous and slightly unstable ripple bed in Silurian limestone (**A**) at the Wren's Nest NNR is marked by metal post (**B**) and explained in several booklets, including these two (**C**), with the one of the right in a populist format, from the late-1970s which were in circulation for many years but now replaced by much shorter and simpler explanations in various leaflets.

In the early 1970s, the 4.5 km Mortimer Forest Geology Trail near Ludlow, was the first purposely-established rural educational geology trail in the UK. It was completed, following cooperation between the Forestry Commission and the Nature Conservancy Council, in 1973. Its booklet (Lawson, 1973, 1977) directed users along a trail way-marked by discrete wooden posts (Fig. 10.12A–C) at several small, mainly roadside, quarries (Hose, 2006, pp 229–237). In the late-1980s, geology-focused visitor centers, commonly providing a range of activities such as talks, identification services, and guided

Geotrails

walks, were opened. The National Stone Centre (Hose, 1994a; Thomas and Hughes, 1993) fully opened in the early 1990s, on the southern edge of the Peak District, was one of the few to have a linked geotrail within its quarry setting. Its six disused Lower Carboniferous Limestone quarries, with their four associated lime-kilns, encompass over 100 old leadmine shafts on some 20 hectares. The short 0.5-km circular geotrail begins outside the Discovery Centre, its visitor center housing the "Story of Stone" exhibition with its emphasis on stone quarrying and construction, with seven stops ranging from a lime-kiln to a coral-reef mound (Fig. 10.13A), with views of one major quarry face from a viewing-platform. At each location, there is an interpretative panel (Fig. 10.13B) and several versions of a trail-guide are available to meet the needs of school parties and general recreationalists.

FIGURE 10.12 The Mortimer Forest Geology (Location 12) Trail, near Ludlow. Note the gently dipping Upper Silurian medium grey to greenish-grey, flaggy siltstones of the Lower Whitcliffe Formation, Ludlow Series in this small redundant roadside quarry (**A**). The difficulties of maintaining access to its rock faces are highlighted by the overgrowth of algae, lichen, and mosses. Ground ivy and bracken incursion on the floor also obscures the section, together with the soil and vegetation creep from the hillside above. However, not obvious is that the quarry's orientation and the deciduous and evergreen tree cover keeps it in shade and shadow for much of the day, making photography difficult. Its entrance is marked by a numbered post (**B**) and its geology is described in an illustrated A5 trail booklet (**C**). The nature of this two-colored printed publication can be briefly gauged from the page shown below the front cover.

FIGURE 10.13 This view of one of the smaller quarries at The National Stone Centre near Wirksworth show a fossil-packed coral reef (**A**) and the associated interpretative panel (**B**), noteworthy for its excellent typography and short content.

Outside of such multipanel provision, an attempt to provide interpretative provision aimed at tourists, also from the early-1990s, was when single, but not trailside, panels were placed by English Nature (Page, 1994) at two popular east coast seaside resort geosite locations; they were located at Hunstanton (Hose, 1994b, 1995) (Fig. 10.14), Norfolk and Scarborough, Yorkshire. It is still a common practice to place a single panel at many geological sites accessible to the public; for example, at Farndon (Fig. 10.15), near Chester. The emergence in the early-1990s of the Geological and Geomorphological Sites (RIGS) initiative (Harley and Robinson, 1991; Harley, 1996), and the consequent RIGS Groups (Burek, 2008a, 2008b) and Earth Heritage Trusts, generated a burgeoning supply of panels, geotrails, and field-guides; for example, in England at Cleeve Common near Cheltenham (in 1998) and Moorfield and Wellfield Quarries near Huddersfield (in 2001).

In the mid-1990s the North-East Wales RIGS (NEWRIGS) group began preparing their series of trail leaflets, such as the innovative 1997 "Steaming Through the Past" (Burek and France, 1998; NEWRIGS, 1997) (Fig. 10.16) based on the route of a preserved heritage railway at Llangollen and an example that, for sustainable geotourism purposes should be more widely adopted. With users of the trail needing to see views of geological and geomorphological interest on both sides of track from the carriage windows, the popularity of such tourist attractions suggests that with such carriage views at a premium, their getting let alone keeping a window seat might be a challenge! Interestingly, it adopts and updates an approach pioneered by the Geologists' Association from 1860 to around 1914 when it mainly employed trains to get its members to their field excursion location. A good example, that underpins a new cyclists' geotrail (Hose, 2017) is for its 1905 excursion to Flitwick and Silsoe (Hopkinson, 1905).

FIGURE 10.14 At the end of marine promenade in a seaside resort lies the Hunstanton Cliffs SSSI, Norfolk, significant for their 'Red' and 'White' Chalk cliffs (**A**) overlying an Upper Greensand Carstone. The panel (**B**) does a fairly good job of explaining to really casual recreationalists why the cliffs are striped, one of the few anywhere that actually answer viewers' most likely question!

The Wildlife Trusts also began to provide limited but welcome geointerpretative provision. For example, at Brown End Quarry (Fig. 10.17A) (Green, 2008; Robinson, 2004), on the south-western edge of the Peak District, the Staffordshire Wildlife Trust in partnership with the North Staffordshire Group of the Geologists' Association. The geotrail originally developed in 1991 was renewed in 2004. It lies around the 300 m or so fenced-off perimeter of the foot of Lower Carboniferous limestone quarry face. This was worked from the mid-18th century to the mid-1960s. Now, several panels have been sited to explain its geology and the use of the stone.

FIGURE 10.15 At Farndon (**A**) on the banks of the River Dee near Chester a single, and much improved over its predecessor, panel (**B**) attempts to explain the sedimentary features in the Triassic desert sandstones. It is just a pity that the obvious question asked by viewers about the mottled red-green appearance of the rock is the last thing mentioned in the text!

Meanwhile, moving away hard-rock geotrails The Great Eastern Pingo Trail (Fig. 10.18), developed by Norfolk County Council, follows the route of an abandoned railway line and other tracks through an area of rough pasture, or Breckland, over some 10 km. Its circular route, which includes several cuttings and embankments, passes close to several (with over 300 in the area) relict Ice Age pingos formed about 20,000 years ago, preserved as flooded ramparts (Fig. 10.18A), some of the most southerly such surviving periglacial features more associated today with areas such as Greenland and Alaska. The route has occasional railway station-like signs (Fig. 10.18B), marker posts and some trailside panels. The panels are a reminder of the railway that ran along part of the route from 1869 to 1965 and the old railway company of the 1920s that once ran the trains has given its name to the trail. In 1971, it was acquired by Norfolk County Council, originally to rebuild a main road route but when that plan changed in 1991 the Trail and its three associated Sites of Special Scientific Interest (SSSIs) were designated. The

Geotrails 269

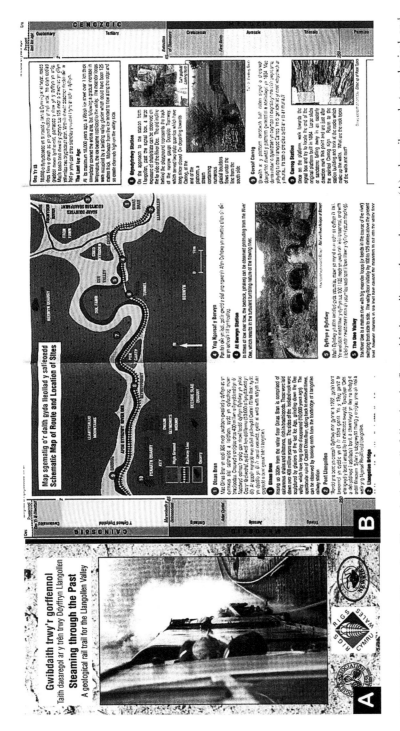

FIGURE 10.16 The "Steaming Through the Past" full-color leaflet, with its railway themed cover (A), describes what geological and geomorphological features can be seen (B) from the carriage windows of, and at the stations along, a steam heritage railway train.

accompanying guide is an A5 bifolded photocopied leaflet—simple but effective! A very new approach to geotrails is the use of downloadable pdf files which can either be printed or read on a smartphone or tablet computer.

FIGURE 10.17 At Brown End Quarry around the lower perimeter of a fenced-off Lower Carboniferous limestone quarry (**A**) several panels in a very populist style, reprieved in the trail leaflet (**B**), have been placed that explain its geology and the uses to which the stone was put.

Geotrails

FIGURE 10.18 Alongside the North East Pingo Trail several relict Ice-Age (20,000-year old or more) pingos (**A**) can be seen. The trail has occasional railway station-like signs (**B**), together with marker posts and some trailside panels, that remind users that part of its route is along an abandoned railway line in use from 1869 to 1965.

Examples of these are two cyclists' geotrails for the Milton Keynes (Hose, 2012) (Fig. 10.19) and Flitwick (Hose, 2017, 2018) areas of eastern England. The former, the "Bradwell to Newport Pagnell Geotrail," is also routed along an abandoned railway line, opened in 1867 and closed in 1967, now part of the cycle route network of the new town. The latter geohistorical trail is based upon the route of a 1905 Geologists' Association excursion (Hopkinson, 1905). Along the Bradwell to Newport, geotrail (Fig. 10.19) are several cuttings and small but either overgrown or built over with housing quarries and pits, with small but locally significant outcrops of Jurassic limestones. These, and the geology of various railway structures and local architecture, including a motorway bridge from the late-1950s, are highlighted at 14 stops on the 5-km route. The "Awheel from Flitwick Station: an 'Edwardian' Geotrail" is specifically aimed at cyclists to recreate and explain a geological excursion route originally undertaken in 1905. Its A4 trifold leaflet, themed along railway lines, is accompanied by an A3 briefing sheet, not needed on the route; both ask users to imagine what it must have been like to cycle and do geology at the beginning of the 20th century—there are references to box cameras, telephones and telegrams, railways, cycling clothing and bicycles, and the geologists on the original trip. The 22-km circular route along quiet country roads visits a dozen locations, mainly the sites of old sandstone quarries and gravel pits but including a purported 17th-century gold mine! Its railway connection is literal in that its start and finish point is a railway station on the mainline to London, some 50 minutes away. On that capital thought, it is time to conclude this chapter.

FIGURE 10.19 The Bradwell Station site (**A**) on the 'Bradwell to Newport Pagnell Geotrail' with its remaining platform and road-bridge, the geology of which are described in the trail leaflet (**B**). The full-color leaflet can be printed onto both A4 and A3 paper and trifolded for carrying in the field; but better still, it can be downloaded as a PDF file and read on a smartphone or tablet computer. Its content is mainly geological but also covers both railway and local history. Note the way the leaflet's headers and footers (their design is based upon the British Railways logo of the mid-1960s—when the railway was closed) reinforces a traditional railway publicity flyer theme.

10.6 CLOSING COMMENTS

This illustrated theoretical account has presented a broad overview of the nature, together with some aspects of the management, history, and current provision of geotrails in the UK. Its focus on panels is a useful starting point in the analysis of trailside media and not an end in itself. Indeed, many smartphones and tablet computers have form factors and screens similar in proportion to panels commonly seen along trails. Hopefully, the future will see a greater encompassing of such technologies, particularly since they are likely to better connect with young and new audiences than traditional media mainly dating in their approach from the 1970s. Whilst obviously not exhaustive it should indicate some fruitful lines of inquiry for both UK, European, and wider afield studies of such an important and widespread aspect of modern geotourism provision.

KEYWORDS

- **geo-interpretation**
- **geotourism**
- **geotrails**
- **interpretative media**
- **geosites**

REFERENCES

Badman, T. Interpreting Earth Science Sites for the Public. In *Geological and Landscape Conservation*; O'Halloran, D., Green, C., Harley, M., Stanley, M., Knill, S., Eds.; Geological Society: London, 1994; pp 429–432.

Baldwin, A.; Alderson, D. M. A Remarkable Survivor: A Nineteenth Century Geological Trail in Rochdale, England. *Geol. Cur.* **1996**, *6* (5), 227–231.

Burek, C. V. History of RIGS in Wales: An Example of successful cooperation for Geoconservation. In *The History of Geoconservation: Special Publications, 300*; Burek, C. V., Prosser, C. D., Eds.; The Geological Society: London, 2008a; pp 147–171.

Burek, C. V. The Role of the Voluntary Sector in the Evolving Geoconservation Movement. In *The History of Geoconservation: Special Publications, 300*; Burek, C. V., Prosser, C. D., Eds.; The Geological Society: London, 2008b; pp 61–89.

Burek, C. V.; France, D. E. NEWRIGS Uses a Steam Train and Town Geological Trail to Raise Public Awareness in Llangollen, North Wales. *Geoscientist* **1998**, *8*, 8–10.

Countryside Commission. *Interpretive Planning. Advisory Series No. 2.*; Countryside Commission: Cheltenham, 1979.

Dowling, R. K.; Newsome, D. The Future of Geotourism: Where to From Here? In *Geotourism: The Tourism of Geology and Landscapes*; Newsome, D., Dowling, R. K., Eds.; Goodfellow: Oxford, 2010; pp 231–244.

Doyle, P. The Lessons of Crystal Palace. *Geol. Today* **1993**, *9*, 107–109.

Doyle, P. *Crystal Palace Park Geological Time Trail: Paxton's Heritage Trail Number Two*; London Borough of Bromley: Croydon, 1995.

Doyle, P. A. Vision of 'deep time': The 'Geological Illustrations' of Crystal Palace Park, London. In *The History of Geoconservation: Special Publications, 300*; Burek, C. V., Prosser, C. D., Eds.; The Geological Society: London, 2008.

Doyle, P.; Robinson, J. E. The Victorian Geological Illustrations of Crystal Palace Park. *Proc. Geol. Assoc.* **1993**, *104*, 181–194.

Green, C. P. *The Geologists' Association and Geoconservation: History and Achievements*. In *The History of Geoconservation: Special Publications, 300*; Burek, C. V., Prosser C. D., Eds.; Geological Society: London, 2008; pp 91–102.

Ham, S. *Environmental Interpretation: A Practical Guide for People with Big Ideas and Small Budgets*; North American Press: Golden, 1992.

Hawkins, B. W. *On Visual Education as Applied to Geology Illustrated by Diagrams and Models of the Geological Restorations at the Crystal Palace*; Journal of the Society of Arts and Republished as a Leaflet by James Tennant, 149 Strand: London, 1854.

Hopkinson, J. Excursion to Flitwick and Silsoe. Saturday, 15th April, 1905. *Proc. Geol. Assoc.* **1905,** *65* (1), 1–10.

Hose, T. A. Telling the Story of Stone-Assessing the Client Base. In *Geological and Landscape Conservation*; O'Halloran, D., Green, C., Harley, M., Stanley, M., Knill, S., Eds.; The Geological Society: London, 1994a; pp 451–457.

Hose, T. A. *Interpreting Geology at Hunstanton Cliffs SSSI, Norfolk: A Summative Evaluation*; Buckinghamshire College: High Wycombe, 1994b.

Hose T. A. Evaluating Interpretation at Hunstanton. *Earth Herit* **1995,** *4*, 20.

Hose, T. A. Geotourism, or Can Tourists Become Casual Rockhounds? In *Geology on Your Doorstep: The Role of Urban Geology in Earth Heritage Conservation*; Bennett, M. R., Doyle, P., Larwood, J. G., Prosser, C. D., Eds.; The Geological Society: London, 1996; pp 207–228.

Hose T. A. *Geotourism-Selling the Earth to Europe.* In *Engineering Geology and the Environment*; Marinos, P. G., Koukis, G. C., Tsiamaos, G. C., Stournass, G. C., Eds.; AA Balkema: Rotterdam, 1997; pp 2955–2960.

Hose, T. A. How Was It for You? Matching Geologic Site Media to Audiences. *Proceedings of the First UK RIGS Conference*; Oliver, P., Ed.; University College Worcester: Worcester, 1998, pp 117–144.

Hose, T. A. European Geotourism - Geological Interpretation and Geoconservation Promotion for Tourists. In *Geological Heritage: Its Conservation and Management*; Barretino, D., Wimbledon, W. P., Gallego, E., Eds.; Instituto Tecnologico Geominero de Espana: Madrid, 2000; pp 127–146.

Hose, T. A. Geotourism in England: A Two-region Case Study Analysis, Unpublished PhD thesis, University of Birmingham: Birmingham, 2003.

Hose, T. A. Writin Stone: A Critique of Geoconservation Panels and Publications in Wales and the Welsh Borders. In *Stone in Wales*; Coulson, M. R., Ed.; Cadw: Cardiff, 2005; pp 54–60.

Hose, T. A. Geotourism and Interpretation. In *Geotourism*; Dowling, R., Newsome, D., Eds.; Elsevier: London, 2006; pp 221–241.

Hose, T. A. *Newport Pagnell to Bradwell Geotrail*. BEHG and Chiltern Archaeology Online Publication, 2012. (available at http://www.bucksgeology.org.uk/pdf_files/Geo-cycle-trail-Bradwell-to-Newport-Pagnell-DRAFT.pdf).

Hose, T. A. *Appreciating Physical Landscapes: Three Hundred Years of Geotourism*; The Geological Society: London, 2016a.

Hose, T. A. *Geotourism in Britain and Europe: Historical and Modern Perspectives. Geoheritage and Geotourism: A European Perspective*; Hose, T. A., Ed.; Boydell Press: Woodbridge, 2016b; pp 153–171.

Hose, T. A. *A Wheel from Flitwick Station: An 'Edwardian' Geotrail*; Rockhounds Welcome: Chalton, 2017.

Hose, T. A. A Wheel in Edwardian Bedfordshire: A 1905 Geologists' Association Cycling Excursion Revisited and Contextualised. *Proc. Geol. Assoc.* **2018,** *129* (in press).

Knudsen, D. M.; Cable, T. T.; Beck, L. *Interpretation of Cultural and Natural Resources*; Venture Publishing: State College, PA, 1999.

Lawson, J. D. New Exposures on Forestry Roads Near Ludlow. *Geol. J.* **1973,** *8*, 279–284.

Lawson, J. D. *Mortimer Forest Geological Trail*; Nature Conservancy Council: London, 1977.

Light, D. Visitors' Use of Interpretative Media at Heritage Sites. *Leisure Stud.* **1995**, *14*, 132–149.

McCarthy, S.; Gilbert, M. *The Crystal Palace Dinosaurs: The Story of the World's First Prehistoric Sculptures*; Crystal Palace Foundation: London, 1994.

NEWRIGS. *Steaming Through the Past: A Geological Trail for Llangollen Valley*; NEWRIGS/Chester College: Chester, 1997.

Page, K. N. Information Signs for Geological and Geomorphological Sites: Basic Principles. In *Geological and Landscape Conservation*; O'Halloran, D.; Green, C.; Harley, M.; Stanley, M.; Knill, S. Eds.; Geological Society: London, 1994; pp 433–437.

Prosser, C. D.; Larwood, J. G. Conservation at the Cutting-edge: The History of Geoconservation on the Wren's Nest National Nature Reserve, Dudley, England. In *The History of Geoconservation, Special Publications, 300*; Burek, C. V., Prosser, C. D., Eds.; Geological Society: London, 2008; pp 217–235.

Robinson, E. Geology Reserve: Wren's Nest revised. *Geol Today* **1996**, *12*, 211–213.

Robinson, E. Briefing. *Geol. Today* **2004**, *20* (5), 162–169.

Thomas, I. A.; Hughes, K. Reconstructing Ancient Environments. *Teach Earth Sci* **1993**, *18*, 17–19.

Thomas, I. A.; Prentice, J. E. A Consensus Approach at the National Stone Centre, UK. In *Geological and Landscape Conservation*; O'Halloran, D., Green, C., Harley, M., Stanley, M, Knill, J., Eds.; The Geological Society: London, 1984; pp 423–427.

Tilden, F. *Interpreting our Heritage*; University of North Carolina Press: Chapel Hill, 1967.

Trapp, S.; Gross, M.; Zimmerman, R. *Signs, Trails, and Wayside Exhibits: Connecting People and Places*; 2nd ed; UW-SP Foundation Press: Stevens Point, 1994.

Ververker, J. A. *Interpretive Master Planning*; Falcon Press: Helena, 1994.

CHAPTER 11

Interpreting Geological and Mining Heritage

ROSS DOWLING

Honorary Professor of Tourism, School of Business & Law, Edith Cowan University, Western Australia.

E-mail: r.dowling@ecu.edu.au

ABSTRACT

Geoheritage are those parts of the geological environment that may be identified as having specific value, whether scientific or otherwise, and therefore worthy of being conserved. One of the key elements of geoheritage is the interpretation of the geological heritage which a visitor is viewing. This is referred to as geo-interpretation which places geology within the environment's **A**biotic (nonliving), **B**iotic (living), and **C**ultural (human) attributes. This approach to environmental interpretation is termed the "ABC" Interpretation Method. In geo-interpretation, the most important element is geology and it is suggested that this be interpreted according to its "form," "process," and "time" aspects. This chapter illustrates geo-interpretation by showcasing interpretation for three geotourism attractions and three mine sites from six countries around the world.

11.1 INTRODUCTION

One of the key elements of geotourism is the interpretation of the geological heritage which a visitor is viewing. This chapter identifies a number of characteristics of sound geo-interpretation. It first begins by defining geological heritage (geoheritage) and geotourism. It then describes interpretation before characterizing geological interpretation (geo-interpretation) through a variety of interpretation methods. A number of essential elements of geo-interpretation are introduced including couching geology within the environment's **A**biotic (nonliving), **B**iotic (living), and **C**ultural (human) attributes, giving rise to the "ABC" approach to interpretation. In geo-interpretation the most important element is geology and it is suggested that this be interpreted according to its "form," "process," and "time" aspects. It is also noted that geological information should be presented in such a way as to provide the reader with some form of emotional connection to the geosite being viewed, rather than just presenting its geological facts. Finally, the "Rule of Thirds" is presented which suggests that geological information presented in brochures or information panels and so on should comprise one-third text, one-third visuals, and one-third space. This makes for better interpretation which will resonate more with the reader than one filled with highly geologic detail. The final part of the chapter presents examples of geo-interpretation showcasing interpretation for three geotourism attractions and three mine sites from six countries around the world.

11.2 GEOHERITAGE

Heritage simply relates to those things which we value. Geological heritage, or more simply, geoheritage, relates to parts of the geological environment that may be identified as having specific value, whether scientific or otherwise, and therefore worthy of being conserved (Gray, 2018). A good example of geoheritage is in the United States where it is viewed as arising from "features, landforms, and landscapes characteristic of the United States, which are conserved in consideration of the full range of values that society places on them, so that their lessons and beauty will remain as a legacy for future generations" (NPS/AGI, 2015: 6).

Thus geoheritage is about understanding and conserving geological formations and their processes due to their esthetic, artistic, cultural, ecological, economic, educational, recreational, and scientific values. This is carried out through geoconservation which is the application of management techniques to protect and conserve geoheritage sites and collections (Newsome and Dowling, 2018). Geoheritage sites provide educational value by offering examples of geological principles and processes for study. They also provide formal and informal settings for instructing people not only about geology but also about history, culture, and other subjects. Finally, geoheritage enhances our science literacy about the Earth by demonstrating natural processes and hazards and how human activities impact the environment (NPS/AGI, 2015).

11.3 GEOTOURISM

At its simplest, tourism comprises either mass tourism or alternative tourism. The former is characterized by large numbers of people seeking replication of their own culture in institutionalized settings with little cultural or environmental interaction in authentic settings. Probably the most well-known types of alternative or special interest tourism are ecotourism and cultural tourism. However, a new type of tourism has emerged called geotourism (Dowling, 2013). A simple and easy to understand synopsis of geotourism is that coined by the US National Park Service. It is:

> "Geotourism is the term used internationally for the practice of hosting visitors at natural areas specifically to enjoy geologic features and processes. Many communities around the world are sustained by geotourism" [NPS/ACI, 2015: 33].

Geotourism comprises both "geo" (geological) and "tourism" elements (Dowling and Newsome, 2018). Geology is the study of the Earth and geomorphology is the study of landforms. Natural resources include landscapes, landforms, rock outcrops, rock types, sediments, soils, and crystals. The "tourism" part means visiting, learning from and appreciating geosites in a sustainable manner (Newsome et al., 2012). Geotourism has a primary focus on experiencing the Earth's geological features in a way that fosters environmental and cultural understanding, appreciation and conservation, and is locally beneficial (Dowling, 2015a). It is about creating a geotourism product that conserves, educates, and promotes geological heritage (Dowling, 2015b). Overall, geotourism comprises the geological elements of "form" and "process" combined with the components of tourism such as attractions, accommodation, tours, activities, interpretation as well as planning and management. Thus geotourism may be viewed both as a form of tourism as well as an approach to it, but one that firmly ties itself first to the geologic nature of an area's "sense of place." Such tourism development generates benefits for conservation (especially geoconservation), appreciation (through geoheritage interpretation), and the economy. An added component of geotourism is that, unlike ecotourism which by definition can only occur in natural environments, geotourism occurs wherever there is rocks or landforms and so on, and hence can take place in both built (urban) as well as natural settings (Riganti and Johnston, 2018).

One particular type of geotourism is tourism related to mining and mine sites (Garofano and Govoni, 2012; Mata-Perelló et al., 2018). Whilst there is a long history of utilizing former quarries and mine sites for geoconservation (Prosser, 2016) through tourism they can play a crucial role in unveiling geological objects which would not normally be seen. Through tourism development mines can bring about considerable economic and social benefits. In Italy a decade ago, it was estimated that there were 35 active tourist mines visited by approximately 215,000 visitors per annum (Govoni, 2007). There a number of examples of former mines being transformed into tourist sites include the Mazarrón District in south east Spain. It has been mined since Roman times (200 BC–300 AD) for lead, and later for aluminum, silver, and zinc. Another mining district in the region is the Cartagena–La Unión region in which lead and zinc have been mined first, in traditional underground methods, and latterly by large open pit mining operations (López-García et al., 2011). Both sites are used for research and education and each year they attract visiting geological lecturers and students from Spain and other European countries.

The Sterling Forest area of New York, the United States, is a former iron smelter complex and today an Iron Mine Trail provides geo-interpretation of the area's mining and smelting heritage (Gates, 2018). The trail illustrates how iron is obtained and processed as well as the importance of geology in everyday life. In 2008, in its first year of operation 1500 visitors in 26 groups undertook tours of the trail as part of an estimated 5000 visitors. Today it is estimated that about 25,000–30,000 visitors use the Trail for geotourism and attendance at the nearby Sterling Forest Park has risen from 120,000 in 2005 to 350,000 visitors per year (Gates, 2018). At a height of over 4000 m, Potosí, the capital of the Bolivian state of the same name, is one of the highest cities in the world. The city was founded as a mining town in 1546 and over the next 200 years, more than 40,000 tons of silver were shipped out of the town, making the Spanish Empire one of the richest the world had ever seen (www.atlasobscura.com/places/potosi-silver-mines). Here the city's operating silver mines are also tourism attractions (Pretes, 2002). The city offers a variety of attractions including well-preserved colonial buildings, interaction with the indigenous Quechua population, and tours of its silver mines. The Quechua miners guide tourists through the mines sharing stories about the area's mining history. Companies don't employ the miners; individuals simply take a contract out on a plot of the mountain and are allowed to do as they please. They supply their own materials and sell any minerals they gather to buyers in town. These enterprising young men (sometimes children) allow tourists into their mines in exchange for supplies, which increases their profit margin. However, the mine tour is not for the faint hearted and is described by the locals as a trip to the "gates of hell" (Castell, 2017).

11.4 INTERPRETATION

Interpretation is used to help a visitor gain a knowledge of, and affinity for, the natural and cultural world. Effective interpretation uses both accurate information and a variety of interpretive techniques which help visitors respond to the environment on both an intellectual and emotional level. Information is not interpretation—interpretation is revelation based upon information. The "father" of heritage interpretation was the American Freeman Tilden (1883–1980). He was one of the first people to set down the principles and theories of interpretation in his book, *Interpreting Our Heritage* (1957), which is still considered to be the definitive text for the discipline. His central message was that interpretation is an educational activity which aims to reveal meanings and relationships through the use of original objects, by firsthand experience, and by illustrative media, rather than simply to communicate factual information. He proposed that interpretation must relate to what is being displayed or described to something within the personality or experience of the visitor (Tilden, 2007). One of his most cited phrases was "Through interpretation, understanding; through understanding, appreciation; through appreciation, protection" (Tilden, 2007).

A particular challenge for interpreters is to engage younger people in taking and maintaining an interest in the natural environment, including its geoheritage. Louv (2005) in his book the *Last Child in the Woods,* identified a new and growing body of research indicating that direct exposure to nature is essential for healthy childhood development and for the physical and emotional health of children and adults. Thus interpretation addressed to children (up to the age of 12), should not be a dilution of the presentation to adults, but should follow a fundamentally different approach and hence requires a separate program. The youth of today are defined by their increased focus on formal education. They spend much of their life behind screens and on digital devices. They live "time compressed" lives which are very busy and full and their lives tend to be "micromanaged" by hypervigilant parents who are "risk adverse" and "over protective" (Dowling, 2017). Our challenge then is to find ways of getting young adults interested in the great outdoors generally, and in geoheritage particularly. This can be achieved by appropriate, effective, and transformative geo-interpretation.

11.5 GEO-INTERPRETATION

Interpretation is essential in the delivery of sound geotourism. Geo-interpretation may be carried out through publications and websites, electronic educational resources, visitor centers, self-guided trails, and guided touring. The US National Park Service undertakes geological education and interpretation of its sites through "guided field trips, interpretive signage, lectures, presentations, publications, documentaries, and online formal and social media" (NPS/AGI, 2015: 46). All are used to facilitate public engagement and appreciation for geological sites. To achieve this, educators, museum staff, and park interpreters all play important roles in promoting geologic heritage education. Before undertaking any interpretation it is important to devise an interpretation strategy which begins with the question "why do we want to interpret this landscape/geopark/reserve…?" and always focuses on keeping the visitor safe (Macadam, 2018). Once the strategy is in place there are a number of essential elements to be included when interpreting geological heritage (Fig. 11.1). These include the use of a variety of interpretation methods, ensuring that geology is presented as the foundational element of the other **A**biotic, **B**iotic, and **C**ultural features of the natural environment. These include information about and interpretation of, the "form," "process," and "time" aspects of geology through clear and easy-to-understand written material, whilst at the same time providing "connection" for the tourist through appropriate language and illustrations.

There are a variety of methods of modern geo-interpretation. Traditional methods include print material such as brochures, guidebooks, information panels, and visitor centers. Face-to-face methods include geology lectures, presentations, or tours with appropriately trained geoguides. Increasingly though electronic methods of interpretation are being generated either by *industry*, through videos, websites, or documentaries, or by *consumers* themselves, through Instagram, You Tube or Trip Advisor. It has been further suggested that such methods can be divided into *on-site* interpretative facilities available to tourists at any time, such as, information panels, and *off-site* facilities such as visitor centers, museums, exhibitions, and websites (Migon, 2018).

Essential to geo-interpretation is the understanding of the identity or character of a region or territory. To achieve this geotourism is viewed as being based on the idea that the environment is made up of **A**biotic, **B**iotic, and **C**ultural components. This **"ABC"** approach comprises the *Abiotic* elements of geology and climate, the *Biotic* elements of animals (fauna) and plants (flora), and *Cultural* or human components, both past

Geo-interpretation is used to help a person gain a knowledge of, and affinity for, the geological world in a safe manner

Interpretation Methods	Traditional	Face-to-Face	Electronic
	Brochures Guidebooks Information Panels Visitor Centers	Lectures Presentations Tours with experts or guides	1. *Industry generated* Videos, websites, documentaries, smartphone applications 2. *Consumer generated* Instagram, You Tube, Trip Advisor
The 'ABC' Approach	**Abiotic** Starts with 'big picture' descriptions of geology and climate as these shape the biota (fauna and flora) of a region	**Biotic** Describes the animals and plants which live in the area, as taken together with the abiotic elements, these shape the cultural landscape	**Cultural** The abiotic and biotic elements (including geological) have shaped how people have lived in the area in the past, as well as how they live there today
Geological Elements	**Form** An understanding of the landforms and other geological features starting from large (landscapes) to small (rocks and minerals)	**Process** A description of the ways in which the geological features were formed, e.g., volcanism, plate tectonics, erosion or weathering, etc	**Time** A description of the geological era and length of time over which the features were formed – always related to human years for ease of understanding
Providing 'Connection'	**Head** *Learning* about the region generally and its geology specifically	**Heart** *Connecting* the visitor to the landscape in an engaging affective manner through storytelling	**Hand** *Motivating* the visitor to do something for the geological environment through transformational interpretation
Print Presentations [The rule of 'thirds']	**Text**[⅓] ABC information including the geological content presented in simple, easy to understand words and / or concepts	**Visuals**[⅓] Appropriate diagrams and photos which reflect and interpret the landscape or feature being viewed	**Space**[⅓] Ensuring that the text and visuals make up no more than two thirds of the available space with the final third being 'free' space

FIGURE 11.1 Essential elements of geo-interpretation.

and present (Dowling, 2015c). Geotourism argues that to fully understand and appreciate the environment we must know about the **A**biotic elements of geology and climate first, as these determine the **B**iotic elements of animals and plants which live there. By extension, the combination of the **A**biotic and **B**iotic components of the environment, determine the **C**ultural Landscape of how people have lived in the area in the past, as well as how they live there today, in the present. Therefore, any form of geo-interpretation should start with an understanding of the geology of an area then use this information to provide a simple account of how this has shaped the plants and animals there, which then informs the reader why and how people have lived there in the past, as well as today. This then presents a holistic picture of the natural and built environment based firmly on the foundation of geological understanding.

Another essential element in the interpretation of geoheritage is to ensure that includes the geological elements of form, process, and time. "**Form**" refers to the shape of an area including its landscape and landforms, that is, what the viewer can see. "**Process**" describes how the landscape and landforms originated, through processes such as volcanism, plate tectonics, or erosion and so on. Finally, the element of "**Time**" refers to when and how long these processes occurred. This last element should always relate geological time to human time by placing the 4.2 Billion years of earth history in the context of the average human life of around 70–80 years. In this way the reader will be better able to relate to geological time and better understand whether the geological time period being described is old or relatively recent.

Geo-interpretation is also about providing the tourist or visitor a form of "connection" with the geological features and geoheritage of an area. It includes *communication*—bringing items and events to life; *learning*—in order to facilitate understanding by revealing meaning; *values*—by stimulating an appreciation of heritage, and finally *action*—by involving or immersing people in their geological encounter. To achieve this it is important to ensure that geo-interpretation includes aspects of the "head," "heart," and "hand." This is carried out by interpreting geological information through the **Head** in which *learning* about the geology of an area and its form, process, and time occurs; the **Heart**—*connecting* the visitor to the landscape in an effective manner that engages their interest; and finally by **Hand**—inspiring the reader so that transformation occurs where they are motivated into *doing* something for the geological environment through their new found understanding and connection to geoheritage.

Interpreting Geological and Mining Heritage   285

This approach provides the basis of a more holistic understanding of the environment and its component parts and thus provides the tourist or visitor with a greater connection to the environment in which they live or are visiting (Hughes and Ballantyne, 2010). Essential to this approach in communicating geoheritage understanding and connection is the role of **storytelling** (Pastorelli, 2003). This is particularly important when geo-interpretation is being undertaken by experts or guides. Stories are about relationships and in a geological setting they connect the visitor to the geology and landforms on view in an interesting and informative manner (Macadam, 2018).

Perhaps the most problematic issue in geo-interpretation is the amount of information presented in a brochure or interpretive panel. Often the reader is daunted by considerable scientific information on the geology of the area which is often unintelligible to the average person. Here is it helpful to remember "**The Rule of Thirds**." That is, any interpretive panel, sign, or brochure should comprise one-third text, one-third illustrations (diagrams or photos), and one-third space. A key aspect is that the text must be written in such a way as to be clearly and easily readable and understandable by members of the general public. Therefore, the geological content should always be written first by expert geologists, then interpreted by professional communicators for consumption by the general public. As has been eloquently stated before "Geotourism is too important to be left to geologists" (Hamilton-Smith, 2008).

11.6 CASE STUDIES

The remainder of this chapter comprises examples of geo-interpretation showcasing interpretation for three geotourism attractions and three mine sites. The geotourism attractions are a cliff site in South Korea, a buried village in Iceland, and a geological visitor center in the United States. The three mine examples are a former coal mine tour in New Zealand, a current rock salt mine, and underground geotourism attraction in Poland, and a proposed phosphate museum and experience in Jordan.

11.6.1 A GEOSITE ATTRACTION, JEJU ISLAND, SOUTH KOREA

Jeju is an oval-shaped island in the southern part of South Korea. It is 73 km from east to west and 31 km from north to south (Jeju Tourism Organisation,

2011). Also a province (Jeju Province), the Island is the largest and southernmost of South Korea and is located south-west of the Korean Peninsula (Jeon and Woo, 2018). The island is a geological wonderland dominated by the volcano Mt Hallasan, which is 1950 m high, which is the highest mountain in South Korea. The island is replete with volcanic landforms including approximately 360 oreums (small volcanic scoria or tuff cones), lava domes, a lava tube system, tuff rings, and columnar joints. The island's beautiful and excellent natural resources have been recognized by its designation as a UNESCO—biosphere reserve (2002), World Heritage Site (2007), and Global Geopark (2010). The island as a whole is a geotourist's paradise (Kim, 2017). A major geotourism attraction is the Jusangjeolli columnar rock pillars near Jungmun Beach on the southern side of the island. The impressive rock pillar formation is the largest in Korea and was formed when the lava from Mount Hallasan erupted into the sea of Jungmun. The cliffs created between 250,000 and 140,000 years ago when Hallasan was still an active volcano, and it offers a rare glimpse at what can only be seen at very few points around the globe. Much like a folding screen, the columnar joints at Jungmun Daepo Coast spread out for about 2 km along the southern coastline (Jeju Province, nd). It was formed when hot lava contracted as it cooled forming cracks perpendicular to the surface of the flow. The cracks grew from the top, contacting the air in a downward direction, and from the base, touching the ground and upward to form long straight columns. While variations on the surface range from between four to seven sides, a hexagonal honeycomb-like pattern is dominant in most columns. More rapid cooling makes narrower columns with thinner surface bandings on the sides. Clinkers can be seen on the upper side of columnar joints at Jungmun Daepo Coast, which were formed by viscous lava flows breaking up the surface of columns. The geo-attraction is situated within a park close to the imposing cliffs which features different types of palm trees and flowers. The park has been created parallel to the cliffs allowing visitors to walk a 2 km trail so the pillars can be seen from as many angles as possible. Its most remarkable aspect is the multi-level observation deck which allows various viewing points of the Jusang jeolli Cliff. The deck and walkway is complemented by a viewing platform and interpretive signage. The signs are written in three languages—Korean, English, Japanese, and Chinese, are easy to understand, and have an excellent balance of text, illustrations (and /or photos) and space, thus being an exemplar of geo-interpretation's "rule of thirds" (Fig. 11.2).

11.6.2 ELDHEIMAR VOLCANO MUSEUM, VESTMANNAEYJAR, ICELAND

Iceland is also a geotourist's paradise with rugged landscapes, glaciers, volcanoes, and geothermal activity. Sitting astride the Mid-Atlantic Ridge, Iceland has 22 active volcanoes, 250 geothermal areas, 780 hot springs, and the world's third largest icecap. It is one of the world's most active hot spots with one-third of all the lava to surface on earth in the last 1000 years being of Icelandic origin. Lying off the southern coast of Iceland, the Vestmannaeyjar Islands were formed by submarine volcanoes 11,000 years ago. On the main island of Heimaey, a volcanic eruption in 1973 created a 1.5 km fissure which split the eastern side of the island. The eruption area formed a new mountain the red cinder cone Eldfell "Fire Mountain." One-third of the town on the island was buried beneath the lava flow and the island increased in size by 2.3 sq km. Today, the resultant cinder cone is a major tourist drawcard and has given rise to the local tourist attraction of a *Volcanic Film Show* which focuses on the eruption (Lopes, 2005).

FIGURE 11.2 Geo-interpretation of columnar rock pillars, Jeju Island, South Korea. The rock pillars of Jusangjeolli (left) are a major tourist attraction on Jeju Island. Their interpretation in the information panel (right) is an excellent example of "The rule of thirds" displaying one-third text, one-third visuals, and one-third space. It also displays simple geological information in four languages—Korean top left, English (lower left), Japanese (center), and Chinese (right).

Source: Ross Dowling (2013).

A new geotourism attraction called "Eldheimar—World of Fire" was opened in 2014. It is a museum of remembrance based on the aftermath of a volcanic eruption on the island of Heimaey (http://eldheimar.is). The eruption started in the early hours of January 23rd 1973 on Heimaey, the only inhabited island of the Vestmannaeyjar islands, and it lasted for five months. Lava and ash destroyed almost 400 homes and businesses, a third of all buildings on Heimaey. While the eruption lasted no one knew if the island would ever be populated again. The highlight of the exhibition is the excavation of a house which was buried in ash and lava for over 40 years, has now been excavated and is displayed as part of the museum experience. The attraction aims to excavate a number of other former buried houses which were buried in tephra by the eruption. Inside the houses are all of their contents which were buried 4 days after the eruption commenced. The excavations so far have revealed much that is well preserved over the past 35 years. At present excavations have commenced and the tops of some houses are exposed. Interpretive signs have been erected and it is already attracting many visitors (Fig. 11.3). The excavation project has been called "The Pompeii of the North."

FIGURE 11.3 Eldheimar Volcano Museum, Heimaey Island, Iceland. Eldheimar, a museum of remembrance, commemorates the 1973 volcanic eruption on the island of Heimaey, Iceland. The eruption lasted for 5 months with the resultant lava and ash destroying almost 400 homes and businesses, a third of all buildings on the island. A central part of the museum is the in-situ exhibition of one of the house which was buried in ash. Photos: Top Left- Museum Logo; Top Right: Eldheimar Museum; Lower Left: House buried by ash (2008); Lower Right: The same house in the museum today (2018).

Source: Museum photos courtesy of Kristín Jóhannsdóttir, Director, Eldheimar Museum (2018), Buried house, Ross Dowling (2008).

11.6.3 A GEOTOURISM VISITOR CENTER, IDAHO, USA

The Teton Geotourism Center is located in Driggs, Idaho, the United States on the western side of the Teton Range, close to the Grand Teton National Park. It is billed as the world's first geotourism center (http://www.tetongeotourism.us). Established by the Driggs City Council, Driggs Urban Renewal Agency, and Teton Valley Chamber of Commerce in association with National Geographic, it opened on 1 August 2014 and has been attracted over 40,000 visitors since that time (K. Eggebroten, personal communication, 26 January 2018).

The center is an excellent example of geo-interpretation with exhibits on the geology, fauna and flora, history and culture of the region. The Center's Welcome signboard statement defines geotourism and introduces the concept of the Geo-traveler.

From there the walk through the centre starts with information about the geology of the Grand Teton Range and nearby Yellowstone National Park. It then provides information about the animals and plants of the region before introducing the visitor to the human (Indian) history and the lives of the early European settlers. The exhibits also show how the area has grown as a tourism destination over the past 100 years. Altogether the visitor leaves the Center with an excellent understanding of how the geology and landscape of the Teton Range has shaped the biotic and cultural characters of the surrounding region (Fig. 11.4).

11.6.4 A FORMER COAL MINE TOUR, WESTLAND, NEW ZEALAND

The Stockton Coal Field is located between 500 m and 1100 m above sea level in the Buller Coalfield, 35 km north of Westport on the West Coast of the South Island. It is being mined in New Zealand's largest opencast mining operation. In recent years, Stockton production has been around 2 million tonnes (mt) a year of high-value bituminous coal. By continuing to develop its available resources in the area, the mine operators expected the operation to continue production at these levels until 2028. However, with a major drop in coal prices in 2012, operations were scaled down to an annual production of around 1 million tonnes.

Geotours of the Stockton Coal Mine were undertaken for 15 years until mid-2016 by Outwest Tours (http://outwest.co.nz). Initially, the tours were run without any financial assistance from the mine but in 2003 a new mine manager recognized their public relations potential. Stockton Mine then fully subsidized the cost of the tours and the only payment by visitors was a gold coin donation to the West Coast Rescue Helicopter. During the 15 years of the tour operations, over 20,000 people undertook the mine tours. The tour was operated as a 5-hour return journey from Westport, the nearest town, to the mine and return. Visitors received a quality, highly informative commentary from local guide who had worked in the underground mines as a school leaver and in later years as a machine operator in the opencast mine (Dowling and Pforr, 2017). The owner/guide, Mickey Ryan, has an excellent knowledge of the mine's history, the mining operation and machinery, and the rehabilitation methods. Besides the many individuals and family groups, large numbers of schools visited the mine as part of their geography studies and university geology and environmental students were also frequent visitors.

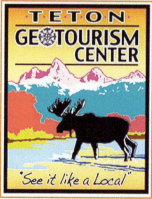

FIGURE 11.4 Teton Geotourism Center, Idaho, the United States. The Teton Geotourism Center in Driggs, Idaho was the first of its type in the world. It opened in 2014 and the center is an excellent example of geo-interpretation with exhibits integrating information about the region's geology, fauna and flora, history and culture. Photos: Upper left, Grand Teton Range; Upper Right: Teton Geotourism Center; Lower left: Center Logo; Lower Right: Center Exhibit.

Source: Upper photos: Ross Dowling, 2016. Lower photos: Teton Geotourism Center.

The mine tours generated many benefits for the mine and the tourists, making it a win-win situation between mining and tourism. First, it gave tourists the opportunity to see opencast coal mining close up and learn about the mining techniques, what the miners were doing, and water management and rehabilitation methods (Fig.11.5). The mine was no longer hidden from the general public and visitors could see for themselves the benefits which the mine brought to the local community. In addition, the tours were undertaken by many school groups who were able to learn about the value of mining and its contribution to the local society and economy. They were also able to view the resultant rehabilitation of the open cast mine which was being conducted in an environmentally sustainable manner.

FIGURE 11.5 Stockton Coal Mine Tour, South Island, New Zealand. Geotours of the Stockton Coal Mine were undertaken for 15 years until mid-2016 by Outwest Tours (http://outwest.co.nz). During this time over 20,000 visitors undertook a tour of the open cast mine site. *Source:* Outwest Tours.

However, with the drop in coking coal prices in 2012 and the financial difficulties of the mining company, support for the fully subsidized cost of the tours was withdrawn. A charge of NZb$25 per person was introduced in 2012 which was further increased to $50 in 2014. However, with the introduction of the fee the numbers visiting Stockton declined markedly and today the tour no longer operates. In 2017, the mine was sold to new owners but as yet the Mine Tours have not been resurrected.

11.6.5 A ROCK SALT MINE AND UNDERGROUND CHAPEL, WIELICZKA POLAND

The Wieliczka salt mine is located on a geological rock salt deposit in southern Poland within the Kraków metropolitan area. The mine is currently one of Poland's official national Historic Monuments whose attractions include dozens of statues and four chapels carved out of the rock salt by the miners, as well as supplemental carvings made by contemporary artists. The mine was worked continuously from the 13th century until the late 20th century and it constitutes one of the earliest and most important European industrial operations. This major industrial undertaking has royal status and is the oldest of its type in Europe.

The Wieliczka Royal Salt Mine illustrates the historic stages of the development of mining techniques in Europe from the 13th to the 20th centuries. Along with the nearby Bochnia mine, they have hundreds of kilometers of galleries with works of art, underground chapels and statues sculpted in the salt, making a fascinating pilgrimage into the past. The mines were administratively and technically run by Wieliczka Saltworks Castle, which dates from the medieval period and has been rebuilt several times in the course of its history (UNESCO, nd).

The Wieliczka rock salt mine has been a royal property and the salt trade provided a significant part of the income for the Polish Kingdom. Until it finally finished operations in 1996, it was the oldest continuously operating underground mine in Europe. After centuries of mining activity the Mine has a 300 km long labyrinth of galleries and chambers, including nearly 3000 open spaces, with a vertical extent from 60 m to 300 m below the surface. The Wieliczka Mine includes a Saltwork Museum (www.kopalnia-wieliczka.com.pl) and it is a world-class geotourism attraction visited each year by over 1 million tourists (Slomka and Mayer, 2010). In 1978, the mine was one of the first sites in the world included on UNESCO's World Heritage list. In 1994, it was officially recognized as the Monument of National History (same as the adjacent Bochnia Salt Mine in 2000). The underground trail is about 3.5 km long and includes shafts, chambers, and galleries, and even an underground pond. The museum offers a unique exhibition: from historical documents and maps through rock-salt sculptures, minerals, and rocks to original mining equipment. At the lower levels, which are closed to the ordinary tourists but may be visited by specialized groups, outcrops of Miocene evaporites and examples of salt tectonics can be examined.

The Wieliczka mine is often referred to as "the Underground Salt Cathedral of Poland." The leading attraction of the underground tourist trail

Interpreting Geological and Mining Heritage

is the St. Kinga Chapel. Located 100 m below the surface, the chapel has been continuously carved in a worked-out rock-salt block between the years 1896 and 1963 although the last sculptures were made in 2003. The chapel is over 50 m long, 20 m wide and up to 12 m high (Fig. 11.6). Today it still plays a religious role but is also a unique exhibition of rock-salt sculptures and a concert hall. A chamber has walls carved by miners to resemble wood, as in wooden churches built in early centuries. A wooden staircase provides access to the mine's 64 m level. A 3-km tour features corridors, chapels, statues, and a lake, 135 m underground. An elevator which holds 36 people returns visitors to the surface in a 30-s trip.

FIGURE 11.6 Saint Kinga Chapel at the Wieliczka Salt Mine, Poland. The Saint Kinga Chapel at the Wieliczka Salt Mine—a chamber cut in a single block of rock-salt and decorated with sculptures and bas-reliefs. The chapel is a major tourist attraction in Poland and one of the leading geotour attractions.
Source: Krzysztof Slomka, with permission.

The other outstanding features of the Wieliczka Mine and Salt work Museum are the "Crystal Caves," two large cavities encrusted with perfect, cubic halite mega crystals and showing the examples of salt karst. In 2000, the caves were granted the status of a nature reserve. Both are permanently closed to the public due to protection of sensitive microclimate in which highly soluble halite crystals require for stability.

11.6.6 A PROPOSED PHOSPHATE MINE MUSEUM AND EXPERIENCE, RUSEIFA, JORDAN

An abandoned phosphate site in Ruseifa, Jordan is being developed as a geoheritage park museum as part of the country's first geopark (Al Rayyan et al., 2017). The city of Ruseifa, in the Zarqa governorate, is located around 15 km to the north of Jordan's capital city, Amman. In 1935, the region was a thriving mining area but today it is the site of abandoned mines and neglected buildings and closed tunnels. Phosphate outcrops display the story of their genesis and paleo-environment with remnants of various fossils and other sedimentary features.

A proposal has been generated to create a geotourism attraction on the site of the abandoned mines to restore the heritage buildings and build a geological educational center. The project's focus is to create local awareness about the history of mining, preserve some of the old mines, explain phosphate formation, composition, and use, and showcase the role the mining industry played in the lives of people. The overall goal is to add to the existing tourism attractions in the city by developing a new type of museum that will generate both economic and social benefits for the city and its surroundings. A point of difference is that it is located in the heart of the modern city of Ruseifa on a 12.6 ha site which is bounded on one side by the Zerqa River.

The main aim of the project is to transform the site into a geopark with emphasis on the geological heritage, the history of phosphate mining in Jordan and its importance to local communities. Architecturally, the site will focus on the user experience through exploration which will create a journey that illustrates the story behind the site in an innovative and interactive way. The key interpretive goal is that it will be designed in such a manner as to allow visitors of all ages to be able to explore the site and learn more about its history in an informed, entertaining way. To facilitate that, the site will be divided into three zones, linked together by various pathways, bridges, and visual points. The first zone created will use a former water tower building as a focal point for activities. From here visitors will be able to choose to walk to either the museum, to the mines or to another section in which various activities will take place. Along the mine tunnel entrances, a meshed wall will be created exposing geological layers and in-situ fossils. One of the mine tunnels will be transformed into an interactive gallery showcasing the historical aspects of mining operations and equipment. Visitors will wear mining clothes and safety mining hats to enhance their experience.

The second zone will focus on the heritage buildings which will become a phosphate museum showcasing the story of phosphate and mining through interactive multimedia and the display of artefacts. The third zone will be a tourist area with restaurants and cafes as well as informative, interactive children's exhibitions. The creation of a phosphate museum in the core of this geopark within the abandoned Ruseifa mines will create a new experience in Jordan, add diversity to its tourist attractions, and provide hands-on educational opportunities for children of all ages to interact with this import resource. It is hoped that by working with the local community to develop the geopark, the associated educational material for local guides and schoolchildren and souvenir products will allow the project to become a source of knowledge, income, and pride in a place that is currently an urban waste (Al Rayyan et al., 2017).

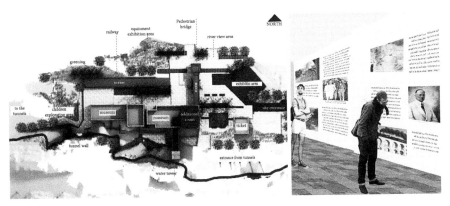

FIGURE 11.7 Proposed Phosphate Mine Museum, Ruseifa, Jordan. Design of the proposed phosphate mine museum and interpretation panels.
Source: Architect Kawthar, with permission.

11.7 CONCLUSION

The interpretation of the geological heritage is a sophisticated medium which comprises part science and part art. Gone are the days when at geological sites an interpretive panel completely filled with geological facts, written by a geological scholar, was the norm. Today, there are a number of essential elements of geo-interpretation which start with simple, easy to read, and understand facts about the geology, but are then complemented by visuals which show the viewer both the geological features of the landscape under view as well as how they were formed. This should be presented by professional communicators who can link the geoheritage information and climatic factors with the area's fauna and flora, so that taken all together these inform the viewer/reader of how people have lived in the area in the past, as well as today. This then presents a more holistic understanding of the way in which a region's geological attributes have contributed to the ecology and culture of the area. This method of interpretation then creates a greater "connection" for the viewer with the geology of the area resulting in an increased propensity for the visitor to want to get involved in increased understanding and appreciation of geoheritage with the end goal hopefully leading to them wanting to do something for geoconservation.

ACKNOWLEDGMENTS

The author wishes to thank the following people for their contribution to the case studies in this chapter. They are: Tobias Dowling—Teacher, Jeju Island, South Korea for information on the languages on the interpretive sign at the rock pillars of Jusangjeolli; Kristie Eggebroten—Program Manager Teton Regional Economic Coalition, Driggs, Idaho, USA, for information and photos of the Teton Geotourism Center; Stu Henley—Consultant Geologist, formerly Senior Geologist Solid Energy, Westport for information on the Stockton Mine Tours, New Zealand; Kawthar Rayyan—architect, Amman, Jordan for permission to republish the drawings of the proposed phosphate mine museum, Ruseifa, Jordan.

KEYWORDS

- geology
- geoheritage
- mining heritage
- geotourism
- interpretation
- geo-interpretation
- the ABC method

REFERENCES

Al Rayyan, L.; Hamarneh, C.; Sukkar, H.; Ghaith, A.; Abu-Jaber, N. From Abandoned Mines to a Labyrinth of Knowledge: A Conceptual Design for a Geoheritage Park Museum in Jordan. *Geoheritage* **2017,** DOI: 0.1007/s12371-017-0266-8.

Castell, M. What Locals Call the 'Gates Of Hell': Inside the Infamous Mines of Potosi, 2017. http://www.news.com.au/travel/world-travel/south-america (accessed Feb 1, 2018).

Dowling, R. K. Global Geotourism—an Emerging form of Sustainable Tourism. *Czech J. Tour.* **2013,** *2* (2), 59–79. DOI: 10.2478/cjot-2013-0004.

Dowling, R. K. Geotourism. In *The Encyclopedia of Sustainable Tourism*; Cater, C., Garrod, B., Low, T., Eds.; CABI: Oxford, 2015a; pp 231–232.

Dowling, R. K. Geotourism. In *Encyclopedia of Tourism;* Jafari, J., Xiao, H., Eds.; Springer: Berlin, 2015b.

Dowling, R. K. Geotourism's Contribution to Sustainable Tourism. In *The Practice of Sustainable Tourism: Resolving the Paradox*; Hughes, M., Weaver, D., Pforr, C., Eds.; London: Routledge, 2015c, pp 207–227.

Dowling, R. K. Engaging Youth in the Conservation and Sustainability of Protected Areas. Paper presented at *The 10th Anniversary Ceremony and Global Forum of Jeju World Heritage Inscription*, Jeju Island, South Korea, 11th September 2017.

Dowling, R. K.; Newsome, D. Geotourism—Definition, Characteristics and International Perspectives. In *Handbook of Geotourism*; Dowling, R. K., Newsome, D., Eds.; Edward Elgar Publishing: Cheltenham, Gloucestershire, 2018.

Dowling, R. K.; Pforr, C. Geotourismus in Australien und Neuseeland. In *Tourismus in Australien und Neuseeland;* Pforr, C., Reiser, D., Eds.; De Gruyter: Oldenbourg, Berlin, 2017; pp 206–223.

Eggebroten, K. Email Personal Communication with Kristie Eggebroten, 2018.

Garofano, M.; Govoni, D. Underground Geotourism: A Historic and Economic Overview of Show Caves and Show Mines in Italy. *Geoheritage* **2012,** *4,* 79–92.

Gates, A. F. Generating Interest in Geotourism near Urban Areas through Integration with Historical Sites: Iron Mine Trail, Sterling Forest, New York, USA. In *Handbook of Geotourism*; Dowling, R. K., Newsome, D., Eds.; Edward Elgar Publishing: Cheltenham, Gloucestershire, 2018.

Govoni, D. Miniere visitabili in Italia: Una risorsa geoturistica da capire. Third Geotourism Course. Geotourism Association: Genova, 2007; pp 14–18.

Gray, M. Geodiversity, Geoheritage, Geoconservation and their Relationship to Geotourism. In *Handbook of Geotourism*; Dowling, R. K., Newsome, D., Eds.; Edward Elgar Publishing: Cheltenham, Gloucestershire, 2018.

Hamilton-Smith, E. Geotourism: Too Important to be Left in the Hands of Geologists. In *Inaugural Global Geotourism Conference*; Dowling, R. K., Newsome, D., Eds.; Proceedings of the Conference held from 17–20 August: Perth, Australia, 2008; pp 5–8.

Hughes, K., Ballantyne, R. Interpretation Rocks. Designing Signs for Geotourism Sites. In *Geotourism: The Tourism of Geology and Landscape;* Newsome, D., Dowling, R. K., Eds.; Goodfellow Publishers Ltd.: Oxford, 2010; pp 184–199.

Jeju Tourism Organisation *Jeju Island: An Island of Sky and Sea.* Jeju Tourism Organisation: Jeju City, South Korea, 2011.

Jeon, Y. M.; Woo, K. S. Geotourism in Jeju Island UNESCO Global Geopark. In *Handbook of Geotourism*; Dowling, R. K., Newsome, D., Eds.; Edward Elgar Publishing: Cheltenham, Gloucestershire, 2018.

Kim, B. H. *Jeju Geotourism Perspective.* Jeju Self-Governing Province Cultural Policy Department: Jeju City, South Korea, 2017.

Lopes, R. *The Volcano Adventure Guide;* Cambridge University Press: Cambridge, 2005.

López-García, J. A.; Oyarzun, R.; López Andrés, S.; Manteca Martínez, J. I. Scientific, Educational, and Environmental Considerations Regarding Mine Sites and Geoheritage: A Perspective from SE Spain. *Geoheritage* **2011**, *3*, 267–275. DOI: 10.1007/s12371-011-0040-2.

Louv, R. *Last Child in the Woods: Saving our Children from Nature-Deficit Disorder*; Algonquin Books: Chapel Hill, NC, 2005.

Macadam, J. Geoheritage: Getting the Message Across. What Message and to Whom? In *Geoheritage: Assessment, Protection and Management*; Reynard, E., Brilha, J., Eds.; Elsevier Inc.: Amsterdam, 2018; pp 267–288.

Mata-Perelló, J.; Carrión, P.; Molina, M.; Villas-Boas, R. Geomining Heritage as a Tool to Promote the Social Development of Rural Communities. In *Geoheritage: Assessment, Protection and Management*; Reynard, E., Brilha, J., Eds.; Elsevier Inc.: Amsterdam, 2018; pp 167–177.

Migoń, P. Geo-Interpretation: How and for Whom. In *Handbook of Geotourism*; Dowling, R. K., Newsome, D., Eds.; Edward Elgar Publishing: Cheltenham, Gloucestershire, 2018.

Newsome, D.; Dowling, R. K. Geoheritage and Geotourism. In *Geoheritage: Assessment, Protection and Management*; Reynard, E., Brilha, J., Eds.; Elsevier Inc.: Amsterdam, 2018; pp 305–322.

Newsome, D.; Dowling, R. K.; Leung, Y. F. The Nature and Management of Geotourism: A Case Study of Two Established Iconic Geotourism Destinations. *Tour. Manag. Perspect.* **2012**, *2–3*, 19–27.

NPS/AGI. *America's Geologic Heritage: An Invitation to Leadership.* NPS 999/129325. National Park Service in association with the American Geosciences Institute: Denver, Colorado, 2016.

Pastorelli, J. *Enriching the Experience: An Interpretive Approach to Tour Guiding.* Frenchs Forest, NSW: Hospitality Press, 2003.

Pretes, M. Touring Mines and Mining Tourists. *Ann. Tour. Res.* **2002**, *29* (2), 439–456.

Prosser, C. D. Geoconservation, Quarrying and Mining: Opportunities and Challenges Illustrated Through Working in Partnership with the Mineral Extraction Industry in England. *Geoheritage* 2016. Published Online. DOI: 10.1007/s12371-016-0206-z.

Riganti, A.; Johnston, J. Geotourism—a Focus on the Urban Environment. In *Handbook of Geotourism*; Dowling, R. K., Newsome, D., Eds.; Edward Elgar Publishing: Cheltenham, Gloucestershire, 2018.

Slomka, T.; Mayer, W. Geotourism and Geotourist Education in Poland. In *Geotourism: The Tourism of Geology and Landscape*; Newsome, D., Dowling, R. K., Eds.; Goodfellow Publishers Ltd: Oxford, 2010; pp 142–157.

Tilden, F. *Interpreting Our Heritage*, 4th ed.; University of North Carolina Press: Chapel Hill, 2007.

UNESCO. Wieliczka and Bochnia Royal Salt Mines. UNESCO World Heritage List, n.d. http://whc.unesco.org/en/list/32 (accessed Feb 1, 2018).

CHAPTER 12

Evolving Geological Interpretation Writings about a Well-Traveled Part of California, 1878–2016

WILLIAM (BILL) WITHERSPOON

PO Box 33522, Decatur, GA, 30033, USA

E-mail: bill@georgiarocks.us

ABSTRACT

Nine nontechnical geology publications cover the route between Sacramento and Reno, and were published between 1878 and 2011. They include bestselling books that have garnered prestigious awards, as well as underrecognized works of high quality. They demonstrate a variety of strengths, from literary polish in accounts of the Old West linked to geology, to lucid explanations of complex debates, to well-designed maps for use at stops along the route. Examining the works in chronologic order, it is possible to trace California's pivotal role in the development of particular geoscience concepts. This includes early uses of plate tectonics to interpret outcrops and landscapes. Three works appeared in 1975, when geologists had only begun to debate the implications of plate tectonics for the geology of particular regions. Of these books, one ignored plate tectonics, another placed it in a final, speculative chapter and tried to blend it with the old paradigm, and only one was ready to apply plate tectonics fully throughout the book. Digital tools, in particular Google Earth® and Google My Maps®, were used to collate the guidebook material, draping geologic maps over the topography of the Sierra Nevada, placing text and images at location markers, and planning out a route with driving times. There is potential for social media to build on this foundation, continuing the public conversation that the nine works represent. The digital possibilities can be explored at http://georgiarocks.us/spots/usca.

12.1 INTRODUCTION

Works offering geological interpretation to the traveler in the United States go back at least to the 1870s. *An American Geological Railway Guide* (MacFarlane, 1878, 1890) undertook to give the "geological formation at each railway station" in the US. Geoscience education for the public was the goal from the beginning, as MacFarlane wrote:

> No person who has the least power of observation can fail to notice the peculiarities in the scenery and the great variety in the formations of rock.... [There is a need] to teach persons not versed in geology...not as in a textbook, but by pointing to the things themselves....*There are some kinds of knowledge too that cannot be obtained from books, but must be gathered by actual observation* (MacFarlane, 1890: 3; his italics).

Beyond curiosity about rocks and "peculiarities in scenery," there are big-picture reasons why the public can be interested in the geosciences. "Civilization exists by geological consent, subject to change without notice," wrote historian and philosopher Will Durant (Durant, 1946). California, where Durant had settled in 1943, exists mainly by consent of the subject of this chapter, the Sierra Nevada. The Sierra foothills yielded the gold that built San Francisco and propelled California to statehood. Its heights trap Pacific moisture as the snowpack after which the Sierra Nevada ("snowy range") is named. Without the snowpack, neither California's multibillion-dollar agriculture nor its ability to sustain 40 million inhabitants would exist.

The plate tectonics revolution boosted both curiosity about the geosciences and the availability of big-picture geological explanations. Because of it, geoscience writers are increasingly able to explain why the Sierra Nevada came into existence, in relatively recent geologic time. The revolution began in the 1960s with revelations about the ocean floor, then moved onto land. In a sense, the revolution continues, as improved tools reveal more and more of the story. Present-day movement of Sierra summits relative to Nevada desert, at rates of millimeters per year, is tracked using global positioning system (GPS) technology. The rising or sinking of crust in the interior of continents is better understood using newfound abilities to peer into Earth's mantle, thanks to improvement in computer processing of seismic waves.

Publishing technology for telling the story has also evolved, with full color books now widely available. Geological interpretation for the public has barely tapped into the potential of digital media, but tools such as Google Earth, Google Maps, and Environmental Systems Research Institute's (ESRI's) ARC-GIS software show promise.

12.2 A PATH ACROSS THE SIERRA NEVADA: SAMPLING THE EVOLUTION OF WRITINGS FOR THE GEOTOURIST

The route of the first transcontinental railroad, and later, Interstate Highway 80 (I-80), passing from Cheyenne, Wyoming through Salt Lake City and Reno to San Francisco, has been a well-trod path for geological interpretation. Where this route crosses the Sierra Nevada, between Sacramento, California, and the Nevada state line, no fewer than eight books, in widely varied formats, have explained geology to the public at points along the way.

The railroad and I-80 follow a route that humans and animals undoubtedly established millennia ago. Beginning in the Great Valley near sea level, it ascends the gradual westward-tilted slope of the Sierra Nevada, crossing one of California's richest gold-producing areas, to the highest elevation of 7000 feet, which is the lowest available crossing of the ice-carved granite crest of the range. From there it abruptly descends down a steep eastern slope, more than 1000 feet to Donner Lake, famed for the tragic end of an 1846 journey, and descends further along the Truckee River to the border.

Because of California gold, the Sierra foothills along this route are in one of the first areas in the West to be well-studied geologically. A century later, because of strong geology research programs in California universities and its location on the tectonically active Pacific Rim, the state was one of the first locations where plate tectonics illuminated geology on land. As geology has advanced, so has understanding of the Sierra Nevada's gradual western slope and abrupt eastern drop-off to the Basin and Range province. Similarly, it is interesting see how the true origins of the region's rock types emerged with plate tectonics, both the dominant granite that makes this the "Range of Light" (Meldahl, 2011, paraphrasing John Muir), and the darker, far less abundant greenstone/amphibolite and serpentine. Finally, there is the story, coming into ever sharper focus, of how gold began in those darker rocks and gradually became concentrated into the source of California's early wealth.

In short, this crossing of California's mountains provides an attractive window into how popular geoscience interpretation of a place evolved, along with geoscience itself, over the past 140 years.

12.3 RAILROAD ERA (1878, 1915)

An American Geological Railway Guide (MacFarlane, 1879, 1890) devotes most of its 200-plus pages to the northeastern US, where the railway network was densest, and geology was best understood. For the transcontinental railway, completed only in 1869, just a small table with footnotes covered the Sierra Nevada crossing (Fig. 12.1). Each station along the way was noted to rest on upon loose rock and soil deposited during the most recent geologic period (Quaternary), but footnotes tell of the neighboring rock outcrops. The east half of the route reveals granite topped with volcanic rocks and glacial deposits, while the west half includes loose sediments overlying gold-bearing Jurassic-age rocks, with additional granite at the end nearest Sacramento. All of this accords with present knowledge, but the granite was misinterpreted as Archaean (the oldest age category available), perhaps on an assumption of that day that all granite must be very old. Actually, the granite is mainly Cretaceous, and its intrusion as magma into the slightly older Jurassic rocks helped produce their gold-bearing veins.

CALIFORNIA. 319

Central Pacific Railroad.

Ms.			Alt.
.....	State Line.	20. Quaternary.	
616	Boca.[4]	"	5531
624	Truckee.	"	5819
638	Summit.	"	6983
652	Cisco.[4]	"	5934
660	Emigrant Gap.[5]	"	5221
665	Blue Cañon.	"	4693
675	Alta.	"	3607
677	Dutch Flat.	"	3395
679	Gold Run.	"	3220
680	Colfax.	"	2422
701	Clipper Gap.	"	1759
707	Auburn.[5]	"	1360
712	Newcastle.[6]	"	956
718	Pino.	"	
721	Rocklin.[6]	"	249
725	Junction.	19 c. Pliocene, "	163
729	Antelope.[7]	Quaternary, above Granite (Arch.?)	154

Central Pacific Railroad—*Continued.*

Ms.			Alt.
731	Arcade.	20. Quater. Alluvial.	55
744	Sacramento.	"	30
.....	Sacramento.	"	30
.....	Elk Grove.	"	53
525	Galt.	"	49
607	Stockton.[8]	"	23
650	Lathrop.	20. Quaternary.	26
706	Banta.	19. Tertiary, Plio., 19 b. Miocene & lignite, 19. Eocene(?)	30
713	Tracy.	20. Quaternary.	
745	Byron.	"	
815	Antioch.	"	
859	Martinez.	18. Cretaceous and 19. Eocene.	
863	Port Costa.	18. Cretaceous.	
877	San Pablo.	20. Quaternary.	
890	Oakland Pier.[9]	"	14
895	San Francisco.[10]	18. Meta. Cretaceous.	

4. *Boca* to *Cisco.* Volcanic and glacial, with 1. Archæan (granite) and metamorphosed rocks of uncertain age. Metalliferous but not rich. Mt. Stanford, northward, is 9,500 feet high.
5. *Emigrant Gap* to *Auburn.* Glacial and detrital above 16. Triassic and 17. Jurassic sandstones, containing most of the gold mined on the western slopes. A fine iron mine seven miles north of Auburn.
6. *Newcastle* to *Rocklin.* Detrital above 1. Archæan granite, surface mining for gold, platinum, telburet of silver and nickel. Diamonds also occur in small quantities.

FIGURE 12.1 List of stations in California and footnotes relevant to Sierra Nevada crossing from MacFarlane (1890: 319).

In 1915, the US Geological Survey (USGS) published *Guidebook of the Western United States* in four volumes (a total of six by 1933), each well over 100 pages in length, with mile-by-mile geologic and geographic descriptions of popular railway trips. Vetter (2008) cited this series as an example of

Evolving Geological Interpretation Writings 303

government scientists serving economic development of the West, in this case helping rail companies promote tourism.

Part B (Lee et al., 1915) covered the Omaha-to-San Francisco "Overland Route," plus a side trip to Yellowstone Park. About 10 pages of text, with a pair of geologic sketch maps (see Fig. 12.2) cover the Sierra traverse. The writing intersperses geology with scenic attractions such as a side-trip to Lake Tahoe, and economic activities such as pulp mills, ice-harvesting stations, fishing, and apple orchards. Unlike later writers, Lee et al. do not supplement the mile-by-mile narrative with a geologic overview section, but the following summarizes points that are mentioned.

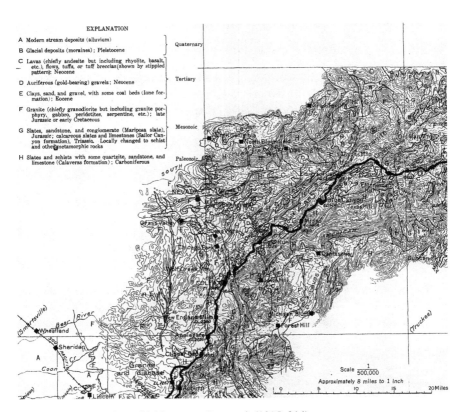

FIGURE 12.2   Portion of foldout map, Lee et al. (1915: 214).

The work recognizes that gold-bearing gravel, the richest and first target of the gold rush, had settled from long-vanished streams. These had eroded gold from a deeply weathered, gently rolling landscape. That countryside was subsequently tilted toward the west in a trap-door-like movement that, on its eastern edge, created the heights of the Sierra Nevada. A fault marks the steep eastern side of the range. Contemporary with the fault movement, volcanoes erupted, and lava flowed down the tilted surface. East of the Sierra, a few miles south of the route, a natural lava dam helps confine the waters of Lake Tahoe.

The gold-bearing gravels rest on top of much older rocks: on granite in the eastern, highest foothills (and also west of Auburn), or atop the "Mother Lode," where gold turned up in bedrock: a complex belt of rock types including slate, marble, serpentine, and amphibolite.

12.4   A *GEOLOGICAL HIGHWAY MAP* WHETS THE APPETITE ALONG I-80

A helpful aid for geotourists, beginning in the 1960s, was the *Geological Highway Map* series by the American Association of Petroleum Geologists (AAPG). The Pacific Southwest Region map (AAPG, 1968), covering California and Nevada, was third in a series with the stated goal to be "clearly understandable to the general public." Prefolded like a road map, it included:

- a full-color geologic map (1 inch = 30 miles scale), with highways and cities clearly marked
- a physiographic map to show landforms
- a tectonic map to show major faults and areas of uplift or subsidence
- a number of time-rock columns to show the age and rock type of strata as they vary by region
- a few cross sections along lines marked on the tectonic map
- an array of thumbnail-sized geological history maps sliced by time
- a few paragraphs of explanation, mainly on how to read the various resources.

The USGS *Geologic Map of the US* (Stose and Ljungstedt, 1932; King and Beikman, 1974) had previously compiled this area in a single geologic map, but the greater detail, additional graphics, and clearly visible highway network provide a boost to geological tourism. By studying the colorful map with a magnifying glass, and referring to the appropriate time-rock column, a user can work out the age and rock type of about 10 geologic units that I-80 crosses. The remaining graphics offer a hint of geologic structure and history. The map whets the traveler's appetite for further explanation, even if, lacking narration, it may fall short of full accessibility to non-geologists.

12.5 THE PLATE TECTONICS REVOLUTION STIRS THE POT

In the 1970s, the plate tectonics revolution was overturning geological thinking. The anthology *Plate Tectonics* (Oreskes, 2001) recounts for the general reader how the evidence unfolded, in the words of the scientists who uncovered and made sense of it.

The first compelling evidence that the continents are moving was the Vine and Matthews (1963) recognition of mid-ocean ridges as pull-apart zones, where new oceanic crust and mantle are continuously generated by rising magma. The discovery flowed from improved locating of worldwide earthquakes, as well as magnetic mapping of the seafloors, both spinoffs of Cold War military research. Soon, a program of deep-sea drilling to determine age and rock type throughout the world's oceans was launched. Astonishingly, in contrast to rock ages of up to billions of years on the continents, the oldest oceanic crust, found in the western Pacific Ocean, was less than 200 million years old. It seemed unlikely that seafloor spreading had begun so recently, so some process must have disposed of the older material. By 1969, a few geoscientists had a good idea of the process, but had not yet settled on a name. In fact, the name "plate tectonics" was not yet a household term.

In Oreskes' book, William Dickinson, of Stanford University (Dickinson, 2001), recounts his January 1969 proposal to the Geological Society of America for an informal conference with the "unwieldy title, *The Meaning of the New Global Tectonics for Magmatism, Sedimentation, and Metamorphism in Orogenic* [Mountain] *Belts*." Dickinson wrote of a "comprehensive theory" developed "in recent months" and pointed out that "far-reaching reinterpretations of geologic history are possible with the theory in mind." The conference was held in December in the coastal town of Asilomar, south of San Francisco. Dickinson recalls, "The group in attendance was largely self-selected, apart from a few invited speakers. Only about 150 geoscientists saw fit at the time to apply, and nearly 100 were invited." One of them was Donald Hyndman, later co-author of one of the works described below.

The process disposing of seafloor acquired the name "subduction" at the Asilomar conference. Borrowed from Alpine geology by Dietrich Roeder, then of Esso Research (Dickinson, 2001) the Latin origin translates as "pulling under." Essentially, at deep sea trenches, a slab of ocean floor begins a sloping descent into the mantle (within a "subduction zone"). When the slab reaches a particular depth, magma rises above it to make a chain of volcanoes called an arc, such as form the "ring of fire" around the Pacific Rim. The Cretaceous age granitic rocks of the Sierra Nevada, along with sediments of that age in the Great Valley and Coast Ranges, found their place in this picture in a paper by Hamilton (1969).

But geologists picked up these ideas at different speeds based on their experience. When I took my first geology class in 1975, few geologists doubted plate tectonics, but articulating its impact on their own area of geology might take time.

There was, at that time, some tug-of-war between "fixists" and "mobilists," over how much geology could still be explained by purely vertical movements of Earth's crust. Three questions in particular had to be answered:

1. How are rocks, deposited as sediments at the surface, transported to the depths needed to account for the heat and pressure at which rocks become metamorphic?
2. How do you explain deformation during metamorphism that indicates great horizontal compression?
3. Why do rocks transformed in this way later get uplifted as mountain ranges?

The fixist answers were (1) There is massive down-folding of the crust, dropping sedimentary rocks into a sort of pocket. (2) The pocket becomes a sort of vice or nutcracker. (3) We don't know why.

One of the reasons these ideas had never quite been satisfying was their inability to explain what would induce such movements, or control their timing. The mobilists correctly interpreted the same geologic history in terms of the mechanics of plates in collision, and in a few years had won the debate.

So, given the flux this debate was still in 1975, it is not surprising that of three works for the public published that year, discussed below, one ignored plate tectonics completely, another placed it in a final, speculative chapter and tried to blend it with the fixist view, and only one was bold enough to apply plate tectonics fully in the book introduction.

12.6 A FULL-COLOR GUIDE TO THE ROCKS ALONG I-80

The American Geological Institute sponsored *Roadside Geology of US Interstate 80 Between Salt Lake City and San Francisco: The Meaning Behind the Landscape* (Hamblin et al., 1975). It is a full-color booklet of 50 notebook-sized pages, which could have been designed to complement the *AAPG Geological Highway Map* along I-80. Unlike the rail guide of 1915, it does not itself include geologic maps, instead simply marking numbers keyed to the text onto a base taken from regional-scale USGS topographic maps.

The introduction includes a generalized six-page geology primer (types of rocks, geologic time), then a single page summarizing how the landscapes and rock types differ between the four provinces of the route: Basin and Range, Sierra Nevada, Great Valley, and Coast Ranges. The rest of the work is road guide, with the above-mentioned maps and numbered descriptions of where particular rock types are best seen. For example, near Auburn, location 4 on the Sacramento to Colfax map has "dark gray-green and greenish-brown metamorphosed Triassic and Jurassic volcanic rocks" (shorthanded by other writers as greenstone), and location 7 near Weimar is where I-80 crosses a "narrow belt of serpentine [that] marks a fault."

The stated goal of the booklet is to reveal the "stories behind the scenery" to the public, but the missing ingredient seems to be the regional geologic history, that is, the steps through time that made the rocks and landscapes what they are. Plate tectonics is not mentioned.

12.7   A PANORAMA OF SIERRA NEVADA GEOLOGIC HISTORY

Geology of the Sierra Nevada (Hill, 1975, 1996; Hill and Faber, 2006) was published as part of the University of California Press' California Natural History Guides series. The first 10 chapter titles in the 1975 edition give an idea of the book's use of plain language to systematically open the panorama of Sierra Nevada geologic history to a general audience: "Sierra Nevada through the Ages," "Of Time and Rocks," "The Range Today," "Seas of Long Ago," "Great is Granite," "Treasures from the Earth," "Rivers of Yesterday," "Days of Fire," "Days of Ice," and "The Mountains Tremble."

Hill builds geologic knowledge gradually while referring to the spectacular locations, whether roadside or accessible by two-day backcountry hike, that best illustrate her points. No locations along the I-80 route receive such attention, but several appear on maps showing examples of features. For example, Donner Lake is a "moraine-dammed lake," and there is "greenstone" at the American River canyon near Auburn, and "conglomerate" at Gold Run.

Plate tectonic theory is introduced in a final chapter, "The Mountains Grow, Unnoticed." Hill is clearly aware of growing interest, as she writes, "No geological 'explanation' (or myth, if you will) captures the American fancy like the current theory of 'global tectonics' of which 'continental drift' is a part" (p 178). After an introduction to the ideas of seafloor spreading and subduction, she speculates on whether and how it might apply to the Sierra Nevada. She provides diagrams that show a continental edge with volcanoes, but without the descending oceanic slab of a subduction zone. She refers to rocks "squeezed and metamorphosed in the crustal nutcracker," holding to the fixist model discussed above. The difference two decades can make is evident in the much-improved explanation of the role of plate tectonics near the front of the second edition (Hill, 1996).

12.8 PLATE TECTONICS UP FRONT AND LABELED GEOLOGIC MAPS: *ROADSIDE GEOLOGY OF NORTHERN CALIFORNIA*

Roadside Geology of Northern California (Alt and Hyndman, 1975) was the first title to follow the authors' popular *Roadside Geology of the Northern Rockies* (Alt and Hyndman, 1972; see Witherspoon and Rimel, this volume (Chapter 13)). The format the authors invented includes road guides along major highways, in this case, I-80 and I-5, as well as four US and eight state highways. Each route segment is accompanied by a geological sketch map, with text and arrows marking geologic points of interest (Fig. 12.3). They grouped the road guides into four regions, each with its own introduction in addition to the book's nine-page master introduction, "The Great Collision."

"The Great Collision" was the *Roadside Geology* series' first foray into plate tectonics. The importance they gave the topic is evident on the book's cover, which is a schematic based on Hamilton's (1969) application of the concept (soon to be called subduction) to California geology. There are three main elements shown (Fig. 12.4):

1. The scraping off of deep-water sediments and bits of oceanic crust at the trench, elevated as land when the trench periodically relocates oceanward;
2. Some distance landward of the trench, the intrusion of magma and its volcanic expression in a chain of volcanoes known as an arc;
3. Between arc and trench, an accumulation of sediments in what was dubbed a "forearc basin."

"The Great Collision" explains that California is made of deep-water sediments and ocean floor added to the west edge of North America beginning around the beginning of the Jurassic Period about 200 million years ago, later intruded by arc magma. The Jurassic-age sediments in the Sierra foothills around Auburn are among the oldest sediments accreted,

Evolving Geological Interpretation Writings   309

and the Cretaceous-age granite, both west of Auburn and in the higher part of the route, formed from the magma that came up in the volcanic arc while later accretion was going on in the Coast Ranges. Cretaceous Great Valley sediments, on which Sacramento is built, are part of the forearc basin from that time.

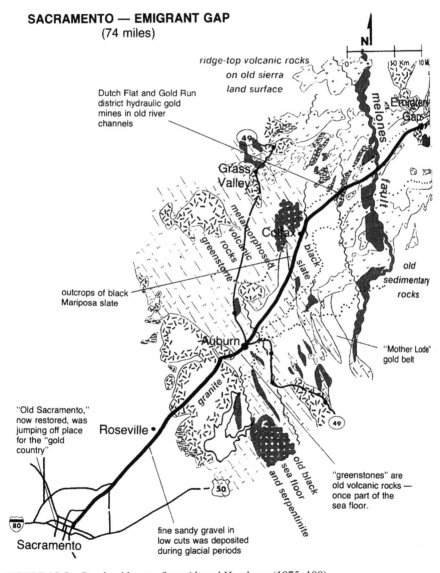

FIGURE 12.3 Road guide map from Alt and Hyndman (1975: 100).

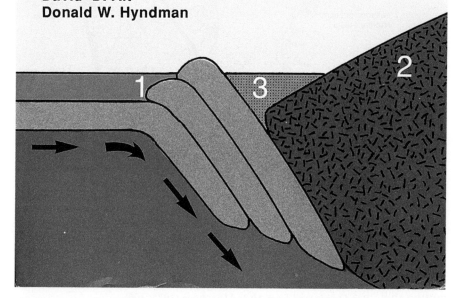

FIGURE 12.4 Alt and Hyndman (1975) book cover; numbers added (see text).

Alt and Hyndman further note a break, beginning around 80 million years ago, in which magma was not rising in California. During this period, erosion removed the volcanoes, exposing the granite that had lain beneath them. On top of this, rivers flowing westward deposited gold-bearing gravels. Then, about 15–20 million years ago, volcanic activity resumed. Lava flowed down the river valleys and covered much of the land surface. The trap-door-like hinged uplift of the eastern Sierra accompanied the later volcanic eruptions and continues to the present. Over the last 2 million years in the Quaternary Period, glaciers have carved the elevated crest, and around it have deposited sediments.

The 12-page Sierra Nevada-Klamath introduction delves deeper into these topics, laying ground for the 9 road guides of the region, including the I-80 guide from Sacramento to Reno, Nevada. The nine pages of that guide are about half illustrations—two geologic maps with labeled locations, four cross sections, and six field photos. An excerpt discussing the older rocks near Emigrant Gap gives an idea of the writing style:

> Bedrock in this part of the Sierra Nevada consists almost entirely of old sedimentary rocks originally deposited in the ocean and then mashed against the edge of the continent about 200 million years ago. Intense deformation and prolonged heating have changed the original sediments almost beyond recognition. Now they are slabby rocks that break easily into slaty plates often glistening with flakes of mica.... [they] look as if they might split into good roofing shingles or stepping stones (p 105).

12.9 WRITER JOHN MCPHEE'S TOME HITS A HOMER

By the 1980s plate tectonics had become the unifying principle of geology, and many readers were eager to grasp the rejuvenated field. John McPhee, a prolific nature and travel writer of articles in the *New Yorker* and several books, set out to learn and write about the new geology as illuminated by crossing the continent from New York to San Francisco. The four-volume series began in 1981 with *Basin and Range*, which runs from Salt Lake City to the eastern edge of the Sierra Nevada, and concluded 12 years later with *Assembling California*, from there west to San Francisco. The entire set was later released in a single tome, *Annals of the Former World* (McPhee, 1998), which remains one of the top selling geology books for the public, according to Amazon (Amazon, 2018). The work is well received by geologists, with many awards from professional societies.

McPhee developed each volume by accompanying one prominent geologist in his or her fieldwork: Ken Deffeyes for *Basin and Range*, Anita Harris (from New York to the Great Plains) for *In Suspect Terrain*, David Love (in Wyoming) for *Rising from the Plains*, and Eldridge Moores for *Assembling California*. The Sierra Nevada segment illuminates about seventy pages of McPhee's (and Moores') big picture geology. McPhee briefly treats the "trapdoor" analogy and the origin of the granite, then allots gold mining history about twenty pages. The remainder follows Moores through his specialty, studying ophiolites.

An ophiolite is an assemblage of rock types that was known in many parts of the world before plate tectonics came along. McPhee describes how Moores participated in the discovery that an ophiolite sequence represents a slice of ocean floor, from mantle through oceanic crust to the thin veneer of deep-sea sediments. With Moores, he visits a "sheeted diabase dike complex" along I-80 near Auburn. A diabase dike is a crack filled with the same type of magma that makes the dark lava rock, basalt, familiar to visitors to Hawaii or Iceland. A sheeted dike complex consists entirely of dikes that have invaded and split earlier dikes, and is considered diagnostic of a place where new ocean floor has been forming, as plates continuously pull apart.

Other components of an ophiolite include pillow basalt, where basalt erupted underwater has formed lumps called pillows; layered gabbro (where magma that fed the sheeted dikes cooled beneath them in layers); peridotite, the main rock of which the mantle is composed; and the green rock serpentine (or serpentinite), a rock formed by the chemical interaction of seawater with peridotite.

The sheeted dikes McPhee visited with Moores near Auburn belong to what was then dubbed the Smartville ophiolite. It was interpreted to represent a Jurassic-age subduction zone abandoned in the Cretaceous Period when the trench location jumped westward. McPhee contrasts this apparent simplicity with the messiness of the Feather River ophiolite they later visit, higher in the Sierra. Geologists had begun to realize that the origin and emplacement of rocks identified as ophiolites were more complicated that they had thought.

McPhee's writing stands out for capturing both the excitement and the complexity of geologic fieldwork and interpretation. There is humor and humanity in it, keeping the reader engaged, while inviting the nongeologist to drill deep into geologic ideas. Its practical limitations as a traveler's reference are that is not systematic enough to support a quick grasp of the big picture, nor detailed enough in its directions for the reader to easily find specific outcrops.

12.10 *ROADSIDE GEOLOGY* MATURES WITH "AN ENTIRELY NEW BOOK"

In 2000, *Roadside Geology of Northern and Central California* (Alt and Hyndman, 2000) appeared. The series that Alt and Hyndman invented was well under way by then, with 18 titles (including their own Oregon, Washington, Idaho, and Montana books and a Hawaii volume Hyndman had co-authored). They wrote, "A lot has happened in California geology during the last twenty-five years; the time has come to reconsider the rocks and write an entirely new book." The new introduction to the Sierra Nevada opens by explaining the ophiolite concept, and identifying the four terranes, or separate pieces of crust, from which this part of California was then thought to have been assembled. The Smartville ophiolite is identified as a "fairly well-preserved slab of Jurassic oceanic crust," and a part of the Western Jurassic terrane, that was "jammed into the Sierran trench... before Jurassic time ended." The Sierra Nevada introduction continues with an updated and expanded discussion of the Sierra Nevada granite batholith, the trap-door uplift of the range, the action of glaciers, and gold.

The road guide to I-80 east of Sacramento is similar to that of the 1975 book, with an improved map (Fig. 12.5) and now with headings: "Granite at Auburn," "Western Jurassic Terrane," "Remnant of the Old Landscape," "Auriferous Gravels," and "Sierra Nevada Batholith."

12.11 GEOLOGIC TRIPS WITH CLARITY AND SIMPLICITY: TED KONIGSMARK

In *Geologic Trips: Sierra Nevada* (Konigsmark, 2002), I-80 between Auburn and Truckee is part of "Trip 5: Northern Sierra," one of nine featured loop drives. Overall, this is a very handy guide for the traveler. The clear and simple graphics are as readable as monochrome publication will allow. The 65 pages of introduction that bring the nongeologist up to speed are well-organized and thorough. In addition to Sierra Nevada topics such as gold, granite, timing of uplift, volcanism, and glaciation, there is a rock and mineral primer with tables including the locations where rock types are seen.

Konigsmark's Northern Sierra loop has an illustration plotting 16 gold mining areas onto the reconstructed path of the Yuba River about 50 million years ago, when it was much larger than today (Fig. 12.6). This includes Dutch Flat and Gold Run, along I-80. The description of the gravels exposed

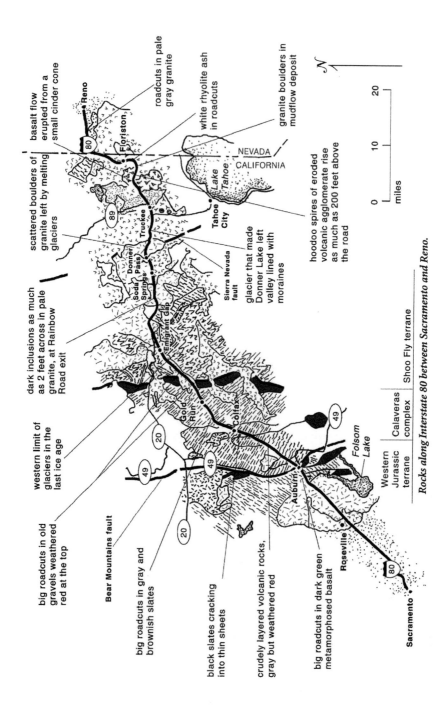

FIGURE 12.5 Road guide map from Alt and Hyndman (2000: 32).

Evolving Geological Interpretation Writings

at the I-80 rest stop at Gold Run is meticulous, explaining that the "blue gravel" (the coarsest, deepest fraction of the stream bed deposit) was where the gold riches lay, and is consequently now gone. The other focus along I-80 is at Donner Pass, with three sites, which separately reveal aspects of the granite, volcanism, and glacial erosion. This approach of delving deeply into specific sites is similar to the vignettes of the *Geology Underfoot* series of Mountain Press Publishing, which had originated in California in the 1990s (Sharp and Glazner, 1993, 1997).

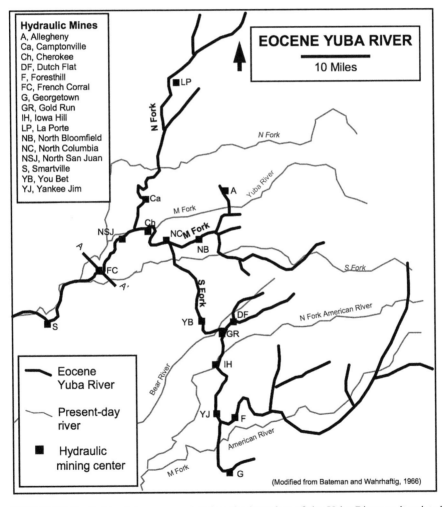

FIGURE 12.6 Gold mining areas related to the branches of the Yuba River as it existed about 50 million years ago, from Konigsmark (2002: 159), with permission.

12.12 THE LITERARY APPROACH AND UPDATED GEOLOGY: *ROUGH-HEWN LAND*

Rough-Hewn Land (Meldahl, 2011) in its literary polish approaches John McPhee's work, but is the work of California geology professor Keith Heyer Meldahl. He covers the same band along I-80 as McPhee's three western volumes, from the Pacific coast to eastern Wyoming. Like McPhee, he is willing to digress to tell stories of the Old West, such as the whole series of mistakes, beginning in Utah, that left the Donner party so vulnerable to starvation at Donner Lake. At the same time, *Rough-Hewn Land* is more systematic in using outcrops to piece big picture geology together than the *Annals* opus. For example, the first outcrop the book visits, pillow basalts exposed at Point Bonita within view of Golden Gate Bridge, conveniently drives the first lesson, that of seafloor spreading.

Meldahl also reaps the benefit of writing about California geology more than four decades after the plate tectonics revolution began. Today, the "quiet period" of no magma following 80 million years is recognized as due to a speeding up of North America's westward movement. This faster overrunning of Pacific seafloor led to the slab descending at a shallower angle (because it did not have as much time to cool and sink). One of the consequences of "flat slab subduction" was that the slab under California was not at the requisite depth to generate rising magma, and the volcanic arc relocated well to the east.

Another more recent concept that Meldahl introduces is the Nevadaplano. Lines of evidence that came together in the 2000s suggest that, 50 million years ago when the gold-bearing gravels were being deposited in those westward-flowing streams, the interior of Nevada was already high, like the Altiplano of Peru and Bolivia today. Rather than the crest of the Sierra Nevada being hinged upward from sea level after about 20 million years ago, the trap-door profile is due to the east side of the fault permitting a previously high area to drop. As Meldahl puts it, the pulling apart of the Basin and Range caused the area now east of the crest—and most of the state of Nevada—to collapse "like a punctured soufflé." As Meldahl illustrates (Fig. 12.7), this means that the gold-rich gravels were deposited from rivers flowing down from an elevated region, not, as previously envisioned, crossing a gently rolling near-sea-level expanse.

For geotourists inclined to use *Rough-Hewn Land* as a travel guide, there is an appendix, "Seeing for Yourself" that includes latitude/longitude and road directions for each site visited in the text. Along the Sierra Nevada crossing, these outcrops reveal green metamorphosed volcanic rocks near Auburn, pillow basalts near Applegate, auriferous gravels at Gold Run, a

Evolving Geological Interpretation Writings

spectacular gradational west edge of the Sierra Nevada batholith on the Loch Leven trail, and glacial erosion with history at Donner Pass.

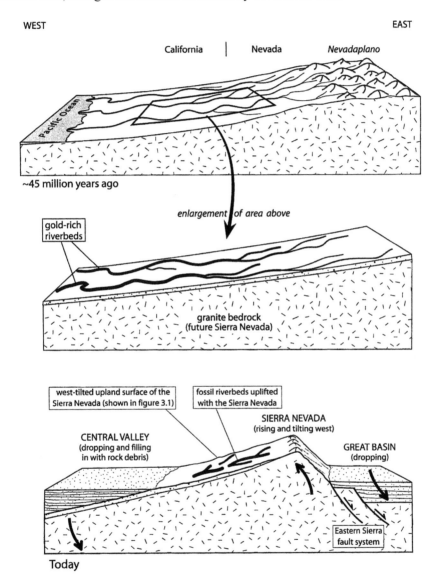

FIGURE 12.7 Gold-bearing riverbeds in relation to the Nevadaplano 50 million years ago, and the uplift of the eastern Sierra with the collapse of the Nevadaplano beginning 5 million years ago. Reprinted with permission from p. 42, *Rough-Hewn Land: A Geologic Journey from California to the Rocky Mountains*.
Source: With permission from Keith Heyer Meldahl, ©2011 by the Regents of the University of California, published by the University of California Press.

12.13 THIRD PASS OF *ROADSIDE GEOLOGY* SERIES: FULL COLOR

The most recent geologic guide for the public on the I-80 Sierra Nevada crossing is the new full color edition of *Roadside Geology of Northern and Central California* (Alt and Hyndman, 2016). For a geology book, color is a big change—compare the new I-80 segment map (Fig. 12.8) to the previous version (Fig. 12.5). The map is adapted from the California Geologic Survey's 2010 digital *Geologic Map of California*.

David Alt had passed away by the time this edition came out, and Don Hyndman had retired (though in 2018, he delivered a new edition of *Roadside Geology of Montana* to MPP). Perhaps as a result, there is little revision to the text in the road guide or the Sierra Nevada introduction. This means that some updated concepts found in Meldahl (2011), such as the flat-slab subduction beginning 80 million years ago, or the Nevadaplano, are not in the book. They can be explored in *Roadside Geology of Nevada* (DeCourten and Biggar, 2017; see discussion in Witherspoon and Rimel, this volume (Chapter 13)).

12.14 DOCUMENTING A CALIFORNIA GEOTOURISM TRIP WITH GOOGLE MAPS

Four of the titles described in this chapter (McPhee, 1998; Alt and Hyndman, 2000; Konigsmark, 2002; Meldahl, 2011) were very useful for a trip I took in 2015 with a geologist colleague. To plan and then report on the trip, I developed a resource in Google Maps (now linked from georgiarocks.us/spots/usca), which has been used by other geotourists, with more than 1500 views. The trip arose as a merger of excursions that Habte Churnet, of the University of Chattanooga, and I had independently planned to lead for our respective audiences.

In 2001, I had led a group of teachers and geology enthusiasts on an eight-day trip in Arizona that I named "Geology on a Grand Scale." I discovered from *Roadside Geology of Arizona* (Chronic, 1983) that along State Highway 87 north and east of Phoenix, it is possible to examine each of the rock units one would see in climbing from the Colorado River to the rim of the Grand Canyon. After that beginning, the trip visited Meteor Crater, Sunset Volcano, cliff dwellings, a dinosaur track way, a Hopi village, the red rocks of Sedona, and, of course, the Grand Canyon.

After the final *Roadside Geology of Georgia* submission to Mountain Press Publishing in 2012 made more planning time available, I decided to

Evolving Geological Interpretation Writings 319

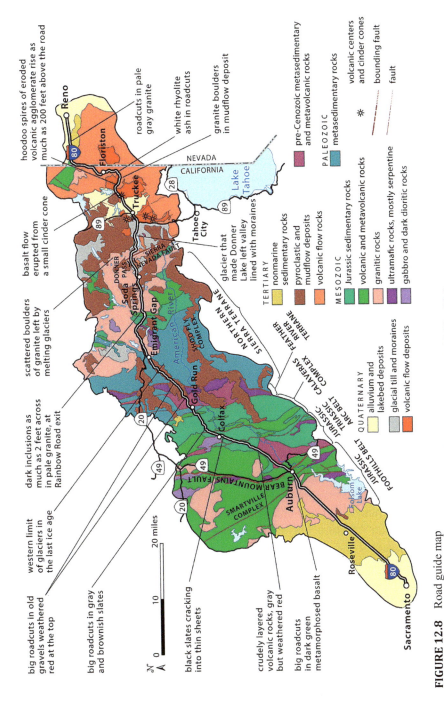

FIGURE 12.8 Road guide map
Source: With permission from Alt and Hyndman (2016: 32); compare with Fig. 12.5.

try for a similar teaching experience, taking Georgia science teachers to see plate tectonics first-hand at the western edge of the North American Plate, which can be defined most strictly as the San Andreas fault, or more broadly as spanning the width of California. I promoted "California: Geology on the Edge" with a slideshow, a Google Earth tour, and a web site that featured Google Maps place marks of all the stops and a photo gallery. Active faults on both sides of San Francisco Bay, as well as active faults and recent volcanoes around Mammoth Lakes, were to be the "Edge" destinations, with Yosemite Park and the Gold Rush country as bonuses. Unfortunately, the 2013 environment was less receptive to free-wheeling professional development trips as compared to 2001, and two funding proposals were unsuccessful.

I was ready to give up on the trip, but my friend Habte Churnet had the idea that we could explore California geology without leading others. He and I share an interest in ophiolites, those slivers of oceanic crust and mantle heaved onto land that John McPhee pursued with Eldridge Moores. Two classic California areas for such rocks, not along the route I had planned, were part of a trip he had been laying out for his students. They are in the Klamath Mountains close to the Oregon border, and in the Sierra foothills just north of I-80. In addition, Habte introduced me to *Rough-Hewn Land* (Meldahl, 2011), and he aimed to visit as many of that book's California localities as possible. It was following Meldahl that took us to the route that is the subject of this chapter.

We also found that there are still some geologically amazing (and often scenic) sites in California that are untouched by any geology books we could find. Thanks to two geologists (and one-time Moores students), Nancy Lindsley-Griffin for the Klamaths and Terence Kato for the Sierra foothills, we managed to find such special locations. Moores himself was kind enough to respond to an email about a guidebook for a geological society field trip he had co-led in the Sierra foothills a few years previous.

When I plan a trip nowadays, I create a map in Google My Maps. The "Add Marker" feature allows placing a pin symbol at each location of interest. Next, I use the "Add Directions" tool to create a proposed route connecting the markers. The markers also permit adding a title and description that appear in a box when the marker is selected. For mobile device reference on the trip, I filled description fields with a summary of what might be seen, and page references from the geology sources.

I share a My Map with others who are helping plan the trip. By the time we were ready to travel, it was apparent that the My Map could be helpful to other geological tourists, so I shared it publicly. On returning home, I thought about how it might be made more useful for others. A screenshot

(Fig. 12.9) shows an example of a finished product for a location not far off I-80, where an older road crests Donner Pass. A bit of text summarizes what *Rough-Hewn Land* (Meldahl, 2011) had to say about this location. I uploaded a picture from our visit. The hashtag #USCA_RHL55 is a modification of the first page reference RHL 55, which stands for *Rough-Hewn Land*, p 55, and the second page reference follows.

Why a hashtag? I would like future visitors to be able to find one another on social media in order to converse about this site and share pictures. If they also use the hashtag shown in the box to the left, #geospot_usca, geological tourists using such hashtags anywhere in California can find one another. As the box indicates, full directions are at georgiarocks.us/spots/usca.

Use of such a plan is theoretical at this writing, although participants on the 2016 Georgia Geological Society annual trip were encouraged to experiment with the social media platform of their choice with the hashtag #geospots. I have also been assisting Katayoun Mobasher of the University of North Georgia as she uses ARC-GIS to equip undergraduates with digital tools for their field explorations (Mobasher et al., 2015). The system she has developed supports:

- Locating oneself in the field on a digital geologic map
- Adding photos and narrated videos
- Putting standardized information such as book references, rock type and age, and size of the outcrop into a database
- Permitting a stream of observations and questions about a specific site, with commenters self-identified as to whether they are professional geologists, or other levels of expertise

Perhaps a popular app that brings such capabilities to anyone with a mobile device lies not too far in the future. (Note as this goes to press: the free app "Rockd," from researchers at the University of Wisconsin, shows promise in many of the above respects.)

ACKNOWLEDGMENTS

This chapter is dedicated in honor of my friend and dissertation advisor Dietrich Roeder, from whom I first heard the word "geotourism." Thank you to John Rimel of Mountain Press Publishing and Don Hyndman for answering my questions about the early history of the Roadside Geology series, to Habte Churnet and Russell Jones for inspiring the 2015 geotourist

322   The Geotourism Industry in the 21st Century

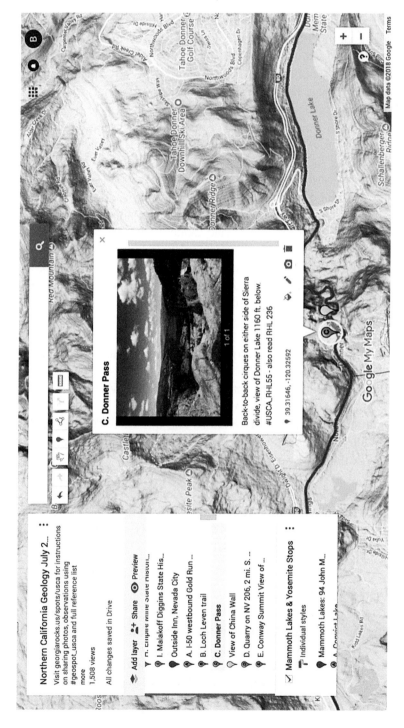

FIGURE 12.9 Screen capture of Donner Pass using Google My Maps, with marker and photo by the author.
Source: Accessed accessible from georgiarocks.us/spots/usca.

trip, and to geologists Nan Lindsley-Griffin and Terry Kato for meeting us in the field and showing us memorable outcrops. Thank you to my wife, Rina Rosenberg, who helped remove some of the jargon that had crept into this piece.

KEYWORDS

- place-based
- education–geosciences
- nature tourism
- history of geology
- Sierra Nevada

REFERENCES

Alt, D. D.; Hyndman, D. W. *Roadside Geology of the Northern Rockies*; Mountain Press Publishing: Missoula, MT, 1972.

Alt, D. D.; Hyndman, D. W. *Roadside Geology of Northern California*; Mountain Press Publishing: Missoula, MT, 1975.

Alt, D. D.; Hyndman, D. W. *Roadside Geology of Northern and Central California*; Mountain Press Publishing: Missoula, MT, 2000.

Alt, D. D.; Hyndman, D. W. *Roadside Geology of Northern and Central California, 2nd ed.*; Mountain Press Publishing: Missoula, MT, 2016.

Amazon.com. Amazon Best Sellers in Geology. https://www.amazon.com/gp/bestsellers/books/13603/ref=pd_zg_hrsr_books_1_4_last#4 (accessed Feb 1, 2018).

AAPG *Pacific Southwest Region Geological Highway Map*; American Association of Petroleum Geologists: Tulsa, OK, 1968.

Chronic, H. *Roadside Geology of Arizona*; Mountain Press Publishing: Missoula, MT, 1983.

DeCourten, F.; Biggar, N. *Roadside Geology of Nevada*; Mountain Press Publishing: Missoula, MT, 2017.

Dickinson, W. The Coming of Plate Tectonics to the Pacific Rim. In *Plate Tectonics: An Insider's History of the Modern Theory of the Earth*; Oreskes, N., Ed.; Westview Press: Cambridge, MA, 2001; pp 264–287.

Durant, W. J. What is Civilization? *Ladies Home J.* **1946**, *63*, 22.

Hamblin, W. K.; Rigby, J. K.; Snyder, J. L.; Matthews, W. H. *Roadside Geology of U. S. Interstate 80 Between Salt Lake City and San Francisco: The Meaning Behind the Landscape*; Varna Enterprises: Van Nuys, CA, 1975.

Hamilton, W. Mesozoic California and the Underflow of Pacific Mantle. *Geol. Soc. Am. Bull.* **1969**, *80,* 2409–2430.

Hill, M. *Geology of the Sierra Nevada*; University of California Press: Berkeley, CA, 1975.

Hill, M. *Geology of the Sierra Nevada*, 2nd ed.; University of California Press: Berkeley, CA, 1996.

Hill, M.; Faber, P. M. *Geology of the Sierra Nevada*, 3rd ed.; University of California Press: Berkeley, CA, 1996.

King, P. B.; Beikman, H. M. *Geologic Map of the United States (Exclusive of Alaska and Hawaii)*; U. S. Geologic Survey: Washington, DC, 1974.

Konigsmark, T. *Geologic Trips: Sierra Nevada*; GeoPress: Mendocino, CA, 2002.

Lee, W. T.; Stone, R. W.; Gale, H. S. *Guidebook of the Western United States Part B: The Overland Route*; U. S. Geologic Survey Bulletin 612; Department of the Interior: Washington, DC, 1915.

MacFarlane, J. R. *An American Geological Railway Guide*; Gies and Co.: Buffalo, NY, 1878.

MacFarlane, J. R. *An American Geological Railway Guide*, 2nd ed.; Gies and Co.: Buffalo, NY, 1890.

McPhee, J. *Annals of the Former World*; Farrar, Straus, and Giroux: New York, 1998.

Meldahl, K. H. *Rough-Hewn Land: A Geologic Journey from California to the Rocky Mountains*; University of California Press: Berkeley, CA, 2011.

Mobasher, K.; Turk, H. J.; Witherspoon, W.; Tate, L.; Hoynes, J. Enhancement of a Virtual Geology Field Guide of Georgia Initiative Using Gigapan© and ArcGIS Online's Story Map. Presented at American Geophysical Union, Fall Meeting 2015, San Francisco, December 2015; ED31C-0917. SAO/NASA ADS Physics Abstract Service. http://adsabs.harvard.edu/abs/2015AGUFMED31C0917M (accessed Feb 1, 2018).

Oreskes, N. *Plate Tectonics: An Insider's History of the Modern Theory of the Earth*; Westview Press: Cambridge, MA, 2001.

Sharp, R. P.; Glazner, A. F. *Geology Underfoot in Southern California*; Mountain Press Publishing: Missoula, MT, 1993.

Sharp, R. P.; Glazner, A. F. *Geology Underfoot in Death Valley and Owens Valley*; Mountain Press Publishing: Missoula, MT, 1997.

Stose, G. W.; Ljungstedt, O. A. *Geologic Map of the United States*; U. S. Geologic Survey: Washington, DC, 1932.

Vetter, J. Field Science in the Railroad Era: The Tools of Knowledge Empire in the American West, 1869–1916. *Historia, Ciencias, Saude—Manguinhos* **2008,** *3,* 597–613.

Vine, F. J.; Matthews, D. H., Magnetic Anomalies Over Oceanic Ridges. *Nature* **1963,** *199,* 947–949.

Witherspoon, W.; Rimel, J. Commercially Successful Books for Place-based Geology: Roadside Geology Covers the U.S. In *The Geotourism Industry in the 21st Century*; Sadry, B. N., Ed.; The Apple Academic Publishers Inc.: 2020; pp 325–353.

CHAPTER 13

Commercially Successful Books for Place-Based Geology: Roadside Geology Covers the US

WILLIAM (BILL) WITHERSPOON[1,*] and JOHN RIMEL[2]

[1]PO Box 33522, Decatur, GA 30033, USA

[2]Mountain Press Publishing Company, PO Box 2399, Missoula, MT 59806, USA

*Corresponding author. E-mail: bill@georgiarocks.us

ABSTRACT

The *Roadside Geology* series from Mountain Press Publishing (MPP) has profoundly improved access to geology for travelers in North America. Over one million copies have connected places to Earth science concepts for readers, many of whom have little or no formal background in geology. The books, widely available in parks, museums, and on the shelves of both independent and large chain booksellers, now cover 38 US states and part of Canada.

The format is characterized by concise introductions to concepts, end-to-end road guides for major highways within each geologic region, and road guide geologic maps with arrows marking outstanding features. All books since 2010 are in full color, which enhances both their visual appeal and the usefulness of the road guide maps. The eight-year process to prepare *Roadside Geology of Georgia* typifies the effort that goes into each volume in the series. Its authors learned to give readers just enough information to understand what is most interesting about a feature, rather than a systematic textbook. MPP staff helped improve conciseness, as well as clarity for non-geologists. The local expertise of six geologist reviewers helped to insure accuracy.

In addition to directly serving travelers, the book is a textbook in Georgia universities, and a springboard for dozens of its authors' geology walks and talks. Tailored programs have tied geology to landscape evolution, plant communities, gold rush and settlement history, Civil War battles, fossil collecting, climate history, and specific travel destinations. Slideshows and web pages supplement illustrations from the book with special uses of Google Earth®, such as geologic map overlays, tours, and pull-up cross-sections.

13.1 INTRODUCTION

Tourists who seek knowledge can be found browsing the bookshelves of park gift shops. They find books specific to the place, including fiction, history, and general travel guides. They also find reference works, mainly nature identification guides. Among the few that blend place with nature are the geology books.

In the USA, perennial sellers in park and museum stores include the *Roadside Geology* (RG) series from Mountain Press Publishing (MPP). More than a million copies of RG books are in the hands of the public today. This chapter traces the surprising commercial success and educational impact of the series. In a country that often neglects Earth science in its schools, these geology books provide an example of how the geotourism industry benefits the world, by deepening travelers' curiosity and knowledge about the workings of planet Earth. In return, tourists' awakened curiosity leads them on new travels to mountains, deserts, and shorelines to investigate the planet with new eyes.

The success of geology books is a surprise to many, because of a common cultural assumption that geology is arcane, or worse, dull. Although deciphering the deep past gained many admirers when it was new in the 1800s, by the time the first RG book appeared in 1972, geology had fallen far behind other sciences in the popular imagination. Physics had produced the bomb and Sputnik, and biology and chemistry had revealed DNA. Meanwhile, in the 1966 Halloween TV special, "It's the Great Pumpkin, Charlie Brown," Charlie is laid low, because other trick-or-treaters get candy, and all he ever gets is a rock. There is a parallel in today's dazzlingly successful TV comedy about scientists, The Big Bang Theory, whose occasional geologist character Bert is played for laughs, as doggedly sniffing out the dullest of all possible trails in science.

Predictably, geology guides must explain the payoff to the general reader in their opening sentences. Geology "is the foundation of the scenery because landscape, after all, is geology with trees growing on it" (Alt and Hyndman, 1972). Learning it can be the antidote to a long drive across Texas that otherwise could amount to "miles and miles of miles and miles" (Sheldon, 1979). These books promise that the reader will see the landscape in a whole new way. Delivering on that claim is both the challenge and the reward of writing a popular geology guide.

13.2 A NOTE ABOUT AUTHORSHIP AND USE OF FIRST-PERSON NARRATIVE

I (first author Witherspoon) am honored that John Rimel consented to be co-author of this chapter. Publisher of MPP since the 1990s, he has been involved with the RG series since his college days in the 1970s. This chapter recognizes that the business John runs has made a key contribution to geoscience education and geotourism in the US. He is the source of the information in this chapter about MPP as a business.

John writes about the value of MPP's work for geoscience education:

> I think perhaps now, perhaps more than at any point in our history, we need to refocus our efforts in making the sciences accessible and understandable. All too often as the sciences have become more specialized and filled with jargon, we tend to lose the ability to translate that understanding of science to the layperson in a way that makes sense to them. Geology is as guilty of this as any of the sciences ... Tying [geology] to the landscape, to the place they live or are visiting, seems to be the best way to make it both understandable and meaningful. Geology and the earth sciences are so vital to understanding the issues we face as a country with regard to global warming, severe weather events, natural catastrophes, and events both man-made and of nature's creation that are impacting our world.

In preparing this chapter, I also spoke to Donald Hyndman, who with the late David Alt started the RG series (*Roadside Geology of the Northern Rockies*, Alt and Hyndman, 1972). Don generously wrote a brief account of its history for the time he was involved and allowed me to quote it extensively. He declined to be listed as co-author.

A large part of this chapter is my own description of working on *Roadside Geology of Georgia* (Gore and Witherspoon, 2013) and the subsequent opportunities its publication has given me to promote geoscience education and geotourism. I decided that writing in the first person worked best, as it also does for Don Hyndman's quoted narrative. Both of us refer to our first author collaborators by first name, as Dave and Pamela. This seemed natural, and no disrespect is intended.

13.3 COVERING A CONTINENT WITH GEOLOGY FOR THE MASSES: MOUNTAIN PRESS PUBLISHING

Don Hyndman writes:

The series began when Dave Alt and I, young professors of geology at the University of Montana, decided to write a book on geology for non-professionals. Our beginning geology students kept asking where they could read more about geology in Montana. Nothing available, we decided to write one ourselves. After discussing various options, we decided that a road-by-road approach would be best.

By 1971, we had a manuscript in hand. I had just finished writing a textbook, *Petrology of Igneous and Metamorphic Rocks,* published by McGraw Hill in New York. Challenges in dealing with a big publisher 2000 miles away (pre-internet/email), especially with drafting the diagrams, suggested that we should try to work with a small local printer/publisher, Mountain Press Publishing (Hyndman, 2018).

Mountain Press, founded as a printing business by David Flaccus in 1948, had published its first book in 1964 after a similar approach from a university professor with *The Psychology of Cornet and Trumpet Playing*. Moderately successful and well reviewed, the book was followed by a series of manuals written by local physicians for coronary care nurses, which also did well and sold throughout the world. Soon the publishing business had outgrown the printing business. Books interested David more than letterheads, so he sold the printing business and reincorporated in 1970 as a book publisher. By 1972, 29 books had been published, including 11 on coronary care.

Hyndman recalls Flaccus being skeptical that there was a market for a book on roadside geology. He continues:

We suggested that the book could be sold, not only in bookstores and rock shops, but also in gas stations to travelers; the latter didn't ever amount to much because Flaccus didn't have a budget for marketing. An additional incentive was that we offered to take no royalties for the first 5000 copies, a sales number that he couldn't imagine ever reaching.

He finally agreed to print 1800 copies on non-reusable plastic printing plates, which were cheaper to produce. When the 1800 books sold out in a few weeks, he was delighted, of course, but mortified that he would then have to re-burn the copy onto reusable metal plates. Word spread quickly, and his subsequent print runs were 5000 copies each.

Neither of us had known anything about geology of the northern Rockies before coming to Montana, but we had since done research with graduate students, and taught field geology for a few weeks each summer. To prepare for writing, we both read everything we could find on the local geology. We used the 1955 U.S. Geological Survey (USGS) geological map of Montana as a base.

We had settled on an audience with little or no background in geology, though we find that many of our readers took a single beginning geology class in college. That audience was not intended to be professional geologists, but many of them do buy the books. Our writing was aimed at people with a ninth-grade education—that is, avoiding words that would be not familiar to those who did not go to college, and avoiding technical language as much as possible. At the same time, we try to be scientifically correct, based on published literature.

Among the challenges for this first book was finding published papers on various topics. Distribution and identification of igneous and metamorphic rocks were covered moderately well. Generalized geological maps, published by the USGS and the Montana Bureau of Mines and Geology helped piece together gaps in coverage. Origin and evolution of various rock types was rather hit and miss. Academic studies tend to be rather esoteric, much of it not suitable for our audience. There was very little published on landscapes and landscape evolution, topics important to our target readers.

Since we wanted to give our non-geologist readers a feel for the rocks and how they originated and evolved, we spent a lot of time and effort to make our own interpretations based on available information, and driving as many of the roads as we could. Cross-sections were either not available, or not where we wanted them, that is, perpendicular or parallel to our roads. We drew our own cross-sections mostly perpendicular to the particular road, and extending several to a few tens of miles from the road.

Dave and I worked well together. I gathered much, but not all, of the published literature, drew all of the initial geological maps, and the initial cross-sections. I took almost all of the photos. Together we had extensive discussions of all aspects of the geology. Dave did most initial writing of the text, and hand-drew most of the final copies of maps and cross-sections. He interpreted most of the landscapes, and I wrote more on igneous and metamorphic rocks, and geologic structures. Together, we sorted out the tectonics. We spent a lot of time inferring evolution of all aspects of the geology, most of which was not available in the published literature. In that sense, we had to do much of our own research (Hyndman, 2018).

Roadside Geology of the Northern Rockies (Fig. 13.1) is out of print today, superseded by the authors' later Montana and Idaho volumes and by *Roadside Geology of the Yellowstone Country* (Fritz, 1985; Fritz and Thomas, 2011). It is hole punched with a plastic binding, and my copy's binding is fragmenting with age. It is the only volume in the series to use fold-out maps. These are confined to the 23-page introductory section, and show the whole region, first without the geology, and then four maps each showing a different age of strata. Subsequent printings would eliminate the fold out maps but employ a more traditional paperback binding.

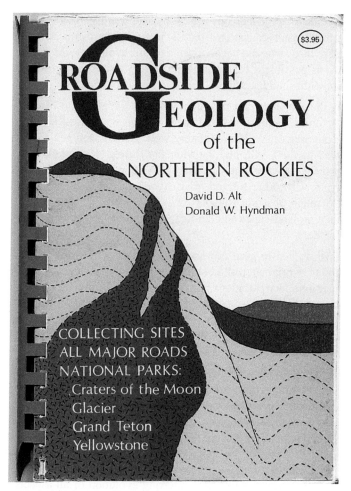

FIGURE 13.1 Cover of the first book in the *Roadside Geology* series.
Source: Adapted from Alt and Hyndman (1972), with permission.

Commercially Successful Books for Place-Based Geology   331

For this book, there was no subdivision into regions, such as separate introductions would be written for later titles of the series. Don Hyndman says this was logical, in that they had chosen mountainous western Montana and northern Idaho, and excluded areas, such as the Montana Great Plains, or volcanic southern Idaho, that would have demanded a separate introduction.

Patterns that would characterize the whole series were established, such as the road segment map with labeled points of interest (Fig. 13.2), cross-sections, and diagrams illustrating geologic principles (Fig. 13.3).

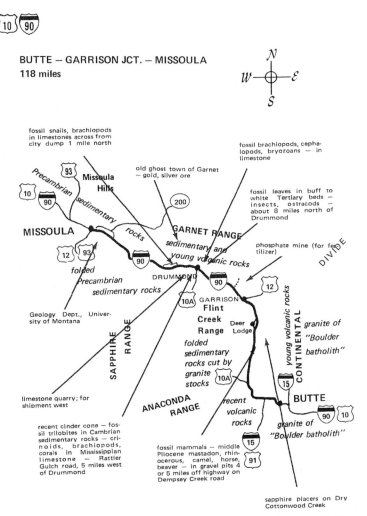

FIGURE 13.2 Road guide map along a segment of I-90, between Butte and Missoula. *Source*: Reprinted from Alt and Hyndman (1972), with permission.

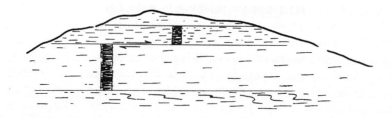

Diagram showing how vertical dikes of basalt are offset by horizontal faults in layers of volcanic material near Wolf Creek.

It is difficult at first to imagine how layers of rock can slide past each other like this because it would seem that there would be too much friction between them. Apparently this happens when the water trapped in the rocks is under sufficient pressure to float most of the load, thus greatly reducing the friction between the layers. The same sort of thing happened on a very much larger scale to form Glacier National Park and the Sawtooth Range between here and Glacier Park.

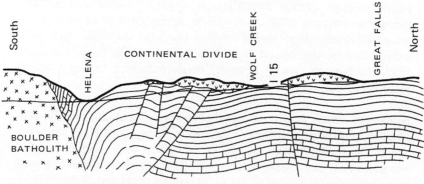

South-north cross section along the line of Interstate 15 between Helena and Great Falls. Gently folded sedimentary formations are intruded by granites of the Boulder batholith near Helena and locally overlain by dark-colored Mesozoic volcanic rocks.

Interstate 15 follows the valley of the Missouri River almost the entire distance between Wolf Creek and Great Falls. Between Wolf

FIGURE 13.3 Example of a diagram and a cross section in RG Northern Rockies.
Source: Reprinted from Alt and Hyndman (1972), with permission.

Roadside Geology of the Northern Rockies is the only series volume with no mention of plate tectonics. Although Hyndman had attended the 1969 Asilomar, California, Penrose Conference that was a milestone in moving plate tectonic insights onto land in general and California in particular (Dickinson, 2001), applying plate tectonics to the Rocky Mountains was barely in its infancy.

The commercial success of this first book led to the authors tackling *Roadside Geology of Northern California* (Alt and Hyndman, 1975; see Witherspoon, this volume). The location was chosen, according to Hyndman, not only for the practical reason that he had completed his Ph.D. at University of California Berkeley, but also to take advantage of the work others had begun in bringing plate tectonics to California geology. Besides plate tectonics as the centerpiece of its introduction, and stable book binding, there were two particular improvements. These were grouping of routes by geologic region, each with its own introduction; and geologic mapping as the base for the road guide maps.

After California, the pair completed Oregon, Washington, Montana, and Idaho (Alt and Hyndman 1978, 1984, 1986, 1989). Meanwhile, other authors began to ask to write for the series. Hyndman recalls,

> For quite a few years we acted as (very part-time) editors for the RG series. This included finding capable new authors for new books. Most potential authors approached us at Geological Society of America (GSA) meetings. Each was asked to provide a book outline and a writing sample, such as a chapter for the book. Most had good background in geology and a record of professional publications, and seemed to be very knowledgeable of the geology of their state. Most thought they could complete a draft of the book in a year or two. Almost all were very slow to provide initial material, and most failed to show any progress at all, in spite of intermittent urging (Hyndman, 2018).

Despite such setbacks, as of 2018, 45 authors have contributed 39 titles to the series, covering all but 12 US states, plus southern British Columbia (Fig. 13.4). MPP's business has also grown. In the catalog, I count 90 history-related titles, 68 in the Earth sciences, and 54 other nature titles. Out of all three categories, 43 titles are written for young readers. The RG series leads the pack. In 2017, the top six selling titles were all in the RG series (Northern and Central California, Oregon, Nevada, Colorado, Washington, and Utah, in that order), as were 18 of the top 30 titles. Another 5 of the top 30 were geology titles.

The general format set in RG Northern California in 1975 has characterized the books, that is, book introduction, regional introductions, and road guides with customized geologic maps. Naturally, individual writing and instructional styles differ, as well as the familiarity of authors with specific subdisciplines. Among geologists, you sometimes hear grumbling about how a particular concept was simplified or misrepresented, yet academic geologists buy and rely on the books when heading into unfamiliar states. The original idea of professors recommending the books to introduce their students to local geology has also been fulfilled many times.

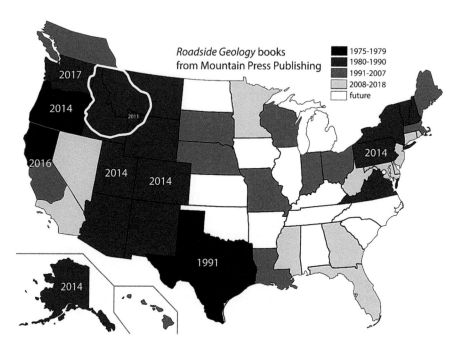

FIGURE 13.4 Coverage of the US and part of Canada by the RG series as of 2018. White outline: RG Northern Rockies (1972). Shading (see legend) refers to the original year of publication. Year of the most recent later edition, if any, is shown in white. Base map by Wikimedia Commons, the free media repository.

Two authors are worth noting in that, beginning in the 1980s, they had the time and energy to create multiple state volumes, as Alt and Hyndman did in the Northwest. These were Halka Chronic, who cornered the "four corners states" of Colorado, Arizona, New Mexico, and Utah (Chronic, 1980, 1983, 1987, 1990), and Bradford Van Diver who was first to bring the series to the East, turning out New York, Vermont/ New Hampshire, and Pennsylvania (Van Diver 1985, 1987, 1990). Chronic, who won a 2004 Geosciences in the Media award from the American Association of Petroleum Geologists, had a more informal style and more emphasis on sedimentary history than tectonics, as compared to Van Diver, whose work tends to be more structured and comprehensive.

Out of the more than one million RG books sold since 1972, two Chronic volumes (*RG Arizona* and *RG Colorado*) are probably the top sellers, followed by Alt and Hyndman's *RG Montana* and *RG Northern and Central California*. Exact sales figures are not available on all four, but RG Montana as of early 2018 is at 79,000 copies.

The series won MPP the 2001 Journalism Award from the American Association of Petroleum Geologists, "for notable journalistic achievement in communications contributing to public understanding of geology." It was the first time a company, rather than an individual, received the honor. In 2015, the Geoscience Information Society awarded MPP its first Outstanding Geologic Field Trip Guidebooks Award for the RG series.

13.4 CREATING A WORK IN THE SERIES: *ROADSIDE GEOLOGY OF GEORGIA*

The most visible improvement to the RG series came in 2008, when the publisher went to full-color editions. More about that below, but first, as one who co-authored an RG book published in 2013, which has sold more than 6000 copies, I would like to describe the authoring process, and what we learned about creating a book for the series. *Roadside Geology of Georgia* (Gore and Witherspoon, 2013) began in 2004, in the fashion Hyndman describes as typical, when Ed Albin, my planetary geologist colleague at Fernbank Science Center, approached Dave Alt at a GSA meeting in Denver.

Georgia had no books on the geology of the whole state, and barely any writings on geology for the public. Two regional geology studies (Butts and Gildersleeve, 1948; McConnell and Abrams, 1984) were useful mainly to professionals, and a series of state park geologic guides (e.g., Abrams and McConnell, 1977; Joyce, 1985) were among very few writings designed to explain Georgia geology to a general audience.

Ed Albin, after his sample chapter had been approved, chose Pamela Gore of Georgia Perimeter College (now part of Georgia State University) as his co-author. She had built an award-winning website for Georgia geology, beginning when few people had begun to use the internet, and, while not a field geologist, was widely connected with the field experts of the state. Pamela obtained college funding for a 6-month sabbatical in 2005 and began work. However, also in typical fashion according to Hyndman, they had both underestimated the time needed to complete a draft. In the fall of 2007, with Ed no further along on his part, he decided to bow out, and the two of them invited me in as Pamela's co-author.

Pamela was working on an update of her well-regarded *Historical Geology Lab Manual* (Gore, 2013), carrying a full-time teaching load, pursuing grants, and serving on committees both local and national. Fortunately, neither she nor I were under typical university pressure to produce a stream of published research. My then-employer, Fernbank Science Center, is an unusual, if not unique, institution. Part of the DeKalb County (GA) School District, it employs instructors with advanced science degrees to develop and teach a variety of programs and courses for K-12 students and their teachers. My work schedule was the same as that of a classroom teacher (but with a smaller grading load), so I controlled my own time outside the 200-day school calendar. Other than Pamela's sabbatical pay, and the promise of less than a dollar per copy apiece in eventual royalties, our effort was self-funded.

My essential first task was to learn Adobe Illustrator®, and start producing road guide maps, based initially on the digitized version of the *Geologic Map of Georgia* (Lawton,1976), which had been imported to Illustrator format for Ed Albin by then-high-school student (now Ph.D. geologist) Steven Jaret.

During her sabbatical, Pamela had managed a healthy start on most of the needed road guides, but she allowed me to rewrite and add to her material where I was most familiar with the geology, in the northern and western parts of the state.

She joked that for the book's outcrop photos, I liked mine and she liked hers. While gracious, as first author, she made the final decisions. We left writing introductory material for the book and the regions to the later part of the project. As her work pressures mounted, including her other book project, it became necessary for me to draft those introductions, except for her draft on the Coastal Plain. Later, we edited one another's introductory material.

At first, I viewed writing introductions as an opportunity to produce a mini-geology textbook, with Georgia's natural attractions as "teachable moments." The book's order, starting with the Sea Islands where sediments are accumulating today, and progressing ultimately to the Piedmont, where many episodes of heat and pressure have transformed rocks most intensely, preserves some of this idea.

However, I was to learn that the nature of the RG series is more to give readers just enough information to understand what is most interesting about a feature, than to burden them by laying a foundation as systematically as a textbook would. Whenever possible, a concise explanation should happen as each feature is encountered in a road guide. Regional introductions are for explaining points that will come up repeatedly in multiple road guides. The book's introduction is written to excite initial interest, and to lay the twin foundations of geologic time and plate tectonics, with a hint of how the two will play out in the state.

In a warm, humid location like Georgia, vegetation and soil annoyingly reduce the availability of rock outcrops. I tell audiences that the name, "Roadside," reflects the series' origin in states where the geology visible from the major highways can fill a book; but in Georgia that geology itself would amount only to a pamphlet. The good news is that when we stretch beyond that title in Georgia, we find popular destinations with eloquent geologic stories, from canyons and waterfalls of the mountains to beaches and marshes of the coast.

Still, expectation piles up on a relatively few outcrops to demonstrate the key conclusions field geologists have drawn. We were grateful to use Georgia Geological Society field trip guidebooks to locate many such outcrops, but on our visits to outcrops from earlier guidebooks, there sometimes was less to see than we hoped. For example, a critically significant ancient fault, separating Blue Ridge from Valley and Ridge rocks, had once been nicely exposed near the base of Carters Dam. I found it overgrown with small trees and fenced off (Fig. 13.5). No sooner had I skirted a concrete moat and thrashed through briars to peer through a chain link fence, than I had to thrash back out to explain myself to a security guard. He had driven up quickly, sensitive to the perceived risk of dam sabotage in a post-September 11 world. There were several examples of key locations that no longer were able to demonstrate important conclusions geologists had reached. One hard rock outcrop on an interstate actually disappeared in the redesign of an interchange, between the first and second printing of the book. Fortunately, Pamela noticed, and removed it from the text.

Our biggest challenge as we neared completion was having nearly twice the expected number of words, requiring much to be cut before a book could be printed at the target price of $24. Four road guides had to be scrapped. One of the deleted road guides is preserved at georgiarocks.us/book, and the remainder may yet appear there. Pamela and I agreed that the assistance from staff provided by MPP was outstanding. James Lainsbury, whose expertise is in technical writing rather than geology, would read a draft that I prided for a clever explanation. When he diplomatically made clear that he did not understand a word I had written, the writing began to show some improvement. At the end of the process, too, he would correct mileages to locations that we had given, having checked them himself using Google Maps®. The reductions needed to get the word count down relied heavily on his editing expertise. Also at the end, the maps I had produced had a colorful and professional makeover, thanks to the skills of Chelsea Feeney.

FIGURE 13.5 Outcrop of Blue Ridge thrust fault near Carters Dam, considered for inclusion in RG Georgia because of its significance, but omitted because of barriers to access.

A final step was reviewing the book for geologic accuracy. Pamela's contact list paid off: we managed to divide the manuscript among six leading geologists active in field research in particular regions of the state. They were able to critique our facts and our simplifications of current thinking.

13.5 THE NEW CROP OF COLOR: MOUNTAIN PRESS GEOLOGICAL INTERPRETATION BOOKS SINCE 2008

Roadside Geology of Connecticut and Rhode Island (Skehan, 2008) was the first of the RG series to be printed in full color. The expense both of color reproduction, and the required heavier, glossy paper, had come down significantly. There was also cost savings because books were printed in Hong Kong. Color makes diagrams, especially the labeled geologic maps that are an RG hallmark, far more readable. Good color photography also can make the books visually stunning. To advertise the color maps inside, many of the book covers switched from the simple but colorful graphics of the early books to a geologic map of the state. These not only catch the eye but also stimulate curiosity, as the browser of a bookshelf realizes that familiar divisions of a state have geologic reasons behind them. Today, 17 states, the District of Columbia, and "Yellowstone Country," have color editions (see lightest shade in Fig. 13.4). Several of these are updates or new works replacing books from the 70s and 80s (dates on states in Fig. 13.4). With advances in technology, the US has once again become competitive in both pricing and quality for color, and more recent color editions are often being printed stateside.

When an older title is replaced in color, the difference between an update and a completely new work depends on the book project. *Roadside Geology of Oregon* (Miller, 2014) totally replaces its predecessor (Alt and Hyndman, 1978), and *Roadside Geology of Washington* (Miller and Cowan, 2017) replaces Alt and Hyndman (1984). Likewise, the second edition of *Roadside Geology of Yellowstone Country* (Fritz and Thomas, 2011) is a new work, with more than twice as many pages as its predecessor (Fritz, 1985), and a far more nuanced understanding of the Yellowstone hot spot. All three new books reflect the remarkable advances in geology in the wake of the plate tectonics revolution.

In northern California, much of the step-up in plate tectonic understanding was reflected in the difference between Alt and Hyndman (1975) and Alt and Hyndman (2000). As mentioned in Witherspoon (this volume), Alt had passed away and Hyndman had retired by the time the full color edition was published (Alt and Hyndman, 2016). That edition is not only much more eye-catching, but also far more useful than its predecessor on account of the improved readability of full-color geologic maps. However, the big differences in text are limited to particular areas such as the Klamath Mountains. As a result, there are better sources to interpret 21st century advances in plate tectonic knowledge to the reader (e.g., in California, Meldahl, 2011).

One of those sources is *Roadside Geology of Nevada* (DeCourten and Biggar, 2017), which filled a longtime hole in the RG series' coverage of western states. Nevada is now known to be about twice as wide as it was 40 million years ago. Its pulling apart followed a period of compression that had uplifted the Nevadaplano, a high region similar to today's Altiplano in Bolivia and Peru. The state has had multiple phases of volcanic activity as well as quiescence over this time, leading up to the present.

The transition from purely describing these events to being able to explain them is a 21st century development. It relates to understanding what happened to oceanic crust (the "Farallon slab") that had been overridden by the westward-moving North American plate. Changes in the geometry of the Farallon slab are thought to explain the timing of events in much of the western US, including the uplift of the Rocky Mountains in Colorado and Wyoming. One of the figures (Fig. 13.6), a collaboration between DeCourten and MPP illustrator Chelsea Feeney, demonstrates both the changes in the Farallon slab and the value of full color.

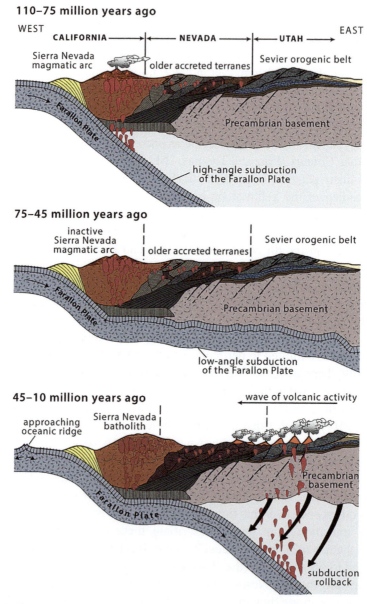

FIGURE 13.6 RG Nevada color illustration of the changes in the Farallon slab through time that drove both Nevada and California geology.

Source: Reprinted with permission from DeCourten and Biggar (2017).

Besides the full-color editions of *Roadside Geology*, color has helped two other MPP series to blossom. The *Geology Underfoot* series differs from the RG series in using vignettes rather than road guides. *Geology Underfoot in Yosemite National Park* (Glazner and Stock, 2010) was the first full-color book in the series, following two other series books also in California (Sharp and Glazner, 1993, 1997). I let the Yosemite book guide me to Pothole Dome, a less-visited feature only about a mile west of Tuolomne Meadows Visitor Center, where I had purchased the book. The dome has features in common with Stone Mountain, Georgia about which Pamela Gore and I had written, related both to the history of magma intrusion and the scaling off (exfoliation) of its surface. It adds marks that a moving glacier made, called striations and chatter marks (the chapter title is "It Went Thataway! The Shaping of Pothole Dome"). By carefully following the authors' directions, my son and I also made our way to one of the person-sized cylindrical cavities that give the dome its name, drilled by swirling gravel, as water poured through a hole in glacial ice.

The book's introduction covers plate tectonics and refers to the Farallon slab in discussing why granite intrusion stopped—"one hypothesis is that the angle of the subducted plate flattened out, so that it no longer reached the depth for melting to occur until it got much farther inland." Appropriately, about half the introduction is devoted to glaciers and their effect on the landscape.

The newest MPP collection is the *Rocks!* series, which chooses geologic sites in a state and devotes (usually only) a single page to the geology of each. It is like having a collection of high-quality interpretive signs that one would hope to see on visiting each location. The series began with *California Rocks!* (Baylor, 2010). One of its 65 sites is Panum Crater, a volcano that last erupted 650 years ago on the shore of Mono Lake, and one of my favorite geologic sites anywhere. I had been guided there by the description in Konigsmark (2002), but did not have *California Rocks!* at the time. Baylor's photo captures one of the site's charms, a view to the steep faulted edge of the Sierra Nevada across a ridge of volcanic ash ejected by the volcano, with obsidian (volcanic glass) that oozed from the center of the volcano in the foreground. The text mentions that Native Americans shaped the obsidian into tools and arrowheads, warning that acquiring the skill would take years of practice and lots of protective equipment.

The *California Rocks!* introduction includes a one-page geologic time scale with California events marked, and a four-page color-illustrated introduction to plate tectonics and its effects on the state.

13.6 APPLYING *ROADSIDE GEOLOGY OF GEORGIA* TO PROMOTE GEOLOGICAL TOURISM

Beginning in the summer of 2012, almost a year before *Roadside Geology of Georgia*'s release, but after most of our writing and drafting duties were complete, its authors began promoting it. I developed the georgiarocks.us web site, and Pamela and I launched the book's Facebook® page. By 2013, we were busy contacting parks, museums, libraries, a book festival, and mineral clubs to arrange talks and book signings for the summer and fall. For the book launch, we planned and helped publicize an event at Fernbank Science Center. We billed it as a rock festival (pun intended) that would include a kid-friendly rock identification clinic, and tables staffed by local geological and mineral societies. We promised door prizes, including a book, for those who pre-registered online.

I created a Google Earth® tour of favorite outdoor spots in Georgia that appear in the book (Fig. 13.7), which was projected in action in the exhibit hall. The tour did double duty for the book talk. Using two projectors and screens, we traded off jobs, one presenting and changing PowerPoint® slides, while the other "flew" the Google Earth® display to each new location in the slideshow.

Eagle Eye Books, a local independent, sold from cases of books that had arrived just in time. We donated our (modest) royalties for the night to the Center. Eagle Eye's initial four cases (96 books) sold out, and we had to lend them a handful of author copies to finish meeting the demand.

That summer, we made a concerted effort to present talks and walks with book signings at as many state parks as possible. On the weekend of July 19–21, I presented four talks and three walks, at Amicalola Falls State Park, Dahlonega Gold Museum State Historic Site, and Fort Mountain State Park. After months of preparation, I had four very different talks ready. "Georgia Geo-Travelogue" was similar to what we had presented at the book launch, but "River Rivalries" was about how stream capture may have determined details of the landscape along the southern edge of the Blue Ridge. The other talks were "Geology, Gold, and the Making of Georgia" and "Fort Mountain Oddities." I was pleased with the publicity the talk hosts had spread (Fig. 13.8), and very satisfied with the attendance at all seven programs that weekend.

In our home town of Decatur, I had approached the library author series, and Pamela was very persistent with the book festival, both of which worked mainly by invitation. The Georgia Center for the Book, based out of the Decatur Library, would fit us into their author series, but preferred a

Commercially Successful Books for Place-Based Geology 343

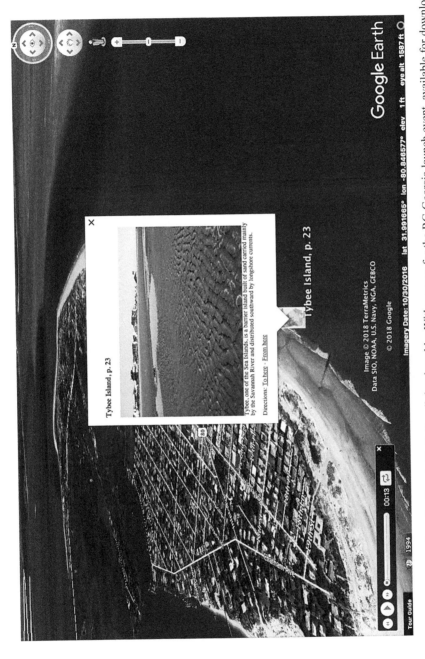

FIGURE 13.7 First "stop" on the Google Earth "tour" developed by Witherspoon for the RG Georgia launch event, available for download at georgiarocks.us/maps.

children's program. Given that our book was "aimed at people with a ninth-grade education" as Hyndman notes, this would seem to dampen attendance and book sales. But we filled the hall, partly by again pairing the event with a rock identification clinic, and it turned out that many parents of precocious children are delighted to present a child with a geologist-autographed book for the child's budding interest to grow into.

FIGURE 13.8 Advance publicity by a state park lodge for a weekend of walks and talks soon after the launch of RG Georgia.

As someone who had frequented elementary school classrooms for the past 16 years, I had two ideas of how to keep children from squirming as we showed the audience around the geological wonders of the state. There was a "matching game" handout for kids to pair locations we would mention with something about each location, and there was the "sock rock" reward. I carefully enclosed each of several special rocks with an identifying tag inside a balled-up, unmated sock from home. I tossed sock rocks to those with correct answers. I joked that the sock was necessary, because if I started throwing unprotected rocks into the audience, I would likely be ejected from the room. The other joke was that a sock rock combined two great mysteries: where do rocks come from, and where did that mate to the sock go?

The Decatur Book Festival bills itself as the largest independent book festival in the US, with around 80,000 attendees (Phillips, 2017). Pamela jockeyed for an invitation to the science track, which generally brings in best-selling authors with a national reputation. We were grateful to get a smaller room adjacent to the ones where the renowned authors were speaking, which, as it turned out, would not be set up for computer projection. All of our talks to date had been using one or two projectors. The sock rock idea came in handy again. The first sock rock question was to guess the location of the closest site mentioned in the book (it was a local soapstone boulder with a historical marker about Archaic Indian bowl manufacturing, outside the courthouse two blocks away). We had about 80 attendees crowded into our room, who quickly exhausted the vendor's book supply as they exited.

To create the various slideshows, figures from the books were supplemented by specialized Google Earth displays. The USGS had converted the digitized Georgia Geological Map into Google Earth. KML format, overlaying its complicated patchwork of colored map units onto Google Earth geography. As when preparing the road guide maps, I simplified the color patterns, grouping rock units by age or dominant rock type.

One advantage of displaying geologic maps in Google Earth is that they drape over its 3D-rendering of topography, highlighting the relationship of rugged terrain to erosion-resistant strata in a particular color. Moreover, once the best angle of view is chosen to show this relationship, it can be saved as a "snapshot view" so a "tour" always returns to that angle to display the point of interest.

Two Google Earth tools that James Madison University professor Steve Whitmeyer has shared online (Whitmeyer, 2018) are particularly useful. One is the ability for a graphic, such as a map legend, to float in a corner of the screen, as the map underneath is moved or zoomed. The other is the ability to fix a graphic, such as a geologic cross-section, perpendicular to the plane of the map (Fig. 13.9).

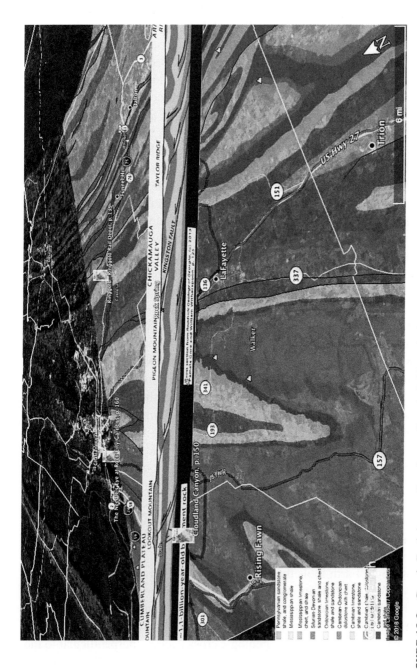

FIGURE 13.9 Google Earth image of northwest Georgia (with geological map overlay simplified from the digital version of Lawton, 1976), showing cross-section by Witherspoon. The technique of floating legend and upright cross-section is from Whitmeyer (2018). Accessible at georgiarocks.us/maps.

Northwest Georgia is an area where the relationship between geology (flat or folded sedimentary strata) and topography (plateaus, valleys, and ridges) is well illustrated using such tools. Relating history to the topography enhances the interest. Henderson (1999) had led part of a Georgia Geological Society field trip to point out relationships between geology and a critical series of Civil War battles along the route from Chattanooga to Atlanta. We had cited some of his observations in the I-75 road guide, and prepared our own figure placing the fought-over railroad and the locations of battles on a shaded-relief map from the USGS (Fig. 13.10). I used this and many other book illustrations, together with Google Earth screen shots, in "Geology and the War in Georgia," first given in September 2013 at the Kennesaw Mountain National Battlefield Park.

Beginning in fall of 2014, the program "Golddiggers, Generals, and Tightrope Walkers" combined elements of three previous programs into what I hoped would be an entertaining introduction to the geology of North Georgia. The "Generals" part incorporates much of the "War in Georgia" program and its points about folded strata determining ridges, valleys, and plateaus. The "Golddiggers" tells the story of the 1829 Dahlonega Gold Rush. Colored polygons in Google Earth represent Georgia's separate slices of crust, called terranes, of which the "Dahlonega Gold Belt" is one. A slide series shows their assembly by plate tectonic movements. "Tightrope Walkers" recalls high wire acts that in 1886 and 1970 crossed northeast Georgia's Tallulah Gorge. It uses Tallulah Gorge, a classic example of stream capture, as a lead-in to discussing the landscape evolution of Georgia's mountains.

The "Golddiggers" program has been handy for groups looking for a generic geology program. This has included mineral clubs, college geology clubs, lifelong learning classes, libraries, a Rotary Club, and a church men's group. To reach nature enthusiasts whose first interest might be birds or plants, finding a co-presenter with ecology expertise has been very successful. By happy coincidence, *Natural Communities of Georgia* (Edwards et al., 2013) appeared at about the same time as our book. It began as an update to the Georgia Geological Survey's *Natural Environments of Georgia* (Wharton, 1978). First author Leslie Edwards is a biogeographer, and worked in the Department of Geology at Georgia State University. Given this origin, accurate geology is not too surprising in this well-organized and beautifully photographed guide about the plants and animals that cluster in particular environments.

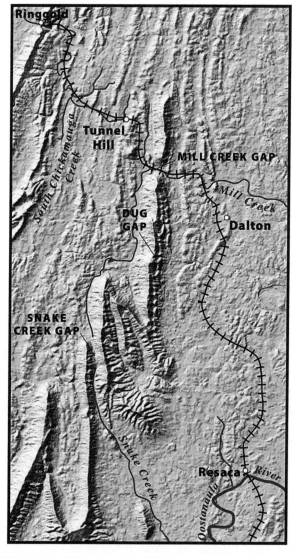

Landscape image showing key locations of May 1864 campaign.

FIGURE 13.10 Illustration from RG Georgia showing the rail line and key Civil War locations in northwest Georgia in relation to the geology-influenced landscape.
Source: Reprinted with permission from Gore and Witherspoon (2013).

After Leslie Edwards and I coincidentally both retired late in 2014, we found many opportunities to lead walks and talks together. This included answering invitations I had received from state parks, nature centers, and environmental education groups, as well as those she received from plant societies and gardening groups. We collaborated on tailoring each slideshow to the geology and natural communities surrounding the talk venue, and alternated speaking during our typically 1-hr long presentations. Wherever possible, we paired each talk with a nature walk, teaching outdoor lessons by pointing out rocks or plants along the way. Our main message, after orienting the viewer to the geologic regions of Georgia and a bit of their origins, is that rock types help determine both soil type and landscape, which together with climate determine the natural communities that can thrive in a given area.

Since recognizing local rock types is not a widely held skill, I developed a game called "Kids Rock!" (Fig. 13.11). As people would arrive for a slideshow or a walk, the game would be spread out on a table, to encourage them to teach themselves their local rocks. The object is to correctly place sample stones on each of five squares labeled with a name such as "granite" or "schist," using extremely simple clues such as "shade: light; pattern: spotted" written in each square. Commonly, even a child too young to read can correctly place the stones when their parent reads them the clues.

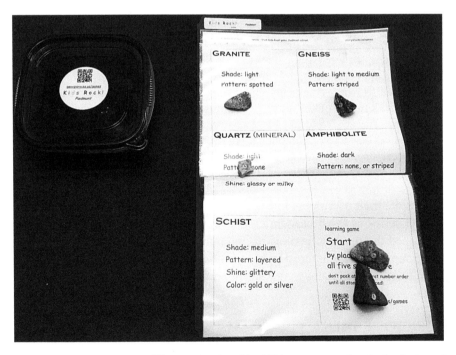

FIGURE 13.11 Kids Rock!™ game developed by Witherspoon, used at the beginning of many walks and talks to introduce local rock types.

Roadside Geology of Georgia is in use as a textbook in "Geology of Georgia" classes in several of the state's colleges and universities. At University of North Georgia at Gainesville, all students enrolled in Introductory Geology and Earth Materials courses are asked to locate an outcrop with the information contained in the book, record photos, populate descriptions in a Geographic Information System (GIS) database, and write a report (Witherspoon and Mobasher, 2018).

Out of the scores of programs for the public since 2013, there have been just a few duds when (virtually) no one came. I learned that profit-making venues (e.g., "gardens" tourist attraction, and a state park hotel run by a private vendor) were more likely to neglect publicity or forget we were coming. Waiting to be invited is good practice, as the host has a larger stake in the outcome and is best at gathering local publicity. From the authors' end, we bring in repeat attendees by putting all public events on Facebook, and issuing a "Walks & Talks E-newsletter" around the beginning of any month with public programs. The e-mail list comes from sign-in sheets circulated at each event, with an optional email address column clearly labeled for the e-newsletter. As of 2018, more than 1200 email addresses are on the list. Unsubscribing continues to be rare.

Looking back, much of what Don Hyndman said in summary is true of Pamela Gore's and my experience as authors of *Roadside Geology of Georgia*:

> In a sense, we wrote these books as a public service, though we certainly enjoyed the research and writing of each book. Over the years we have received glowing comments about our books from individual readers. Some, but not all of our university colleagues, however, didn't view these books as real academic contributions. Our own academic advancement was based on our active professional research, research grants, publications, and teaching. Not lacking in our own self-confidence, we ignored those critical views, and carried on (Hyndman, 2018).

Despite their usefulness in college classes, it is perhaps fair not to regard the RG series as "real academic contributions." Yet they have profoundly broadened the reach of geology to the public, revealing that there is not an age group, nor many interest groups, who are unable to be captivated by the Earth science that relates to our surroundings.

ACKNOWLEDGMENT

This chapter is dedicated in memory of David Alt, whom I did not have the good fortune to meet. From his work, it is clear he had the "ability to convey and translate complex geological and natural science into exciting and understandable concepts, often with an artistic and colorful flare" (Missoulian, 2015). Many thanks to Donald Hyndman for taking time to talk and write about his recollections and giving permission to publish them here. I thank John Rimel for the honor of accepting my invitation to be listed as co-author, for information on the MPP business, and for providing copies of five RG titles to supplement my already long bookshelf of them, along with three examples of the *Rocks!* series. I also thank Pamela Gore for inviting me into her book project and patiently collaborating with me many times, before and since.

KEYWORDS

- **place-based**
- **geoscience education**
- **nature tourism**
- **history of geology**
- **guidebooks**

REFERENCES

Abrams, C. E.; McConnell, K. I. *Geologic Guide to Sweetwater Creek State Park*; Georgia Geologic Survey Geologic Guide 1; Georgia Department of Natural Resources: Atlanta, 1977.

Alt, D. D.; Hyndman, D. W. *Roadside Geology of the Northern Rockies*; Mountain Press Publishing: Missoula, MT, 1972.

Alt, D. D.; Hyndman, D. W. *Roadside Geology of Northern California*; Mountain Press Publishing: Missoula, MT, 1975.

Alt, D. D.; Hyndman, D. W. *Roadside Geology of Oregon*; Mountain Press Publishing: Missoula, MT, 1978.

Alt, D. D.; Hyndman, D. W. *Roadside Geology of Washington*; Mountain Press Publishing: Missoula, MT, 1984.

Alt, D. D.; Hyndman, D. W. *Roadside Geology of Montana*; Mountain Press Publishing: Missoula, MT, 1986.

Alt, D. D.; Hyndman, D. W. *Roadside Geology of Idaho*; Mountain Press Publishing: Missoula, MT, 1989.

Alt, D. D.; Hyndman, D. W. *Roadside Geology of Northern and Central California*; Mountain Press Publishing: Missoula, MT, 2000.

Alt, D. D.; Hyndman, D. W. *Roadside Geology of Northern and Central California*, 2nd ed.; Mountain Press Publishing: Missoula, MT, 2016.
Baylor, K. J. *California Rocks!* Mountain Press Publishing: Missoula, MT, 2010.
Butts, C.; Gildersleeve, B. *Geology and Mineral Resources of the Paleozoic Area of Northwest Georgia*; Georgia Geologic Survey Bulletin 54; Georgia State Division of Conservation: Atlanta, 1948.
Chronic, H. *Roadside Geology of Colorado*; Mountain Press Publishing: Missoula, MT, 1980.
Chronic, H. *Roadside Geology of Arizona*; Mountain Press Publishing: Missoula, MT, 1983.
Chronic, H. *Roadside Geology of New Mexico*; Mountain Press Publishing: Missoula, MT, 1987.
Chronic, H. *Roadside Geology of Utah*; Mountain Press Publishing: Missoula, MT, 1990.
DeCourten, F., Biggar, N. *Roadside Geology of Nevada*; Mountain Press Publishing: Missoula, MT, 2017.
Dickinson, W. The Coming of Plate Tectonics to the Pacific Rim. In *Plate Tectonics: An Insider's History of the Modern Theory of the Earth*; Oreskes, N., Ed.; Westview Press: Cambridge, MA, 2001, pp. 264–287.
Edwards, L.; Ambrose, J.; Kirkman, L. K. *The Natural Communities of Georgia*; The University of Georgia Press: Athens, GA, 2013.
Fritz, W. J. *Roadside Geology of Yellowstone Country*; Mountain Press Publishing: Missoula, MT, 1985.
Fritz, W. J.; Thomas, R. C. *Roadside Geology of Yellowstone Country*; Mountain Press Publishing: Missoula, MT, 2011.
Glazner, A. F.; Stock, G. M. *Geology Underfoot in Yosemite National Park*; Mountain Press Publishing: Missoula, MT, 2010.
Gore, P. J. W. *Historical Geology Lab Manual*. Wiley: New York, 2013.
Gore, P. J. W.; Witherspoon, W. *Roadside Geology of Georgia*. Mountain Press Publishing: Missoula, MT, 2013.
Henderson, S. W. The Geology of Civil War Battlefields in the Chattanooga and Atlanta Campaigns in the Valley and Ridge of Georgia. *Georgia Geol. Soc. Guideb.* **1999**, *19*, 53–78.
Hyndman, D. W. University of Montana, Missoula, MT. Personal Communication, 2018.
Joyce, L. G. *Geologic Guide to Providence Canyon State Park*; Georgia Geologic Survey Geologic Guide 9; Georgia Department of Natural Resources: Atlanta, 1985.
Konigsmark, T. *Geologic Trips: Sierra Nevada*; GeoPress: Mendocino, CA, 2002.
Lawton, D. E. *Geologic Map of Georgia*; Georgia Geologic Survey Map SM-3; Georgia Department of Natural Resources: Atlanta, 1976.
McConnell, K. I.; Abrams, C. E. *Geology of the Greater Atlanta Region*; Georgia Geologic Survey Bulletin 96; Georgia Department of Natural Resources: Atlanta, 1984.
Meldahl, K. H. *Rough-Hewn Land: A Geologic Journey from California to the Rocky Mountains*. University of California Press: Berkeley, CA, 2011.
Miller, M. B. *Roadside Geology of Oregon*. Mountain Press Publishing: Missoula, MT, 2014.
Miller, M. B.; Cowan, D. S. *Roadside Geology of Washington*; Mountain Press Publishing: Missoula, MT, 2017.
Missoulian Obituaries—David Alt. *Missoulian*, [Online], May 17, 2015; p. B2. http://missoulian.com/news/local/obituaries/david-alt/article_cc3198dc-aaaf-5607-9198-ed339d41a1ad.html (accessed Feb 1, 2018).
Philips, R. (PHOTOS) 2017. Decatur Book Festival. *Decaturish.com*, [Online], Sept 2, 2017 http://www.decaturish.com/2017/09/photos-2017-decatur-book-festival/ (accessed Feb 1, 2018).

Sharp, R. P.; Glazner, A. F. *Geology Underfoot in Southern California*; Mountain Press Publishing: Missoula, MT, 1993.

Sharp, R. P.; Glazner, A. F. *Geology Underfoot in Death Valley and Owens Valley*; Mountain Press Publishing: Missoula, MT, 1997.

Sheldon, R. A. *Roadside Geology of Texas*; Mountain Press Publishing: Missoula, MT, 1979.

Skehan, J. W. *Roadside Geology of Connecticut and Rhode Island*; Mountain Press Publishing: Missoula, MT, 2008.

Van Diver, B. B. *Roadside Geology of New York*; Mountain Press Publishing: Missoula, MT, 1985.

Van Diver, B. B. *Roadside Geology of Vermont and New Hampshire*; Mountain Press Publishing: Missoula, MT, 1987.

Witherspoon, W.; Mobasher, K. The Roadside Geology Series as a Geoscience Education Resource. Presented at Geological Society of America, southeastern Section-67th Annual Meeting-2018, Knoxville, March 2018; https://gsa.confex.com/gsa/2018SE/meetingapp.cgi/Paper/312902 (accessed Dec. 23, 2019).

PART IV
Geoparks and Community Developments: A Base for Geotourism Promotion

CHAPTER 14

Community Engagement in Japanese Geoparks

KAZEM VAFADARI and MALCOLM J. M. COOPER*

*Tourism and Hospitality Management Cluster,
Ritsumeikan Asia Pacific University, Beppu 874-8577, Oita, Japan*

Corresponding author. E-mail: cooperm@apu.ac.jp

ABSTRACT

This chapter assesses the processes and patterns to be found in the Japanese community's engagement with geoparks. It is based on trends identified by the authors over the past 15 years and longer, but is a commentary based on the activities of actual communities rather than surveys of tourists. The themes covered are the threads that make up the current community interest in geoconservation, rural decline and revitalization, globally important agricultural and other forms of heritage conservation (geoparks, GIAHS and cultural heritage, Satoyama), and geotourism in Japan. These themes help us identify the impact of geoparks on the community, and the Japanese community's engagement with them.

14.1 INTRODUCTION

This chapter provides a brief overview of the processes and patterns to be found in the Japanese community's engagement with geoparks. It is based on trends identified by the authors and their colleagues over the past 15 years and longer but is a commentary based on the activities of actual communities rather than surveys of tourists (Cooper and Funck, 2013; Vafadari, 2013; Tian et al., 2016). The themes covered are the threads that make up the current community interest in geoconservation, rural decline and revitalization, globally important agricultural and other forms of heritage conservation (geoparks, globally important agricultural heritage system (GIAHS) and cultural heritage, Satoyama), and geotourism in Japan (Chen et al., 2016). These will help us identify the impact of geoparks on the community, and the Japanese community's engagement with them.

The idea of what is now known as geoheritage conservation changed considerably in the 20th century. Prosser (2013) has observed that in the UK understanding sites of national importance came first, but now interest is in community interaction with locally important sites of considerable touristic value. More recently, the focus has shifted to how the geological evaluation of diverse landscapes can help to educate communities about their impact on the natural environment and in the process contribute to geoconservation (Erfurt-Cooper and Cooper, 2013; Boys et al., 2017). The parallel use of the term "geodiversity" for the geological and geomorphological resources of an area first became prominent during the 1990s when communities were reassessing their importance to nature conservation (Gray, 2013).

Geodiversity is defined as: "...the natural range (diversity) of geological (rocks, minerals, fossils), geomorphological (landforms, topography, physical processes), soil, and hydrological features, along with their assemblage, structures, systems, and contributions to landscapes." Importantly, for the later evolution of geoparks, this definition brought the underlying geology into focus (Erikstad, 2013; Chakraborty et al., 2015). Three points should be noticed:

1. Diversity covers the range of natural landforms in an area and their interaction with communities;
2. Geological diversity is analogous to biodiversity but chooses to focus on the abiotic forms of nature; and
3. Geological diversity informs "geoconservation" and therefore geoparks by communities.

Geoconservation is critical for community sustainability: "Given their importance to science and society, geological and geomorphological features and processes are critical parts of our natural environment and are worthy of conservation and sustainable management" (Prosser et al., 2011). The problem is that the required resource backup for conservation activities is generally lacking unless it can be generated through an industry that uses that environment. Geoconservation needs therefore to find a holistic commitment from its constituent communities to be viable (Gordon et al., 2012). In this chapter, these issues are discussed using by placing an important industry that can generate the required resource back-up, tourism, in the context of two specific geoparks in Japan.

14.2 THREATS TO GEODIVERSITY AND THE IMPORTANCE OF TOURISM AND THE GEOPARK CONCEPT

The threats to geological/natural environments depend on the location of those environments and the nature of the geosites involved (Glasser, 2001; Gordon and MacFadyen, 2001). MacFadyen (1999) created a typology of sites subject to human impact as follows:

1) "robust" (resources extensive and of lower value, impact can be managed);
2) "vulnerable" (geoheritage is extensive but has high value and should not be disturbed), and (3) "very vulnerable" (limited heritage resources, with considerable value).

These situations result from land use conflict pressure and change, thus sociocultural processes (Chakraborty et al., 2015) are not always supportive of the need for geoconservation.

Finally, Gray (2013) noted that the "fragmentation" of nature and of the integrity of its underlying geomorphological processes results from urbanization, and we note that this urbanizing force is in fact the fundamental nature of all tourism.

Tourism is therefore now seen as another variable affecting the integrity of geological processes, especially where destinations are sensitive ecosystems (Wright and Price, 1994). Where there are many tourists, natural sites are inevitably degraded. The management of tourism and geological diversity within the geopark context should therefore acknowledge this problem. Another problem that arises in these situations is the lack of knowledge (lack of interest?) about how the landforms present in a destination evolved. While it may be difficult for the layperson (resident and tourist) to understand the importance of geological processes operating across a large area, these processes typically offer economic profit to land users in a specific area, and this often results in community members seeking to "consume" geodiversity rather than protect it, usually through mining and tourism. Thus, identifying geological heritage sites for tourism is not going to be enough to protect them (Erikstad, 2013; Chakraborty et al., 2015) in the absence of an elevated level of awareness of this form of the natural environment in the subject community (including its tourists).

14.2.1 THE CONCEPT OF THE GEOPARK

To offset these problems, the Geopark concept was mooted. A geopark is a nationally protected area usually containing several heritage sites of geological importance and having a visual/informational appeal to the observer (Newsome and Dowling, 2010). These heritage sites are increasingly being seen throughout the world as important resources requiring an integrated management framework of protection, education, and sustainable tourism development if they are to be of benefit to the local, national, and international community (Erfurt-Cooper and Cooper, 2013). The system of geoparks is divided into global (GGN) and national (in Japan, JGN) networks, and achieve their goals through addressing the conservation of the underlying geology, community education, and geotourism.

The first national park to seek to conserve natural heritage was Yellowstone National Park in the USA in 1872, and this was followed by the many now to be found across the world, including in Japan (beginning in 1934; Martini, 2009). The national parks concept gained importance during the second half of the 20th century and continues to be the mainstay of the natural area conservation effort in most countries. Additional foci were established as time went on, including man and biosphere reserves (MABs, in 1976; globally enshrining the joint conservation of community and geological/biodiversity heritage), World Heritage Sites (from 1978, a global outlook), geosite initiatives in Europe, and all refined the national parks concept, bringing it closer to the point where the underlying ecological and physical structure of the earth would become of interest to many visitors in natural surroundings. In 1991, the global geoparks concept (Global Geoparks Network or GGN) was introduced to allow concentration on these geological resources. The GGN was established in 2004 with the support of UNESCO and now has 147 members in 41 countries, including 9 in Japan.

Thus, the geopark concept should be identified as the latest point on a continuum in the development of natural area conservation. The push to establish geoparks in Japan started in 2005 and sped up in 2007 with cooperation from earth scientists and local government. The Japan Geopark Committee (JGC) was established in 2008 and decided on the first three candidate areas in Japan to apply for GGN status in October 2008. The first seven national (domestic) geoparks, including three candidates for the GGN, were then proposed in December 2008. The Japanese Geoparks Network (JGN) was established in February 2009 by these seven national geoparks. Four more domestic geoparks were accepted in October 2009 (Chakraborty et al., 2015).

By January 2020, the JGN had 44 geoparks, including nine global geoparks (Table 14.1), plus 18 associate parks seeking full membership. In Japan, Itoigawa City (located on the Japan sea coast of Honshu) was the first area designated as a geopark. The activities of this geopark have covered planning for tourism, the conservation of geosites, and community education efforts. It became one of the first global geoparks in Japan in August 2009 (the others are Shimabara Peninsula Geopark on Kyushu and Toya-Usu Geopark on Hokkaido). Given the country's geological history, Japanese geoparks have a common volcanic theme. The most important of these being the Shimabara Peninsula (Mt. Unzen), Toya-Usu, Mt. Aso, and Kirishima Geoparks (Shinmoedake), with the first being also classified as a decade volcano [the most hazardous volcanoes on the planet; Erfurt-Cooper and Cooper (2013)]. However, there are some that have a broader focus, covering ancient cultures (Itoigawa), MAB sites (Hakusan Tedorigawa, Southern Japan Alps), and new themes such as "natural hazards" (after the Tohoku earthquake and Tsunami in 2011; Oike et al., 2011), and the rapidly depopulating and aging countryside (Eades and Cooper, 2007) that is now a prominent management concern covers regional redevelopment by attracting new residents or supports a sustained tourist presence through geopark activities. The Japan Geopark Committee (JGC) oversights the formation and operation of geoparks (Chakraborty et al., 2015). This committee is based in the offices of the Geological Survey of Japan.

14.3 GEOTOURISM: THE "COMMUNITY INTEREST" FACTOR

Chakraborty et al. (2015) note that the term geotourism is relatively new. Based in rural tourism, it is based in the recent rise in public awareness geophysical hazards.

Newsome et al. (2012) defined geotourism as "…geologically focused, environmentally educative… and fostering local community beliefs," following on from the work by Hose (1995, 2000), Dowling and Newsome (2006), Joyce (2010), and Newsome and Dowling (2010). Geotourism is thus "Tourism based on an area's geological or geomorphological resources that attempts to minimize the impacts of this tourism through geoconservation management" (Gray, 2008). Geoparks are therefore a prime venue for geotourism (Prosser, 2013; Cooper, 2014).

UNESCO defines a geopark as "…a nationally protected area containing a large number of geological heritage sites of particular importance, rarity, or esthetic appeal" (UNESCO, GGN, updated 2010). The JGN website defines them as:

TABLE 14.1 Geoparks of Japan as in 2020.

Global Geoparks	Japanese Geoparks (+ 18 Associate Members)
Toya Caldera and Usu Volcano	Minami-Alps (Mtl Area)
Itoigawa	Dinosaur Valley Fukui Katsuyama
San'in Kaigan	Shirataki
Unzen Volcanic Area	Izu Oshima
Muroto	Kirishima
Oki Islands	Oga Peninsula-Ogata
Aso	Mt. Bandai
Mt. Apoi	Shimonita
Izu Peninsula	Chichibu
	Hakusan Tedorigawa
	Yuzawa
	Hakone
	Happo Shirakami
	Choshi
	Mikasa
	Hagi
	Sanriku
	Sado
	Shikoku Seiyo
	Oita Himeshima
	Oita Bungo-ono
	Sakurajima Kinkowan
	Tokachi Shikaoi
	Tateyama Kurobe
	Nanki Kumano
	Amakusa
	Naeba-Sanroku
	Mt. Kurikoma Area
	Mine-Akiyoshidai Karst Plateau
	Mishima Kikai Caldera
	Shimokita
	Mt. Tsukuba Area
	Asama North
	Mt. Chokai, Tobishima Island
	Shimane Peninsula and Shinjiko Nakaumi Estuary

Source: The authors. (*See also* https://geopark.jp/en/)

"Geopark is a single unified geographical area where people conserve important geological heritage and landscapes, and utilize this heritage for education, disaster mitigation activities, and geotourism, all with the aim of sustainable development for local communities" (Chakraborty et al., 2015). The interest in geological processes in the Japanese community is high because it lives in a very tectonically active environment.

The next section covers the issues for Japanese geoparks in terms of conservation and community involvement.

14.3.1 ISSUES IN GEOHERITAGE CONSERVATION AND COMMUNITY INVOLVEMENT: THE EXPERIENCE OF THE IZU PENINSULA GEOPARK

Community involvement is critical for Japanese geoparks, but exactly how is this to be achieved, and does it include tourists? These questions can be addressed using the experience of the Izu Peninsula Geopark, Shizuoka Prefecture (Chakraborty et al., 2015). The whole Peninsula is designated a geopark and has a total area of 1585 km^2. In tourist terms, Izu is a famous hot spring resort area on the Pacific coast of Japan and a favored destination for tourists from the Tokyo Metropolitan Area. Up to 19 million people visit the peninsula annually (Ministry of Land, Infrastructure, Transport, and Tourism, 2012), but it is unclear how many of these visit the area with a view to gaining knowledge about its geoheritage rather than just using it. Also, in the Izu Peninsula geopark context, 15 local government areas jointly manage the entity, which means that gaining stakeholders consensus on management policy is not always easy (ibid). A further complication is that the geopark also has a wide variety of geoheritage sites (114). Unfortunately, each of these sites may be claimed by its local community as a premier site over others in the same geopark.

The Izu Geopark seeks to conserve and promote its geodiversity in the face of these pressures through three approaches: using existing legal regulations (Fuji-Hakone-Izu National Park); encouraging "good practice" such as the protection of important landscapes by each local community; and through promoting awareness of geoheritage. To effectively deliver its message, this geopark, as do others, seeks the full participation of locals, tries to involve tourists in the flow of information, and trains local operating staff. Above all, it is seeking to ensure that geoconservation is appreciated as widely as possible and properly enforced in all parts of the Geopark. The greatest challenges for conservation in this geopark are that it must remain careful of visitor impacts on sites and heritage, and deal with the low level of knowledge about geoheritage and its conservation in the community. Complicating this situation is the fact that Izu is a popular tourist destination close to Tokyo. In the north of the peninsula, where most of the population and infrastructure are concentrated, visitors are essentially urban and resort minded. Thus, mass tourism dominates the northern Izu Peninsula, and "hot spring bathing tourism" and visits to important cultural sites are major tourism pull factors. The geological underpinning of these attractions is still not well known despite the existence of the geopark (Chakraborty et al., 2015).

Izu Geopark's Promotion Council (the geoparks are critically dependent on local leadership, or a bottom-up approach to their development and resilience) ran three workshops in 2014, and nearly all the attendees preferred to increase tourist numbers to the park (ibid). The problem with this attitude is that the greatest challenge to effective geoconservation in the park is managing the mass tourism impact on its geoheritage. Due to its location near Tokyo, the landscape and ecological components of the park are in constant risk of damage from excessive human activity and a disregard for the approach taken by the local community, even though that community wishes to see more of them. Compounding the tourist impact is the large number of geosites (114) in the peninsula and the lack of trained staff to manage all of them. These sensitive ecosystems can easily be affected by visitors, but few specific conservation plans for the park have been formulated, despite their importance to sustainable tourism. Fortunately, most of the major sites are protected by either the Fuji-Hakone-Izu National Park or the coastal conservation plan of Shizuoka Prefecture. However, as the conservation policies in these areas were made for purposes other than protecting geoheritage in mind, it is necessary to formulate a more comprehensive conservation framework that can integrate the visual aspects, the ecosystems, and the geological resources in these landscapes.

The next problem is the pervading disinterest in geoconservation, even within the local communities in the park. The council deserves praise for trying to raise interest through geotours and field trips, but community involvement remains low to moderate (Chakraborty et al., 2015). The population of the peninsula is 690,000 (plus 19 million tourists), but there are significant variations in terms of its distribution. Depopulation has occurred in the south and west, and what population remains is a rapidly aging one, whereas the northern part is densely populated (although not much younger). In addition, while awareness about the geopark's activities is comparatively high in the coastal areas, it is not further inland. Of course, this might seem positive for community involvement since the major geosites are concentrated along the coast, but it is also an expression of the problem that sites having visual appeal for tourists are deemed "more valuable than those which do not." As Chakraborty et al. (2015) note, this hinders the development of a global understanding that values this heritage because all parts of the local geoheritage are equally important as they convey essential parts of the "story" of the geopark. In addition, residents, politicians, and economic planners often see the geopark concept as a new "tool" to revitalize depopulated areas, without much thought for the resulting impacts on the geology. The depopulation problem is common to most Japanese geoparks, thus using the geopark for "branding" purposes is seen to be a policy direction that may ultimately aid repopulation and economic redevelopment, but only in fringe areas to urban settlements. However, in the parts of the geopark that are away from the geosites on the coast and in the north, knowledge of the park's purpose falls quickly in the declining local population.

In addition, some local communities with deep traditions in the Izu Geopark landscape are not part of the planning activities, and this reduces the potential impact of community involvement. For example, the female "Ama" divers who collect shellfish in the area comment that the health of the coastal ecosystem has declined over recent years (ibid). However, their voice is not adequately represented in the activities of the geopark. A more visible role for this sort of community in geopark-related activities is thus important for coastal conservation if the park managers want to understand these changes in coastal ecosystems. Finally, the examples quoted above are evidence that the evaluation of geodiversity is still predominantly seen from a top-down scientific perspective, not from the perspective of community (and tourist) involvement in understanding georesources.

14.3.2 A SOLUTION: THE KUNASAKI PENINSULA AND GIAHS

The FAO initiative of GIAHS is one of the leading initiatives for the dynamic conservation of geopark landscapes that explicitly includes the value-added potential of tourism (Vafadari, 2012). The Kunisaki Peninsula of Oita Prefecture, Kyushu, Japan, was nominated as a GIAHS site for its traditional resource circulation landscape in 2013 (Fig. 14.1). The peninsula used to be a religious and cultural hub in historical times but is currently facing rapid decline mainly due to depopulation and aging. Vafadari (2013) explored the potential of geotourism based on agricultural heritage in the Peninsula by analyzing the components of the resource circulation landscape and saw the importance of cultural heritage in persuading visitors to come to experience the geopark like resources of this area (part of the area is a geopark—Himeshima Island, just off the coast). The key findings were: the resource circulation landscape of this GIAHS site can be divided into

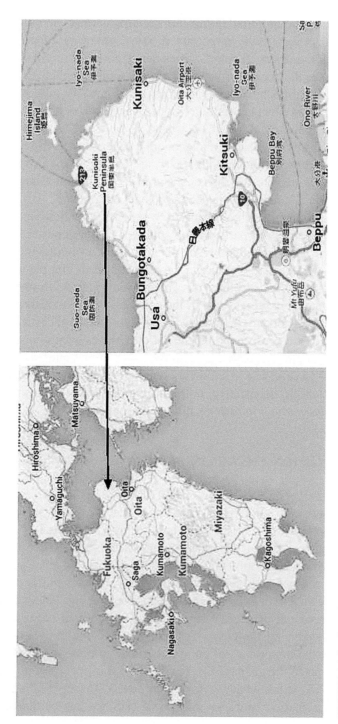

FIGURE 14.1 Location of Kunasaki GIAHS and Himeshima Geopark, Oita Prefecture, Kyushu, Japan.
Source: Google Maps.

"core" and "value-added" components; core components being the small scale interlinked "tameike" reservoirs based on the water courses of the area and the Sawtooth Oak forests, and the endemic shichitoui grass (for Tatami Mat production) and shiitake mushrooms are value-added products. These social ecological production landscapes are presently underutilized, but if geotourism is planned it can generate renewed stakeholder interest in these landscapes, and thus pave the way for revitalizing this declining region, which otherwise has immense value as a low-carbon, sustainable rural society based on traditional farming knowledge.

Kunisaki is located within the West Japan climatic zone. Average annual rainfall is 1570 mm, and in summer temperatures climb above 30°C in many places, but in winter snowfall is rare except at high altitudes. A mixture of landscape, climatic, and settlement patterns are found because of the mainly high seasonal variation in precipitation and the porous volcanic soil that absorbs water quickly. This environment led to the development of a unique and dynamic traditional water management system in the peninsula through the creation and maintenance of multiple "tameike" reservoirs. Although such ponds are observable throughout Japan, Kunisaki offers a remarkable density of them, and their location ensures that they manage the circulation of resources from the mountain to the sea, by acting like valves on the waterways of the area. This intricate, complex, and dynamic resource circulation system has been managed over centuries through traditional knowledge and has created an array of secondary nature landscapes high in production and biodiversity values.

The farming of shiitake mushrooms as a special, value-added local product is a unique type of agriculture that partially utilizes the tameike system. A special type of secondary forest of Japanese Sawtooth Oak or Kunugi (*Quercus acutissima*) is utilized to grow the mushroom in this area. The high density of this deciduous forestry peal so plays a crucial positive role in groundwater retention, soil replenishment, and supports a variety of species and ecosystem services. A rare type of straw called Shichitoui is also cultivated and is used to prepare high quality mats. Kunisaki is the only location where this weaving culture still exists in Japan. Finally, wet rice cultivation was central to the spread of agriculture and sedentary societies in the peninsula, which began as far back as the Yayoi Period (300 BC–AD 300) and could not exist without the Tamieke.

This complex and dynamic system has its core the concept of society's co-evolution with nature and this is perfectly suited to the fusion of geoparks and tourism discussed in this chapter. To see the importance of the tourism attractions of Kunisaki in the geopark context, they should be classified into two main types, natural and cultural. The natural attractions center around the agricultural and forested landscapes, and they are comparatively underused. Tourists visiting this area come instead to see the cultural relics, the most famous of which is the nationally important Usa Hachiman Shrine. This cultural site is also deeply related to resource circulation and the former agricultural society. However, the main attraction for agri-tourism is the Kunisaki landscape itself. The landscape embodies a culture of resource circulation, through the several components discussed above. The FAO also recognized the "integrated" nature of these landscapes, which together make Kunisaki an excellent example of a social ecological production landscape. In other words, the main tourism capital of Kunisaki is the peninsula itself, which is more than the sum-total of its specific components. Agri-tourism (Sidali et al., 2011), or more specifically, agricultural heritage tourism, is based on "experiencing" the landscape and local culture (Vafadari, 2012) and in this sense, the potential for community engagement with the attractions is much more diffuse in nature than conventional tourism to the cultural sites. Typically, GIAHS tourism is associated with the landscape and local lifestyles, and the aim of GIAHS tourists is to be educated about the resilience of local societies and the functional components of their socioecological production landscapes. The Kunisaki Peninsula is a dynamic landscape of resource circulation and co-evolution of nature and agrarian societies. This area was a hub of cultural activities in historic times but declined during the last century mainly due to demographic factors like rural–urban migration, aging, and depopulation. However, the remaining farming community retains an excellent example of community resilience, and the landscape is rich in tourism capital.

The nomination of this landscape as a GIAHS/geopark has opened new opportunities to re-evaluate the traditional resource circulating society and indigenous resource conservation culture of the Kunisaki Peninsula, while protecting the geopark values of the natural environment and promoting the tourism based on them. The GIAHS accreditation committee recognized "core" areas such as the broadleaf Kunugi forests which are increasingly rare in Japan, and the indigenous "tameike" water reservoirs. These were considered vital for sustaining indigenous agriculture, which in turn now sustains the tourism potential of the area. In the 21st century, when the peninsula has steadily receded into the background of life and activities in Japan, the successful nomination raises an exciting opportunity to rediscover, and eventually revitalize this landscape as a commentary on sustainable communities and geotourism.

The strategy for basing geotourism in this GIAHS/geopark area was refined during the process of developing the first community action plan in 2013. This is a bottom-up approach. Villages that were suffering most from a lack of human and institutional resources in the Peninsula were selected for community-building projects. One, the Ryoai Village Project in USA city, is an example where community initiatives gained local government support and created a showcase of community development through geotourism (Figs. 14.2 and 14.3). This small community enjoys a rich heritage of stone structures that are environmentally friendly, technically appropriate, and economically useful in a wet climate. This serves to keep soil and landscape together when water flow is high, but also assists in the maintenance of the unique response to geological conditions known as the Tamieke system (Vafadari, 2013) found in the area.

FIGURE 14.2 A group of students (mixed from elementary to university) participating in a local event organized by the community (note the stone walls and bridge).
Source: The Authors.

FIGURE 14.3 Community rice planting activity.
Source: The Authors.

The community organizes local events. The village enjoys a unique landscape made by stone walls that protect paddy fields form landslides, and a historical bridge that connects the communities living on both sides of the river. Geological resources combined by traditional agriculture landscape has made the village attractive and community has been trained to utilize these for tourism. Furthermore, the local governments in the area have provided abandoned houses to support research and development by local schools and Universities, and to accommodate visitors who come for training and preservation activities. Thus, existing traditional resources have been and can be turned into centers for local planning and internal/external community engagement and knowledge transfer but are also capable of bringing to life the geology of an area while they do so.

14.4 CONCLUSIONS

For effective community involvement in geoparks, activities within them must be linked based on their shared geolandscapes, and not seen as possibly interesting but isolated sites within a larger mass tourism orientation centered on "rural" resort tourism. As a coherent offering, geotourism activities with diverse stakeholder participation can help a geopark like Izu and a GIAHS/geopark site like Kunisaki Peninsula to evolve as role models for popularizing geoheritage. The geopark implementation model of addressing geoconservation that includes legal protection, community site management, conservation awareness-raising among residents and tourists, and coordinating all these with ecological restoration initiatives is effective in creating multi-stakeholder community understanding (Chakraborty et al., 2015). Sustained community participation in educating tourists, managing landscapes, and imparting traditional geophysical and cultural knowledge is a major strength of Japanese geoparks. Together, these influences make up the form of social capital that will develop a better understanding about long-term ecological processes and insights into human impact on geoheritage. Such community knowledge can play a vital part in safeguarding the natural environment. Community knowledge of the natural environment is also helpful for managing geoheritage when it is difficult to achieve planning consensus in other forums.

KEYWORDS

- **GIAHS**
- **community involvement**
- **geoparks**
- **geotourism**
- **Kunisaki Peninsula**
- **participation**
- **knowledge**

REFERENCES

Boys, K. A.; DuBreuil-White, K.; Groover, G. Fostering Rural and Agricultural Tourism: Exploring the Potential of Geocaching. *J. Sustainable Tourism* **2017,** *25* (10), 1474–1493.

Chakraborty, A.; Cooper, M.; Chakraborty, S. Geosystems as a Framework for Geoconservation: The Case of Japan's Izu Peninsula Geopark. *Geoheritage* **2015,** *7,* 351–363.

Chen, B.; Qui, Z.; Nakamura K. Tourist Preferences for Agricultural Landscapes: A Case Study of Terraced Paddy Fields in the Noto Peninsula, Japan. *J. Mountain Sci.* **2016**, *13* (10), 1880–1889.

Cooper, M. Volcanic National Parks in Japan. In *Volcano Tourism: Dynamic Destinations*; Erfurt, P., Ed.; Springer: London, 2014; pp 231–247.

Cooper, M.; Eades, J. S. Landscape as Theme Park: Demographic Change, Tourism, Urbanization, and the Fate of Communities in 21st Century Japan. *Tourism Rev. Int.* **2007**, *11*, 9–18.

Cooper, M.; Funck, C. *Japanese Tourism*; Berghahn Books: Oxford, 2013.

Dowling, R.; Newsome, D., Eds.; *Geotourism*; Elsevier/Heineman: Oxford, 2006.

Erfurt-Cooper, P.; Cooper, M. *Volcanic Tourism—Geo-Resources for Leisure and Recreation.* Earthscan: London, 2010.

Erikstad, L. Geoheritage and Geodiversity Management: The Questions for Tomorrow. *Proc. Geol. Assoc.* **2013**, *124* (4), 713–719.

Glasser, N. F. Conservation and Management of the Earth Heritage Resource in Great Britain. *J. Environ. Plann. Manage.* **2001**, *44*, 889–906.

Gordon, J. E.; MacFadyen, C. C. J. Earth Heritage Conservation in Scotland: State, Pressures, and Issues. In *Earth Science and the Natural Heritage*; Gordon, J. E.; Leys, K. F., Eds; Stationary Office Books: Edinburgh, 2001; pp 130–144.

Gordon, J. E.; Barron, H. F.; Hanson, J. D.; Thomas, M. F. Engaging with Geodiversity: Why it Matters. *Proc. Geol. Assoc.* **2012**, *123* (1), 1–6.

Gray, M. Geodiversity: Developing the Paradigm. *Proc. Geol. Assoc.* **2008**, *119* (3–4), 287–298.

Gray, M. *Geodiversity: Valuing and Conserving Abiotic Nature*, 2nd ed. Wiley Blackwell: Chichester, 2013.

Gray, M.; Gordon, J. E.; Brown, E. J. Geodiversity and the Ecosystem Approach: The Contribution of Geoscience in Delivering Integrated Environmental Management. *Proc. Geol. Assoc.* **2013**, *124* (4), 659–673.

Hose, T. A. Selling the Story of Britain's Stone. *Environ. Int.* **1995**, *10* (2), 16–17.

Hose, T. A. European Geotourism: Geological Interpretation and Geoconservation Promotion for Tourists. In *Geological Heritage: Its Conservation and Management*; Barretino, D.; Wimbledon, W. A. P.; Gallego, E., Eds; Instituto Tecnologico Geominero de Espana: Madrid, 2000, pp. 127–146.

Hose, T. A. The Significance of Esthetic Landscape Appreciation to Modern Geotourism Promotion. In *Geotourism: The Tourism of Geology and Landscape*; Newsome. D.; Dowling, R. K., Eds.; Goodfellow: Oxford, 2010; pp. 13–26.

Joyce, E. B. Australia's Geoheritage: History of Study, A New Inventory of Geosites and Application to Geotourism and Geoparks. *Geoheritage* **2010**, *2*, 39–56.

Larwood, J. G.; Prosser, C. D. Geotourism, Conservation and Society. *Geol. Balcan.* **1998**, *28*, 97–100.

MacFadyen, C. *Fossil Collecting in Scotland: Information and Advisory Note 110.* Scottish Natural Heritage: Edinburgh, 1999.

Martini, G. Geoparks: A Vision for the Future. *Geologia USP* **2009**, *5*, 85–89.

Ministry of Land, Infrastructure, Transport and Tourism (MLIT). *Izu Kanko-ken Seibi Keikaku (Tourism plan for Izu area)*, 2012. http://www.mlit.go.jp/common/000221544.pdf

Newsome, D.; Dowling, R. K., Eds.; *Geotourism: The Tourism of Geology and Landscape*; Goodfellow: Oxford, 2010.

Newsome, D.; Dowling, R. K.; Yeung, Y. F. The Nature and Management of Geotourism: A Case Study of Two Established Iconic Geotourism Destinations. *Tour. Manage. Perspect.* **2012,** *s2–3,* 19–27.

Oike, K.; Kato, H.; Watanabe, M. *Nihonnojiopaku (Geoparksin Japan)*. Nakanishiya Shuppan: Kyoto, 2011.

Prosser, C. D. Our Rich and Varied Geoconservation Portfolio: The Foundation for the Future. *Proc. Geol. Assoc.* **2013,** *124* (4), 568–580.

Prosser, C.; Bridgland, D. R.; Brown, E. J.; Larwood, J. G. Geoconservation for Science and Society: Challenges and Opportunities. *Proc. Geol. Assoc.* **2011,** *122* (3), 337–342.

Sidali, K. L.; Spiller, A.; Schulze, B., Eds. *Food, Agriculture and Tourism: Linking Local Gastronomy and Rural Tourism: Interdisciplinary Perspectives*; Springer: New York, 2011.

Tian, Mi.; Min, Q.; Tao, H.; Yuan, Z.; He, L.; Lun, F. Progress and Prospects in Tourism Research on Agricultural Heritage Sites. *J. Res. Ecol.* **2014,** *5* (4), 381–389.

UNESCO Global Geoparks Network. *Guidelines and Criteria for National Geoparks Seeking UNESCO's Assistance to Join the Global Geoparks Network (GGN)*; New York: UNESCO, 2010.

Vafadari, K. Rural Tourism and Preservation of Landscapes in Japan: A Spotlight on Satoyama in the Noto Peninsula. *J. Hospitality Tourism* **2012,** *10* (2), 87–103.

Vafadari, K. Exploring Tourism Potential of Agricultural Heritage Systems: A Case Study of the Kunisaki Peninsula, Oita Prefecture, Japan. *Issues Soc. Sci.* **2013,** (1), 33–52.

Wright, J. R.; Price, G. Cave Conservation Plans: An Integrated Approach to Cave Conservation. In *Conserving our Landscape*; Stevens et al., Eds.; English Nature: London, 1994; pp. 195–197.

CHAPTER 15

The Role of Volunteer Management Programs in Geotourism Development

CRISTIAN CIOBANU* and ALEXANDRU ANDRĂȘANU

Hațeg Country UNESCO Global Geopark, University of Bucharest, Romania

*Corresponding author. E-mail: cristian.ciobanu@unibuc.ro

ABSTRACT

Youths are a reservoir of creativity and energy for any community. If properly engaged and empowered, they can have a very high impact on the growth of a geopark. Since 2013, Hațeg UNESCO Global Geopark has developed a Volunteer Program to use and inspire the young people. The result is the most successful Volunteer Group in Romania. This chapter shows why volunteers are needed, how a volunteer program works in a geopark, and what results were reached for local development, geoconservation, and geotourism promotion.

15.1 INTRODUCTION

Hațeg Country is a territory in Transylvania, western Romania, where, since the year 2000, the University of Bucharest coordinates a local partnership for the implementation of the geopark concept.

In 2005, Hațeg Country Geopark was accepted as a member of the European and Global Geoparks Networks, and since 2015, it is part of the UNESCO International Geoscience and Geoparks Programme (Andrăşanu, 2015; Ciobanu, 2016).

To the visitors, the Geopark offers a unique experience, a journey through time from Precambrian to the present day. The stories about Earth and Mankind are supported by geosites with dinosaur fossils and outcrops, natural reserves, geoproducts, interpretation points, and many historical monuments.

All this heritage was created and shaped in thousands of years by the communities living in Hațeg area, who, in return, developed in a special unique way. Inhabiting a mountain depression with clear boundaries gave the people here a special sense of place, which helped them keep their identity for hundreds of years. In recent history, however, the local community of Hațeg region passed through difficult times. Half a century of communism changed the landscape, the local economy, and the social fabric and values. After the fall of communism, the people were affected by the many transformations inherent to a shift toward democracy and free market. The result is a community with many challenges, where local traditions and the sense of place are overwhelmed by the "dream" of working abroad or at least in a big city. The present social model begins with moving away from the Hațeg Country.

This is the social landscape in which the Geopark operates. It has to find solutions to reconnect the communities with the land, both at a spiritual level and at a very practical economic reality (Martini, 2009; UNESCO, 2016). It also has to create economic value in the form of geotourism. One of the answers is working with youth and involving them as responsible Geopark partners in community development and geotourism promotion.

15.2 THE VOLUNTEER PROGRAM NEED AND FUNCTIONALITY

In 2014, only one year after its creation, "Volunteers for the Geopark" program received the highest distinction at the Romanian National Volunteer Gala, followed by a series of other national and regional prizes. Now it is recognized as a model and best practice example (Istrate and Baciu, 2015). The reason for this quick success is the fact that it satisfied a great need in the local community as well as in the Romanian society as a whole—an active youth model.

The philosophy behind the Geopark's decision of creating the volunteer program is captured in Figure 15.1.

Every geopark has a team tasked with many activities in a very wide range of domains. The geopark team has to be creative enough as to successfully promote the geopark outside and raise awareness inside the territory, an activity which is very difficult, and energy and time consuming to sustain. But not all the geoparks are aware that they are standing on a reservoir of energy and creativity—the local youth. If this resource is identified and properly engaged and empowered, then it can release great amounts of imagination, innovation, and passion for the geopark and for the community.

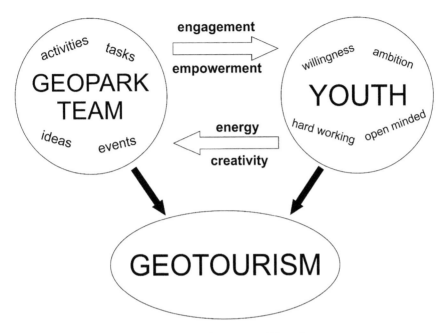

FIGURE 15.1 Volunteers for the Geopark program philosophy.

Usually, the geopark team is engaging the community in a direct manner. The innovative approach proposed by our volunteer program is engaging the community also through the volunteers, who are part of this community. In this way, geotourism development and promotion benefits from three key elements:

1. The creativity of a young team adds to any initiative a component of innovation and sets it in line with the new ways of communication;
2. The community is more favorable to geotourism, having its youth directly involved in decision making;
3. The volunteers are trained in the entrepreneurship approach of the geopark and could be easier involved in further geotourism activities and geotourism promotion.

From the point of view of the youth, the volunteer program also fulfills a need. The youth of Hațeg Region doesn't have a valid manner of actively using spare time, with the exception of the Red Cross and some religious communities. In an internal survey conducted in March 2017, 80% of the volunteers answered that if the geopark didn't exist, they wouldn't volunteer for something else. Contributing to this situation is the atmosphere of the volunteer group, perceived as different from an ordinary system, like school. The program is centered on the individual, his or her skills, competences, personality, and direction of development.

A specific characteristic of the volunteer program is the "captive" target group, as most of the volunteers are (or were) attending the only high school in the region—"I. C. Brătianu."

This particular situation makes recruiting easier and shapes the strategy of group targeting. As the program developed, other factors appeared which I didn't foresee at the beginning. First, there is the lack of an associative form of representation for young people in the community, then the discrepancies between the workforce market demands and youth training, the deep fault between school and real life, and the need of young people to belong to a group that appreciates and defends them. All these new factors raised the importance and bettered the results of the program.

The volunteer program follows the Romanian volunteering law (78/2014), using all the respective documents. Volunteering is a new trend in Romania, so the legislation is not fully covering the issue. This is the reason why the assistance of the major national organization in this field—VOLUM Federation and ProVobis is essential. The financial aspects of the activities are covered by participating in national grant competitions. Most of the funding comes from the Foundation for Partnership and MOL Romania through the Green Spaces Program.

The functionality of the program can be synthetized by discussing innovation and engagement.

In Figure 15.1, we showed the innovative approach on geotourism using the young volunteers for creativity and community acceptance. Another innovative element is the role of volunteers inside the organization. Each volunteer has to find or create his or her own place in the organization. There are no pre-assigned positions, but young people shape their activities according to their own personality, skills, and objectives. Engaging the volunteers is a three-step process (Fig. 15.2). Recruitment is made mostly through recommendations from one volunteer to the others, and also through regular public meetings. After the recruitment, the volunteer is added to the data base and receives all the information about undergoing projects and activities. He or she then is able to choose to participate in one or more activities.

The key component of the engagement strategy is that the volunteers are not only informed and consulted regarding the Geopark's activities, but they are also involved in decision making.

The Role of Volunteer Management Programs

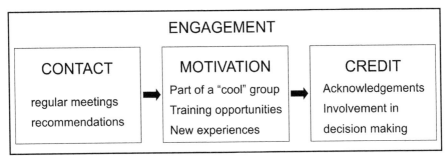

FIGURE 15.2 Structure of volunteer engagement.

15.3 GEOPARK AMBASSADORS—A FURTHER STEP IN YOUTH INVOLVEMENT

In 2016, a project managed by the Geo-media Association and financed by the Foundation for Partnership and MOL Romania through the Green Spaces Program, created the next step for the Volunteers for the Geopark initiative. It developed the concept and trained the first "Geopark's Ambassadors."

Ambassadors are a group of volunteers trained to represent their community and Hațeg Country Geopark everywhere and to everyone on the national and international level. The Ambassadors have the knowledge about the land and the geopark concept and also the abilities to communicate this information and to create bridges of cooperation.

The first generation of Ambassadors, trained in 2016, numbered 17 volunteers. From these, only six were revalidated for 2017 and 13 more were trained in the second generation. In 2018, a third generation joined the group after the usual training and selection process. Ambassadors are defined as (Geomedia, 2016):

- Enthusiastic young persons, intelligent and charismatic, capable of representing their community in projects and national and international exchanges;
- They are able to contribute to raising the understanding, appreciation, conservation, and promotion of the natural and cultural values of Hațeg Region; and
- They are partners in geopark projects and events and could be considered as part of the management geopark team.

Involving representatives of local communities in promoting the geoparks' values is a concept used by different geoparks. Fforest Fawr UNESCO Global Geopark is organizing workshops and field trips annually for its Ambassadors as a contribution to sustainable tourism and its engagement with local businesses (EGN Newsletter, 2017). Other geoparks are recruiting and training local people as ambassadors for international conferences or events.

Hațeg Country Geoparks Ambassadors is a new and innovative approach aiming to train and select already experienced geopark volunteers to become partners of the Geopark's administration and official representatives of their territory for national and international missions. Every year, from May to September after training sessions on subjects ranging from leadership, public speaking, heritage interpretation, the history of the EGN and GGN, and leading field excursions, some of the young volunteers pass the evaluation and validation process and become Geopark Ambassadors. The 10 total steps of youth involvement in Hațeg Country Geopark are described in Figure 15.3. The last two steps are at the ambassador stage.

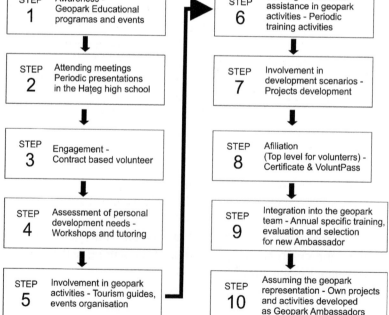

FIGURE 15.3 The steps of youth involvement in Hațeg Country Geopark.

The ambassadors are in charge with local, national, and international communication of the geopark, and presentations in local schools, fairs, and study visits. For example, the ambassadors made their first study visit in Bakony Balaton Global Geopark in 2016.

In 2018, they helped setting up the exhibition Stone Made Objects during the 8th International Conference on UNESCO Global Geoparks in Adamello Brenta. Since 2016, they are coordinating the international project Andi Andezit's Journey and from 2018, the #GGNyouth Instagram account project. Thus, the Ambassadors are helping the geopark team with key activities. Based on their experience, a similar project is going to be developed in Tumbler Ridge Geopark, Canada, Hațeg's sister geopark. To become a geopark ambassador is a logical step forward in the evolution of a geopark volunteer. Their role is to represent officially the geopark territory and its communities and to be accepted as having this position by the geopark team.

15.4 RESULTS FOR GEOTOURISM DEVELOPMENT AND PROMOTION OF THE TERRITORY

Over the years, the volunteers for the Geopark became the driving force of the Geopark. Most of the activities are made with the help of the volunteers and, more than that, some of the most successful projects were created by them.

One example out of many initiatives is the Bridge over Time, built in 2015 to facilitate the access over a river toward one of the Geopark's interpretation points—the House of Volcanoes (built also by the volunteers in 2013). A simple infrastructure element was transformed by the volunteers' creativity into geoheritage interpretation.

On the other side of the river, an important geosite is located, showing both Cretaceous traces of volcanic activity and marine sedimentary rocks. The bridge was imagined as connecting the present to the Upper Cretaceous, guiding the visitor through key moments in the Earth's history. The scientific information was provided by the specialists from the Geology and Geophysics Department, University of Bucharest, and then transformed and translated by the volunteers, to address a certain target group (Fig. 15.4). This way, the volunteers' actions contributed to:

- Geoheritage conservation and promotion;
- Geointerpretation and geoeducation;
- Geotourism infrastructure and promotion.

Puntea peste Timp
The Bridge Over Time

Atenție, tocmai ești pe punctul de a face o călătorie în timp! Pe malul celălalt vei descoperi o lume fascinantă cu vulcani, mări preistorice și viețuitoare ce populau Pământul în urmă cu 72 de milioane de ani.

Be careful, you're about to travel back in time! On the other bank of the river you'll discover a fascinating world with volcanoes, prehistoric seas and creatures that inhabited the Earth 72 million years ago.

FIGURE 15.4 Example of interpretation panel on the Bridge over Time.
Source: Hațeg Country UNESCO Global Geopark interpretation panel, 2015.

Generally, the volunteering results are threefold:

1. The volunteer work is used to develop geotourism infrastructure, organize events, and developing educational activities, which all contribute to local welfare by creating new opportunities, especially for geotourism (Andrășanu, 2017);
2. Investing in local youth, in their training and experience, we invest in the community's future generation. Training the youth in areas connected to the market increases the chances for them to return to the region and work in tourism, local development, environmental protection, education, and other connected domains;
3. Raising awareness for problems like geoheritage conservation, sustainable tourism development, risk management, climate change, and others on the Geopark's agenda received a boost along with the volunteer program. The young volunteers are an important vector for the key messages to get across to the community. We can say that the Geopark has a team member in about 300 households all over the region.

The results for the community can also be expressed quantitatively: If we take into account only the 16 volunteers who received the Volunt Pass certificate, their work hours, relative to the Romanian minimum salary, value 12,000 euros. This is equivalent of the work done by these 16 young persons for the community.

All the above contributions were not only felt in the Geopark team, but also meant a boost for other local organizations to use volunteering as a form of training and reaching out to the community. At least three NGOs from Hațeg Region started working with young volunteers for the first time in 2017, in activities not directly related to the Geopark. Even if these organizations don't have a proper volunteer program and don't follow all the law requirements so far, they still are the sign of a positive trend set by the Geopark.

15.5 CONCLUSIONS

Hațeg Country Global Geopark is the result of a grass-root effort that began in 2000 and was initiated and coordinated by the University of Bucharest. Previous experiences and research analysis indicated the need for a strong partnership with local schools and direct involvement of local young people. The geopark management team needed to find the best solutions to successfully promote the geopark outside and raise awareness inside the territory. This important objective requires energy, creativity, and time, thus being difficult to sustain on long term.

The Volunteers for Geopark program, established five years ago, was a new approach aiming to involve young people as Geopark partners to fulfill their needs for the social recognition of their skills and creativity and to support their personal and professional development. The innovative approach proposed by our volunteer program is engaging the community also through the volunteers as partners in developing geotourism activities, to become ambassadors of their territory, and, some of them, tourism guides or entrepreneurs.

More than 300 local students were enrolled as volunteers in different Geopark projects. According to Romanian legislation they signed a contract, succeeded in training stages and were involved in several projects (see Fig. 15.5). Some of these projects were initiated by the volunteers themselves. Based on an evaluation process, volunteers are receiving, since 2015, an international VoluntPass issued by the University of Bucharest. In 2016,

FIGURE 15.5 The Volunteers for the Geopark at the House of Science and Art.
Source: Hateg Country UNESCO Global Geopark archive, 2016.

a new program was initiated aiming to train the best volunteers to be able to represent their communities in national and international projects and to contribute to increasing the level of understanding, appreciation, conservation, and promotion of Hațeg Country Global Geopark.

Volunteers and Ambassador are key partners in organizing and promoting geotourism by contributing to geopark projects and annual events: *Geopark Challenge*, in April, a half marathon competition; *European Geoparks Week*, May–June; *Dinosaurs Festival*, in July; *Vulcano Day* in August. They have been involved in the construction of interpretation points and installations: House of Volcanoes, House of Stones, House of Dwarf Dinosaurs, The Fragrant Time Garden; they are guiding tourists and help in educational activities.

The volunteer program is one of the most successful activity of the Geopark, being known as the best volunteering program for protected areas in Romania and having gained numerous national prizes. It is also a very efficient way to transmit the Geopark's values directly to the local community through its youth and to organize and promote geotourism.

ACKNOWLEDGMENTS

The authors are grateful to the numerous (more than 300) young volunteers who participated in the program since 2013, thus making Hațeg Country Geopark a more successful project and creating a dynamic and creative working environment for the team.

Our gratitude goes also to John Macadam for his priceless contribution in shaping this text.

KEYWORDS

- **volunteering**
- **geotourism**
- **tourism**
- **geodiversity**
- **education**

REFERENCES

AER Asociația de Ecoturism din România. *Tourism Monitoring Results* 2016-2017. Database collected as part of the project: ȚaraHațegului–Retezat: destinațiaecoturisticădeexcelențăa României. Un model de dezvoltaredurabilăprinecoturism, financed through a Swiss grant as part of the Swiss Contribution for the extended European Union. 2017.

Andrășanu, A. Geoparcul UNESCO—model de dezvoltarecomunitarășiconstrucție de brand. In Societate, Publicitate, Consumator; Rujoiu, O., Editura; ASE, 2017; pp 185–212.

AsociațiaGeomedia. *Ambasadorii Geoparcului*, Promotional Leaflet, 2016.

Ciobanu, C. *Space and Time Perception and the Geopark's Communities. From Mythical Geography to Heritage Interpretation. Internat. Rev. Social Res.* **2016,** *6* (2). Available at https://www.degruyter.com/view/j/irsr.2016.6.issue-2/irsr-2016-0013/irsr-2016-0013.xml?format=INT 2016.

EGN Newsletter, no 8, **2017**. Available at http://www.europeangeoparks.org/wp-content/uploads/2014/09/newsletter-8F.pdf (accessed Dec 15, 2018).

Istrate, L.; Baciu, V. *VoluntariatulpentruNatura. Studiu*, 2015. Available at http://voluntariatnatura.ro/downloads/Voluntariatul%20pentru%20natura.pdf.

Law no. 78/2014. Available at http://federatiavolum.ro/wp-content/uploads/2014/11/LV_promulgata.pdf.

Martini, G. Geoparks ... A Vision for the Future. *RevistadoInstit. Geociênc. USP* **2009**, *5,* 85–90.

UNESCO. *UNESCO Global Geoparks*, Paris, 2016.

CHAPTER 16

Geotourism and Proposed Geopark Projects in Turkey

GÜLPINAR AKBULUT ÖZPAY

Education of Social Science and Literature Department, University of Sivas Cumhuriyet, 58140, Sivas, Turkey

E-mail: gakbulut58@gmail.com

ABSTRACT

Geoparks and geotourism concepts are emerging trends in the tourism industry and protection of geological sites. The aim of this chapter is to explore proposed geopark projects and the progress of geotourism in Turkey. At the same time, this chapter describes some general information about the geographical and geological characteristics of Turkey.

16.1 INTRODUCTION

"Geoparks were defined as a geographical area where geological and geomorphological heritage sites are part of a holistic concept involving the conservation of all natural and cultural heritage" (Akbulut, 2009: 264). As of 2020, 75 European Geoparks and 147 Global Geoparks have been declared. The geopark process takes a long time (Fig. 16.1), starting with the determination of the boundaries of the geopark area. Secondly, characteristics of geosites in geoparks should be evaluated according to UNESCO criteria for rare, esthetic, value, and landscape properties. Thirdly, "geoparks require a management plan provided for the social and economic needs of the local population protect the landscape in which they live and conserve their cultural identity" (UNESCO, 2016: 5). Fourthly, "this management body should include all relevant local and regional actors and authorities" (ibid: 5). Fifthly, geoparks have played an important role in sustainable rural development through local investments (Farsani et al., 2014: 185). Lastly, information or education seminars related to the protection process must be provided to local people, students and visitors.

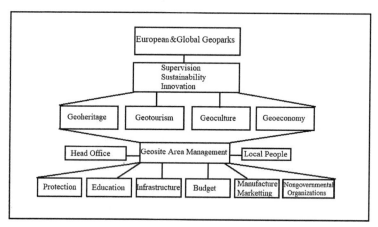

FIGURE 16.1 The process of geoparks.
Source: Gülpınar Akbulut, 2014.

There are three aims in geopark establishment: protection, education, and geotourism (UNESCO, 2006: 1). Geotourism is a niche form of tourism focusing on the visitation of geological and geomorphological sites in rural areas and aiming at educating visitors about the Earth's heritage (Akbulut, 2016: 87). Newsome and Dowling noted that:

> Geotourism is a form of natural area tourism that specifically focuses on geology and landscape. It promotes tourism to geosites and the conservations of geo-diversity and an understanding of earth sciences through appreciation and learning. This is achieved through independent visits to geological features, use of geo-trails and viewpoints, guided tours, geo-activities and patronage of geosite visitor centers (Newsome and Dowling, 2010: 4).

Turkey has various natural and cultural landscapes, but it is a typical Mediterranean country, which is developing the 3S tourism (sea, sand, and sun). One of the most important tourist attracting countries of the world, Turkey is visited by over 38.6 million tourists, generating an income of more than USD 26 billion involved in beach-based tourism, which has significantly improved in 2017 (Ministry of Culture and Tourism, 2017). Furthermore, there are outstanding resources such as volcanic mountains, karstic, fluvial, and glacial features of Turkey. This chapter includes general information about the geographical and geological characteristics of Turkey and evaluation of its geopark potential for geotourism. Consequently, the chapter has discussed proposed geopark projects of Turkey.

16.2   INTRODUCING TURKEY

Turkey is a country located in Asia and Europe and close to Africa. "It has an area of 814,578 km^2 and a total coastline of 8333 km along the Black Sea in the north, the Mediterranean Sea in the south, the Sea of Marmara in the northwest and the Aegean in the west" (Akbulut, 2016: 88). Turkey shares borders with Azerbaijan, Georgia, Iran, Iraq, Syria, Greece, and Bulgaria and has a population of over 80 million people (Ministry of Culture and Tourism, 2017).

Turkey has a very rich geodiversity due to its geological and geographical evolution. Bozkurt and Mittwede (2001) highlighted that "just as Anatolia has had a rich cultural history as home to numerous and diverse civilizations, the geology of Turkey is a colorful, fascinating mosaic" (Bozkurt and Mittwede, 2001: 578). The complex geology of Turkey, which is part of the Alpine-Himalayan orogenic belt, is a product of the collision of two mega-continents, Gondwanaland in the south and Eurasia in the north (Angus et al., 2006: 1336). Bozkurt and Mittwede noted that:

> The geological framework of Turkey comprises many lithospheric fragments that were derived from the mega-continental margins and then were amalgamated during the Alpine orogeny when the Arabian plate collided with the Anatolian plate in the Late Cretaceous-Tertiary. As the rifts widened the Tethyan oceans developed; when, subsequently, the mega-continents collided, these oceans closed sequentially (Bozkurt and Mittwede, 2001: 578).

Due to the closure of an ocean located between these plates around 65 million years ago, the main geological and geomorphological features of Turkey occurred. "Two mountain ranges lie on the northern and the southern parts of Anatolia reflecting the tectonic and stratigraphic features of the country: The Northern Anatolian Mountains and The Taurus Mountains" (Atalay, 2002). The northern Anatolian Mountains constitute the major part of western Himalayas. The Taurus Mountains are a wide and long mountain range that lies paralleled to the Mediterranean coast (Atalay et al., 2008). Besides these, fault-block mountains with uplifted blocks along the normal faults occurred in the western part of Anatolia (Atalay, 2002). Anatolia demonstrates evidence of extensive volcanic activity, particularly during the Neogene to Quaternary periods; it has the dormant strato-volcanoes of Mt. Ağrı, Mt. Nemrut, Mt. Süphan, Mt. Erciyes, and Mt. Hasandağı, calderas, volcanic cinder and tuff cones and layers of pyroclastic deposits (Hose and Vasiljevic, 2016: 184, cited by Kazancı et al., 2005). "Many of Anatolia's geological structures are in carbonaceous and evaporate rocks; hence, Karst landforms and underground features such as Karrens, Dolines, Poljes, and caves are very common. Turkey as a whole has over 2500 caves and more than 300 lakes" (Hose and Vasiljevic, 2016: 184). Anatolian geomorphology is characterized by high relief contrast and its average elevation of 1132 m. "In this geomorphology, plateaus and mountains are deeply cut by short rivers with powerful currents, deeply cut valleys and eroded surfaces" (Gümüş, 2014: 38).

16.3 THE HISTORY OF GEOTOURISM AND GEOHERITAGE IN TURKEY

Though geotourism and geopark are new concepts in Turkey, there are many scientific or popular studies on nature protection, ecology, wildlife, geosite, and nature-tourism related to these concepts. The most important reason is that there have been many studies on the geology and geography of Turkey from the 19th century until today. These studies have been carried out by General Directorate of Mineral Exploration and Research (MTA), earth science societies, institutes, and universities.

MTA was founded in 1935 (Doğanay, 2011). This institute published thousands of scientific reports and articles and contributed to the determination of formations, fossils, landforms and tectonics of Anatolia. Kazancı noted that the number of the geological research projects, meetings, panels, and workshops about natural monument increased significantly in the late 1960s and 1970s. In the late 1980s, the government had to pay more attention to the protection of the natural environment and to raise awareness of the society. In 1997, a national action plan was prepared for geoheritage protection (Kazancı et al., 2012: 367–368), and in 2003, Turkish Geological Heritage Research Project (TUJEMAP) was started by MTA. The main aims of this project were to prepare an inventory of geosites, draw up site maps and to organize research in the field of geotourism. Moreover, the project was supposed to determine methods for the conservation and use of the geosite (Gürler and Timur, 2007). According to these aims, 29 areas have been identified so far. Nine of them were proposed as high priority potential geopark areas: Karapınar, Mut Miocene Basin, Kula Volcanic, Cappadocia, Dilek Peninsula, Biga Peninsula, Tuz Lake, Artabel Basin, and Sivas geoparks (Gümüş, 2014: 77).

Other important natural conservation associations are the Turkish Geography Society (1941) and the Turkish Geological Society (1945). Both of them have published many studies with different aims such as the geology and geomorphology of a place, its tectonics, natural tourism assets, and the protection and conservation of geographical features (or landscapes). Moreover, Sea Sciences and Geography Research Institute was founded by Geography Department of Istanbul University for the purpose of developing earth sciences. In connection with the same department, the *Journal of Turkish Geography* started publishing articles about geosites, nature protection, ecology, geomorphology, and botanics in 1943.

The first Turkish populist geo-journal, named *Yeryuvarı ve İnsan* (The Earth and Human), was published by the Turkish Geological Society to draw attention to natural monuments (Arpat, 1976, cited by Kazancı et al., 2012). Furthermore, they offered that some outstanding places such as Cappadocia, Ararat Mountain, and Pamukkale's Travertine Terraces should be evaluated as georoutes and tourism areas. Many articles about geoparks, geosites and geotourism were published by Geological Society.

Other important protected areas are National Parks owned by the state. The first national park in Turkey named Yozgat Çamlığı was established in 1958. Today, there are 44 national parks in Turkey. The majority of them having a rich flora and fauna. They have been declared as national parks to protect their biodiversity, but include many geosites as well. Geosites in national parks are protected; particularly, in registered archaeological sites, which are generally fortunate enough to be protected; for example, the Pamukkale travertines near Hierapoli (Denizli), and the tuff cones and maars in Cappadocia (Nevşehir) (ibid: 185). Thousands of people visit these places every year and are interested only in viewing natural and cultural-historical sites (Akbulut, 2016). In Turkey, some social associations have been established with the purpose to protect the country's geological heritage. One of them is The Turkish Association for Conservation of the Geological Heritage (JEMIRKO). In 2000, JEMIRKO was officially established by a small group of scientists and volunteers. This organization has played a significant role in the development of geopark and geotourism-related concepts (Kazancı, 2010). It has collected a list of hundreds of geosites in Turkey, nine of which have been proposed as high priority potential geoparks, which are Karapınar, Mut Composite, Kula Volcanic, Cappadocia,

Narman Red Happiness Valley, Pamukkale, Nemrut Volcano, Van Lake and Çamlıdere geoparks.

The first scientific studies on geotourism and geoparks in Turkey were published in the 2000s by Koçman and Koçman (2004), Yıldırım and Koçan (2008), Gümüş (2008), Akbulut (2009), and Kazancı (2010). These studies have increased awareness among scientists about Turkey's geoheritage and the opportunities for the development of geotourism in Turkey (Akbulut, 2016: 88). Upper Kızılırmak Gypsum Karst Geopark, Levent Valley Geopark, and Van Lake and Nemrut Caldera Geopark have been started officially, based on the criteria of UNESCO.

Some of the universities in Turkey have been paying attention to geoparks in recent years. For the first time in Turkey, İnönü University Geopark Society was established by the students of İnönü University in 2012. This society organized meetings, conferences, logo studies, field excursions, national television and radio programs, geopark education, and conservation in secondary schools. A Geopark Bulletin was published by this society in 2012. A second geopark society was established by Cumhuriyet University in Sivas in 2017. This society aims to provide information to people and students about geoparks.

The Geopark Research and Application Center (JARUM), which is the first geopark research and application center in Turkey, was founded by the Manisa Celal Bayar University in 2013. The aims of this center are close the cooperation with the Kula Geopark, to constitute academic coordination, to serve as a mediator between the EGN and GGN, and to promote Turkish geopark attempts (Gümüş, 2014: 76–77).

Also, a geopark meeting, which is another important progress, was organized by UNESCO in October 2015. At this meeting, people who are studying geoparks, suggested 26 new geopark areas in Turkey. However, no decisions were taken at the meeting. Today, there are several geopark projects such as Kızılcahamam-Çamlıdere Geopark, Levent Valley Geopark, Narman Red Happiness Valley Geopark, Nemrut Geopark, and Upper Kızılırmak Geopark. Some of these geoparks are organized by municipalities, while the others are organized by governorship.

Although there are developments in the fields of geopark and geotourism in Turkey, there are also some problems. The studies in Turkey have shown that scientists in Turkey have been aware of the geological heritage resources for a long time. The most common terms employed in their studies are natural matter, natural object, natural monument, and national park. On the other hand, geopark and geotourism terms are not specifically mentioned in Turkish legislation (Hose and Vasiljevic, 2016: 185). Thanks to the unique geological and geomorphological characteristics of Turkey, there is an abundance of geosites. Hence, it is challenging to determine the inventory of the geosites and existing geopark potential. The government, some universities, and civil organizations are working to create an inventory of geosites in Turkey. For example, the Sivas Geosite Inventory Project has been studied by Akbulut since 2013, but is not completed yet, although 300 geosites have been determined until today. Another issue is lack of scientific information and basic data related to geosites that can be easily understood by the public (Tongkul, 2006: 27). In this sense, special educational programs should be developed to educate children about the geology and geography of Turkey. Published materials, visitor centers, and national history museums could also increase public awareness (Akbulut, 2016).

16.4 TURKEY'S GEOPARK POTENTIAL AND ITS GEOTOURISM

16.4.1 KULA VOLCANIC GEOPARK

The Kula Volcanic Geopark in the Manisa province of West Anatolia is Turkey's first Global Geopark recognized by UNESCO in 2013. This geopark has played a vital role as a model for the organization of other geoparks and it increases awareness about geotourism activities in Turkey. Indeed, after opening, many tourists have started to visit the geopark to see its landscapes and to understand the geological history of Turkey.

"Katakekaumene" or Kula volcanic geopark is an area between the Kula and Karatas settlements and the Demirköprü dam-lake, which covers an area of 300 km². The Kula Geopark Project has been identified and classified under three main categories: volcanic geosites (Sandal, Kula, and Elekçitepe Divlit cinder cones, Saraçlar horseshoe and Kula Divlit parasite cone, Kula, Sandal and Kaplan Divlit lava flow, Çukuradamaar, Saraçlar base surge, Dereköy volcanic canyon, Suuçtu waterfall, and lava tubes, and Çakırca columnar basalts), erosional geosites (Karapınar and Börtlüce cave, Harmanyeri sinkhole, and Kula fairy chimneys), and mixed geosites (footprints) (Gümüş, 2012; Gümüş and Zouros, 2012: 2; Gümüş, 2014).

There are many non-geological sites in Kula such as Kula Town Urban Architecture, Gölde Village Rural Architecture and Greek Heritage, Menye Village Historical Site, Kız Bridge and the Palankaya Roman Settlement, rock tombs, Sacred Tomb of Tabduk Emre and Yunus Emre, and Carullah bin Süleyman Mosque and churches (Gümüş, 2014).

Furthermore, this area has also a cultural heritage, which includes Kula houses, traditional handicrafts, a historical bazaar, and special foods. This geopark is one of the youngest volcanic regions of Turkey, which continued from the middle Miocene to the early Holocene periods (Koçman, 2004). It has been formed by extensive tectonics and represents a landscape with volcanic elements, such as volcanic cones, maars, lava flows, volcanic canyons, lava caves, columnar basalts, badlands, hot springs, fairy chimneys, and tuff layers with is the collection of footprints made by prehistoric humans, (Gümüş, 2014) (see Fig. 16.2). The most interesting feature in the Kula volcanic area is the collection of footprints made by prehistoric humans, which were discovered during the construction of the Demirköprü Dam (1954–1960). In the first stage of research, 200 footprints were found and 60 of these prints were removed and taken to the depository of the Natural History Museum in Ankara (Akdeniz, 2011). The most important cultural heritage consists of the sacred tombs of Tabduk Emre and Yunus Emre, who are Sufi; and the Carullah bin Süleyman Mosque. This mosque is quite different from most other mosques in Turkey, because it is characterized by dense wall paintings including many figures of daily life that are valid even today, such as apartments, boats, trees, and flags. Kula houses and foundations are popular for geotourism. These Ottoman-style houses, despite being a protected site, unfortunately were destroyed by fires and other human activities, because wood was used as the main building material of the houses. The background of the Kula Geopark Project has 10 years of history. The manager of this geopark on behalf of a municipality is Erdal Gümüş, who prepared the management and action plan of the Kula Volcanic Geopark, and determined the inventory of geosites in the Kula. The Kula Geopark Visitor Center was established to serve as tourism information office, a laboratory for geo-education activities, and to function as a basic museum (Gümüş and Zouros, 2013: 82–84). Geopark walking trails and bridges were built in the geopark area, and geopark direction panels were prepared (see Fig. 16.3). This geopark also has a Geopark Kula Municipality football and volleyball team.

The Kula Geopark has significant geotourism potential. According to the latest census announced by the *Turkish Statistical Institute (TurkStat)* in 2017, the population of the geopark area is 44,403. Its closeness to large cities such as İzmir (4279,677 people) and Manisa (1413,041 people) is a great advantage, with more than one million tourists visiting the Kula Volcanic area every year, the majority of whom are local people.

Geotourism and Proposed Geopark Projects in Turkey

FIGURE 16.2 A view of the Sandal volcanic cone locally called devlit. There are over 80 volcanic cones.

FIGURE 16.3 A view of the bridge, "Kula Geopark Area."
(Photo: Gülpınar Akbulut).

16.4.2 KIZILCAHAMAM-ÇAMLIDERE GEOPARK PROJECT

The Kızılcahamam-Çamlıdere proposed geopark area is located north of Ankara. The first studies on this proposed geopark started with a common project in 2006. Ankara University, JEMIRKO, MTA and National Park General Manager studied "The Geological Heritage in National Parks" in 2006–2009. While this project was underway, the "Kızılcahamam-Çamlıdere Geopark and Geotourism Project" was developed by/under the scientific leadership of Ankara University and JEMIRKO (Gürsay, 2014). The first formal geopark of Turkey in a small part of the Galatian Volcanic Province, "Kızılcahamam-Çamlıdere Geopark" was registered by the national authorities in 2011 (Kazancı, 2012). At the same time, this proposed geopark was the subject of the first master thesis on geoparks of Turkey (Gümüş, 2008). The proposed geopark covers a surface area of 2000 km^2 with at least 23 geosites, some of which are endangered (Kazancı, 2012). It is located within the Galatian Volcanic Province, which is known as the Köroğlu Volcanic Belt (Gümüş, 2014). According to the census (2017), the population of the geopark area is 32,336. This area is very close to big cities such as Ankara (5,445,026 people) and Istanbul (15,029,231 people). It has been identified and classified under three main categories: natural geosites, historical geosites, and archeological geosites. Some of the natural geosites include the Çamlıdere fossil forest, Uzunkavak petrified trees, Güvem-Beşkonak fossil site, Sarıkavak tephra, Güvem columnar lava, Işıkdağı stratovolcano, Abacı fairy chimneys, Kızık fault, Kızılcahamam hot spring, Acısu mineral water springs and travertines, and the Taşlıca village andesite boulders.

The core of natural geosites consists of the Fossil Forest in Pelitçik Village (Çamlıdere), which is very different from other fossiliferous localities in Turkey, thanks to silicified tree stems, branches, and roots covering the surface (Akkemik et al., 2009: 90). The original forest was composed predominantly of Taxodium with few Sequoias, with subsidiary junipers, oaks and pine cupressus. The trees have been preserved as a result of volcanic activity that took place during the lower Miocene (ibid: 96). This fossil forest-bearing zone is exposed at Çerkeş, Kurşunlu, Ilgaz, Bolu, Çankırı, Şabanözü, Beypazarı, Kazan, and Çubuk (Gümüş, 2008). The silicified fossil layer is 15–20 m thick but not every layer is filled with fossils (Kazancı, 2012). Other important fossil forests are also located in Kızılcahamam within the Soğuksu National Park and Güvem-Beşkonak. The Uzunkavak petrified fossil-bearing zone is 1–4 m thick and lies 180 m in E–W direction. Güvem-Beşkonak, which is extremely rich in Pliocene age fishes, insects, and leaf fossils, is only narrow (Gümüş, 2014). Kazancı states that the "Güvem-Beşkonak fossil site is one of the most disturbed geosites in Turkey, particularly by fossil hunters, despite having been a nature conservation site since 1985" (Kazancı, 2012: 258). The Güvem columnar lava is located in a place 1.5 km north of Güvem Village. Columns are very apparent along the two sides of the Sabunkaya creek because of natural erosion and anthropogenic activities (ibid: 257). Hot springs within the proposed geopark are important sources for tourism applications. There is also a national park within the geopark. Soğuksu National Park, which was proclaimed in 1959. It is known as the most famous Aegypius monachus and endemic tulip and fossil forest (Koçan, 2011).

The cultural history of the Kızılcahamam-Çamlıdere Geopark is very interesting. The region has been of great interest since ancient times, when it was the land of Hittites, Phrygians, and Galatians (Kazancı, 2012). Some of the historical and archeological geosites are Karagöl Cave (man-made), Seyhamamı Turkish bath, Roman chapels, Alicin monastery, and rock-carved churches of Mahkemeağcin village (Gümüş, 2014).

The Kızılcahamam-Çamlıdere Geopark has a very important geotourism potential with these characteristics. The education studies and the classification of geosites have been made by JEMIRKO. It organized activities such as some meetings, workshops, and summer education programs. Many booklets have been published and distributed to people and children from different age groups. Most of the geosites have been grouped in four georoutes and three geotours using the high quality of roads in the area in order to encourage both visitors and also local people to engage in geotourism (Kazancı, 2012). The Kızılcahamam-Çamlıdere Geopark has some advantages over the other proposed geoparks. Firstly, it has two protected areas; Soğuksu National Park in Kızılcahamam (nominated in 1959), and Pelitçik Fossil Forest, nominated as a natural site in 2007 under the auspices of the Ministry of Forestry and Water Affairs (Gümüş, 2008). Secondly, the proposed geopark has a formal management plan. Despite all advantages, this geopark has many threats, particularly to geological monuments. There are increased threats from collectors, fossil hunters, and local people. Local people have sold fossils to gemstone shops. In the past, the petrified forest was used as a material for garden walls or houses (Gümüş, 2014). Even though Kızılcahamam-Çamlıdere and Çeltikçi municipalities have a formal management plan, and its popularity has been increasing recently, there are some problems with the application of the management plan such as budget, infrastructure, manufacture, and marketing. Numerous people visit this area, but it is not enough for the development of this geopark. The Kızılcahamam-Çamlıdere Geopark has no international status, so it should apply to the UNESCO Geopark Network and EGN.

16.4.3 NARMAN, OLTU, AND TORTUM (NOT) GEOPARK PROJECT

Narman, Oltu, and Tortum Proposed Geopark Projects (NOT Geopark Project) are located in the Erzurum Province in Eastern Anatolia. The total population of the geopark in 2017 was 58,791. It has been identified and classified under three main categories: natural, historical, and archeological geosites.

The geopark is already well-known for its geological monuments such as Narman pillars, Tortum Waterfall, Tortum Gorge, Zökün Lake and floating islands, Dayks, Colorful Organogenic Clusters and Pillow Lava, and Oltu Gemstones. Part of the important geoheritage is Narman Canyon, which covers an area of 63 km^2 (Garipoğlu Farımaz, 1996). "It is covered with Pliyo-Quaternary sediments... Narman Canyon is divided with the Oltu stream into two parts. It is famous for its red fairy chimneys of various sizes" (Güngör et al., 2012: 143). Seventeen geosites, 56 unique inventories, and 15 georoutes were determined, and the proposed geopark was mapped (Azaz and Güngör, 2017). The Oltu basin is one of the predominant tectonic features with its overturned strata, faults, and folds. "In this area, 13 geosites and 7 georoutes were identified" (ibid: 209).

Karahan et al. (2011) state that Tortum Valley and its surroundings have the potential for a geopark with landscape and ecosystem characteristics. "Tortum river has formed a narrow valley with a canyon cut about 1000 m along its watercourse. The lithology of the valley and its environs are composed of Mesozoic and Tertiary rocks" (Karahan et al., 2011: 397–398). It formed a valley by crossing anticlines and synclines. The valley and its surroundings contain important geological inventories, such as the Tortum landslide barrier lake and waterfall, seven lakes, floating islands (Zökün lake), and caves (ibid: 395). In this area, "there are six geotrails in 17 geosites" (Azaz and Güngör, 2017: 209). The Tortum Landslide Barrier Lake is the biggest barrier lake in Turkey. "The drainage area of the lake is 1835 km^2 and the surface area of the lake is 6.5 km^2" (Koparand Sevindi, 2013: 55). The landslide occurred as a rapid-rock slide in the Cretaceous and interbedded limestone with clastics. The volumes of the displaced mass and the landslide dam were estimated as 180 million m³ to 223 million m³ which came from Kemerli Mountain (2770 m) (Duman, 2009) (see Fig. 16.4).

"There are some lakes of different sizes on the landslide material near Ulubağ village. The area is called Seven Lakes" (Karahan et al., 2011: 401). At the same time, Zökün Lake and the floating islands are very interesting landscapes (Bulut et al., 2009). Furthermore, there is Tortum waterfall, which is also one of the tallest at a height of over 48.5 m and a width of 22 m (Doğanay, 2001). "It then runs into Hatka stream and, after that, it reaches its old valley and finally Çoruh River" (Karahan et al., 2011: 401). In addition, there are many non-geological sites in the NOT Geopark Project such as a church, a castle, and an old gemstone quarry. One of the best examples of Georgian style is the Öşvan Monastery. The origin of the monastery goes back to the 11th century and was converted into Üngüzek castle during the 14th-15th century, and then used for different purposes by the people (Gümüş, 2014). Another important geological monument is Oltu gemstone (Turkish black amber). "Oltu stone, which gained importance in the 19th century, is one of the symbols of cultural identity in Erzurum. This stone is a kind of low calories lignite, which was formed as the result of sedimentation of resinous conifer plants" (Doğanay, 1997: 1). Prayer beads, which are used as a meditation tool, are made with of Oltu Stone.

FIGURE 16.4 The view of Lake Tortum.
(Photo: Gülpınar Akbulut).

However, NOT Geopark Projects have some problems. One of the key problems is related to the geopark border and the name, as there are two different geoparks: Narman Geopark and Tortum Geopark. Narman geopark was proposed by the Narman municipality. Tortum geopark has been studied by Karahan and his team for a long time. It will be useful for the geoparks to combine both of their studies and continue as NOT Geopark. Both studies have been undertaken to determine the geosite inventory and geotrails. The second problem is that none of the proposed geoparks in Turkey has a budget. Thirdly, there is no education program concerning this geopark. Actually, this geopark is a natural laboratory for the study of tectonics, with its folded structure, valleys, fairy chimneys, and pillow lava. Despite the negative factors, numerous local and international tourists visit popular geosites such as the Tortum waterfall and lake every year. The proposed geopark area is also close to Erzurum, which has a population of 760,476 and is one of the most important winter tourism centers.

16.4.4 LEVENT VALLEY GEOPARK PROJECT

The Levent Valley Geopark is located within the Tohma Stream Basin in the Malatya Province in East Anatolia, 40 km from the center of Malatya, 8 km from the center of Akçadağ, and just north of Yalınkaya village. According to the census (2017), the population of the geopark area is 26,058. This area is very close to the cities such as Malatya (786,676 people), Elazığ (583,671 people), and Sivas (621,301 people). The age of the geological units in the proposed geopark can be traced back to Mesozoic and Quaternary periods. The oldest unit in the geopark is of Upper Cretaceous age. This unit is overlain conformably by Eosene sediments. The formation of Miocene age is located at the top of this formation. The youngest unit is Yamadağ volcano, which is of Pliocene age (MTA, 2008, cited by Akbulut, 2014).

The first official article on the Levent Valley Geopark was presented by Gülpınar Akbulut in 2012. The first survey studies on this proposed geopark started in 2009 and these studies continued at different times. The proposed geopark covers a surface area of 400 km^2 with at least 34 natural geosites and 15 archeological and historical sites (Akbulut, 2014). It has been identified and classified under three main categories: natural geosites, historical, and archeological geosites. Natural geosites are classified as volcanic, erosional, morphological, and fossil. An important feature is the horizontal bed structure as well as pillars, buttes, caves, valleys, and karstic bridges. The Levent valley is an outstanding valley in Malatya and is also known as the "Little Grand Canyon" (see Fig. 16.5). The total length of the valley is 28 km. There is also a significant historical settlement in the Levent Valley, where some of the houses were built of stone (Akbulut, 2016). The high limestone walls and buttes have mounted on the sides of the valley in which the joints, fractures, and erosion have created beautiful scenery units. The second prominent geological monument is the caves. It shows that "Almost all the caves, which were formed as a result of the fault lines cutting through the valley and vertical to the main fault line, were used by people and reshaped on purpose" (Güngör et al., 2012a: 322). Küçükkürne Caves, which is one of the best examples of human–geology relation, is still inhabited by people (see Figs. 16.6 and 16.7).

There are fossil cemeteries in the proposed geopark, with most of them generally composed of larger benthic for aminifera such as Nummulites, Alveolina, and Discocyclina (Kaygılı et al., 2017). In addition, there are Karadağ volcanic mountain, Ozan Canyon, Dipsiz Valley, and İkizdere waterfall.

Some of the historical and archeological geosites are cave houses, fountains, rock tombs, rock paintings, tumuli, and an observation terrace. It is the first observation terrace in the proposed geoparks in Turkey. The 180 m high deck was initially opened to visitors in 2012 (see Fig. 16.8). The tumuli are located on the Karacadağ volcanics. There are three tumuli, but one of them was destroyed. The Bağköy rock cemetery used to be five rock cemeteries. The entrance door in the cemetery was destroyed by people hunting for treasure. The cemeteries have some symbols on the walls of the rock cemetery rooms (Akbulut and Ünsal, 2013) and over one million people visit this area every year, including school students.

FIGURE 16.5 The view of Levent Valley.
(Photo: Gülpınar Akbulut).

FIGURE 16.6 The cave at Küçükkürne Village in Dipsiz Valley. (Photo: Gülpınar Akbulut).

FIGURE 16.7 The view of inhabited caves in Levent Valley. (Photo: Gülpınar Akbulut).

FIGURE 16.8 The view of the observation glass terrace in Levent Valley. (Photo: Gülpınar Akbulut).

The Levent Valley Proposed Geopark was surveyed several times, which resulted in the determination of six main georoutes. Most of georoutes are hiking trails, but are also useful for bikes, buggies and animals (see Fig. 16.9). The georoute is classified according to difficulty rating (easy, middle, and hard), age (child, young, and old), package program (one day, three days, and a week), and activities (car tour, walk, and rock climbing). The Levent Valley Geopark has an education program; for example, İnönü University Geopark Association organized several meetings and media programs for different age groups of students.

This geopark was suggested as one of the priority geopark areas of Turkey by Akbulut in the meeting organized by UNESCO in 2015 and it was accepted. Recently, the number of different activities has been increased to promote the Levent Valley. One of them is International Malatya Photo Camp, which was organized by the Malatya Photography and Cinema Art Association (MAFSAD). Ten Turkish and foreign photographers have taken photographs in the valley and provided photography education to students and volunteers for two years. Fashion shoots were made in the proposed geopark, especially in the Hasanağa Stream Valley. Despite all of the positive progress, there are some problems in the proposed geopark. First, there is no management plan, even if the inventory of geosite has been completed in geopark studies. It should have a budget available, infrastructure, and marketing. Secondly, it should have a visitor center providing information about the geopark project for visitors. Thirdly, it should have a museum about the geosite, fossils, and other essential features. Consequently, if the process of candidature for the UNESCO and the EGN geopark network is completed, this geopark in process can apply for membership of these networks and can be accepted by them.

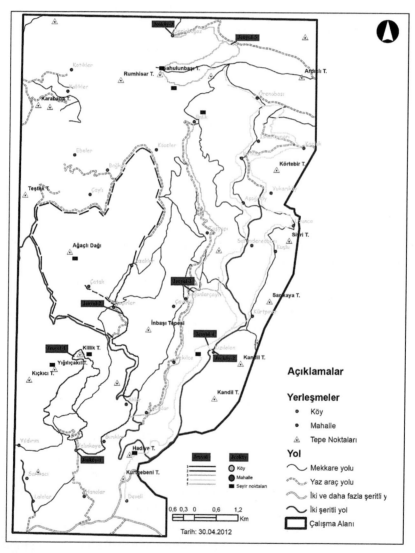

FIGURE 16.9 The georoutes on Hasanağa stream in the east of Levent Valley Geopark
Source: Reprinted with permission from Akbulut and Ünsal, 2013.

16.4.5 BITLIS VOLCANIC GEOPARK PROJECT

One of the first articles on Lake Van as a suggested geopark was presented by Gülpınar Akbulut in the Proceedings of the 10th European Geopark conference of Norway (2011). In that article, she discussed the importance of geological and geomorphological heritages at Lake Van and its surrounding in Eastern Anatolia (Akbulut, 2012), and in 2013 a geopark workshop was organized by the governorship of Bitlis. The proposed geopark mainly focused on the Nemrut Caldera and Mountain (2935 m), a part of Lake Van and Mount Süphan, an area covering 1000 km^2 (Gümüş, 2014: 58). The total population of the geopark in 2017 is 1,448,365. In the geopark area, different units from Paleozoic to Quaternary can be observed, including some young volcanic activities, such as from Mt. Nemrut, which is a stratovolcano. The most recent eruptions of this volcano were in 1441 and 1597 (Aydar et al., 2003).

The most prominent geological monument is the Nemrut Caldera, which was formed by the collapse of the peak of the volcano (Gürbüz, 1995) and which is one of the most beautiful and largest calderas of the world. Fumarole activity from several outlets takes place in the caldera, which is approximately 450–500 m deep. The caldera is characterized by 10 maars, 12 lava domes, 3 lava flows, and mineral springs (Ulusoy et al., 2008). "The western half of the caldera is filled by a freshwater lake with a surface area of 12.36 km^2, and a smaller lake with hot springs" (ibid: 269).

The second prominent geo-monument is Lake Van, which is the biggest lake in Turkey, occupying 3738 km^2 with a depth of 451 m. Furthermore, it is the largest soda lake on earth. The altitude of the lake surface is 1648 m above sea level. Lake sediments have supplied a continuous Paleo-climatic and Paleo-ecological record since the late Pliocene until today for the last 15,000 year of the lake's history (Gümüş, 2014). It is a closed lake without any significant outflow in a tectonically active zone in Eastern Anatolia (Degens et al., 1984; İzbırak, 1996).

The third prominent geo-monument is Mount Süphan, is a stratovolcano formed during the Quaternary and located to the north of Lake Van. The peak of Mt. Süphan is 4058 m. above sea level. Other geo-monuments are Lake Aygır on the eastern slope of Mt. Süphan, the Muradiye Waterfalls in the northeast of Lake Van, obsidian layers, Başkale travertine, and vapor chimneys (Akbulut, 2012). Among these, the number of people visiting the Muradiye waterfall (18) is above of other small monuments. This waterfall is on Bendimahi stream in the Muradiye district and has seven vertical drops and flows over young basalt lava (Doğanay, 2000).

The settlement history of the proposed geopark dates back to Neolithic civilizations and continues with Urartu, Persians, Assyrians, Meds, Byzantine, Seljuk, and the Ottoman State. Non-geological sites are strongly connected with geological structures. Blocks of red ignimbrite (Ahlat stone) have been used as raw material for construction purposes for thousands of years. One of the outstanding examples is the Seljuk Cemetery, which was built between the 11th and 13th centuries. This cemetery is the biggest open-air Turkish-Islamic graveyard of the world (Fig. 16.10). Some of the grave stones are of very large dimension up to 3.5–4 m in height (Akbulut, 2012). Obsidian is the other important stone, which was used as weapons and material for sculptures in ancient time, due to its sharp edges and a smooth surface when polished.

FIGURE 16.10 Headstones at the Seljuk Cemetery (Ahlat, Lake Van) are beautifully decorated with Persian–Islamic floral and geometric patterns.
(Photo: Gülpınar Akbulut).

The proposed geopark area has many archaeological sites. In the construction of mosques, tombstones, such as Hüseyin Temur Tomb and Emin Bayındır Bridge and other bridges. Ahlat stone was used as a raw material. Also, volcanic tuff in the proposed geopark has been excavated and the remaining rock converted into caves. These caves have been used for different purposes since ancient times. One of the most important archaeological sites is the Tilki area in the vicinity of Van castle. Another archeological site is the Akdamar church, which is located on Akdamar island in Lake Van.

The proposed geopark area had a feasibility report prepared by the governorship of Bitlis in 2013. According to the report, it has identified and classified under three main categories: natural geosites, historic geosites, and archeological geosites, but a detailed geosite inventory has not been finalised yet. Moreover, most of the geosites are under threat from anthropogenic and natural factors, and there is no effective management plan for the geopark. Also, there are no education studies about the geopark and geotourism concepts either. Despite the negative factors, over a million of tourists visit the proposed geopark every year, which is one of the most interesting volcanic areas in Turkey from esthetic, scientific, natural, and educational points of view. This geopark is very important for geotourism and understanding the development of the volcanic heritage of Turkey (Akbulut, 2012).

16.4.6 UPPER KIZILIRMAK GEOPARK PROJECT

The Upper Kızılırmak proposed geopark area is located in the center of Sivas city, in the Hafik, Zara, and Imranlı districts. The proposed geopark covers a surface area of 5618 km^2. According to the census (2017) the population of the geopark area is 408,865. This area is very close to the cities Malatya (786,676 people) and Sivas (621,301 people).

This proposed geopark is quite different from the other geoparks, because there are some projects related to the geopark such as "The Geosite Inventory of Sivas" (2013–2020), "Upper Kızılırmak Cultural and Natural Way" (2013–2014), and The Upper Kızılırmak Geopark Project (2016–continue). A formal management plan for the geopark and education programs have been prepared. The first official article on the proposed geopark was presented by Akbulut in 2012 (Akbulut and Ünsal, 2015).

The geological history of the geopark consists of various lithostratigraphic units of Paleozoic–Quarternary ages. It is located within the Sivas Tertiary Basin in northeastern and southwestern direction, length and width of which are about 250 km and 50 km, respectively (Ayaz, 2013). During the

Late Oligocene–Miocene, the basin transformed into a lagoonal and detrital clastic basin (Doğan and Özel, 2005).

It has been identified and classified under three main categories: natural geosites, historical, and archeological geosites. Natural geosites are classified as karstic springs, gypsum doline lakes, dolines; canyon, gorge, caves, natural bridge, travertine, pillar, erosion features, fault, fossil, salt marsh, hogback, Emirhan cliff and Eğribucak rock.

The core of natural geosites consists of massive gypsum series between Sivas and Imranlı. Gypsum covers an area of approximately 5% in Turkey and is mainly distributed in the Sivas Basin (Alagöz, 1967; Akbulut, 2011). The gypsum karst area in Sivas is approximately 300 km long and 40–50 km wide (Keskin and Yılmaz, 2016). These series occurred with the high rate of evaporation in the lagoonal environment (Pekcan, 1995; Yalçınlar, 1997), especially the cross stratification and wave traces on sandstone show that it is comprised of a shallow marine facies formation covered by alluvium in the Quarternary. Many karst structures of various size are observed in the region, however; different types of dolines that have different morphometric characteristics are the main karst type. Collapse and dissolution dolines are widely observed in the basin. The dolines in the basin are generally dish-shaped, however, cone-shaped dolines are also observed, although they are rare (Keskin, 2011). There is water mass within the dolines and some of the doline lakes include Tödürge, Hafik, East and West Lota, Akgöl, Karagöl, Mağara, Çimenyenice, Sarıgöl, and Cankız lakes (Fig. 16.11).

Left: Karagöl Lake   Right: Akgöl Lake

FIGURE 16.11 It is possible to see various and different gypsum doline lakes within the proposed Upper Kızılırmak geopark.

(Photo: Gülpınar Akbulut).

Tödürge (Demiryurt) Lake is the largest collapsed gypsum lake in Turkey and covers an area of approximately 3 km². The lake is generally shallow but the deepest part reaches 30 m (Alagöz, 1967; Yazıcı and Şahin, 1999). Local and migratory birds live in the surroundings of this lake, which are of so much value to the visitors. Hafik Lake is dish-shaped on a gypsum plateau, while West and East Lota lakes are cone-shaped and their water surfaces are at the same level. The depth of the lakes varies between 5 and 35 m (Akpınar and Akbulut, 2007). One of the examples of a collapsed doline is the Kızılçam, which contains a lake 220 m in diameter surrounded by steep rock slopes that rise 30–50 m forming a rim about 350 m diameter (Waltram, 2002). "Kurudeniz, Osmaniye, Başıbüyük and Bedirvan lakes, of which the smallest has a diameter of at least 1 km" (Günay, 2002: 394).

The second prominent natural monuments are caves and hollows. There are numerous cave entrances opened within gypsum rocks. Most of them are important in terms of nature, geo-archeologic, and reflect ancient times. Kalemköy, Ambarkaya, Durulmuş, Köroğlu, Kocabey, Satırlar, Hüyüğünçayır, Mağara, Dışkapı, and Güngörmez are some of the interesting caves.

The Emirhan cliffs are another prominent natural monument. It is a natural geosite, of sandstone, which was overturned in to a horizontal position during the Eocene–Miocene periods due to tectonic movements and created giant layers of 90° vertical structures, covering an area of approximately 15.68 km². Emirhan cliff is a hogback with erosional surfaces made mostly of sandstone, mudstone, and gypsum in colors of red, brown, gray, and yellow formed in its front (Fig. 16.12). Behind the hogback, there is a natural scenery with temporary waterfalls, different forms as tafoni, wild goats and rock partridges, and endemic species on the gypsum rocks. Some of the sandstones with gypsum were quarried and used as watch-towers and a church in ancient times (Akbulut et al., 2017; Akbulut and Ocak, 2017).

In the red-sandstone, mudstone and gypsum formation in the area surrounding Eğribucak village, one can find scenic types of erosion such as Eğribucak rocks and pillars. These interesting forms are known as "Bridge's Rock" and turtle. The weathered units have been eroded by seasonal rains, which created attractive erosion forms (see Fig. 16.13). Furthermore, there is a small valley in this area.

There are many non-geological sites in the Upper Kızılırmak Geopark Project area such as Eğri and Boğaz historical bridges; Seyfe, Deliktepe, and Dışkapı mount; Emirhan, Gökdin, and Tuzhisar churches (Fig. 16.14). Demiryurt, Kalemköy, Ambarkaya, and Köroğlu cave settlements;

Paşapınarı foundation; Hafik old government house and teacher house; and the Ahmet Turan tomb. Furthermore, this area has also a cultural heritage including Yarhisar houses; and Sivas knifes, carpets, and special foods.

FIGURE 16.12 A view of hogbeck of Emirhan cliff. (Photo: Used with permission of Ozan Özpay).

FIGURE 16.13 Erosion block and tafoni of Eğribucak rocky (Photo: Used with permission of Ozan Özpay).

The Eğriköprü bridge is located on Kızılırmak in the center of Sivas. This bridge represented the east entrance to the city for a long time. The Boğaz bridge is also located in Kızılkavraz village. The main characteristic of the bridge are the shelter and observation rooms in the bridge piers (Pürlü et al., 2011; see Fig. 16.15).

FIGURE 16.14 A view of Church of Gökdin Village.
(Photo: Used with permission of Ozan Özpay).

FIGURE 16.15 A view of Eğri Bridge.
(Photo: Gülpınar Akbulut Özpay).

The proposed geopark has several churches such as Tuzhisar, Gökdin, Günyamaç, and Düzyayla. One of the best known orthodox churches is Gökdin, which is located 30 km from the center of Sivas, near Gökdin village. A family has the ownership of the church. Tuzhisar church is located in Tuzhisar village, which was an old Armenian settlement. The majority of churches were destroyed by local people. Among the most important cultural heritage is the sacred tomb of Ahmet Turan. Parents, who want children, pray at this tomb and make a wish.

The territory of the proposed Upper Kızılırmak geopark is an open laboratory to study gypsum karst morphology and aims to protect and manage the natural heritage by introducing the alternative use of geological and geomorphological monuments to sustain these for future generations. Other objectives include:

- providing geopark education at schools, to local people, and to the general public;
- increase the sustainability of the economic sources in rural areas;
- complete the necessary procedures such as forming a management body, management plan, and other relevant activities to apply to the European Geoparks Network (EGN) and UNESCO Global Geopark Network (GGN) for approval and to be declared as a global geopark.

The proposed geopark is under the umbrella of Sivas governorship. Its management plan was prepared by Akbulut. Until now, 300 geosites have been found, some of which are very outstanding, rare, scientific, and esthetic. There are eight georoutes in this proposed geopark. The main route, route 3 and route 6 more outstanding than other georoutes. The length of the main route is about 80 km. Some of the monuments on these georoutes are listed in the Table 16.1. A map of this main georoute, which is named as Upper Kızılırmak Culture and Natural Way, has been prepared.

TABLE 16.1 Summarized List of Monuments along the Main Georoute.

Eğriköprü (Eğri Bridge);	Dışbudak tumulus;
Paşapınarı fountain;	Kızılırmak fault;
Seyfe spring and Seyfe tumulus;	Yarhisar Ambarkaya cave;
Kalemköy cave;	Taşlıgöl, Akgöl, Karagöl, Çimenyenice and
Deliktaş tumulus;	Kızgölüdoline lakes;
Gökdin church and spring;	Demiryurt caves;
Hafik lake;	Tödürge lake;
West and East Iota lakes;	Kızıl mountain.

Source: Author.

The third georoute is 54 km long. One can see the Kızılkavraz landscape, Kızılırmak gorge, Bingöl salt basin, Emirhan cliff and hogback, Eğribucak pillar, erosion surfaces, valley, formation borders, tafoni, waterfall, caves, church and cave settlements, numerous mountain goat, long-legged buzzard, and bear. The georoute is known as Emirhan georoute. Emirhan georoutes have been visited by hundreds of secondary school students, local people, and scientists supported by non-government associations. On both of these georoutes, tent camps have been organized by volunteers. Geopark management estimates that over one million people visit the geopark area.

The length of sixth georoute is about 60 km. There are Tavra stream and valley, Historical Paşa Garden, travertine, waterfall, Köroğlu cave, big church, Yıldız mountain, and Yakupoğlan dam on the direction of this georoute. Yıldız mountain is a stratovolcano, with 2252 m altitude at the sea level. This mountain was used as a ski resort by people, as well as for trekking and climbing. The historical Pasha Garden was designed based on London's Hyde Park as a model in the 19th century by Halil Rıfat Pasha. Furthermore, hobbit houses were built by the Sivas municipality within the garden for touristic purposes. Also, the majority of local people use it for picnics during the summer season. The garden also includes a small bird area and a small waterfall.

The Upper Kızılırmak proposed geopark is developing step by step. However, the most important problem is the budget. If the budget necessary for the infrastructure could be found, the process of the geopark development would be faster.

16.4.7 THE OTHER PROPOSED GEOPARKS

Turkey has a number of proposed geopark areas; many of them are still not a project, but only an idea. The most important proposed geopark at the idea stage is *Mount Ağrı (Ararat) and its surrounding Geopark*. The most prominent geological monument is Mount Ağrı, which is a young stratovolcano formed during the Pliocene. Its 5137 m summit is the highest peak in Turkey. Furthermore, this mountain is regarded as sacred by followers of different religions. Around this mountain, it is possible to see other heritages such as Noah's Ark, Cehennem Valley, Doğu Beyazıt meteor, Diyadin spring, the İshak Pasha Place, and Bayazıd Mosque. Mountaineers from all over the world also come to climb. Mount Ağrı, which was declared as a National Park in 2004 (Akbulut, 2014a, 2016).

The second proposed geopark idea is *Cappadocia*, which is located in the triangle of the Kayseri-Aksaray-Niğde Provinces in the Central Anatolia. Cappadocia is characterized by volcanic landforms and the deposition of ignimbrites during the Neogene–Quaternary periods, which covers an area of 7000 km^2 (Aydan and Ulusoy, 2003). There are many special morphological features in this area such as volcanic mountains, fairy chimneys, badlands, and caves. Among these, Mount Erciyes is the highest peak in Central Anatolia. In the area surrounding Mount Erciyes are parasite cones, as well as a caldera and feature lava tunnels, which have been used for stone mining. There are also several maars which include the Meke maar, Cora maar, Nar Maar, and the Gölcük Maar (Kopar, 2007). Göreme in Cappadocia also is currently only a national park, although Ürgüp-Göreme was declared a UNESCO World Heritage site in 1985 (Akbulut and Gülüm, 2012). Millions of tourists visit this area to see the special features of Cappadocia.

Another proposed geopark is Karapınar, which is located within the boundaries of Konya in the center of Anatolia with 1500 km^2. There are maars, maar lakes, pyroclastic cones, stratovolcanoes, hot springs, sinkholes, dolines and caves, castles, a caravansary, and underground cities in this area. The geology and geomorphology of Karapınar proposed geopark have been studied by MTA, but there is no management body established as yet.

16.5 CONCLUSIONS

Turkey has great opportunities to develop geotourism due to over 38.6 million tourists visiting this country every year and very rich geoheritage values. It is important to determine these natural and cultural values, therefore geoparks are significant opportunity for the protection and conservation of these values. The geopark attempts in Turkey are at different stages. Some geopark attempts were shallow and remote from the confines of planning and strategical developments such as Yatağan geopark project. Some of the proposed geoparks had studies on only one of the three key concepts of geopark, for example, some of these were confined to the attempts to increase the infrastructure necessary for the development of geotourism, but they ignore the education and conservation. In fact, there are some problems related to geoparks. The biggest problem for the development of geoparks in Turkey is the lack of a formal recognition of the concepts of geotourism and geoparks. The second problem is that geosite inventories of Turkey has been studied by different institutions and organizations and therefore a data center for geosite inventory should be established. For geoparks, it is important to find financial support and qualified human resources for the development of geotourism.

The Kula Volcanic Geopark is the first geopark of Turkey with international status. Several local geopark projects also exist, which are not included in the Global Geoparks Network, such as Bitlis Volcanic Geopark, Kızılcahamam-Çamlıdere Geopark, Levent Valley Geopark, Narman Geopark, and Upper Kızılırmak Geopark. These proposed geoparks have a scientific value as they include different types of stratigraphic, structural, and geomorphological geosites, as well as cultural sites. The majority of geoparks have completed their geosite inventory. On the other hand, some geosites in Kızılcahamam-Çamlıdere, Upper Kızılırmak, and also Kula Volcanic geoparks have been destroyed by people. Some proposed geoparks have no official activities, budget, management plan, and body. Despite this, the number of geoparks in Turkey should be increased. Geoparks should raise awareness, provide education and encourage research on geosites and geoheritage for the people. Turkey should apply to the UNESCO Global Geoparks Network for the approval of more geoparks. In this sense, management plans and bodies, particularly budgets should be prepared. It is hoped that the proposed geoparks will achieve their targets in a short period of time.

ACKNOWLEDGMENTS

I would like to thank Ozan Özpay for permission to use the photos, Patricia Erfurt and Sedat Bay for correcting the text and improving the English.

KEYWORDS

- **geopark**
- **geotourism**
- **environmental education**
- **georoutes**
- **sustainability**
- **Turkey**

REFERENCES

Akbulut, G. Suggested Geoparks ın Turkey: Volcanic Mountains. In *New Challenges with Geotourism*, Proceedings of The VIII European Geoparks Conference Idanha-a-Nova, 4–6 September, 2009; Portugal, 2009; pp 264–269.

Akbulut, G. *A Suggested Geopark Site: Gypsum Karst Topography Between Sivas-Zara. Natural Environment and Culture in the Mediterranean Region-II*; Cambridge Scholars Publishing: Newcastle, UK, 2011; pp 137–148.

Akbulut, G.; Gülüm, K. A Suggested Geopark Site: Cappadocia. *J. Balkan Ecol.* **2012**, *15* (4), pp 415–426.

Akbulut, G. Volcanic Features in the around Lake Van in the Eastern Anatolia Region of Turkey as a Suggested Geopark. In *The 10th European Geoparks Conference*, 16–18 September, 2012; Norway, 2012; pp 12–19.

Akbulut, G.; Unsal Ö. The Geopark Potential of Dipsiz Creek Valley (Akçadağ/Malatya). *The Science and Education at the Beginning of the 21st Century in Turkey*; St. Klıment Ohridski University Press: Sofia, Bulgaria, 2013; pp 986–994.

Akbulut, G. Önerilen Levent Vadisi Jeoparkı'nda Jeositler. *Cumhuriyet Üniversitesi Sosyal Bilimler Dergisi* **2014**, *38* (1), pp 29–45.

Akbulut, G. Volcano Tourism in Turkey. In *Volcanic Tourist Destinations Geoheritage, Geoparks and Geotourism*; Erfurt-Cooper, P., Ed.; Springer-Verlag: Berlin, Heidelberg, 2014a; pp 89–102.

Akbulut, G.; Ünsal, Ö. Yukarı Kızılırmak Kültür ve Doğa Yolu. *Coğrafyacılar Derneği Uluslararası Kongresi*, Özet Basım, 21–23 May 2015; Ankara, 2015.

Akbulut, G. Geotourism in Turkey. In *Alternative Tourism in Turkey Role, Potential Development and Sustainability*; Egresi, I., Ed.; Springer: Switzerland, 2016; pp 87–107.

Akbulut Özpay, G.; Erdem, Özgen N.; Ayaz, E.; Ocak, F. Yeni Bir Jeoturizm Sahası: Emirhan Kayalıkları (Sivas). *SOBİDER* **2017**, *14*, pp 15–29.

Akbulut Özpay, G.; Ocak, F. Sivas İlinde Bir Jeosit Alanı: Eğribucak Kayalıkları. *SOBİDER* **2017**, *18*, pp 77–93.

Akdeniz, E. Some Evidence on the First Known Residents of Katakekaumene (Burned Lands). *Mediterranean Archaeology and Archaeometry* **2011**, *11* (1), pp 69–74.

Akkemik, Ü.; Türkoğlu, N.; Poole, I.; Çiçek, İ.; Köse, N.; Gürgen, G. Woods of a Miocene Petrified Forest Near Ankara. *Turk. J. Agric. For.* **2009**, *33*, pp 89–97.

Akpınar, E.; Akbulut, G. Hafik Gölü ve Yakın Çevresinin Turizm Olanakları. *Erzincan Eğitim Fakültesi Dergisi* **2007**, *9*, 1–24.

Alagöz, C. A. *Sivas Çevresi ve Doğusunda Jips Karstı Olayları*. Ankara DTCF Yayınları No: 175: Ankara, 1967.

Angus, D. A.; Wilson, D. C.; Sandvol, E.; Ni, J. F. Seismology Lithospheric Structure of the Arabian and Eurasian Collision Zone in Eastern Turkey from S-Wave Receiver Functions. *Geophys. J. Int.* **2006**, *166*, 1335–1346.

Atalay, İ. Türkiye'deki Dağlık Alanların Oluşumu, Yapısal ve Ekolojik Özellikleri. In *Türkiye Dağları, I. Ulusal Sempozyumu*; 25–27 Haziran, 2002; Orman Bakanlığı Yayınları183: Ankara, 2002; pp 12–23.

Atalay, İ.; Efe, R.; Soykan, A. Mediterranean Ecosystems of Turkey. In *Ecology of the Taurus Mountains. Environment and Culture in the Mediterranean Region Part I*; Efe, R., Cravins, Öztürk, I., Eds.; Atalay, Cambridge Scholars Publishing: Newcastle, UK, 2008; pp 3–37.

Ayaz, E. Sivas Yöresinin Karmaşık Jeolojik Yapısına Bağlı Olarak Gelişen Önemli Maden Yatakları ve MTA'nın Sivas Yöresindeki Yeni Bulguları. *MTA Doğal Kaynaklar ve Ekonomi Bülteni* **2013**, *16*, pp 65–87.

Aydan, Ö.; Ulusoy, R. Geotechnical and Geoenvironmental Characteristics of Man-Made Underground Structures in Cappadocia Turkey. *Eng. Geol.* **2003**, *69* (3–4), pp 245–272.

Aydar, E.; Gourgaud, A.; Ulusoy, I.; Digonnet, F.; Labazuy, P.; Sen, E.; Bayhan, H.; Kurttaş, H.; Tolluoglu, A. Ü. Morphological Analysis of Active Mount Nemrut Stratovolcano, Eastern Turkey: Evidences and Possible Impact Areas of Future Eruption. *J. Volcanol. Geotherm. Res.* **2003**, *123*, pp 301–312.

Azaz, D.; Güngör, Y. The Geopark Potential of Oltu Narman Basin (Eastern Turkey: Erzurum). In *70. Türkiye Jeoloji Kurultayı*, 10–14 Nisan, 2017; Ankara, 2017; p 209.

Bozkurt, E.; Mittwede, S. K. Introduction to the Geology of Turkey. *Intro. Geol. Rev.* **2001**, *43*, 578–594.

Bulut, I.; Kopar, I.; Zaman, M. Karadeniz Bölgesindeki Yüzen Adalara Yeni Bir Örnek: Zökün Gölü Yüzen Adaları (Tortum-Erzurum). *Atatürk Üniversitesi Fen-Edebiyat Fakültesi, Sosyal Bilimler Dergisi* **2009**, *8* (41), 215–230.

Degens, E. T.; Wong, H. K.; Kempe, S.; Kurtman, F. A Geological Study of Lake Van, Eastern Turkey. *Geol. Rund.* **1984**, *73* (2), 701–734.

Doğan, U.; Özel, S. Gypsum Karst and Its Evolution East of Hafik (Sivas, Turkey). *Geomorphology* **2005**, *71*, 373–388.

Doğanay, H. Fitolojik Kökenli Bir Fosil: Oltutaşı. *Doğu Coğrafya Dergisi* **1997**, *3* (2), 1–22.

Doğanay, H. Türkiye'de Az Tanınan Üç Doğa Harikası Tomara-Sırakayalar ve Muradiye Çağlayanları. *Doğu Coğrafya Dergisi* **2000**, *3*, 1–25.

Doğanay, H. *Türkiye Turizm Coğrafyası*. Çizgi Kitabevi: Konya, 2001.

Doğanay, H. *Türkiye Ekonomik Coğrafyası*. Pegem Akademi Yayınları: Ankara, 2011.

Duman, Y. The Largest Landslide Dam in Turkey: Tortum Landslide. *Engineer. Geol.* **2009**, *104*, 66–79.

Farsani, N. T.; Celeste, O. A.; Coelho; Costa, C. M. M.; Amrikazemi, A. Geo-Knowledge Management and Geoconservation via Geoparks and Geotourism. *Geoheritage* **2014**, *6*, 185–192.

Garipağaoğlu, Farımaz, N. Doğal Bir Anıt Olarak Narman Havzası'nda Kuesta Reliefi, *Türk Coğrafya Dergisi* **1996**, *31*, 291–304.

Gümüş, E. *Yeni bir doğa koruma kavramı: UNESCO jeoparklar çerçevesinde Çamlıdere (Ankara) fosil ormanı fizibilite çalışması*. Ondokuz Mayıs Üniversitesi Sosyal Bilimler Enstitüsü Yayınlanmamış Yüksek Lisans Tezi: Samsun, 2008.

Gümüş, E. Türkiye'nin ilk Jeoparkına Doğru-Kula Volkanik Jeoparkı Projesi. In *I.Ulusal Coğrafya Sempozyumu*; 28–30 Mayıs, 2012; Erzurum, 2012; pp 1081–1088.

Gümüş, E. *Geoparks: Multidisciplinary Tools for the Protection and Management of Geoheritage in Turkey. Kula Volcanic Area (Manisa) and Çamlıdere Fossil Forest (Ankara) as Case Studies*. University of Aegean, Doctorate Thesis: Greece, 2014.

Gümüş, E.; Zouros, N. Geosite Identification in Kula Volcanic Geopark/Turkey. In *11th European Geopark Conference*;19–21 September, 2012; Portugal, 2012; pp 1–2.

Gümüş, E.; Zouros, N. Aspiring Kula Volcanic Geopark—Earth Heritage Protection and Promotion. *Online Soc. Geol. It.* **2013**, *09*, 82–84.

Günay, G. Gypsum Karst, Sivas, Turkey. *Environ. Geol.* **2002**, *42*, 387–398.

Güngör, Y.; Direnç, A.; Çelik, Y.; Yalçın, M. N. Investigate The Potential of Being Jeopark of Narman Canyon (Narman-Erzurum) and Prepare of Jeopark Inventory. In *International*

Multidisciplinary Scientific GeoConference: SGEM: Surveying Geology & Mining Ecology Management; Sofia, 2012; Vol 1, pp 143–153.

Güngör, Y.; İskenderoğlu, L.; Azaz, D.; Güngör, B. Levent Vadisinin (Akçadağ- Malatya) Jeopark Envanter Çalışması. In *Türkiye Jeoloji Kurultayı*; 2–6 Nisan, 2012; Ankara, 2012a; pp 322–323.

Gürbüz, O. Turizm Coğrafyası Açısından Nemrut Kalderası. *Türk Coğrafya Dergisi* **1995**, *30*, 255–265.

Gürler, G.; Timur, E. Jeoparkların Koruma ve Kullanım Yöntemlerinin Belirlenmesi; Karapınar Potansiyel Jeopark Alanı İçin Bir Değerlendirme, Türkiye. In *Proceedings of the Second International Symposium on Development within Geoparks Environmental Protection and Education*; 12–15 June, 2007; Lushan, Jiangxi Province, China, 2007; pp 1–13.

Gürsay, M. S. *Kızılcahamam—Çamlıdere Jeoparkında Jeoturizm ve Sürdürülebilirlik*. Atılım Üniversitesi Sosyal Bilimler Enstitüsü Basılmamış Yüksek Lisans Tezi: Ankara, 2014.

Hose, T. A.; Vasiljevic, D. A. Protecting and Promoting the Geoheritage of South-Eastern Europe. In *Geoheritage and Geotourism: A European Perspective*. Hose, T. A., Eds.; Boydell Press: The UK, 2016; pp 173–194.

İzbırak, R. *Türkiye I*. MEB Yayınları: Istanbul, 1996.

Karahan, F.; Kopar, İ.; Orhan, T.; Çakır, T. *The Geopark Potential of Tortum Valley (Erzurum-Turkey) and Its Surroundings. Natural Environment and Culture in the Mediterranean Region-II;* Cambridge Scholars Publishing: Newcastle, UK, 2011; pp 395–406.

Kaykılı, S.; Avşar, N.; Aksoy, E. Paleontolojik Bir Jeosit Örneği: Hasanağa Deresi, Akçadağ, Malatya. *Türkiye Jeoloji Bülteni* **2017**, *60* (1), 93–105.

Kazancı, N. *Jeolojik koruma kavram ve terimler*. JEMİRKO: Ankara, 2010.

Kazancı, N. Geological Background and Three Vulnerable Geosites of the Kızılcahamam-Çamlıdere Geopark Project in Ankara, Turkey. *Geoheritage* **2012**, *4*, 249–261.

Kazancı, N.; Şaroğlu, F.; Doğan, A.; Mülazımoğlu, N. S. Turkey. In *Geoheritage in Europe and Its Conservation*; Wimbledon, W. A. P., Smith-Meyer, S., Eds.; ProGeo: Oslo, 2012; pp 367–377.

Keskin, İ. *Jipslerde Dolinlerin Oluşum Mekanizmaları Açısından Süreksizlik Özelliklerinin Etkilerinin Araştırılması: KD Sivas Örneği*; Cumhuriyet Üniversitesi Fen Bilimleri Enstiütsü Basılmamış Doktora Tezi: Sivas, 2011.

Keskin, İ.; Yılmaz, I. Morphometric and Geological Features of Karstic Depressions in Gypsum (Sivas, Turkey). *Environ. Earth Sci.* **2016**, *75* (1040), 1–14.

Koçan, N. Jeoturizm Planlaması ve Peyzaj Mimarlığı Açısından Bir Değerlendirme: Kızılcahamam-Çamlıdere Jeoparkı. *Ege Üniversitesi Ziraat Fakültesi Dergisi* **2011**, *48* (1), 47–53.

Koçman, A. Yanık Ülkenin Doğal Anıtları: Kula Yöresi Volkanik Oluşumları (Natural Wonders of the "Burnt Land (Katakekaumene): Volcanic Features of Kula Area. *Ege Coğrafya Dergisi* **2004**, *13*, 5–15.

Koçman, A.; Koçman, Ö. Yanık Ülke (Katakekaumene) Kula Volkanik Yöresinde Jeoturizm Üzerine Değerlendirmeler. In *II. Uluslararası Turizm, Çevre ve Kültür Sempozyumu Bildiriler*; 10–11 May, 2004; İzmir, 2004; pp 91–103.

Kopar, İ. *Hasan Dağı ve Yakın Çevresinin Fiziki Coğrafyası*. Gündüz Yayıncılık: Ankara, 2007.

Kopar, İ.; Sevindi, C. Tortum Gölünün (Uzundere-Erzurum) Güneybatısında Aktüel Sedimantasyon ve Siltasyona Bağlı Alan-Kıyı Çizgisi Değişimleri. *Türk Coğrafya Dergisi* **2013**, *60*, 49–66.

Ministry of Culture and Tourism (2017). Turkey Tourism Statistic Data, Ankara.

Newsome, D.; Dowling, R. Setting an Agenda for Geotourism. In *Geotourism: The Tourism of Geology and Landscape*; Newsome, D., Dowling, R. K., Eds.; Goodfellow Publishers: The UK, 2010; pp 1–12.

Özpay, O. *The Photograph of Upper Kızılırmak Geopark*, Sivas, 2017.

Pekcan, N. *Karst Jeomorfolojisi*. Filiz Kitabevi: İstanbul, 1995.

Pürlü, K.; Altın, Y.; Aygün Canan, A.; Cebecioğlu, M.; Özkanat, M.; Çetindağ, E.; Bedir, A.; Kaya, A.; Çavuş, İ.; Babacan, S. S. *Sivas Kültür Envanteri I-II*. Kadir Pürlü, Eds.; Es-Form Ofset: Sivas, 2011.

Tongkul, F. Geotourism in Malaysian Borneo. In *Geotourism*; Dowling, R. K.; Newsome, D., Eds.; Elsevier/Heinemann: Oxford, 2006; pp 26–41.

Ulusoy, İ.; Labazuy, P.; Aydar, E.; Ersoy, O.; Çubukçu, E. Structure of the Nemrut Caldera (Eastern Anatolia, Turkey) and Associated Hydrothermal Fluid Circulation. *J. Volcanol. Geotherm. Res.* **2008**, *174*, 269–283.

UNESCO Global Geoparks. http://unesdoc.unesco.org/images/0024/002436/243650e.pdf, 2016.

Waltham, T. Gypsum Karst near Sivas, Turkey. *Cave and Karst Sci.* **2002**, *29* (1), 39–44.

Yalçınlar, İ. Sivas Çevresinin Strüktüral Jeomorfolojisi Üzerine. *Türkiye Coğrafyası Araştırma ve Uygulama Merkezi Dergisi* **1997**, *6*, 407–410.

Yazıcı, H.; Şahin, İ. F. Demiryurt (Tödürge-Sivas) Sulak Alanı ve Yakın Çevresinde Coğrafî Gözlemler. *Türk Coğrafya Dergisi* **1999**, *34*, 19–30.

Yıldırım, T.; Koçan, N. Nevşehir Acıgöl Kalderası Kalecitepe ve Acıgöl Maarlarının Jeoturizm Kapsamında Değerlendirilmesi. *Ege Üniversitesi Ziraat Fakültesi Dergisi* **2008**, *45* (2), 135–143.

CHAPTER 17

Geotourism Development in Latin American UNESCO Global Geoparks: Brazil, Uruguay, Mexico, and Peru

JOSÉ LUIS PALACIO PRIETO[1*], CÉSAR GOSSO[2], DIEGO IRAZÁBAL[2],
JOSÉ PATRÍCIO PEREIRA MELO[3], FRANCISCO DO O´DE LIMA, JÚNIOR[3],
CARLES CANET[4], MIGUEL A. CRUZ-PÉREZ[4],
ERIKA SALGADO-MARTÍNEZ[4], JUAN CARLOS MORA-CHAPARRO[4],
KRZYSZTOF GAIDZIK[5], JERZY ŻABA[5], and JUSTYNA CIESIELCZUK[5]

[1]*Instituto de Geografía, UNAM, Scientific Coordinator of the Mixteca Alta UNESCO Global Geopark, Mexico*

[2]*Grutas del Palacio UNESCO Global Geopark, Uruguay*

[3]*Araripe UNESCO Global Geopark, Brazil*

[4]*Comarca Minera UNESCO Global Geopark, Mexico*

[5]*Silesian University of Technology, Poland*

Corresponding author. E-mail: palacio@unam.mx

ABSTRACT

The Latin American and Caribbean Geopark Network (GeoLAC) was founded in May 2017 with the participation of four UNESCO Global Geoparks: Araripe (Brazil), Grutas el Palacio (Uruguay), and Comarca Minera and Mixteca Alta (both in Mexico).

GeoLAC shares the objectives of other regional networks as well as those of the Global Geoparks Network, namely "to conserve and enhance the value of areas of geological significance in Earth history, including landscapes and geological formations, which are key witnesses to the evolution of our planet and determinants for our future, and to promote sustainable development for example through geotourism and education" (GGN, 2018).

Geotourism is the main activity promoting the economy of Latin American geoparks, which focuses on a cultural and educational perspective for all audiences.

Latin American geoparks, both those recognized by UNESCO and others in development, maintain close collaboration through joint geotourism activities.

Geotourism displays a growing dynamism and acceptance in various Latin American territories, although with striking differences in their development. Latin American Geoparks are contrasting territories, some with a long tourism tradition, extensive infrastructure while others are socially and economically marginalized.

17.1 INTRODUCTION

On May 5, 2017, the Mexican territories *Comarca Minera* and *Mixteca Alta* were listed as UNESCO Global Geoparks. On May 26th of the same year in the City of Chivay (Arequipa, Peru), these two geoparks, together with the UNESCO Global Geoparks of Araripe (Brazil, listed in 2006) and *Grutas del Palacio* (Uruguay, in 2013), agreed on the foundation of the Latin American and Caribbean Geopark Network (GeoLAC) with the sponsorship and collaboration of the UNESCO International Geoscience and Geoparks Programme and the General Secretariat of the Global Geoparks Network (GGN). GeoLAC shares the objectives of other regional networks as well as those of the Global Geoparks Network, namely "to conserve and enhance the value of areas of geological significance in Earth history, including landscapes and geological formations, which are key witnesses to the evolution of our planet and determinants for our future, and to promote sustainable development for example through geotourism and education" (GGN, 2018). Culture is a key element of the Latin American Geoparks, given the traditional indigenous knowledge of the local people across the region. In this way, education and geotourism are two fundamental and closely related elements to achieve the objectives stated.

Education in UNESCO Global Geoparks, and particularly in those recognized in Latin America, comprises formal and informal programs. The first includes those that are part of academic programs of educational institutions at different levels, from basic to college education. In these programs, students are introduced to geopark subjects including geodiversity, geological heritage, and geoconservation, as well as their relationship with society and culture. These formal programs, developed in educational institutions with the support of the scientific committees of geoparks, include activities that are carried out on a daily basis in the territory through school practices. On the other hand, informal education programs include the dissemination of Earth Sciences through printed materials (brochures, maps) and activities at geopark interpretation centers, which are open to all interested visitors and offering conferences, courses, and recreational activities. Informative panels located along geotrails are also informal educational strategies that deliver information to visitors on topics of interest related to each particular territory.

The educational offer is, thus, the core element to promote geotourism, which has been conceptualized from two different perspectives. On the one hand, Hose (1995) defined geotourism for the first time as "the provision of interpretive and service facilities to enable tourists to acquire knowledge and understanding about the geology and geomorphology of a site (including its contribution to the development of the Earth sciences) beyond the level of mere esthetic appreciation." More recently, geotourism (as a modality of geographical sustainable tourism) was defined as "tourism which sustains and enhances the identity of a territory, taking into consideration its geology, environment, culture, esthetics, heritage and the well-being of its residents" (Arouca, 2011). This chapter does not provide an in-depth discussion of these two perspectives, although geotourism activities in Latin American geoparks are compatible with both. Undoubtedly, geological heritage is a key component of the geotourism offer of the geoparks referred to herein, with the relationship between geological heritage with other types of heritage—both natural and cultural—as major factors.

The distinctive geoproducts of each geopark are also key elements to promote geotourism; these include handcrafts and gastronomic experiences, where the geodiversity-biodiversity-culture relationship is amalgamated.

Although this chapter focuses on geoparks registered with UNESCO in Latin America, there are several projects currently underway that, with different extent of progress, promote geotourism in territories aiming to be inscribed at UNESCO in the near future. UNESCO Global Geoparks are single, unified geographical areas where sites and landscapes of international geological significance are managed under a holistic concept of protection, education and sustainable development (www.unesco.org). Aspiring geoparks refer to those who have officially applied to UNESCO and are currently under evaluation. Geopark projects have not yet applied for inscription at UNESCO, although they are expected to do so in the near future (Table 17.1 and Fig. 17.1).

TABLE 17.1 UNESCO Global Geoparks, Aspiring Geoparks, and Projects in Latin America.[1]

UNESCO Global Geoparks		
Araripe	Brazil	2006
Grutas del Palacio	Uruguay	2013
Comarca Minera	Mexico	2017
Mixteca Alta	Mexico	2017
Aspiring Geoparks		
Imbabura	Ecuador	2015
Tungurahua	Ecuador	2015
Colca y Volcanes de Andagua	Peru	2016
Tacaná	Mexico	2017
Río Coco	Nicaragua	2017
Chirripo	Costa Rica	2017
Kütralkura	Chile	2017
Geopark Projects		
Viñales	Cuba	
Volcánico del Ruiz	Colombia	
Seridó	Brazil	
Napo Sumaco	Ecuador	
Torotoro	Bolivia	
Litoral del Biobio	Chile	
Pillán Mahuiza	Argentina	

[1] In 2019, during the preparations for printing of this book, three of the geoparks listed in Table 17.1 joined the Global Geopark Network: these are Colca y Volcanes de Andagua UNESCO Global Geopark, in Peru, Imbabura UNESCO Global Geopark, in Ecuador, and Kütralkura UNESCO Global Geopark, in Chile.

These geoparks and projects are framed within a tectonic context resulting from the interaction of six large plates—North American, Pacific, Cocos, Nazca, South American, and Antarctic—which explains the presence of regional mountainous structures of volcanic and structural origin and the lithological diversity spanning across 2500 million years, from the Proterozoic to date. The wide regional climatic diversity results in the action of modeling geomorphological processes that ranges from glacial to tropical environments, including deserts and temperate environments.

The extraordinary geological and geomorphological diversity of the Latin American region is closely related to the equally outstanding biological diversity, explained by the former; 6 of the 17 countries considered as mega biodiversities worldwide are located within this region (Llorente-Bousquets and Ocegueda, 2008). A distinctive feature of the region should be added, that is, the presence of numerous indigenous people, whose ancestral knowledge in the use of resources, and in particular geodiversity, completes a complex and interesting framework that characterizes Latin American Geoparks.

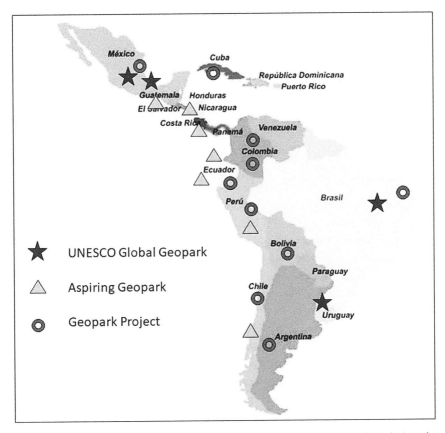

FIGURE 17.1 UNESCO Global Geoparks, Aspiring Geoparks, and projects in Latin America.

17.2 LATIN AMERICAN UNESCO GLOBAL GEOPARKS

The complex tectonic, geological, biological, and social settings constitute the distinctive foundations of the Latin American Geoparks and set the ground for geotourism projects that, although under different stages of development, show a significant growth potential. Four UNESCO Global

Geoparks and one Aspiring Geopark, the latter in Peru, are referred to in this text to illustrate the development of geotourism in Latin America.

17.2.1 ARARIPE UNESCO GLOBAL GEOPARK, BRAZIL

The Araripe UNESCO Global Geopark is located in the Araripe Basin, which is considered to be the largest sedimentary basin in northeastern Brazil, reaching southern Ceará, northwestern Pernambuco and eastern Piaui. The Geopark was created in 2005 from an initiative of *Universidad Regional de Cariri* (URCA) and registered with the GGN in 2006 during the 2nd UNESCO Conference on Geoparks held in Belfast, Northern Ireland. It involves the territory of the municipalities of Barbalha, Crato, Juazeiro do Norte, Nova Olinda, MissãoVelha, and Santana do Cariri, comprising an area of 3441 km^2.

In terms of relief, the Araripe Plateau is its main distinctive landform (Fig. 17.2). Key geological records from the Lower Cretaceous between 90 and 150 million years ago characterize this Geopark's geological heritage, especially regarding its paleontological content. The preservation of this vast paleontological fossil heritage in the region is the result of unique conditions during the geological evolution of the Araripe Basin. Paleobiology research works reveal a huge diversity, including silicified trunks, impressions of ferns, conifers, and flowering plants, as well as foraminifera, mollusks, arthropods (ostracoda, spiders, scorpions, and insects; Fig. 17.3), fishes (sharks, rays, bony fishes, and many coelacanths), amphibians, and reptiles (turtles, lizards, crocodilians, and pterosaurs). Sedimentary deposits preserve a high diversity of rocks (limestones, claystones, sandstones, thick gypsum deposits) that serve as records of past geological environments.

There are nine geosites, each characterizing a different period of geological time in this region. Geosites worth mentioning for their scientific importance are Parque do Pterossauros, Pedra Cariri, and Floresta Petrificada do Cariri. Other geosites are important from historical and cultural perspectives in addition to their geological value, such as Colina do Horto, Ponte de Pedra, Cachoeirade MissãoVelha, and Pontal de Santa Cruz; others have high ecological interest, such as Riacho do Meio and Batateiras geosites.

The municipality of Juazeiro do Norte is home to the largest population in the Cariri region, with 261,289 inhabitants, and includes the greatest cultural heritage, receiving the highest number of visitors per year—approximately 2 million. The entire city is considered as a sacred site by the thousands of pilgrims who visit it four times a year.

In the past, indigenous people settled in this region of the Geopark and were subsequently conquered, followed by an extensive colonization and settlement of the first villages and cities of the early Caririenses. The miscegenation of various people (indigenous, European, and African) and the relative isolation of Cariri from major Brazilian cities led to a distinctive cultural identity with particular folk dances, songs, religious and artistic expressions. Therefore, Cariri became known as a "cultural melting pot" where ancestral traditions are kept alive.

FIGURE 17.2 The Araripe Plateau, a distinctive landform of the Araripe UNESCO Global Geopark.
Source: http://www.globalgeopark.org/aboutGGN/list/Brazil/6404.htm.

The Geopark has infrastructure that supports tourism and educational uses. Based on the historical and cultural identity of the Cariri people, the Environmental Education and Interpretation Center of the Araripe UNESCO Global Geopark was created in 2010 (Fig. 17.3). This Center organizes pedagogical activities for public and private schools, higher education institutions, and the general public. The purpose is to disseminate the concept of UNESCO Global Geoparks and environmental education practices in an interactive way, providing an integrated understanding of the historical, cultural, socio-environmental, paleontological and landscape aspects of geological sites. Also, the Geopark has raised interest from several governmental institutions and NGOs, as well as from the productive and private sectors, establishing partnerships and jointly developing projects with the ultimate aim of stimulating tourism in the region from a sustainable development perspective.

FIGURE 17.3 Interpretation Center at Araripe UNESCO Global Geopark, Brazil.
Source: http://geoparkararipe.org.br/museu-de-paleontologia-da-urca/.

17.2.2 GRUTAS DEL PALACIO UNESCO GLOBAL GEOPARK, URUGUAY

The *Grutas del Palacio* Geopark comprises 3600 km² and is located in the center-southern region of Uruguay, in the Department of Flores. With extensive plains and low hills, this Geopark displays numerous round-shaped granite outcrops and other remarkable features that confer a unique beauty to the natural landscape.

Geological and morphogenetic processes during the past thousands of years under different climate conditions (alternating humid and dry periods) shaped the actual landforms. Chemical and physical weathering processes produced relictual inselberg areas on thick granite soils that support fertile prairies. Uruguay is a water-rich land and the Geopark has large bodies of water. Uruguay has extensive hydrological basins and large artificial lakes abound in the Geopark.

FIGURE 17.4 Grutasdel Palacio, a main geosite within the Geopark (Photo: D. Irazábal).

In *Grutas del Palacio* geological ages are represented by various rocks and sediments (Proterozoic, Palaeozoic, Mesozoic, and Cenozoic). The site includes two distinctive geological areas; one comprises central and southern Proterozoic crystalline basements composed of granites, gneisses, and metamorphic belts with magmatic intrusions. In the Northern portion, a small and isolated area of Permian conglomerates and sandstones represent old glacial environments. Mesozoic sedimentation of conglomerates deposited in Cretaceous fluvial environments is exposed, as well as Cenozoic inter-bedding with nondepositional sediments like palaeo-soils. Some Holocene sand terrace deposits and fluvial erosion processes that have occurred over the last four decades are exposed in a coastal lake. This geodiversity within a small territory allows visitors to acquire a broad vision of the evolution of history.

From the local development perspective, the Geopark was established as a strategy to reactivate and invigorate the territory through various activities, including geotourism. "It enables visitors to understand a past spanning 4600 million years, making it possible to view the present from a different perspective and project ahead a likely shared future for the Earth and humankind" (Arouca, 2011).

The *Grutas del Palacio* Geopark is closely linked with primary and secondary schools in the Department of Flores. Several indoors and field geo-education and environmental projects have been carried out. A special publication targeting students was prepared in 2013. There are highly active local Fair of Science Clubs, and some projects related to the Geopark are selected each year for the National Fair of Science. Further education programs are also carried out for university students. These curricular activities focus mainly on geological mapping, geosite characterization, and inventory activities. Talks are underway with other UNESCO Global Geoparks in Spain, Portugal, and China to promote the relationship between local students. The Geopark is the means to promote local development based on geological and historical heritage, with geotourism, geo-education, sports, and culture as core instruments.

Thanks to an agreement with *Universidad de la República* (School of Sciences, Bachelor's Degree in Geology), the Geopark has achieved to involve about a dozen advanced students. They are provided with information and tools such as magnifying glasses, hammers and lamps, and invite tourists to "travel back in time" by explaining the processes and phenomena that occurred millions of years ago. Besides, this is not an isolated process either; it has to be closely linked to the traditions and practices of the area. Local communities become hosts, opening the doors to old country houses and indulging people with age-old recipes and flavors.

Since 2016, *Grutas del Palacio* holds the Geotourism Week at the same time as the traditional "Holy Week," as it is one of the periods with the largest influx of tourists in the country (Fig. 17.5).

Finally, geotourism has become a project that promotes local and sustainable development for the following reasons:

First, it has entailed a new tourist product at a national level, giving the territory a competitive edge and comparative advantage when attracting visitors and tourists. For instance, the *Grutas del Palacio* geosite (after which the geopark is named, Fig. 17.4) experienced a 30% annual increase in the number of visitors. Second, and closely related to the fact just mentioned, this geosite has created opportunities for local inhabitants who had been previously excluded from the tourism experience. Today, geodiversity-based tourism also creates jobs, although they may be initially temporary. Third, in *Grutas del Palacio* we call ourselves the *"Tierra del Geoparque"* (Geopark Land), and have developed and improved the food and handcraft product lineup under the name *"Producto de la Tierra del Geoparque"* (Geopark Land Products), in relation with geotourism routes.

FIGURE 17.5 Visitors during the "Geotourism week" in Grutas del Palacio Geopark (Photo: D. Irazábal).

In *Grutasdel Palacio* we are aware that the success of the activities in place is indissolubly linked to the level of participation and involvement of the local community. The Geopark lives and is shaped by the empowerment of the population, once the people understand the concept and become familiar with it.

Since the early development stages of these activities, *Grutas del Palacio* encouraged the creation of a "Promoter Group" composed of representatives of the territory's live forces, mainly civil society organizations, chambers of commerce, entrepreneurs, and students. This "Promoter Group" was involved in awareness-raising and dissemination by the Intendancy of Flores and the Faculty of Sciences at *Universidad de la República*, aimed at facilitating the understanding of the Geopark concept, particularly as regards its holistic approach.

When leaders serving as guides regarding social, cultural, and economic development aspects of the territory understand this dynamics, they are capable of participating in a strategy that benefits the population as a whole.

Geotourism activities, characterized by community tourism in the beginning, as well as the talks that were delivered in all educational centers of the territory, rapidly boosted enthusiasm in the population, as well as the recognition that the "Geopark" title was a territorial denomination worth valuing and promoting. The latter, in addition to communication and awareness campaigns, made the *Grutas del Palacio* Geopark a true departmental strategy focused on integrated development based on social and political legitimacy.

The Latin American experience indicates that geopark projects are being developed by municipal, local, and provincial authorities with strong support from the national government. In turn, this geopark development involves the participation of universities and NGOs, which are strategic partners of geoparks under governance schemes that still need to be strengthened.

Under this scheme, the participation of local communities and the general public is a challenge for geopark designers and developers alike, since building geopark structures is not feasible without social participation, particularly of communities inhabiting the territories where geosites and sites of interest are located.

To achieve this objective, multiple strategic actions should be implemented, as was the case in *Grutas del Palacio*:

- Involvement of university students in geotourism activities.
- Community tourism programs in sites and geosites across the Geopark.
- Outreach and social awareness in public spaces, official propaganda, and in sports, cultural and artistic events.
- Training programs for entrepreneurs and development of handcrafts, including the Geopark logo.
- Technical talks targeted to specific social groups, to explain the Geopark objectives and scope.
- Publication of contents for primary education throughout the territory.
- Participation in local, national, and regional exhibitions and fairs.

17.2.3 MIXTECA ALTA UNESCO GLOBAL GEOPARK, MEXICO

The *Mixteca Alta* UNESCO Global Geopark is located in the *Sierra Madre del Sur* physiographical province, a mountainous region bordered to the north by the *Sierra Madre Oriental*; it encompasses 415 km². From the geological standpoint, this is considered to be the most complex region of Mexico. It consists of Precambrian (more than 500 million years old) and Paleozoic metamorphic and plutonic rocks (250–500 million years old) that make up the complex basement. A layer of Mesozoic—largely marine—rocks (60–250 million years old) and Cenozoic volcanic rocks and continental sediments (less than 60 million years old) overlay the basement lithology.

Most of the 35 geosites identified are related to erosion-deposition processes and landforms associated with the millenary intensive use of land for farming purposes (Palacio-Prieto et al., 2016). A number of geosites were selected to explain these links, including gullies and badlands (Fig. 17.6), mass-wasting features, ancient agricultural terraces locally known as *lamabordos*, and palaeosols (fossil soils). Additional geosites include plutonic and tectonic structures (dikes and sills, faults) and outcrops featuring spheroidal weathering.

FIGURE 17.6 Vista Hermosa Geosite, gullies, badlands and dykes, Mixteca Alta Geopark (Photo: J. L. Palacio)

The name of the Geopark derives from the Mixteca civilization that flourished between the 2nd century BC and the 15th century AD, ending in the early 16th century AD with the arrival of the Spaniard conquerors (Spores, 1969). The territory has some of the most important traces of the Mixteca culture. The presence of human settlements in the region (circa 3400–3500 years BP) has been established based on radiocarbon dating of soil organic carbon in *lamabordos* (valley-bottom terraces; Leigh, et al., 2013). *Lamabordos* were a local farming innovation for food production to meet the demands of a population that reached 50,000 inhabitants during the Post-classic period, around 1000–1520 AD, living in 111 towns (Spores, 1969).

The *Mixteca Alta* in general, and in particular the nine municipalities included in the Geopark, are characterized by low population density and demographic growth. Due to the lack of economic opportunities, emigration to urban areas and abroad (United States of America) is significant and contributes to low population numbers. The total population is about 7000 inhabitants. Most local inhabitants are indigenous and include groups such as the Chinantecos, Mixes, Mazatecos, Zapotecos, and mainly, Mixtecos.

The *Mixteca Alta* region is a popular touristic destination for both Mexicans and foreigners. Tourism is based on extraordinary attractions including pre-Columbian, Colonial, and contemporary sites, although geotourism is relatively recent. However, tourists who visit these attractions stay only a few hours at the most. The start-up of activities in the Geopark, in operation since 2014, has contributed to increase the number of visitors, who now stays an average of 4 days. This trend has increased in recent years, reaching to about 1400 visitors in 2017 specifically seeking geotourism sites, about 10% being foreigners. Although these numbers are still low, the trend is on the rise and the recognition by UNESCO has largely contributed; a boost in the number of visitors is expected in the near future.

There is a set of facilities available in the Geopark, which include a visitor's center, museums, and historical monuments; geotourism information is also available in all nine municipality halls. Ten geotrails have been established to allow people to visit geosites and other locations of interest. Visitors are guided along these geotrails by duly trained local people, who provide information about the geological heritage and its relationship with the ecological, historical, archaeological, and cultural heritage (Fig. 17.7).

The Mixteca Alta Geopark Project was launched in 2013 from the proposal of the municipal authorities of Santo Domingo Yanhuitlán, where the Geopark is located, to which other eight municipalities joined. This initiative also involved various groups in the community, including business owners, schools from preprimary to senior high school, hoteliers, restaurateurs, transporters, artisans, and so on. Community work is preserved, being a cornerstone to ensure the Geopark success; all activities and proposals are discussed in meetings, where specific agreements are reached to assign persons/areas in charge of each of them. This governance modality, derived from local traditions and practices, is particularly common in indigenous communities in Mexico, as in the case of the Geopark territory.

FIGURE 17.7 Local guides of the Mixteca Alta Geopark (Photo: J. L. Palacio).

17.2.4 COMARCA MINERA, HIDALGO UNESCO GLOBAL GEOPARK, MEXICO/SECRETARIAT OF TOURISM, AND CULTURE OF THE STATE OF HIDALGO

The Geopark is located in Central Mexico, in the State of Hidalgo. It consists of nine municipalities that jointly cover an area of 1600 km^2. It is characterized by a rugged relief, with altitudes in the range of 1300–3200 meters above sea level, and is crossed by the volcanic ranges of Pachuca and *Las Navajas*. The mountainous area is limited to the north by the Meztitlán canyon, a breathtaking gorge that drains into the Gulf of Mexico. The physiographic variation and altitudinal range of the Geopark territory results in a mosaic of diverse landscape and vegetation types. On the one hand, mountain ranges covered by oyamel fir (*Abies religiosa*) and subtropical pine-oak forests, whereas the northern gorges exhibit columnar cacti-rich shrub lands.

The Geopark has an outstanding geodiversity. It encompasses the junction of two geological provinces, the Trans-Mexican Volcanic Belt and the *Sierra Madre Oriental* thrust and fold belt, as well as the overlap of two magmatic suites. World-class silver and gold deposits occur, which were exploited from the 16th century to the late 20th century (Probert, 1987). It is estimated that almost 6% of the world's historical silver production was mined from this area.

Besides the ore deposits, another remarkable element is *Cerro San Cristobal*, the type locality for two key minerals: tridymite and cristobalite, discovered by Gerhard vom Rath in 1868 and 1887, respectively. In addition, this Geopark is well-known for the *Santa María Regla* basaltic columns, which are among the longest reported worldwide (Sánchez-Rojas and Osorio-Pérez, 2008), with heights above 40 m (Fig. 17.8). The first traveler that came from afar to admire and study this natural wonder was Alexander von Humboldt, who stayed in *Comarca Minera* between 1803 and 1804. Volcanism in *Comarca Minera* produced a variety of structures including cinder cones, dacite domes, lava flows, obsidian deposits, strato-volcanoes, and a caldera structure. Fracturing and erosion have sculpted the volcanic deposits into scenic rock formations, such as those of *El Chico* National Park and *Peñas Cargadas*, both popular touristic destinations.

FIGURE 17.8 Basaltic columns at Santa María Regla in Mexico (Photo: C. Canet).

Human presence in *Comarca Minera* dates back to the pre-Columbian period, when the Xaltocan Kingdom expanded between 1220 and 1380 A.D. The Otomí people dominated the Metztitlán, Atotonilco el Grande and Tutotepec areas. During the 15th century, the Aztecs settled down in Huejutla and Patlachihuacán, now Pachuca, the State capital city. The Spaniard conquerors arrived in the State of Hidalgo shortly afterwards. In 1552, once the mining potential of the Real del Monte mines was discovered, a community settled down nearby. A five-century mining history left a cultural legacy that is deeply rooted in the identity of the region. The imprint of mining is evident not only in the modified landscape, but also in the rich industrial heritage that remains, including superb mining *haciendas* (i.e., colonial metallurgy plants). Moreover, the flourishing mining industry drew the attention of scientists, artists, and explorers, and prompted the migration of miners, mostly from Cornwall, United Kingdom of Great Britain, and Northern Ireland. Cornish miners arrived in the late 19th century and introduced locals to football—which was played for the first time in the country in the towns of the *Comarca Minera*—and *pastes*, a local version of the Cornish pasty.

Nowadays, the *Comarca Minera* Hidalgo UNESCO Global Geopark territory is home to over 500,000 inhabitants, most of them concentrated in the City of Pachuca.

The territory of this Geopark has a well-established tourism infrastructure (Canet et al., 2017). There are three *Pueblos Mágicos* (magical towns, Fig. 17.9)—a denomination granted by the Federal Secretariat of Tourism aimed at the conservation and tourism promotion of towns with an outstanding cultural heritage. One of these towns, Huasca, is the administrative seat of the Geopark and the main promotion office for geotourism activities, including tours to visit sites of geological, natural, and cultural interest. The presence of important cities in the territory and its proximity to Mexico City (about 70 km to the south) make of the Geopark a major tourist destination, being visited by more than two million tourists each year.

The management structure of this UNESCO Global Geopark has been designed to warrant the participation of citizens, with a 22% representation in the steering committee. This participation is represented by *ejido* (communal land) communities, who actually manage most of the 31 geosites identified. The Comarca Minera Hidalgo UNESCO Global Geopark is supported by the Geophysics Institute at *Universidad Nacional Autónoma de México* (UNAM). The Institute considers the Geopark as a tool to carry out public outreach and awareness activities on geohazards and environmental geosciences delivered to the general public.

FIGURE 17.9 Mineral el Chico, one of the magical towns within the Comarca Minera Geopark (Photo: C. Canet).

Hidalgo's *Comarca Minera* (mining region) has a long tourism tradition, which was consolidated in 2001 when Huasca de Ocampo was recognized as the first "Pueblo Magico" (Magical Town) at the national level, a successful program developed by the Secretariat of Tourism of Mexico that includes 111 such nominations throughout the country (http://www.pueblosmexico. com.mx/). Today, *Comarca Minera* has two additional towns also included in the program: *Real del Monte* and *Mineral del Chico*, recognized as *Pueblos Mágicos* in 2004 and 2014, respectively.

In this context, the main strategies for local development in the Geopark municipalities are based primarily on tourism and aim to improve quality and diversify offer, with geotourism as an emerging option since the declaration of the UNESCO Global Geopark in 2017. Some examples are:

(a) sustainable trade with the participation of local communities (sustainable orchards and trade of handcrafts made from recycled materials, both in *Mineral del Chico*; pottery workshops in *Huasca de Ocampo*; (b) recovery and valorization of the cultural heritage (*"Vigías del Patrimonio Cultural"* (Cultural Heritage Watchdog) Program of the Secretariat of Tourism of Hidalgo); (c) new geotourism tours (cultural tours in *Omitlan of Juarez*; visits to the mining neighborhoods of the city of Pachuca, which values the industrial mining heritage); and (d) citizen science and geo-education projects (involving the participation of the *Omitlan of Juarez* high school in the GGN Youth Instagram project).

17.2.5 COLCA Y VOLCANES DE ANDAGUA ASPIRING GEOPARK

This Geopark project is located in the Central Andes, southern Peru, in the northwestern Department of Arequipa. This area is visited by more than 250,000 tourists annually (Autocolca, 2018) and is one of the most important tourism destinations in Peru. It covers an area of 6010 km^2 and includes 19 districts in the provinces of Caylloma, Castilla, and Condesuyos. The elevation in the Geopark ranges from about 800 m above sea level (a.s.l.) at the confluence between the Colca River and the Andamayo River in the southwest to more than 6200 m (a.s.l.) at Ampato Volcano. The Geopark includes "the deepest canyon on Earth" with the famous *Cruz del Condor* site (Fig. 17.10), large and active volcanoes, and the extensive and breathtaking Colca Valley (see Fig. 17.12) with abundant hot springs and active crustal faults some 270 km east of the trench where the Nazca Plate is sub-ducted beneath the South American Plate.

FIGURE 17.10 Majestic condor seen in the famous *Cruz del Condor* geosite (Photo: K. Gaidzik).

The Geopark area is divided into three main zones with different landscapes and characteristics:

(1) Colca Canyon, (2) Colca Valley, and (3) Valley of Volcanoes. It highlights a high geodiversity mainly represented by volcanic landscapes with stratovolcanoes, volcanic domes, and lava flows, thermal springs, glaciers, lakes, scarps associated with active tectonic faults and landslide processes, deeply incised river valleys and gorges (Żaba et al., 2009; Paulo et al., 2014; Gałaś et al., 2015, Rendón and Bidwell, 2015; Zavala and Churata, 2016; Zavala et al., 2016). The high geodiversity of the area is corroborated by the recognition of 119 geosites with high geodidactic, scientific, and esthetic values classified as geomorphological, volcanic, structural, tectonic, sedimentological, geodynamic, mining, paleoclimatic, hydrogeological, stratigraphic, paleontological, neotectonic, and geochronological (Zavala and Churata, 2016; Zavala et al., 2016). The undisputable scientific and esthetic values of these geosites were confirmed by the selection of a photograph of one such geosites (fold structures from Ayo) as the picture of the month by the Journal of Structural Geology (Żabaet al., 2014). The local dry climate has resulted in the unparalleled exposure of the geological record from Precambrian rocks to the present, especially on deeply incised slopes of the Colca Canyon, making of the area a perfect spot for geoeducational purposes.

Pre-Columbian and colonial remains (26 archaeological sites: *chullpas*, citadels, agricultural platform, and terraces from Inca and pre-Inca times, mainly related to ancient Collaguas and Cabanas communities (Fig. 17.11); 20 colonial churches and mining sites, irrigation systems (Zavala and Churata, 2016), high biodiversity (including the Andean Condor—an emblematic national bird—Andean cat, guanaco, vicuñas, and so on), and intangible cultural heritage (cultural-religious practices and traditions, such as clothing, agriculture, folk festivals, handicraft, and dances, represented by Qamiliand Wititi or Wifala dances recognized by UNESCO as intangible cultural heritage of the humanity) significantly enrich the Geopark offer.

The geotourism and geo-educational potential of the Geopark, represented by its geodiversity, geoheritage, biodiversity, and cultural heritage, has been broadly documented (Paulo et al., 2014; Gałaś et al., 2015, Zavala and Churata, 2016; Zavala et al., 2016). However, the need for geoconservation and tourism development has frequently led to conflicts (Rendón and Bidwell, 2015) mostly related to mining, construction of a road network, transmission of high-voltage electricity, water management, and non-regulated tourism. All these issues could severely affect geotourism and geoconservation in the Geopark. Therefore, sound and shared management are imperative to deal with these issues. On the other hand, its location in a seismically active zone makes this area vulnerable to powerful earthquakes and catastrophic land-waste processes that might threaten not only local communities but also tourism activity in this area.

FIGURE 17.11 Colca Valley in the Achoma area, view to the north (Photo: K. Gaidzik).

The worldwide fame and recognition of the Colca Canyon owes to the Polish team Canoandes, who in 1981 rafted down the Colca River. The story of this achievement, together with the recognition of the Colca Canyon as the deepest canyon worldwide (3223 m deep), was later published in the National Geographic Magazine and the Guinness Book of Records (1984). Its slogan as "the deepest canyon on Earth" boosted the number of visitors and tourists seeking for beautiful landscapes and a taste of adrenaline in the deepest canyon. As a result, the Colca Canyon became one of Peru's most popular tourist destinations in the 1990s and still ranks among the top destinations, along with Machu Picchu and Cuzco. Local governments of the districts located along the Colca River acknowledged the potential of this region and in 1986 created AUTOCOLCA (*Autoridad Autónoma del Colca y Anexos*; Autonomous Authority of Colca and Adjacent Regions), an autonomous public entity responsible for managing tourism development in the region. This organization was set to focus on restoration, protection, development, use, and promotion of natural, archaeological and historical heritage, and to manage economic resources within the newly created Tourist Circuit (Autocolca, 2018). Since 2003, members of the Scientific Expedition to Peru (PSEP) launched a research project aimed at creating a national park in this region ("the Colca Canyon and Valley of the Volcanoes National Park"). This work confirmed the high geodiversity and unique landscapes in the area, pointing to the importance of the dry climate that unveiled the geological and morphological structures across an extensive area devoid of vegetation (Paulo and Gałaś, 2008; Żaba et al., 2009; Gałaś et al., 2015, Zavala and Churata, 2016). In 2006, INGEMMET (*Instituto Geológico, Minero y Metalúrgico*; Geological Mining and Metallurgical Institute of Peru) launched a Heritage and Geotourism program focused on documenting geological heritage resources and identifying valuable areas for promotion and protection. Members of this research team also acknowledged the unique values of the territory, but they opted for a geopark instead of national park for this area. The idea of establishing the Colca Canyon and Valley of the Volcanoes Geopark was first announced in the workshop "Geoparks and geoheritage; promoting geoheritage in Latin America" held in Mexico City (Zavala, 2015). The event focused on and was related to the organization and establishment of the Latin American Geopark Network, but it also provided an opportunity to show the geotourism potential of the Colca Canyon and Valley of the Volcanoes area to the international scientific community. *Cañón de Colca and Volcanes de Andagua* is currently under evaluation by UNESCO.

17.3 REGIONAL INITIATIVES FOR THE PROMOTION OF GEOTOURISM

The *Grutasdel Palacio* UNESCO Global Geopark, together with the School of Sciences at *Universidad de la República*, Uruguay, jointly launched the "Geotourism Week" and the "Geotourism Day" initiatives. Starting in the *Grutas del Palacio* Geopark in 2016, these events currently involve all Latin American Global UNESCO Geoparks, Aspiring Geoparks and several similar projects in Ecuador, Chile, Nicaragua, Mexico, Argentina, Brazil, Bolivia, and Peru (Fig. 17.12).

These initiatives seek to promote knowledge on Earth Sciences among the general public and include activities for children and youngsters on issues related to each of the geoparks, as well as for the promotion of local products. These events reach several thousands of visitors and are considered to be one of the main promotional activities of geoparks in the region.

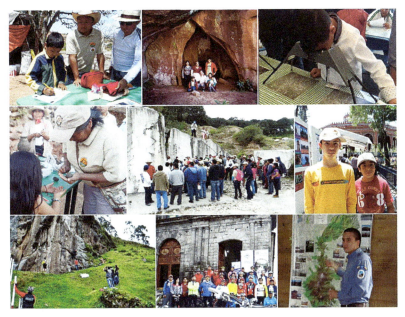

FIGURE 17.12 Geotourism Week 2017; Promoting geoactivities in Latin American Geoparks (Grutas del Palacio Newsletter No. 4/June/2017).

17.4 CLOSING REMARKS

Geotourism displays a growing dynamism and acceptance in various Latin American territories, although with striking differences in their development. Latin American Geoparks are contrasting territories, some with a long tourism tradition, extensive infrastructure and visitors that reach millions (e. g., Araripe, Comarca Minera, and Grutas del Palacio geoparks), while others are socially and economically marginalized, with scarce tourism infrastructure and a number of visitors barely reaching a few thousands per year, as in the case of the Mixteca Alta Geopark.

Nonetheless, geotourism is uniquely relevant to Latin America by representing a true development strategy in rural territories currently characterized by limited social and economic progress that have led to emigration. Geotourism contributes to the well-being of residents and raises awareness about the importance of the geological and geomorphological elements among visitors. In territories already consolidated as tourism destinations, the dissemination of knowledge on Earth Sciences as a recent added value is worth noting, which has expanded and strengthened the traditional tourism offer in these territories.

On the other hand, despite the limited number of Geoparks in the vast Latin American region, the projects underway in a dozen countries evidence the growing interest that identifies geotourism as a key element for development. Therefore, it can be affirmed that this tourism modality has enormous perspectives.

The region currently faces a number of challenges, such as the lack of a legal framework for the creation of geoparks and the protection of nature focused almost exclusively on biodiversity. This scenario, on the other hand, is not unique to Latin America, although some countries have taken important steps in this regard. The scarce recognition of the importance of Earth Sciences in formal educational schemes in various Latin American countries is another major challenge that should be addressed; in this regard, Geoparks undoubtedly represent a valuable alternative to manage these limitations.

Finally, the incorporation of Latin American territories into the Global Geoparks Network represents a relevant contribution to the UNESCO initiative. The geoparks model, originally developed in Europe, was not only imported but has been tailored to the unique conditions of the Latin American region, and has resulted in its conceptual enrichment. This is a key characteristic of the UNESCO model, through which various social contexts contribute to consolidate a truly global program.

KEYWORDS

- geotourism
- Latin American UNESCO global geoparks
- Latin American and Caribbean geopark network
- aspiring geoparks
- geoheritage

REFERENCES

Arouca. *Arouca Declaration on Geotourism*, Arouca Geopark, Portugal. November 12, 2011.

Autocolca. La Autoridad Autónomadel Colca y Anexos; The Autonomous Authority of Colca and Annexes Regions (2018), http://www.autocolca.pe/ (accessed Jan 08, 2018).

Canet, C.; Mora-Chaparro, J. C.; Iglesias, A.; Cruz-Pérez, M. A.; Salgado-Martínez, E.; Zamudio-Ángeles, D.; Fitz-Díaz, E.; Martínez-Serrano, R. G.; Gil-Ríos, A.; Poch, J. Cartografía geológica para la gestión del geopatrimonio y la planeación de rutas geoturísticas:

aplicación en el Geoparque Mundial de la UNESCO Comarca Minera, Hidalgo. Terra Digitalis, 2017, v. 1; pp 1–7.

Gałaś, A.; Paulo, A.; Gaidzik, K.; Kalicki, T.; Ciesielczuk, J.; Żaba, J.; Radwanek-Bąk, B.; Gałaś S.; Krupa, J. Propuestas de geositios y atracciones geoturísticas del futuro Parque Nacional Cañón del Colca y el Valle de los Volcanes. In *Primer Simposio de Geoparques*. INGEMMET, Arequipa, Peru 14–17.07.2015.

GGN. Global Geoparks Network Mission (2018). http://globalgeoparksnetwork.org/?page_id=202 (accessed Feb, 2018).

Hose, T. A. Selling the Story of Britain's Stone. *Environ. Interpret.* **1995**, *10* (2), 16–17.

Leigh, D. S.; Kowalewski, S. A.; Holdridge, G. 3400 Years of Agricultural Engineering in Mesoamerica: Lama-bordos of the Mixteca Alta, Oaxaca, Mexico. *J. Archaeol. Sci.* **2013**, 40, 4107–4111.

Llorente-Bousquets, J.; y S. Ocegueda. Estado del conocimiento de la biota, in Capital natural de México, Vol. I: Conocimiento actual de la biodiversidad. Conabio, México, 2013; pp 283–322.

Palacio-Prieto José Luis; Emmaline Rosado-González; Xóchitl Ramírez-Miguel; Oralia Oropeza-Orozco; SilkeCram-Heydrich; Mario Arturo Ortiz-Pérez; José Manuel Figueroa-Mah-Eng; Gonzalo Fernández de Castro-Martínez. Erosion, Culture and Geoheritage; the Case of Santo Domingo Yanhuitlán, Oaxaca, México Geoheritage, 2016. DOI: 10.1007/s12371-016-0175-2.

Paulo, A.; Gałaś, A.; Gałaś, S. Planning the Colca Canyon and Valley of the Volcanoes National Park in South Peru. *Environ. Earth Sci.* **2014**, *71*, 3, 1021–1032.

Probert, A. En pos de la plata. Episodios Mineros en la Historia Hidalguense: Pachuca, Hidalgo, Colección Hidalguense, Dirección General de Publicaciones e Impresos del Gobierno del Estado de Hidalgo, 1987; pp 446.

Rendón, M. L.; Bidwell, S. Success in Progress? Tourism as a Tool for Inclusive Development in Peru's Colca Valley. In *Tourism in Latin America*, Panosso Netto, A., Trigo, L., Eds.; Springer: Cham, 2015; pp 207–233.

Sánchez Rojas, E.; Osorio Pérez, M. Geología y petrogénesis de los Prismas Basálticos. Santa María Regla, Hgo. *GeoCiencia, Rev. Serv. Geol. Mex.* **2008**, *3*, 5–24.

Spores, R. Settlement, Farming Technology, and Environment in the Nochixtlan Valley. Science **1969**, *166*, 557–569.

Zavala, B. Presentación de propuesta de geoparque cañón del Colca y valle de los volcanes de Andagua en el Workshop: Geoparques en Latinoamérica, proomoviendo la creación de geoparques en Latinoamérica. México, 28 y 29 de mayo 2015. INGEMMET.

Zavala, B.; Churata, D. Colca y Volcanes de Andagua Geopark, Arequipa, Perú: Application dossier fornomination as geopark. Geological Heritage, Document process for UNESCO Global Geoparks aspiring, information prepared by INGEMMET, 2016; pp 9–26.

Zavala, B.; Mariño J.; Peña F. Guía geoturística del valle de los volcanes de Andagua. (in Spanish) INGEMMET, Boletín, Serie I: Patrimonio y geoturismo, **2016**, *6*, 424.

Żaba, J.; Ciesielczuk, J.; Gaidzik, K. Structural Position of Huambo River valley (Central Andes, Peru) and its Geoeducational Aspects. StudiaUniversitatis Babes-Bolyai, Geologia, Special Issue—MAEGS **2009**, *16*, 72–74.

Żaba, J.; Gaidzik, K.; Ciesielczuk, J. Photograph of the Month. *J. Struct. Geol.* **2014**, 65, *123* https://doi.org/10.1016/j.jsg.2014.01.003 (accessed Feb, 2018).

PART V
Globalization and the Future of Geological Attraction Destinations

CHAPTER 18

Dinosaur Geotourism: A World-Wide Growing Tourism Niche

NATHALIE CAYLA

Docteur en Sciences de la Terre, de l'Univers et de l'Environnement, EDYTEM, Université Savoie Mont-Blanc, CNRS Pôle Montagne, 73376 Le Bourget du Lac, France

E-mail: Nathalie.cayla@univ-smb.fr

ABSTRACT

Dinosaurs are a very popular science topic throughout the world, and paleontological tourism highlighting dinosaur paleontological sites is booming. Dinosaur outcrops, "hot spot" of paleontological discoveries, are valorized into geotourism destinations contributing to local economic development based on a specific territorial resource: the paleontological resource. Thus, since the 2000s, no less than 30 new dinosaurian destinations have emerged. The main infrastructures of this touristic offer are open-air or field museums, interpretive centers, local museums, or palaeontological parks that value all these discoveries as closely as possible. Stakeholder network graphs analysis shows that many factors fosters the success of these tourism projects. Beyond the scientific and/or spectacular quality of the paleontological remains, the networks of actors involved, and the scale of these networking are key factors to trigger a sustainable tourism development. Several case studies are presented illustrating the diversity of strategies used to develop these projects. Whatever the strategy adopted, the tourist trajectory of dinosaurian sites must above all respect the "sense of place" allowing the visitor to apprehend a fascinating world forever gone.

18.1 INTRODUCTION

Dinosaurs are considered as an interesting scientific topic, and paleontological sites around the world have led to the promotion of paleontological tourism. About 30 different dinosaur-based tourism destinations have appeared since 2000. Those, whether they are redeveloping geosites that have been known for a long time or showcasing new discoveries are very attractive. For example, since its inscription in 2000 on the UNESCO World Heritage List for its rich triassic fauna, the number of tourists visiting Ischigualasto Regional Park in Argentina has steadily increased from barely 10,000 to nearly 90,000 in 2016. This kind of geotourism is already ancient. The first palaeontological site valorized for tourism was that of Fossil Grove, which since 1887 welcomes the public in the heart of Glasgow to discover the remains of a 330 million year old forest (Gordon, 2016). Geotourism can take a wide variety of forms, hence the need to delineate its boundaries (Reynard et al., 2003; Dowling and Newsome, 2006; Hose, 2008; Cayla, 2009; Sadry, 2009). When dinosaur replicas were first shown to the public at Crystal Palace Park in London in 1854, the aim was to enhance the latest scientific discoveries from a world that already fascinated a wide public. Gradually, throughout the 20th century, the major great cities of the world have incorporated paleontological exhibitions into their tourism offer, developing impressive scenography such as the 10 skeletons of iguanodons presented at the Royal Belgian Institute of Natural Sciences in 1902, which have since attracted no less than 300,000 visitors a year. These two examples illustrate the ex situ valorization of paleontological heritage. They can attract a wide audience and engaged new public to dinosaur science but they don't create new tourism destinations. The birth of dinosaurian tourism itself is more recent and mainly based on the in situ development of a geosite for tourism purposes. It can be dated to 1957, when the Carnegie Quarry, the first in situ exhibition dedicated to this theme, was inaugurated in Utah (Fig. 18.3). Since then, numerous dinosaur remains, "hot spot" of paleontological discoveries, have been valorizated into geotourism destinations (Kadry, 2011) thus contributing to local economic development based on specific territorial resources: the paleontological resources (Landel and Senil, 2009; Peyrache-Gadeau et al., 2016; Hobléa et al., 2017). This chapter aims to understand how, over time, is elaborate the attractiveness of these destinations. After presenting a typology of dinosaurian outcrops, several case studies will illustrate the geotourism trajectories of these sites. The methodology used is based on the analyses of stakeholders networks graphs and their diachronic evolution (Mounier, 2015). The web crawler Hyphe is used to harvest the content and hyperlinks of the web pages of these actors then a graph analysis is done with the software Gephi. These investigations show that geotourism clusters have been developed creating business ecosystems whose spontaneous or constructed origin may explain the greater or lesser success and/or sustainability of these destinations (Fabry, 2009).

18.2 THE SCOPE AND TYPOLOGY OF THE DINOSAURIAN GEOTOURISM

Midway between ecotourism and cultural tourism, widening the heritage and tourist offer, geotourism is a tourism niche that has been asserting itself since the 2000s in territorial development strategies (Cayla and Duval, 2013). According to the authors, the definition of geotourism may vary. Thus, Newsome and Dowling identify it with traditional tourism practices applied to specific sites: "*a specialized form of tourism in that the focus of attention is the geosite*" (Newsome and Dowling, 2006; Newsome and Dowling, 2010), on the contrary, Pralong highlights the importance of the creation of specific interpretive tools: "*A set of practices, infrastructures and products to promote Earth sciences through tourism*" (Pralong, 2006) while Sadry (2009) synthesizes both approaches in: "*Geotourism is a knowledge-based tourism, an interdisciplinary integration of the tourism industry with conservation and interpretation of abiotic nature attributes, besides considering related cultural issues, within the geosites for the general public.*"

The key elements of this paleontological geotourism are in situ valorization (Cayla, 2009) also qualified of external (Perini and Calvo, 2008) as opposed to the ex situ or classical valorization of palaeontological geoheritage. The main infrastructures of this touristic offer are the field or open air museums, interpretive centers, local museums, or palaeontological parks that value all these discoveries as closely as possible. They must enhance the "sense of place" of the local paleontological heritage in an in-depth touristic experience.

18.2.1 A GEOTOURISM PROMOTING IN SITU DINOSAURS GEOSITES

Open air or field museums are facilities that provide in situ interpretation of heritage. The first geological open air museum was opened in Lucerne, Switzerland, where the "glacier garden" was inaugurated in 1874. It presents giant's kettles witnesses of the last glaciation (Cayla, 2010). Since that time, this site, in the heart of the city, has welcomed an average of 100,000 visitors every year.

Three types of deposits characterize the dinosaurian sites:

The dinosaur graveyards, enclosing many skeletal parts, are enhanced by the selective uncovered of the most impressive bones. Discovered in 1909 by Earl Douglas, it took nearly 50 years for the Carnegie quarry, on which we will focus later in this chapter, to inaugurate its visitor center. More recently, in 1972, in a building site of the suburb of Zigong, Sichuan, an outstanding fossil deposit was discovered. It delivered more than 20 new dinosaur species, and immediately mobilized scientific communities and public stakeholders for a geotourism development project. The Zigong dinosaur museum was inaugurated in 1987, barely 15 years after the discovery of this exceptional paleontological site. Today, it receives more than 1 million visitors a year. This example illustrates the increasing interest of policy-makers in the geoheritage resources for tourism development.

The nesting areas offer the opportunity to observe nests and/or eggs of dinosaurs, mostly dating from the Upper Cretaceous. The first discoveries were made in France in 1846 (Taquet, 2001), the Sainte-Victoire National Nature Reserve, created in 1994, protects the Roques-Hautes fossil deposit, but it is forbidden for public access. The famous site of Flamming Cliffs in Mongolia, is now protected in the Bayanzag park and already valorized by an original project: the moveable dinosaur museum. In 2010, in the Sanagasta Valley, northwestern Argentina; a dinosaur hatchery composed by many nests containing eggs associated to a paleo-hydrothermal field was discovered. The "Sanagasta Parque de Dinosaurios" opened in 2014 and has just inaugurated a theme trail where a dinosaur nest is presented in situ. In these cases, dinosaur geosites provided new drive tourism services to sustain economic development of remote regions.

The footprints and track ways left by dinosaurs are numerous throughout the world and new sites are frequently discovered. The main interest of these tracks is to revealed part of the behavior of these animals. The Lark Quarry site in Queensland, Australia presented more than 3000 footprints left by nearly 150 animals 95 million years ago at the very beginning of the Upper Cretaceous. Initially interpreted as a herd fleeing a predator, the new analyses of all footprints reveal more banal scenes of animal movements in a more or less deep lake (Romilio and Salisbury, 2011; Romilio et al., 2013). Discovered in 1971, the site was first developed for tourism in 1979. The interpretive center, which was renewal in 2002, receives an average of 10,000 visitors per year. In central Bolivia, the site of Cal Orck'o, near the town of Sucre, was discovered in 1985, it is the largest current deposit of dinosaur footprints in the world with more than 5000 footprints left by about 15 species 68 million years ago (Meyer et al., 2001; Lockley et al., 2002). The "Parque Cretacico," which was inaugurated in 2006, has become in a few years the most popular tourist attraction in the country, welcoming more than 170,000 visitors in 2015. The mentioned study cases demonstrate that the scientific interest of an outcrop is only a small part of its touristic success.

In situ valorization of geoheritage contributes to the economic development of new tourism destinations, sometimes isolated and remote from major tourist centers (Laws and Scott, 2003). These sites require substantial infrastructures, in terms of accessibility, for example, as well as the creation of a secondary tourist offer that is essential to ensure attractiveness of primary supply (Reynard, 2003). These infrastructures, as well as new functionalities attributed to these areas, which are sometimes already used for other purposes, can lead to local conflicts (Prentz, 2015).

A special category must be made (Perini and Calvo, 2008) for active excavation sites where the visitor observes live work of paleontologists and may even participate in it. One can then speak of experiential tourism since the visitor will be able to interact directly with the researchers.

18.2.2 TERRITORIAL RELAYS ESSENTIAL TO CAPTURE TOURIST FLOWS

Open air museums, main topics of the previous paragraph are, in most cases, well identified by dinomaniac tourists. However, in order to attract a wider public, it is necessary to capture a greater flow of local visitors by creating edutainment infrastructures like interpretive centers and/or museums at strategic points in the territory. Their purpose will be to showcase the interest of the remote sites and to redistribute part of the tourists in transit.

The diversity of interpretive centers, exclusively dedicated to dinosaurs, is very important, revealing, beyond the common intention, the range of financial means mobilized as well as the spatial scale of touristic attractivity considered. These infrastructures, most of the time supported by various levels of public authorities, represent substantial investments which are justified only in a perspective of tourism development that must, to succeed, involved all the actors of the local economy. The two most famous interpretive centers of this type are the Fukui Prefectural Dinosaur Museum in Japan, which has attracted close to 800,000 visitors annually since it inauguration in 2000, and the Royal Tyrrell Museum in Alberta, which has welcomed an average of 450,000 visitors annually since 1985. The first one is the gateway to the Fukui Katsuyama Dinosaur Valley Geopark, which joined the Japanese Geopark Network in 2009. The second is the repository of fossils discoveries done in the Provincial Dinosaur Park, which was inscribed on UNESCO's

World Heritage List in 1979. Throughout the 20th century and even more importantly for the last 20 years, paleontological tourism destinations have emerged worldwide (Figs. 18.1 and 18.2). This touristic segment market led also to the establishment of entertainments dinosaurs theme parks, which are excluded from the study presented here.

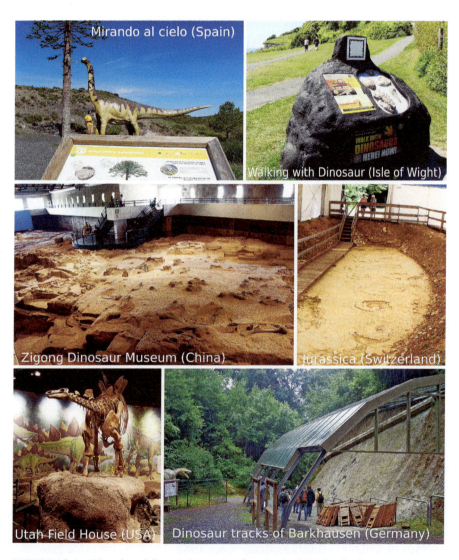

FIGURE 18.1 Diversity of dinosaurian geotourism destinations (N. Cayla).

All these projects emerged from networks of public and private stakeholders at different territorial levels. The relationships created between these actors offer the opportunity to understand the success factors of these projects gathering actors from different backgrounds: researchers, politicians, and tourism actors. A web-crawler software has been used to collect data complemented by more traditional face to face inquiries that highlighted a variety of strategies.

18.3 PALEONTOLOGICAL DESTINATIONS EVOLVING WITH SCIENTIFIC DISCOVERIES

Palaeontological discoveries, resulting from scientific investigations, natural erosion of land, mining and quarry activities, or infrastructure construction due to growing urban development, are regular (Tomic et al., 2014). They can initiate new geotourism destinations projects, expand the reach of existing destinations, or even reduced the interest of a paleontological site in favor of more interesting ones. The stakeholders must therefore constantly adapt their tourism strategies.

Since the 1870s, numerous palaeontological discoveries have been made in a vast territory which extends from New Mexico to Canada. This North American Jurassic Park is linked to the outcrop of the Morrison Formation, which was deposited in the upper Jurassic, around 150 million years ago in an ecosystem dominated by the dinosaurs. On August 17, 1909, the paleontologist Earl Douglas, discovered, on the border of Utah and Colorado, an almost complete dinosaur skeleton. Described by William Holland, director of the Carnegie Museum in Pittsburgh, he became the Holotype (CMNH 3018) of *Apatosaurus Louisae*, named after the wife of the museum's founder, and his mounted skeleton was presented to the public in 1936. As soon as the discovery was announced and throughout the excavation period, many local visitors came to discover the site, first locally and then from neighboring states (Carpenter, 2018). For its "great scientific interest and value," President Woodrow Wilson protected the quarry as a National Monument on October 4, 1915. Dinosaur fossil excavations undertaken by the Carnegie Museum, ended in 1922, while the universities of Utah and Michigan, as well as the Smithsonian, continued to collect fossils for another 2 years. In the 1930s, supported by national funding, a first draft of the future field museum was initiated by the paleontologist Barnum Brown of the National Museum of Natural History. The current center, built above the fossil outcrop, was

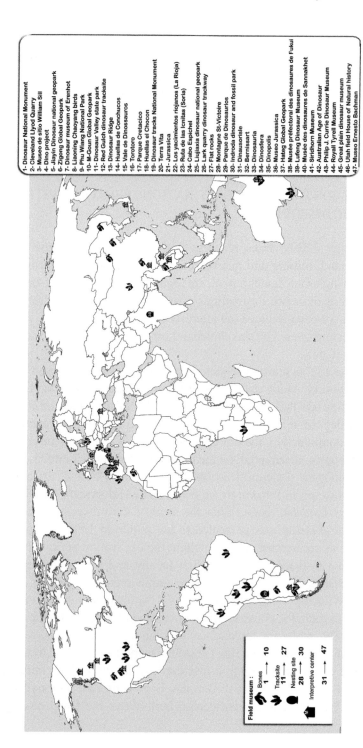

FIGURE 18.2 Overview of main worldwide dinosaurian geotourism destinations (N. Cayla).

inaugurated in 1957. It cost $309,000 ($2.77 million in 2018)[1] financed by the Mission 66, a project of the National Park Service (Allaback, 2000). Since that time, despite a closure from 2006 to 2011 due to a damage on a part of the building, the field museum has welcomed more than 20 million visitors (NPS stats 2016).

FIGURE 18.3 Dinosaur National Monument—Carnegie quarry visitor center, Utah, the United States. (a) First permanent visitor center (Library of Congress, 1958); (b) Quarry exhibit hall (N. Cayla, 2016).

In 1946, as the collections from the Carnegie Quarry expanded, the Governor of Utah Herbert B. Maw allocated $200,000 ($2.58 million in 2018—http://www.in2013dollars.com) to establish the Utah Field House of Natural History in Vernal, some 20 miles away. G.E. Untermannaranger of the Dinosaur National Monument was appointed director of the new museum which was achieved in 1948. Initially managed by the Department of Industrial and Commercial Development, it became part of the Utah State Parks and Recreation Department in 1959. After changing its name several times, it is called the "Utah Field House of Natural History State Park" since 1984. A close partnership between the Dinosaur National Monument and the museum has resulted in the creation of the "Uintah Research and Curatorial Center" in 2012, the museum becoming the official repository for DNM collections.

Simultaneously, the many discoveries made on the border of Utah and Colorado during the 20th century gathered new actors expanding the network of stakeholders.

In Utah, Cleveland-Lloyd's quarry was excavated in the 1920s by paleontologists at the University of Utah. It became a Natural National Landmark in 1965 and its administration was entrusted to the Bureau of

---
[1]http://www.in2013dollars.com

Land Management, which manages the interpretive center. The discoveries made in the quarry led to the creation of the Utah State University Eastern Prehistoric Museum in Price, a nearby locality. It opened in 1961 and displays mainly the mounted skeleton of an Allosaurus, the main specie found in the quarry.

Shortly afterwards, in Colorado, around Grand Junction and Fruita, a museum of natural and cultural heritage opened in 1965. In 2000, a dinosaur-specific building moved to Fruita, named "Dinosaur Journey." It offers numerous palaeontological activities relating to nearby excavation sites like Dinosaur Hill.

Faced with the growing numbers of geotourism points of interest on the theme of Dinosaurs, the local authorities of the two states scheduled, since the beginning of the 1990s, the creation of a "National Scenic Byway," a program launched in 1992 by the "U.S. Department of Transportation, Federal Highway Administration." In 1996, the Dinosaur Diamond Byway partners' association was created with more than 60 stakeholders in both states and the label was awarded in 2000 for a road of more than 500 miles which took the name of "Dinosaur Diamond Prehistoric Highway."

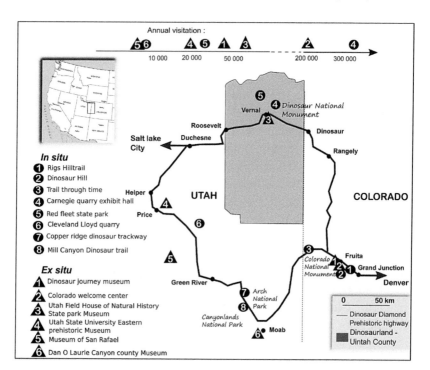

FIGURE 18.4 Map of Dinosaur Diamond Prehistoric Highway (N. Cayla).

It includes dinosaur geosites but more generally natural sites such as the Arch National Park or archaeological sites (Fig. 18.4). The partnership is gradually running out of steam and although it remains quite active in Colorado, it is no longer active in Utah and a marketing study is requested in 2007. It analyses the project's difficulties and recommends a few improvements. Despite this, little is then done at least on the scale of the entire road.

At smaller scale, however, a few projects emerged. For example, the county of Uintah in northeastern Utah, although home of the Dinosaur National Monument, is only in twentieth place among the 29 counties of the state regarding tourism activities (Leaver, 2017) and the unemployment rate reaches 8.3% of the population. In 2010, to promote the Vernal region as a family tourist destination, the public authorities set up Dinoland, a new dinosaur destination which gathered all the touristic actors. The associated marketing campaign was rewarded with numerous prizes and the county's tourist attendance increased by more than 11% between 2015 and 2016.

The web mining done with the software *Hyphe* crawled all the websites of the actors of the "Dinosaur Diamond Prehistoric Highway" clearly highlighting the importance of the cluster "Dinoland" in close connection with the Dinosaur National Monument. This key cluster gathers public and private actors linked to the local tourism economy. The others main nodes of the network show less relationship between policy-makers and tourism service providers. This may no doubt explain the current redistribution of dinosaurian tourism in these territories. Originally composed of isolated range destinations, the geotourism that emerged on the Utah-Colorado border has gradually progress into local networks. These networks then merged into a National Scenic Byway in the early 2000s. This spatial and thematic enlargement is now running out of steam and new local networks appear gathering all the tourism service providers to improve territorial marketing strategy and promote a sustainable tourism (Fig. 18.5).

18.4 GEOTOURISTIC DESTINATIONS RELYING ON AN ALREADY EXISTING NETWORK

Paleontological discoveries can be made on or near touristic sites. Their tourism development then benefits from the networks of actors already established and then participate in the diversification of the offer widening the scope of possibilities for the tourist.

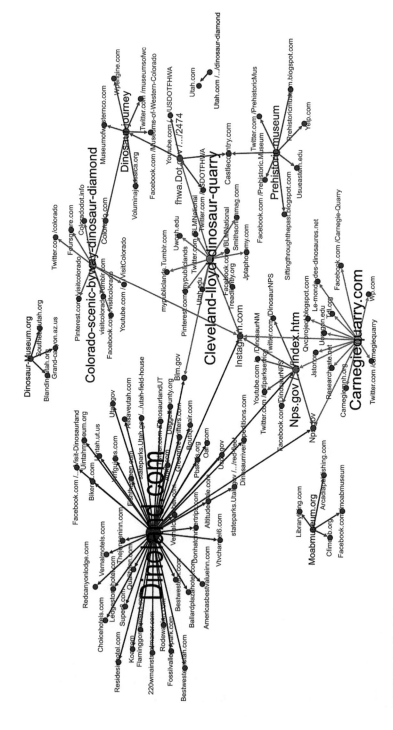

FIGURE 18.5 "Dinosaur Diamond Prehistoric Highway" stakeholders graph network (N. Cayla).

18.4.1 THE ISLE OF WIGHT GEOTOURISM, A NEW PERSPECTIVE FOR A FAMOUS SEA-SIDE DESTINATION

The Isle of Wight, originally a seaside destination in the south of the United Kingdom became a famous dinosaurian destination as progressed the fossil discoveries (Munt, 2008; Munt, 2016).

The tourism attractivity of the island began in 1826 with the first edition of the "Week of Cowes," the world's prestigious sailing event initiated by Georges IV. A few years later, East Cowes became the summer residence of the British royal family. The island was initially enjoyed, during the summer time, by the gentry and in the 1950s became a popular seaside destination for British people. Tourism is a key component of the island's economy, and a recent study by the NGO Natural England (Webber, 2006) showed that the 2.4 million tourist arrivals generate £352 million income. Alongside this touristic trajectory, the island progressively became one of the most famous dinosaurian destination in Europe and the incomes due to geotourism has been estimated at between 2.6 and 4.9 million, thus supporting the employment of 324–441 people. However, since the 2008 crisis, tourist numbers on the island have been declining, with fewer tourists staying for shorter periods (Simpson, 2018).

The south coast of the Isle of Wight exhibits several outcrops of the lower Cretaceous Wealden Group (140–125 Ma). The Wessex formation deposited in freshwater and floodplain environments, is overlying by the Vectis formation which shows very shallow lacustrine or lagoonal deposition with fluctuating salinities (Hopson, 2011). Throughout the 19th century, many fossil collectors such as Reverend William Fox discovered remains of new dinosaur species: *Aristo suchus*, *Hypsilophodon foxii*, *Polacanthus*. Most of these fossil bones have been dispersed in museum collections outside the island, so the Wight Island Council began its own collection which in 1914 became the first geological museum of the island located in Sandown. It is currently home to 40,000 specimens of which nearly 200 are types (Munt, 2008). In the 1970s, the management of the museum has moved to a professional team. At the same time fossicking became very popular on the island. New discoveries such as that of *Neovenator salerii* in 1996 and *Valdosaurus canaliculatus* in 2012 testify to the territory's ever-potential wealth. It explains the development of more commercial prospecting by owners of fossil shops. The growing importance of conflicts of interest between the common heritage of the island and commercial activities led to the establishment of guidelines for collecting fossil on the Isle of Wight in 1995. However, fossil collecting is not the only threat, cliff erosion is currently a major issue in protecting outcrops but it is also this phenomenon that allows new discoveries.

To support island geotourism, an action plan was developed in 2010 (Price and Jakeways, 2010). The central element of this action plan is Dinosaurisle, the new museum, still located in Sandown but closer to the discovery sites, allowing geosites visits in addition to the museum visit. Funded by the Millennium Commission, for a budget of £2.7 million, it opened in 2001 and attracts an average of 70,000 visitors per year. The museum team is also in charge of the "Regionally Important Geological/Geomorphological sites (RIGS)" of the Isle of Wight, a planning policy for the geoheritage protection. Another important actor is the "Dinosaur Expeditions, Conservation and Palaeoart Centre" which was founded in 2013. This private structure, succeeding the "Dinosaur Farm Center," presents an exhibition on the discovery of the island's dinosaurs which attracts a little less than 10,000 visitors but above all organizes about a hundred field trips each year in search of fossils but in the respect of a deontological approach. Several fossil shops take advantage of this tourist expectation to offer not only the sale of fossils and minerals but also numerous fossil hunts on many island sites.

The stakeholders network graph highlights the importance of the three above-mentioned actors but it also relies on many other public and private actors in the field of tourism as well as management and preservation of the island's natural heritage (see Fig. 18.6). This networking, which is undoubtedly based on long-standing tourism, explains the current success of this geotourism destination.

18.4.2 THE GEOPARK OF HAȚEG, A GLOBAL LABEL TO PROMOTE AN EMERGING TOURISM DESTINATION

With the end of the communist regime in the early 1990s, the tourism sector became an important economic activity in Romania. The country, with a population of some 20 million inhabitants, welcomes about 7 million tourists every year, mainly on cultural and seaside sites linked to mass tourism (Turtureanu, 2014). A more widespread tourism highlighting the country's natural and rural resources is gradually emerging in search of attractivity (Iorio and Corsale, 2010).

Being aware of the importance of the geoheritage to develop new tourist offer, the Progeo South-Eastern Europe working group meeting was held in 2003 in Romania. The drawing up of the Hațeg Declaration was the culmination of the meeting, underlining the role of governments in developing

Dinosaur Geotourism: A World-Wide Growing Tourism Niche   463

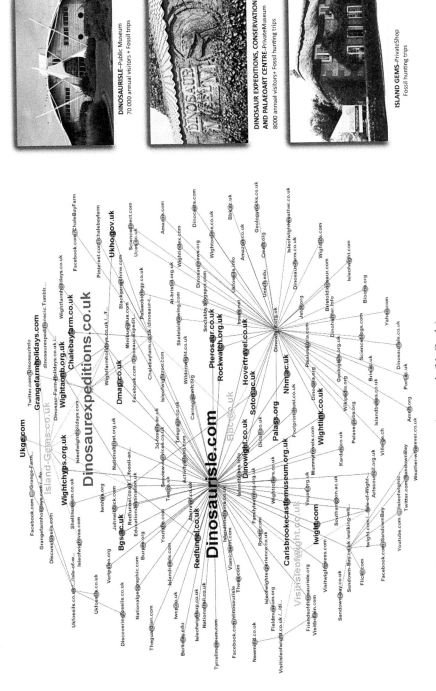

FIGURE 18.6 Isle of Wight geotourism actors graph network (N. Cayla).

environmental education that would find effective support with the emergence of new geoparks (Andrasanu and Grigorescu, 2012). The first discoveries of dinosaur bones were made by the Transylvanian aristocrat Franz baron Nopcsa and published in 1897 (Grigorescu, 2005). The Sanpetru and Densus-Ciula formations, of the Hațeg Basin, are considered to be Maastrichtian, the latest age of Upper Cretaceous (Van Itterbeeck, 2005). At that time, the Hațeg region was an island of about 80,000 square kilometers located near the equatorial belt. The strange dinosaurs found there such as the theropod *Balaur bondoc* or the titanosaur *Magyarosaurus dacus*, are of a spectacular scientific interest because of their smaller size compared to other contemporary dinosaurs of their genus living elsewhere on the planet. This reflects the particular evolution process of insular dwarfism (Benton et al., 2010). The scientific research that resumed in the 1970s allowed in 1988, the discovery of nesting sites similar to those observed in the south of France. The presence of newborn bones associated with these eggs made it possible to attribute the hatchery to *Telmatosaurus transylvanicus*, a hadrosaur (Grigorescu, 2010).

In 2004, the Hațeg Country Dinosaurs Geopark was officially designated. In the Romanian National Spatial Plan, geoparks are considered as territories that must sustain local identity for a sustainable development and to achieve these goals, 2 million euros were invested (Stoleriu, 2014). The management of the geopark is carried out by the University of Bucharest. First, a lot of projects focused on tourism infrastructures improvement. The main goal of the geotouristic experience is to emphasize the authenticity of the "Land of Hațeg" enhancing both natural and cultural heritage as already promote by the Retezat National Park created in 1935. Tourism activities have been diversified respecting the development of the eco-tourism destination Țara Hațegului – Retezat in order to increase the duration of the stays (Vesa, 2017). Seven interpretation centers and a few thematic trails disseminated around Hațeg provide knowledge about local geoheritage. The first visitor center opened in Hațeg in 2010, the "house of the geopark" has welcome 5321 visitors in 2017, nearly 25% more visitors than in 2016 (see Fig. 18.7).

With the UNESCO Global Geopark label, Hațeg country integrates a network of international level but one of the main difficulty is that the international tourists represent, at the national level only 22.6% of the total tourist arrivals registered in 2016 and only 18% of the total overnight stays (Iatu et al., 2018). The stakeholders network graph highlights that if numerous private and public actors work on tourism in Hațeg country there would be a low level of connectivity between the different spatial scale they represent. This can explain that the international level obtained by joining the global

Dinosaur Geotourism: A World-Wide Growing Tourism Niche 465

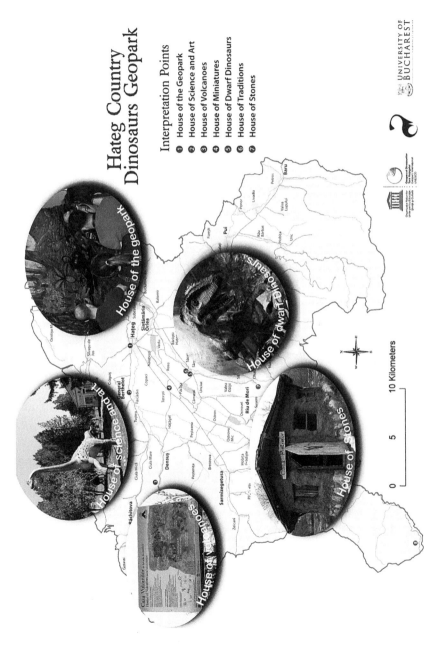

FIGURE 18.7 Interpretive centers of Hateg Country Dinosaurs Geopark (N. Cayla).

geopark network don't already produce the funnel effect that could attract international tourists at the level of the local tourism actors. But this dynamic set up fruitful cooperation initiatives, such as the European Interreg program Danube Geotour 2017–2019 for the development of sustainable geotourism between the geoparks of central Europe. This intermediate level cluster of actors aims to create innovative geoproducts to improve geotourism. The success of the Hațeg Geopark is due in part to a community of local volunteers mobilized by all the projects developed by the management team. Youth activity on social networks contributes to the effective communication of the geopark (see discussion in Ciobanu and Andrășanu, this volume (Chapter 15)).

18.5 A GEOTOURISM DESTINATION WHERE EDUCATION AND PLAY GO HAND IN HAND

Spain is home to palaeontological sites of great importance. Most of them are located in poor rural areas far from the tourist circuits. In the 1990s, a project for inclusion on UNESCO's World Heritage List: Icnitas de Dinosaurio de la Península Ibérica triggered important scientific investigations for the evaluation, conservation, and enhancement of proposed sites. Although the project failed, it is at the origin of several geotourism destinations focused specifically on the theme of dinosaurs melting fun and education. For example, in the Rioja region, better known for its wine production, the El Barranco Perdido paleo-park has welcomed, since 2010, in the small village of Enciso, which has hardly more than 160 inhabitants, more than 30,000 visitors per year. Beyond the recreational practices linked to a leisure park, the discovery of several of the dinosaur tracks sites is proposed as well as the visit of the palaeontological center located in the heart of the hamlet and which has just been completely renovated (Fuertes-Gutierrez et al., 2015).

Further east, the province of Teruel, located in the south of Aragon, focused its tourist development on the creation of the "Teruel-Dinopolis Palaeontological Site," a thematic park located in the provincial capital, a city of 30,000 inhabitants. Territorio Dinopolis is a public scientific foundation founded in 1998 and a private and public management company employing around 100 people year-round. The project was financed by the Instituto Aragonese de Fomento, which invested 15 million euros to support the economic development of the province, one of the least populated in Spain (Bosquet, 2005).

The theme park opened in 2001 and has grown ever since. A study conducted in 2011 by the Fundación de Economía Aragonesa shows that from 2001 to 2011, more than 1,750,000 visitors discovered Dinopolis. The incomes generated by all this tourist activity amount to more than 1560 million euros including more than 184 million euros of direct revenues (Perez y Perez and Gomez Loscos, 2011). In 10 years, the number of tourist beds around Teruel has increased from 700 to 2000. Dinopolis is above all a family leisure park that offers many recreational activities, from 4D cinema to TRex discovery (a species that is only known in North America). It also offers a museographic section dedicated to the history of life where visitors can improve their knowledge particularly about the dinosaurs discovered in the province as *Turiasaurus riodevensis*, a titanosaur of the end of the Jurassic. One of the main objectives of the project was to increase the length of the tourist stay in order to promote the development of the entire province by creating five thematic interpretation centers in the heart of the hamlets with the aim of distributing tourist flows and economic income. For that, the Fondo de Inversiones de Teruel has invested more than 3 million euros, each center being dedicated to a specific palaeontological theme.

Throughout the year, numerous events support the attendance of Dinopolis (Alcalà, 2011). At the same time, the paleontological excavations still carried out by the foundation's researchers have led to new discoveries. Since 2002, many dinosaurs' tracks have been unearthed around the village of El Castellar. There are, in particular, two new ichnogenres: *Deltapodus ibericus*, footprint left by a stegosaurus and *Iberosauripus grandis* made by a theropod. Here again, in order to attract tourist in this remote village, a geotrail was inaugurated in 2015 (see Fig. 18.8). It shows within the village a great number of footprints casts and leads to the deposit of El Castellar which presents tracks of the end of the Jurassic between 150 and 145 million years. The creation of the geotrail and the development of the deposit cost 130,000 euros financed by the government of Aragon (Cobos and Alcalà, 2018).

The strategy chosen by the provincial authorities is a success. The analysis of the network graph of actors involved in this project shows the balance between the two main pillars of "Territorio Dinopolis." On the one hand Dinopolis manages the park and its development with the objective to attract visitors in a concern for legitimate economic profitability. On the other hand, the foundation guarantees active research, the preservation of the provincial palaeontological heritage and the quality of the scientific mediation offered to visitors. The current project responds well to these two expectations, thus demonstrating the importance of a thoughtful offer at several levels in order to reach the widest possible audience.

FIGURE 18.8 The geotrail "Dinopaseo" in El Castellar (N. Cayla).

18.6 CONCLUSION

Meeting the worldwide growth of tourism, there is an increasingly specialized segmentation of practices. Palaeontological tourism is an interesting way of diversifying geotourism, as it is not delocalizable and attracts tourists closer to geosites, sometimes far from the world's leading tourist destinations. However, whatever the scientific interest of the sites, the success of these tourism projects must be based on a structured network of actors, gathering scientists, tourism actors supported by local policy-makers. Although until now, projects were often initiated by scientists, local stakeholders aware of their heritage can drive change by developing innovative geotourism projects. In France, the recent discovery of the longest dinosaur track of the world near the village of Plagne led to the development of an ambitious project of valorization supported by local authorities. The analysis of the actors involved in the various projects mentioned in this chapter shows that there is no standard scheme, but that the scale of reflection and mobilization of these actors with sometimes divergent expectations is an essential asset for the success of a dinosaur destination. Whatever the strategy adopted, the tourist trajectory of dinosaur sites must above all respect the "sense of place" and allow the visitor to apprehend a fascinating world forever gone.

ACKNOWLEDGMENT

I warmly thank Alex Andrasanu, Luis Alcala, Jose luis Barco, Diego Castanera, Angelica Torices, Cristian Ciobanu, Martin Munt, Dmitriy Prykhodchenko for their constructive comments.

KEYWORDS

- **geotourism**
- **dinosaur deposits**
- **stakeholder network**
- **territorial development**
- **Paleontological tourism**
- **in-situ, ex-situ valorization**
- **Global Geopark Network**

REFERENCES

Alcalà, L. Unmodelo de desarrollo regional fundamentado en recursos paleontológicos (Dinópolis-Teruel). *Paleontol. Dinosaurios Am. Latina* **2011,** 253–260.

Andrasanu, A.; Grigorescu, D. *Romania in Geoheritage in Europe and its Conservation*; Wimbledon, W. A. P., Smith-Meyer, S., Eds.; PROGEO: Oslo, 2012; pp 275–287.

Allaback, S. Mission 66 Visitors Centers the History of a Building Type. U.S. Department of the Interior National Park Service Cultural Resources Stewardship and Partnerships Park Historic Structures and Cultural Landscapes Program: Washington, 2000.

Benton, M.; Csiki, Z.; Grigorescu, D.; Redelstorff, R.; Sander, M.; Stein, K.; Weishampel, D. Dinosaurs and the Island Rule: the Dwarfed Dinosaurs from Haţeg Island. *Palaeogeogr. Palaeoclimatol. Palaeoecol.* **2010,** 438–454.

Bosquet, R. L. Un nuevoconcepto de parque tematico: origen e impactos de dinopolis. *Cuadernos de Turismo* **2005,** *15*, 149–167.

Carpenter, K. Rocky Start of Dinosaur National Monument (USA), the World's First Dinosaur Geoconservation Site. *Geoconservation Res.* **2018,** *1* (1), 1–20.

Cayla, N. Le patrimoine géologique de l'arc alpin. De la médiation scientifique à la valorisation géotouristique. Thèse de doctorat de l'Université de Savoie, 2009.

Cayla, N. Les processus de construction du géotourismealpin. *Téoros* **2010**, *29* (2), 15–25.

Cayla, N.; Duval M. Le géotourisme. Patrimoines, pratiques, acteurs et perspectives marocaines. *Collection EDYTEM* **2013**, *14*, 101–116.

Cobos, A.; Alcàlà, L. Palaeontological Heritage as a Resource for Promoting Geotourism in the Rural Setting: El Castellar (Teruel, Spain) *Geoheritage* **2018**, *10* (3), 405–414.

Dacos, M.; Mounier, P. Humanités numériques: Etat des lieux et positionnement de la recherche française dans le contexte international, 2015.

Dowling, R.; Newsome, D., Eds.; *Geotourism*; Oxford: Elsevier/Heineman: Oxford, 2006.

Dowling, R.; Newsome, D. Geotourism a Global Activity. In *Global Geotourism Perspectives*; Dowling, R. K.; Newsome, D., Eds.; Goodfellow Publishers Limited: Oxford, 2010; pp 1–17.

Fabry, N. Clusters de tourisme, compétitivité des acteurs et attractivité des territoires, *Rev. Int. d'intelli. Econ.* **2000**, *1*, 55–66.

Fuertes-Gutierez, I.; Garcia-Ortiz, E.; Fernandez-Martinez, E. Anthropic Threats to Geological Heritage: Characterization and Management: A Case Study in the Dinosaur Tracksites of La Rioja (Spain). *Geoheritage* **2016**, *8* (2), 135–153.

Gordon, J. E. Geoheritage Case Study: Geoheritage and Geoparks in Scotland. In *Geoheritage and Geotourism: a European Perspective*; Hose T. A., Ed.; Boydell & Brewer: Woodbridge, 2016; pp 261–278.

Grigorescu, D. Rediscovery of a "Forgotten Land": the Last Three Decades of Research on the Dinosaur-Bearing Deposits from the Haţeg Basin. *Acta Paleontol. Roman.* **2005**, *5*, 191–204.

Grigorescu, D. The "Tustea puzzle": Hadrosaurid (Dinosauria, Ornithopode) Hatchlings Associated with Magaloolithidae Eggs in the Maastrichtian of the Haţeg Basin (Romania) *Rev. Assoc. Paleontol. Argent.* **2010**, *47* (1), 89–97.

Hobléa F., Cayla N., Giusti C., Peyrache-Gadeau V., Poiraud A., Reynard E., Les géopatrimoines des Alpes occidentales : émergence d'une ressource territoriale. Annales de géographie 2017/5 (N° 717).

Hopson, P. The Geological History of the Isle of Wight: an Overview of the 'Diamond in Britain's Geological Crown'. Proc. Geol. Assoc. **2011**, *122*, 745–763.

Hose, T. A. Towards a History of Geotourism: Definitions, Antecedents and the Future. In *The History of Geoconservation*; Burek, C. V., Prosser, C. D., Ed.; The Geological Society: London, 2008; pp 37–60.

Iatu, C.; Ibanescu, B. C.; Stoleriu, O. M.; Munteanu A. The WHS Designation—a Factor of Sustainable Tourism Growth for Romanian Rural Areas? *Sustainability* **2018**, *10*, 626.

Iorio, M.; Corsale, A. Rural Tourism and Livelihood Strategies in Romania. *J. Rural Stud.* **2010**, *26*, 152–162.

Jacomy, M.; Girard P.; Ooghe-Tabanou, B.; Venturini, T. Hyphe, a Curation-Oriented Approach to Web Crawling for the Social Sciences Proceedings of the Tenth International AAAI Conference on Web and Social Media (ICWSM 2016).

Kadri, B.; Reda, Khomsi M.; Bondarenko, M. Le concept de destination Diversité sémantique et réalité organisationnelle. *Téoros* **2011**, *30* (1), 12–24.

Landel, P. A.; Senil, N. Patrimoine et territoire, les nouvelles ressources du développement, *Développement durable et territoires Ed.* [On line] 2009, 12. http://developpementdurable.revues.org/7563 (accessed Feb 28, 2018).

Laws, E.; Scott, N. Developing New Tourism Services: Dinosaurs, a New Drive Tourism Resource for Remote Regions? *J. Vacat. Mark.* **2003**, *9* (4), 368–380.

Leaver, J. The State of Utah's Travel and Tourism Industry. Kem C. Gardner Policy Institute University of Utah, 2017; p 18.

Lipps, J. H. Paleoparks: Our Paleontological Heritage Protected and Conserved in the Field Worldwide. In *Paleoparks—The Protection and Conservation of Fossil Sites World Wide*; Lipps, J. H., Granier, B. C. R., Eds.; International Palaeontological Association, France, 2009.

Lockley, M.; Schulp, A. S.; Meyer, C. A.; Leonardi, G.; Mamani, K. D. Titanosaurid Track Ways from the Upper Cretaceous of Bolivia: Evidence for Large Manus, Wide-Gauge Locomotion and Gregarious Behaviour. *Cret Res.* **2002**, *23*, 383–400.

Meyer, C. A.; Hippler, D.; Lockley, M. G. The Late Cretaceous vertebrate ichnofacies of Bolivia—Facts and Implications Asociación Paleontológica Argentina. Publicación Especial 7VII International Symposium on Mesozoic Terrestrial Ecosystems: Buenos Aires, 30-6-2001; pp 133–138.

Munt, M. C. History of Geological Conservation on the Isle of Wight. In *The History of Geoconservation*; The Geological Society, London, Special Publications 2008, 300; pp 173–179.

Munt, M. C. Geoheritage Case Study: The Isle of Wight, England. In *Geoheritage and geotourism: a European Perspective*; Hose, T. A., Ed.; Boydell Press: The UK, 2016, 195–204.

Perez y Perez, L.; Gomez Loscos, A. Efectoseconomicos y sobre el empleodel parque Territorio Dinopolis de Teruel, Documentos de trabajo Fundacion Economia Aragonesa 2011; p 56.

Perini, M. M.; Calvo, J. O. Paleontological Tourism: an Alternative Income to Vertebrate Paleontology *Arquivos do Museu Nacional*, Rio de Janeiro 2008, *66* (1).

Peyrache-Gadeau, V.; Perron, L.; Janin, C. La spécificité territoriale comme alternative à la généricité des produits-ressources. In *Pecqueur Bernard et GlonEric (sous la direction de) Au coeur des territoires créatifs–proximités et ressources territoriales*; Presses Universitaires de Rennes, 2016, pp 226–236.

Pralong, J. P. Géotourisme et utilisation de sites naturels d'intérêt pour les sciencesde la Terre: Les régions de Crans-Montana-Sierre (Valais, Alpes suisses) et Chamonix-Mont-Blanc(Haute-Savoie, Alpes françaises), Thèse de doctorat de la Faculté des Géosciences et del'Environnement—Université de Lausanne Géographie, 2006.

Prats, M.; Thibault, J. P. Qu'est-ce que l'esprit des lieux? Symposium scientifique ICOMOS La mémoire des lieux: préserver les sens et les valeurs immatérielles des monuments et des sites, Victoria Falls, Zimbabwe, 28–31 octobre 2003. En ligne: http://www.international.icomos.org/victoriafalls2003/papers/A1-4%20-%20Prats%20-%20Thibault.pdf.

Prenz, M. Gigantes. La guerra de los dinosaurios en la PatagoniaTusquets, 2015.

Price, T.; Jakeways, J. Isle of Wight Local Geodiversity Action Plan (IWLGAP), 2010; p 87.

Reynard, E.; Holzmann, C.; Guex D. Géomorphologie et tourisme: quelles relations? In *Géomorphologie et tourisme*; Reynard, E., Holzmann, C., Guex, D., Summermatter, N., Eds.; Actes de la Réunion annuelle de la Société Suisse de Géomorphologie (SSGm), Finhaut, 21–23 septembre 2001, Lausanne, Institut de Géographie, Travaux et Recherches n 24, 2003; pp 1–10.

Romilio, A.; Tucker, R.; Salisbury, S. Reevaluation of the Lark Quarry Dinosaur Tracksite (late Albian–Cenomanian Winton Formation, Central-Western Queensland, Australia): no Longer a Stampede? *J. Vertebr. Paleontol.* **2013**, *33* (1), 102–120.

Romilio, A.; Salisbury, S. A Reassessment of Large Theropod Dinosaur Tracks from the Mid-Cretaceous (Late Albiane Cenomanian) Winton Formation of Lark Quarry, Central-Western Queensland, Australia: A Case for Mistaken Identity. *Cretac. Res.* **2011**, *32* 135–142.

Sadry, B. N. *Fundamentals of Geotourism: With Special Emphasis on Iran*; SAMT Publishers: Tehran, 2009 (English summary available online at: http://physio-geo.revues.org/4873?file=1 ; Retrieved: Jan. 01, 2020) (in Persian).

Simpson, M. Geotourism and Geoconservation on the Isle of Wight, UK: Balancing Science with Commerce. *Geoconservation Res.* **2018**, *1* (1), 44–52.

Stoleriu, M. O. National Approaches to Geotourism and Geoparks in Romania, 14th SGEM Geo Conference on Ecology, Economics Education and Legislation 2014-06 Conference-Paper.

Taquet, P. Philippe Matheron et Paul Gervais: deux pionniers de la découverte et de l'étude des os et des oeufs de dinosaures de Provence (France). *Geodiversitas* **2001**, *23* (4), 611–623.

Tomic, N.; Markovic, B.; Korac, M.; Cic, N.; Hose, T. A.; Vasiljevic, A; Jovicic, M.; Gavrilov, B. Exposing Mammoths: From Loess Research Discovery to Public Palaeontological Park, *Quat. Int.* **2015**, *372*, 142–150.

Turtureanu, A. Harnessing Tourism Product. *Int. J. Sci. Res.* **2014**, 3 (4), 537–539.

Van Itterbeeck, J.; Markevitch, V.; Vlad, C. Palynostratigraphy of the Maastrichtian Dinosaur-and Mammal Sites of the Râul Mare and Barbat Valleys (Hațeg Basin, Romania). *Geologicacarpathica* **2005**, *56*, 2, 137–147

Vesa, L. Microregional Tourism Branding an Overview of the Brand "The Land of Hațeg-Where Legends Live. *Quaestus* **2017**, *10*, 35–46.

Webber, M.; Christie, M.; Glasser N. The Social and Economic Value of the UK's Geodiversity. English Nature Research Report No. 709. English Nature, Peterborough, 2006.

CHAPTER 19

Accessible Geotourism: Constraints and Implications

MAMOON ALLAN

Department of Tourism Management, Faculty of Archeology and Tourism, the University of Jordan Amman, Jordan

E-mail: m.allan@ju.edu.jo

ABSTRACT

Accessible tourism involves offering free-barrier geological tourism experience for tourists with special needs at geosites. Despite the advancement in the nature and scope of geotourism experiences in the last years, visitors with special needs are usually excluded from different types of geological tourism activities because of different physical and health barriers. Therefore, there is an urgent necessity to mitigate the potential constrains and provide an accessible tourism experience to tourists with special needs in the context of geotourism. As a result, this study aims to clarify the concept of the accessible geotourism and identifies the main constrains for promoting this concept. Moreover, it identifies the implications for such concept in the geotourism context.

19.1 INTRODUCTION

It is becoming increasingly difficult to ignore the importance of tourists with special needs segment in the overall tourism market (Allan, 2015). Deeper understanding of the different needs of individuals with disabilities is crucial for providing proper information and helping them to involve in different tourism experiences (Eichhorn et al., 2008). Accessible tourism may involve people with mobility, vision, hearing, cognitive, and developmental disabilities (Darcy, 2010). Darcy and Dickson (2009: 34) suggest that "Accessible tourism enables people with access requirements, including mobility, vision, hearing and cognitive dimensions of access, to function independently and with equity and dignity through the delivery of universally designed tourism products, services and environments. This definition is inclusive of all people including those travelling with children in prams, people with disabilities and seniors."

Overall, geotourism represents a holistic natural, cultural, and recreational experience for tourists. Thus, a geotourist can have a holistic experience, which distinguishes and recognizes the quality of this geotourism experience from other forms of tourism (Dowling and Allan, 2018). Despite the developments in the nature and scope of geotourism experiences in the last decades, people with special needs are commonly excluded from different types of geological tourism activities because of different physical and health constraints. Therefore, there is an urgent need to tackle the potential constrains and provide an accessible tourism experience to geotourists with special needs. Consequently, it is significant to promote the concept of "Geotourism for all" in different geological tourism sites and enhance the ethical practices in geotourism business. To date, geological tourism tours for people with special needs are still very scant. This is commonly due to several barriers that hamper the supply and demand sides for geotourism experiences.

19.2 THE CONCEPT OF ACCESSIBLE GEOTOURISM

Generally speaking, the concept of "accessible geotourism" involves enabling visitors with special needs to undertake their tourism experiences at different geological tourism sites. It includes providing accessible services, facilities, and requirements for all types of visitors engaging in geotourism experiences. Different types of geological sites, mainly geoparks, should provide barrier-free tours including accessible facilities, amenities, and services. Visitors with special needs may include:

- people with different types of disabilities;
- senior people; and
- people with different health conditions requiring special care, such as diabetic and obesity, and families with children having different kinds of disabilities.

19.3 CONSTRAINTS FOR ACCESSIBLE GEOTOURISM

A big challenge for geotourism is to make it accessible for all types of visitors including people with special needs. Thus, it is essential to create accessible geological tourism sites for people with special needs such as blind, deaf, handicapped, seniors, and children (Lima et al., 2013). Henriques et al. (2018) indicate that "Geoparks' websites generally lack information concerning facilities for special groups, and the web design does not meet the needs of people that have difficulties with written word and speech."

Collectively, many constrains have confronted the development of the accessible geotourism in theoretical and practical levels. It is apparent that there is a clear lack of attention in the geotourism literature for exploring different aspects of the accessible geotourism. Training and qualification for geotourism employees to deal with tourists with special needs is another constraint to enhance the accessibility of geotourism destinations. Also, the availability of qualified geoguides who have a good experience to guide tourists with special needs is still questionable in many geosites. Moreover, the cost of the infrastructure and superstructure pertaining for tourists with special needs is still a real constrain for developing accessible geotourism destinations.

19.4 THE INTERNATIONAL ASSOCIATION FOR GEOSCIENCE DIVERSITY (IAGD)

The purpose of the IAGD is to ensure the inclusion for people with disabilities in the geosciences (see Fig. 19.1). It promotes "different efforts of inclusion through collaboration in research, dissemination of instructional best practices, and professional development opportunities." However, despite the significance of the role of the IAGD in providing accessible experience for people with disabilities in the different dimensions of the geosciences, their main targets are the geosciences students and experts. Thus, it does not pay full attention to the geotourism participants. By reviewing the IAGD resources, the efforts for exploring the accessibility of the geosites for tourists with special needs do not exist. Despite this issue, the IAGD plays vital role in supporting the research to facilitate the practical experiences for geosciences students and experts to undertake their geosciences tasks. Overall, the IAGD visions are as follows:

- Celebrate the diverse abilities of all geoscience students, faculty, and working professionals by fostering student engagement in geoscience career pathways.
- Develop a community of resources for faculty and student support.
- Advance knowledge of access and accommodation within the geosciences through scientific research.
- Promote efforts of inclusion through collaboration in research, dissemination of instructional best practices, and professional development opportunities (IAGD, 2018).

FIGURE 19.1 Geo-tour for part of IAGD members (IAGD, 2018).

19.5 DISCUSSION

Taken together, concerns have been raised to enhance the accessibility of tourism destinations for all types of tourists (Al-Tell et al., 2017). In the context of geotourism, the status quo of the barrier-free geotourism experiences and activities is still very limited. Thus, it is vital to provide a range of activities to meet the needs of different types of visitors with special needs and help to develop their physical and intellectual abilities. It is vital and a priority to create environment for accessibility and enhance the existing ones in order to promote geotourism as an inclusive activity opens to all visitors (Lima et al., 2013). "Geology for Everybody" should be promoted by creating strategies and provide material for the support of various types of tourists (children, elderly people, the disabled, etc.), in order to allow the direct and indirect fruition of geological sites, which are of high scientific significance (Panizza and Piacente, 2008).

Toward providing accessible geotourism activities and experiences and promoting the concept of "Geotourism for all" at geosites, several steps should be done. Geotourism managers, policy-makers, planners, and promoters should provide accessible facilities and services, improve the quality of offering services, and develop marketing and promotion campaigns to enhance the awareness toward the significance of accessible geotourism activities. It is also highly recommended to offer appropriate training programs for geotourism staffs to be able to deal with different kinds of visitors with special needs. Accordingly, media can play a significant role to promote the concept of "Geotourism for all" and attract the attention of public to support the right of people with special needs to travel or experience geotourism activities.

Conway (2009) indicates that several elderly and people with disabilities are not able to visit the real geosites; therefore, it is necessary to develop virtual geopark to provide geotourism experiences for disabled people with mobility problems. As a result, the advancement of Information and Communication Technologies (ICTs) and Internet of Things (IoT) could provide many opportunities to improve the accessible tourism facilities and services at geosites.

UNESCO could play an important role in empowering the geotourists with special needs by including the accessibility of geopark for special needs as a requirement for the nomination of geoparks.

19.6 CONCLUSIONS

Geotourism has a significant role in reinforcing the commitment to apply notions of geoethics among its stakeholders and tourists, and to arouse the public awareness to the value of geodiversity and environment (Allan, 2015). Providing accessible geotourism facilities and services for people with special needs is an ethical practice and a primary responsibility for geotourism authorities and organizations to help visitors with special needs to practice their basic rights and enhance their wellbeing.

KEYWORDS

- **geotourism**
- **geotourists**
- **accessible geotourism**
- **accessibility**
- **barrier-free tourism**

REFERENCES

Allan, M. Accessible Tourism in Jordan: Travel Constrains and Motivations. *Eur. J. Tour. Res.* **2015**, *10*, 109–119.

Allan, M. Geotourism: An Opportunity to Enhance Geoethics and Geoheritage Appreciation. In *Geoethics: The Responsibility of Geoscientists*; Peppoloni, S., Di Capua, G., Eds.; Special Publications, Geological Society: London, 2015.

Al-Tell, Y; Allan, M.; Al-Zboun, N. Investigating Perceived Leisure Constraints for Senior Tourists in Jordan. *Aust. J. Basic Appl. Sci.* **2017**, *11* (2), 67–75.

Conway, J. S. Developing a Virtual Geopark—How Google Software Can Be Used To Provide a Geopark Experience for Disabled People with Mobility Problems. In *New Challenges with Geotourism*; Proceedings of 8th European Geoparks Conference; Neto De Carvalho, C., & Rodrigues, J., Eds.; Portugal, 2009; p 55.

Darcy, S.; Dickson, T. A Whole-of-Life Approach to Tourism: The Case for Accessible Tourism Experiences. *J. Hospital. Tour. Manage.* **2009**, *16* (1), 32–44.

DARCY, S. Inherent Complexity: Disability, Accessible Tourism and Accommodation Information Preferences. *Tour. Manage.* **2010**, *31* (6), 816–826.

Dowling, R.; Allan, M. Who Are Geotourists? A Case Study from Jordan. In *Handbook of Geotourism*; Dowling, R., Newsome, D., Eds.; Edward Elgar Publishing: Cheltenham, Gloucestershire, 2018.

Eichhorn, V.; Miller, G.; Michopoulou, E.; Buhalis, D. Enabling Access to Tourism Through Information Schemes. *Ann. Tour. Res.* **2008,** *35* (1), 189–210.

Henriques, M.; Canales, M.; García-Frank, A.; Gómez-Heras, M. *Accessible Geoparks in Iberia: A Challenge to Promote Geotourism and Education for Sustainable Development.* Geoheritage 2018, pp 1–14.

Lima, A.; Machado, M.; Nunes, J. Geotourism Development in the Azores Archipelago (Portugal) as an Environmental Awareness Tool. *Czech J. Tour.* **2013,** *2* (2), 126–142; DOI 10.2478/cjot-2013-0007.

IAGD. International Association for Geosciences Diversity (IAGD); https://theiagd.org/, 2018 (Accessed Oct 12, 2018).

Panizza, M.; Piacente, S. Geomorphosites and Geotourism. *Rev. Geogr. Acadêmica* **2008,** *2* (1), 5–9.

CHAPTER 20

Space and Celestial Geotourism

BAHRAM NEKOUIE SADRY

Adjunct Senior Lecturer and Geotourism Consultant

E-mail: Bahram.Sadry@gmail.com

ABSTRACT

Since the latest decade of the 20th century and specially in the early 21st century, geotourism, the geoparks movement and consideration, emphasis and conservation of abiotic nature and geoheritage have been seriously proposed in academic and executive circles. The concept and practice are expanding ever-increasingly and a new topic has been added to the aspects of environment and conservation of human heritage (i.e., abiotic nature). During the last 15 years, about 150 global geoparks have been registered around the globe and all this helps sustainable development.

Attention to the "heavens," which has long been considered by humans, in novels, and movies during the last 200 years or so, reveals human desire for travel to the moon and into space. Traveling from our planet is no longer just a science fiction story and space tourism is gradually developing and expanding and one day will become one of journeys that people will be able to option. In terms of the aforesaid concept and with due regard to space or planetary geology, being in the scope of human attention, the abiotic aspect of nature and heavenly geosystems like celestial bodies, clouds, light, Aurora, stars, the Moon (lunology), Mars, etc., give us cause to think that in developing and expanding geotourism concepts such attractions and preserving them have been neglected or underestimated and may have no main relationship with geotourism. However, in fact there are some geo-celestial activities in Australia, Iceland, Iran, and elsewhere, and/or within geoparks and space trips; here, the words celestial geotourism and space geotourism are coined and proposed for developing these new sub-sectors. Also, considering the concepts of primary and secondary geosites in geotourism for categorizing these new sectors, here those concepts can be applied and classified for proposed celestial and space tourism.

Therefore, from the proposed theoretical aspects in space geotourism, the earth itself is considered as a secondary space geosite and, being in a weightlessness condition and visiting space in sub-orbit, on the orbit and beyond earth's orbit is proposed as a primary geosite, of which these geosites would require to be protected in the 21st century and future centuries regarding sustainable development and avoiding space contamination. From a celestial geotourism point of view, planetariums, the Meteorite Museums and similar, Astronomy scientists and astronauts' residences and gravesites, archaeo-astronomical heritage, etc., are proposed as celestial secondary geosites. Celestial geotourism is also applicable in the cities and suburbs (just in free of light pollution areas) and in nature and also adjacent to villages and in polluted cities of the world in order to develop urban geotourism for gazing at sunrise and sunset, etc., affected by pollution, and participating in related interpretive provisions. Such celestial geotourism could help awareness of the public in the cities of developing countries and also help to decrease city air pollution.

From the point of view of managing geotourism, considering the space and time of development of appropriate geosites for tourism, celestial geotourism, both for gazing and for viewing cosmic-like beauties on Earth, and the proper interpretation of the space geology landscape at these geotourism sites, can be used as a platform for the expansion of future public space geotourism.

Geoconservationists can consider new perspectives for developing celestial geotourism in their geotourism development evaluation models and cooperation of geologists, directors of geoparks, institutes of astronomy, observatories, planetariums, universities, and enthusiasts and policy making directors can help the development of celestial geotourism inside geoparks and other urban and natural regions and be a factor for poverty alleviation and income generation for local communities in remote areas of different countries—especially developing countries. Sustainable development of space geotourism provides the study and attentions to atmospheric environmental issues and above atmosphere and in space, when parts of the universe as a geotourism attraction, to be used as space geology in the future. Then the development of celestial geotourism can be a platform for the public development of space geotourism in the world—especially in developed countries.

20.1 INTRODUCTION

Nowadays, protecting life and conservation of natural and cultural heritage, both are vital issues. The existence of humanity (the physical quality, mental conditions, genetics, and environmental awareness of humanity which are inherited from ancestors) are also a legacy of previous generations.

In the 21st century, the leisure times and usable holidays are important matters due to using life such that John Urry (1989, cited by Li, 2015) refers to as "to take holiday-making 'seriously'." People observe the world through sets of ideas, qualifications, interests and expectations which formed by social class, gender, nationality, age and education" (Urry and Larsen, 2011: 2). In the concept of the gaze (tourist gaze), "anything is potentially an attraction" (MacCannell, 1999, cited by Urry and Larsen, 2011: 12). But "looking is a learned ability" (Urry and Larsen, 2011) and gazing is included social work that contains senses beyond sight (ibid). Since humans have a great interest in astronomical observations and phenomena, astronomical observations can become a tourism resource. Astronomical observations and records of extraterrestrial phenomena started at least 5000 years ago (Chen et al., 2015). On the other hand, space travel is also one of humankind's dreams from ancient eras so that in the last 200 years various films and novels have appeared, such as *A Trip to the Moon*, a 1902 French adventure film directed by Georges Méliès (see Fig. 20.1). Inspired by a wide variety of sources, Jules Gabriel Verne published *From the Earth to the Moon* in 1865; it seems that since those times everybody has had a dream of traveling into space, a dream which is now going to be a reality. According to van Pelt (2005), space tourism is not science fiction anymore. When the American multimillionaire, Dennis Tito, went in the Russian Soyuz capsule in 2001 to travel to space, he made space tourism a reality for the present day (Reddy et al., 2012). In the near future, humans will tour the moon and other celestial bodies as tourist areas. Furthermore, there are many tourists on Earth who are interested in gazing at annular, total solar and lunar eclipses, oppositions of Venus and Mars, comets, etc. (ibid: 77). Thus, the main purpose of this chapter is to show the recognition of space and celestial tourism activities as a distinct and remarkable branch of geotourism, which can be a center of attention by academics and relevant institutions as the activity that overmatches tourism's emerging engagement with the heavens.

In the first section of this chapter, defining the term and proposing a grading system, the structure of celestial geotourism, and space geotourism are explained. Then a review of the relevant literature is discussed and relevant management issues are explained. The concept of this issue, management style, and its nature are the topics covered in this chapter.

FIGURE 20.1 The iconic image of the Man in the Moon.

Source: Screenshot from *A Trip to the Moon* (1902), which was named one of the 100 greatest films of the 20th century.

Source: https://en.wikipedia.org/wiki/A_Trip_to_the_Moon (Permission: *as Public domain*)

20.2 DEFINING THE BOUNDARIES

Recently, according to Coratza et al. (2018 and all references therein), two terms, which are being utilized more and more regularly, have been made known in scientific terminology: "geoheritage" and "geodiversity"; and also more recently, geotourism as related tourism to geoheritage and geodiversity (e.g., see Hose, 1995, 2016; Sadry, 2009; Reynard et al., 2009; Newsome and Dowling, 2010; Farsani et al., 2012; Reynard and Brilha, 2018; Dowling and Newsome, 2006, 2010, 2018), and all in all, abiotic nature. The concept of geoheritage was introduced in the 1970s (Martini, 1994, cited by Coratza et al., 2018), whereas geodiversity has been defined as a concept worth investigating since 1990s. The increase in recognition is verified through scientific meetings—conferences, workshops, and sessions regarding geoheritage and geodiversity issues, which have been held during the last two decades. Geotourism was first defined and published by Hose (1995, cited by Hose, 2006) in the modern era; geotourism is a facilitator for developing geoconservation, realizing geological heritage (geoheritage), and comprehending geological diversity (geodiversity). The idea is that geotourism is based on concept of sustainability, while geodiversity is together with biodiversity as a significant element of landscape appreciation and conservation (Dowling, 2015). "Through the investigation of the geological elements of form, process and time, we can gain an understanding of the complexity of process systems and earth history" (ibid). Geotourism has appeared as a new global phenomenon trying to develop sustainable tourism (Dowling and Newsome, 2006). Since humans have great interest in astronomical observations and phenomena, this has been resulted in astronomical observations becoming a tourism resource (Chen et al., 2015). Weaver (2011) argued that "wildlife-based ecotourism depends on the presence of charismatic mega-fauna and vegetation-based ecotourism depends on the presence of charismatic mega-flora and finally geologically based ecotourism depend on the presence of charismatic megaliths." But, nowadays, it is obviously clear that the so-called geologically based ecotourism of Weaver (2011), called geotourism as "tourism of geology and landscape" (e.g., Dowling and Newsome, 2006, 2010, 2018); "tourism of geology and geomorphologic landscape" (Reynard and Brilha, 2018); and "abiotic nature-based tourism" (Sadry, 2009), and also "tourism-earth science" (Chen et al., 2015) and "tourism of geology and geomorphology" (Hose, 1995, 2012, 2016), in the 21st century. According to Chen et al. (2015: 260) with the rise of geoparks in the 21st century, a global wave of geotourism is rapidly increasing. Previously, Sadry (2009) simply included celestial phenomenon such as Diurnal and Crepuscular in an inventory of geotourism activities. Hence, the emerging concept of abiotic nature-based tourism (Sadry, 2009) as the geotourism and geoparks movement in the 21st century, on the one hand, and the concept of the new voyageur/explorer tourists (see Fayos-Sola et al., 2014) on the other hand, bestows us with the new concept of astrotourism toward planetary/space geological tourism, even climatological–astronomical observation tourism, geomorphological landscape tourism, and/or previously mentioned nightscape tourism as new sub-segment of sustainable tourism, so called "Geotourism." According to Carter et al. (2015, 457), space tourism resembles adventure tourism, "space tourism presents an important philosophical challenge that can be harnessed for sustainability, forcing participants to consider their place in the universe, relationship to other beings, and especially concepts on time" (Toivonen, 2017: 24); he (Toivonen, 2017) also added space tourism as a new sector of adventure tourism, which, in the near future, will be fast becoming a new opportunity for experiencing the unknown universe. Astronomical tourism provides a less-studied part of sustainable tourism, in which a dark night sky is the main resource (Collison and Poe, 2013) and recently the universe has been considered a resource for tourism promotion and space tourism would not be considered as just "fly into space for recreation," due to environments and atmospheres and beyond Earth-space sustainability. Because the trends in space and celestial tourism have blurred the boundaries between adventurous activities and celestial ecotourism and astrotourism, and based on the conceptualized astrotourism (Jafari, 2007; Collison and Kevin, 2013; Fayos-Sola et al., 2014) and also space tourism (e.g., Carter et al., 2015; van Pelt, 2005) and celestial ecotourism (Weaver, 2008), and due to the lack of a crystal clear definition or doctrines about space and celestial tourism activities; hence, it is proposed to consider it in the geotourism field, which coheres to the theoretical correspondence of the concept, respectively, as "space geotourism" and "celestial geotourism." So, a consequence of not proposing a clear consensus of currently celestial tourism activities, it may be a delicate and doubtful progress regarding its geological ethics; also, the legitimacy of this newly developing industry would be in peril. Previously, astrotourism was defined as: "From the original significance of 'leisure activities of travelers paying to fly into space for recreation', to 'tourism using the natural resource of unpolluted night skies', and appropriate scientific knowledge for astronomical, cultural or environmental activities" by Fayos-Sola and Marín (2009, cited by Fayos-Sola et al., 2014), which can be defined under the sustainable framework of the newly emerged geotourism industry. Geotourism is defined by Sadry (2009) as *"Geotourism is knowledge-based tourism, as an interdisciplinary integration of the tourism industry with conservation and interpretation of abiotic nature attributes, besides considering related cultural issues, within the geosites for the general public"*; indeed, astronomical objects are geo-objects (space/planetary geological objects) and are abiotic nature-based objects in the universe that here are proposed as underlying resources for celestial geotourism and space geotourism. Fayos-Sola et al. (2014) stated that "Astrotourism had to wait for a new generation of citizens, concerned with the great dilemmas of the 21st century, and interested afresh in scientific answers to progress and development." According to Hose (2006: 238), "Geology contextualizes issues of place in the universe and scheme of life," and, therefore, it is proposed here that astrotourism be considered as a sub-segment of geotourism.

The typology of space tourism proposed by Carter et al. (2015) includes terrestrial space tourism, atmospheric space tourism, and astrotourism. In the schematic chart (see Fig. 20.2), terrestrial space tourism from the viewpoint of the geotourism industry is divided into the terrestrial space geo-landscape (e.g., Valley of the Moon Geosites (Chile), Martian Mountains and also Valley of the Stars (Qeshm Island, Iran), village of Lonar (India), etc.) and secondary geosite (such as Planetarium, astro- and geoscientists biographies, research, publications, notes and artwork, correspondence, diaries, collections, workplaces, residences and their grave sites)—that is, the "geoheritage" (i.e., space and planetary geological heritage) that, according to Hose (2006), encompass an examination and understanding of the physical basis of geosites and even includes the Kennedy Space Center (USA) and "non-site-specific space tourism consisting of virtual gaming environments, space movie-related travel and eclipse tours and also space attractions and launch sites" (Toivonen, 2017), which involves specific space tourism sites on the earth that potentially are applicable to development as a Geosite (hence, geoheritage sites) and geotourism sites and within geoparks in the 21st century. Carter et al. (2015) stated that "the atmospheric space tourism currently involves high-altitude jet flights up to 20 km with Russian MIG flights and weightless flights offered by the Zero G corporation." In Figure 20.2, atmospheric space tourism from the viewpoint of the geotourism industry is divided into space geotourism (up to 20 km) and celestial geotourism, encompassing related activities (e.g., gazing on the stars, planets, the moon, meteor showers, northern or southern lights, galaxies, aurora, comets, nebula, clouds, rainbows, solar and lunar eclipses, "sun dogs," sunrises, sunsets, the midnight sun, and cloudscapes, etc.) within geoparks, celestial geosites, and other national parks and also into celestial urban geotourism (celestial geotourism activities in unpolluted (light) cities for gazing on nightscapes and also related activities in day-time sky) (see Figs. 20.2 and 20.3). Actually, according to Chen et al. (2015), tourism resources of atmosphere includes meteorological tourism resources, climatological tourism resources and clean air (within a proper urban geosite an industrial air pollution story could be one of the interpreting programs in the celestial geotourism development in developing countries; see Table 20.1).

TABLE 20.1 Summarized Atmospheric Resource for Celestial Geotourism.

Geotourism Resources of Atmosphere	The aurora landscape
	Buddha's Halo (or Ratnaprabha) landscape
	"Mirage" landscape
	Landscape of cloud and fog
	"Misty rain" landscape
	"Snow" landscape
	Rosy clouds landscape
	Summering, Wintering, Sunny climate
	And Polar "white night"

Source: Adapted from Chen et al. (2015).

According to Carter et al. (2015), astrotourism, hitherto, has been limited to a "handful of wealthy elite trips" to the International Space Station, and occur both in in-Earth orbit and beyond Earth orbit. Hence, this activity is proposed to be categorized as space geotourism in-Earth and beyond Earth orbits (see Figs. 20.2 and 20.3).

20.3 LITERATURE REVIEW

Space geotourism and celestial geotourism is a neglected subsector of geotourism industry and there is no published academic material on topics. And there is a lack of attention on geotouristic aspects of astronomical tourism. Recently, Chen et al. (2015) emphasized that universe, astronomical, climatological, meteorological tourism resources and Star Tourism Resources are "Tourism earth-science" resources, and, Sadry (2009) simply included daytime celestial phenomenon, such as vivid sunset and cloudscapes gazing, in an inventory of geotourism activities. The original concept of the "abiotic nature tourism as geotourism" was proposed by Bahram Nekouie Sadry in 2009. Also, there is geotourism's contribution to sustainable tourism (Dowling, 2015). "Whilst ecotourism and biodiversity have been described in the academic literature for over thirty years, geotourism and its attendant concepts of geoheritage, geodiversity, geoconservation, and geotours, is relatively new" (Dowling, 2013). Geotourism is a new form of tourism based on the geological environment (ibid) that Sadry (2009) in broader sense, called it as abiotic nature environment's combination with tourism industry. Indeed, space tourism deals with vertical abiotic nature attributes; space tourism is a type of tourism in which the tourist travels with an aircraft or spacecraft

Space and Celestial Geotourism 489

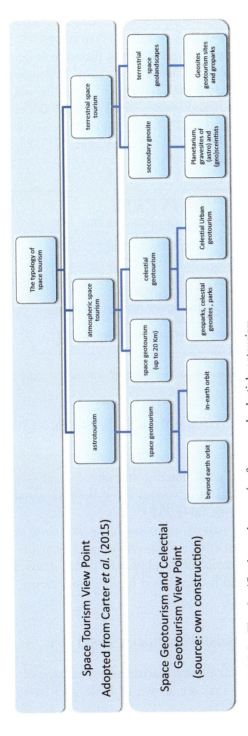

FIGURE 20.2 The classification and synopsis of space and celestial geotourism.
Source: Adapted from Carter et al. (2015) and own construction).

to locations with high height where the tourist can experience the suborbital effects, for example, zero-gravity or earth curvature observing (Weaver and Lawton, 2014); thus, nowadays the academic literature in the field of "space tourism," has been provided as a favored topic of theoretical and experimental research (e.g., Ashford, 1990; Bell and Parker, 2009; Brown, 2004; Crouch and Laing, 2004; Prideaux and Singer, 2005; Smith, 2000, cited by Weaver, 2011; van Pelt, 2005; Jafari, 2007; Reddy et al., 2012; Weaver and Lawton, 2014; Fayos-Sola et al., 2014; Collison and Kevin, 2013; Chen et al., 2015; Carter et al., 2015; Toivonen, 2017).

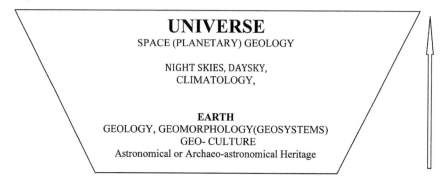

FIGURE 20.3 Schematic overview on the content of celestial and space geotourism.

20.4 CELESTIAL GEOTOURISM: DESTINATIONS, RESOURCES, AND MEANINGS

Because humans have great interest in astronomical observations and phenomena, astronomical observations have become a tourism resource (Chen et al., 2015: 77) (see Table 20.2).

This type of tourism activity, according to Fayos-Sola et al. (2014), "... is probably one of the most effective ways to bring tourism and tourists closer to nature for a comprehension of the physical world systems and dynamics." Celestial geotourism can be developed in urban and nature areas, such as geoparks (see Table 20.3).

20.4.1 CELESTIAL GEOSITES (NIGHTSCAPES)

The starry sky dimension attracts tourists as a landscape of night sky. The world's best stargazing sites are shown in Table 20.3 based on Tafreshi's (2016) account.

TABLE 20.2 Synopsis of Space and Celestial Geotourism Phenomena.

				Examples	
Secondary space geosites	On Earth	Ex situ	Meteorite Museum	-	Celestial geotourism
			Planetarium	-	
			Space Geolandscapes* on Earth, also craters, remains of meteorites	*Valley of the Moon* Geosites (Chile), Martian Mountains (Iran), village of Lonar (India), etc.	
		In situ	Night-time skies	Stars	
				Planets	
				The moon	
				Meteor showers	
				Northern or southern lights	
				Galaxies	
				Nebula	
				Comets	
				Aurora	
			Daytime skies	Clouds	
				Rainbows	
				Solar and lunar eclipses	
				"sun dogs"	
			Twilight skies	Sunrises	
				Sunsets	
				Midnight sun	
				Comets	
Primary space geosites	On orbits	Naked eye	—	All celestial bodies, even Earth	Space geotourism
		Assisted eye	—		

*This could be considered as a new value in the assessment of geodiversity on our planet by geologists and geomorphologists (educational and geotouristic values for space geotourism activities such as space and celestial geotourism).

Source: Some materials adapted from Weaver (2011), Dowling and Newsome (2006), Hose (2006), and Sadry (2009).

TABLE 20.3 The World's Best Stargazing Sites.

	Country name	Region
1	Chile	Atacama Desert
2	US State—Hawaii	High Volcanoes of Hawaii
3	USA	National Parks of the Southwest United States
4	Spain	La Palma, Canary Islands
5	Australia	Western Australia Outback
6	Nepal	Sagarmatha National Park, Nepal
7	Portugal	Alqueva Dark Sky

Source: Adapted from Tafreshi (2016) from an online article available at: https://www.nationalgeographic.com/travel/top-10/worlds-best-stargazing-sites/ (retrieved Dec. 1, 2018).

Also, there is some relationship between geological environment and the free of light pollution condition where tourists or amateur astronomers set up their telescopes. For example, Tafreshi (2016) pointed out that usually in summer, amateur astronomers arrange telescopes at Glacier Point in Yosemite (western USA), and also there are astronomy walks by the rangers in Death Valley, which is one of the largest dark-sky places in world with places altered with sky-friendly lights. In Spain, Astro-tourism is provided on islands with many tours for viewing stars to the border of Caldera de Taburiente, near a major observatory. In Australia, Nambung National Park north of Perth (national park in Western Australia), with the unbelievable Pinnacles rock formations is one of the favorite places for camping under the stars.

In some countries, celestial activities accompanied by geotours have started; for instance, an Iranian geotourism institute recently in order to expand their geotour (two-day tour) experience for visitors added a number of stops starting a journey from the Salt lake close to Damghan and providing some interpreting in the Verkian Observatory, using night sky for sky-gazing activities, then next day with adding other stops in a travertine mine, and finally a natural–historical site of a spring geosite (Damghan Cheshmali spring) to end their geo-journey.[1] In addition to Iran, also in Iceland, celestial activities combined with geotours specialized high-latitude destinations attract thousands of visitors; according to Weaver (2011), aurora-viewing is the most articulated as a specialized commercial tourism industry, and obviously, Iceland is a perfect place to see the Northern Lights, Aurora Borealis. Therefore, the Reykjavík

---

[1]www.parsgeotourism.com

Peninsula day tour of geology, culture, and history offers one such tour to natural wonders, such as Pursuing the Northern Lights around Reykjavík by HIT ICELAND.[2]

20.4.2   CELESTIAL GEOSITE (DAYTIME-SCAPES)

It is obvious that breathing *clean air* can contribute to human's health and well-being. Howard (2003: 1) suggests that "Heritage is taken to include everything that people want to save, from clean air to morris dancing, including material culture and nature." For example, clean air causes energy saving, better health, and other social and economic benefits and, on the other hand, the air polluted with industrial contaminants causes the unhealthy conditions for local communities. According to Chen et al. (2015), clean air is also "tourism earth—science" resources. For example, for the daytime, the corona can be an attraction, the corona is the Sun's rim, which, during a solar eclipse, is observable with the naked eye from Earth (Christopherson and Birkeland, 2018); also clouds and sky cloudscapes are geotourism attractions (Sadry, 2009). Several important components of the energy—atmosphere system work together to determine climate conditions on the Earth and form the Earth's climate system. Indeed, general climate types, which sometimes called climate regimes, like tropical deserts (hot and dry), polar ice sheets (cold and dry), and equatorial rain forests (hot and wet) are combination of two main climatic components—temperature and precipitation (Christopherson and Birkeland, 2018). As mentioned above, clouds landscape are an attraction of geotourism, one of which is a Rosy clouds landscape. Such a Rosy clouds landscape is a reflection of sunlight passing through clouds at sunrise and sunset. The combination of rosy clouds with sunrise and sunset forms a beautiful picture. These can be mentioned as one of the significant natural landscapes in tourist areas. From the managerial view point, all aforementioned geo-stories can be conveyed to geotourists via geo-interpretation provision within cities and geoparks.

20.4.2.1   SUNSET AND SUNRISE AS GEOTOURISM SITE ATTRACTIONS

The story behind sunset and sunrise involves geosystems and space/planetary geological stories. It was proposed by Sadry (2009) to include these

---

[2]https://hiticeland.com/iceland/tours

as geotourism attractions. According to James et al. (2006: 68), Uluru, the world's largest monolith, is the most well-known icon in Australia and is a geological landform. Tourists come to Uluru–Kata Tjuta National Park, in the Northern Territory of Australia, to observe an unbelievable light show at sunset and color changes, specially at sunrise and sunset.[3]

20.4.3 TERRESTRIAL SPACE GEO-LANDSCAPES

There are some geosites with an impressive range of color and texture, looking somewhat similar to the surface of the moon or Mars, etc., that can be considered as space geosites on Earth, such as *Valley of the Moon* Geosites (Chile), Martian Mountains or valley of stars (Iran), etc., that look somewhat similar to space geo-landscape (see Fig. 20.2). All-in-all, observatory activities, skygazing, and walking in a place similar to Moon or Mars on the Earth (space geo-landscapes) can be considered as geotourism activities.

20.4.4 NOCTURNAL ACTIVITIES WITHIN GEOPARKS

This activity uses the natural resource of nights, which are protected in a good condition, for opportunities and sciences relevant to the astronomy. Celestial geotourism is a neglected activity in geoparks where a dark night sky is a resource for observing and to know more about planetary geology and our life on the earth. Excluding geopark's global networks, some other parks started their own nocturnal and day-time activities for tourism, for example: One of the well-developed astronomy programs to serve visitors belongs to Bryce Canyon National Park (BCNP) in the southwestern United States. The program includes solar gazing during the day, multimedia programs during evening, and viewing stars with telescopes during the night. It seems that the BCNP has done well for educating visitors about significance of protecting and/or increasing the darkness of night skies (Collison and Poe, 2013).

---

[3]In 2017, the board of the *Uluru-Kata Tjuta National Park* voted just to end the climb because of the spiritual significance of the site (www.bbc.com), Uluru is well worth visiting with or without the climb (https://www.tripadvisor.com/).

20.4.5 URBAN CELESTIAL GEOTOURISM

Nocturnal Geotourism sites—Urban Night Sky Places—is also a neglected activity in urban geotourism development; geotourism deals with the natural and built environments (e.g., Hose, 2006; Sadry, 2009; Dowling, 2013; Chen et al., 2015; Riganti and Johnston, 2018). International Dark Sky Association (IDA) (http://darksky.org) defines Urban Night Sky Places as:

> An IDA Urban Night Sky Place is a municipal park, open space, observing site, or other similar property near or surrounded by large urban environs whose planning and design actively promote an authentic nighttime experience in the midst of significant artificial light. By virtue of their characteristics, these sites do not qualify for designation within any other International Dark Sky Places category.

Nowadays, in most cities, there are few stars in the sky. About 80% of North Americans live under skies, that are polluted with light, and there is not enough darkness to see the Milky Way (Tafreshi, 2016). Nevertheless, due to increase of light pollution, the dark skies are becoming a hard to find resource. Nowadays, for the first time in human history, most of the celestial objects cannot be seen from cities and wide areas around the cities (Fayos-Sola et al., 2014). The Geotourism industry as tourism of geology and geomorphological landscapes and sky-scapes can help to reclaim the starry nights.

20.4.6 SECONDARY CELESTIAL GEOSITES

It is proposed here (see Fig. 20.2) that the astronomical tourism, such as planetariums, should be considered as a kind of secondary space geotourism sites on Earth; also, the cultural heritage associated with planetary geology (and astronomy) is also an important resource for geotourism.

The conceptual proposed framework of this chapter also includes centers, such as "NASA's Kennedy Space Center" (USA), Space World in Japan, or Space City and Futuroscope in France and also astronomic sites in urban areas such as the Royal Observatory in Greenwich (UK), and the Griffith and Lowell Observatories (USA) that could be considered as secondary space geosite (celestial geotourism sites).

20.4.6.1 ANCIENT SOLAR OBSERVATORY AND CALENDAR STRUCTURE

Astronomical observations have been carried out for millennia by Indigenous cultures around the world, such as Indigenous Australians, with that knowledge passed down through oral tradition (Hamacher and Noon, 2021). About 5000 years ago, Chinese astronomers first recorded astronomical observations in writing (Chen et al., 2015), and more than 2000 years ago, Iranians began astronomical studies and records (Ghiasabadi, 2010). However, the most ancient records go back 5000 to 6000 years ago belonging to the Sumerians (D. Hamacher, pers. comm., Jan.11, 2020). The cultural heritage associated with astronomy is also an important resource for geotourism. Astrotourism requires places for gazing at stars, and often heritage sites, observatories, or natural dark sky areas with distinguished beauty (Fayos-Sola et al., 2014). The other one is astronomical heritage consisting of both cultural heritage and cultural landscapes relating to the sky and, in general, is deemed as a vital component of heritage and a significant resource for astrotourism and archaeoastrotourism (ibid). Additionally, some cultural resources, including archaeological sites at Stonehenge, Chichen Itzá, Giza, Chankillo, Mesa Verde, Persepolis, Almendres, Gochang, or Chaco Canyon, proved to have an astrotourism potential, which leads to the development of archaeoastronomy experiences for the society (Fayos-Sola et al., 2015). Ancient solar observatories and calendar structure such as the functionality of the Persian tetrapylons as solar structures that works with sunlight turn back to about 2000 years ago (Ghiasabadi, 2010, 2015) named as Chartaqi of Niasar (a calendar monument) (Isfahan Province, Iran); in summer and winter solstice, the sunrays enter the monument from different angles, which determines the time of the year. Every year Chartaqi of Niasar hosts a large number of Iranians that gather to cherish their ancestral way of life and to mark the observance of sunrise in summer and winter solstice. This ideology (proposed here as secondary celestial geosites) substantially develops the notion of geoheritage, which includes the built environment, or anthropogenic objects, which to some extent are made of local geological materials and involving planetary geology and astronomical contents and story to be interpreted to the public.

20.4.6.2 METEORITE GEOTOURISM

Our planet is also a sample of the universe. Also, *meteorites* are rocks from space that strike the Earth's surface. Before man provided the lunar rocks, meteorites were considered as the only samples of extracelestial bodies. In present times, more than 1700 kinds of meteorites have been collected and maintained in the world. There are dozens of meteorite museums in the world, which are attracting so much tourists every year, and have become a main tourism resource for scientific research and spreading astronomic knowledge. A meteorite museum in Jilin and an iron meteorite weighing 30 tons in Xinjiang are Meteorite Tourism Resources in China, both important for attracting tourists (Chen et al., 2015: 77–78). There are also other meteorite museums (e.g., The Natural History Museum of Vienna owns one of the largest meteorite collections in the world or a small fantastic Meteorite Museum in New Mexico[4] and so on could be considered as secondary geosites in geotourism industry. On the other hand, meteorite finding field tours are also main interest of youths and adults; for example, in Iran recently, some geological tours for exploring and finding meteorites in day time accompanied by celestial activities (night sky—star gazing) occurs in Khur and Biyabanak region (www.nightsky.ir/tour).

20.4.6.3 PLANETARIUMS: AS SECONDARY GEOSITES

According to Chen et al. (2015: 77), every year Beijing Planetarium attracts thousands of tourists to reach the aim of propagating astronomic and scientific knowledge. Greenwich Observatory, which is near London, UK, is a world-class tourist attraction. Planetariums are astronomical observation tourism resources (ibid). A planetarium is a theatre, which in essence, is built for providing educational and entertaining displays regarding astronomy and the night sky or for training in celestial navigation (Wikipedia).

20.5 SPACE GEOTOURISM: DESTINATIONS, RESOURCES, AND MEANINGS

It is perceived that the universe is unlimited as regards space and time and this is a universal concept for anything scattered and is altering in the space. "Earth, is a small rocky planet hurtling through space. From the edges of the solar system, it is barely visible as a small, blue dot. But this small, blue dot is special. Very special. For 4600,000,000 years, as it orbits the Sun, the surface of this small planet has altered innumerable

---

[4]https://www.tripadvisor.com/Attraction_Review-g60933-d559831-Reviews-Meteorite_Museum-Albuquerque_New_Mexico.html

times. How do we know all this? We know because the memory of these things are written in the stones and rocks all around us in situ and non situ such" (McKeever, 2015: ix). There are at least 125 billion galaxies in the universe. One of them with about 300 billion stars, is the Milky Way Galaxy (Christopherson et al., 2016: 44). The universe is also our heritage. A little point of this universe is our galaxy and our planet. Therefore, to understand the story of our universe and our planet, a precious sample available for us is planetary/space geology.

According to Chen et al. (2015: 76), in 1993, a tour with an itinerary of 4–5 circles around the earth in the space orbit with 8–12 h was planned to be arranged with Space Adventures Corporation in Seattle of the US. This tour, which costs $52,220, includes two space meals and tour expense, and has been reserved by 250 persons up to now. It is noteworthy that for the first time it was Yuri Gagarin, a Soviet citizen, who traveled to the space outside the earth by the spaceship in early 1960s and sighted very exciting spectaculars of the earth and the moon. Later on, in the early 1980s, the crown prince of Saudi Arabia traveled to outer space, at his own expense. It is proposed here that through geotourism, a better understanding of the Earth's geological wonders and the universe wonders can be achieved (see Fig. 20.2). As Chen et al. (2015) indicated, accomplishments of modern astronomy express the idea that the Earth where humankind lives and breathes is a regular planet in our solar system, which itself is a usual astronomical system in the Milky Way Galaxy, which finally, in combination with other galaxies, frames the Meta galaxy. Although the Meta galaxy is quite huge, compared to the Universe it is small. And our planet is a small geo-object in the universe such as a grain of sand on the Earth. Regarding space travels with due regard to definition of tourism, the WTO (1980: 89) defines tourists as "travelers who stay at least one night at a place other than their usual place of residence." From this perspective, the identified space geotourist should stay in space one night and considering future development of space tourism, floating some hours in space and returning to earth, they would not be considered as tourist and it is noteworthy that they would be deemed as space geo-excursionist.

20.5.1 PRIMARY SPACE GEOSITE

Humans were first interested in earth tourism and now their interest has developed into universal tourism (Chen et al., 2015) and gradually the commercial uses of space taking place and space is about to be opened up to more and more people, and the drive behind this is one of the most powerful economic forces: tourism. Today, the competition for taking tourists to suborbital space is increased and some private companies intend to operate a luxury space hotel to orbit in future. But, space tourism has different scales from "atmospheric space tourism" (up to 20 km) to "space tourism" (beyond earth orbit and in-earth orbit) (see Fig. 20.2) for classifying as a geotourism industry concept (e.g., Lunar Geotourists, Suborbital Geotourist, etc.). In the "atmospheric space tourism," passengers would experience 3–6 min of weightlessness, and also suborbital flights and other space tourism experiences is the possibilities of holidays to destinations far, far away in the 21st century. And at the NewSpace conference held (June 26–28, 2018) in Seattle, Blue Origin—governed by Amazon founder Jeff Bezos—declared that it intends to sell tickets to tourists who are interested to travel to space as early as next year (in 2019). Some companies have comprehensive plans for investigating on human space travel (and then, of course, there's SpaceX, which first will concentrate on shuttling astronauts to and from the space station). Blue Origin and Virgin Galactic intend to provide quite the same experience. They will not take tourists into orbit; they will only touch the edge of space, while passing an artificial boundary known as the Kármán line 62 miles up. Developing suborbital space tourism vehicles is also being worked. SpaceX (a manufacturer in the field of aerospace) in 2018 declared that it is planning on sending two space tourists to the moon in a path they are free to return (WIRED website, 2018).[1] Nevertheless, achieving accomplishment for commencing commercial astrotourism is very close. Virgin Galactic provides suborbital flights up to 100 km and Space Adventures travels to the International Space Station up to 350 km. Virgin Galactic and Space Adventures both drive inside the in-earth orbit. Lunar and Martian journeys, which are traveling beyond the earth's orbit, are not yet available, but Google, for financially motivating private companies to move toward special Lunar X, has planned a prize award for it (Davenport, 2015; Google Lunar X Prize, 2017, cited by Toivonen, 2017).

20.5.1.1 SPACE GEOTOUR GUIDES

Similar to the former miners' incorporation into the tourism industry as informed guides (e.g., at the Cape Breton Miner's Museum in Canada; Lemky

---

[1] Amy Thompson; Science page, WIRED website: https://www.wired.com/story/the-race-to-get-tourists-to-suborbital-space-is-heating-up/ (Retrieved Nov. 29, 2018).

and Jolliffe, 2011: 148), spaceflight program operators can be involved, and former astronauts or cosmonaut or space (planetary) geologists as volunteers in interpreting and guiding geo-space visitors. Some of these space geoguides may, as a crew member of a spacecraft, be capable of commanding, piloting, or serving visitors (something like a driver-guide on Earth).

20.6 MANAGEMENT CONSIDERATIONS

20.6.1 MINDFUL VISIT FOR SPACE AND CELESTIAL GEOTOURISTS

Since the 21st century when commencing new types of tourism have arisen, the original prevailing mass travel pattern demonstrates flexibility and regenerated stimulator of the more educational and seasonal passengers who are asking for more intelligent experiences, and the industry distinguishes the need for specialization as an element of destination competitiveness (Fayos-Sola et al., 2014).

Leisure times provide the possibility and opportunity for non-formal educators (under management of the site managers) and some learning opportunity for visitors based on recreational education or edutainment and subsequently leads to improvement in quality of life and the environment—along with increase in awareness about the earth and the universe, in which we are living in a small island of the space, called planet Earth. Indeed, managing leisure time is an important subject and people may undertake non-serious behaviors in their leisure time (e.g., like gazing stars in the desert). On the other hand, sky gazing and stargazing and astronomical observation activities may act as a potential instrument for development, both among the visitors and within the host community (Fayos-Sola et al., 2014). Therefore, managing public education in tourism is an important issue for educational management of societies to achieve sustainability. Leisure time and apparently non-serious behaviors during visiting and mindful gazing may have valuable effects for conservation of individual, national, and international heritage and legacy of this universe, namely, the Earth planet for human kind. According to Fayos-Sola et al. (2014), two key human urges must be properly addressed in the management of destinations for filling the leisure time of visitors: the life drive and the knowledge drive urges that, considering this point, the geotourism industry is a knowledge-based tourism, hence, it is extended to geo-destinations (see Table 20.4).

"Geoconservation" [Geological heritage conservation] was defined by Hose (2003) as *"The dynamic preservation and maintenance of geosites,*

Space and Celestial Geotourism

together with geological and geomorphological collections, materials and documentation" (Hose, 2006: 222). Hence, geoconservation in broad definition to encompass astronomical or planetary geological heritage can be defined as "*The dynamic conservation and protection of all components of abiotic nature and beyond space of the Earth together with geological and geomorphological and climatological and astronomical collections, materials and documentation and the cultural heritage associated with them.*"

TABLE 20.4 Two Key Human Urges in Geo-Destinations

Two key human urges in destinations (e.g., geotourism sites and in geoparks)	The life drive	The appetite for: • Content • Enjoyment • Sharing • Satisfaction	"Traditional/mass" (psychocentric) tourists
	The knowledge drive	The compulsion for: • Geo-information • Geo-education • Geo-understanding • New solutions to deep existential questions	New "voyageur/explorer" (allocentric) tourists (e.g., space geotourists and celestial geotourists)

Source: Materials adopted from Fayos-Sola et al. (2014).

Gordon (2012: 74, cited by Dowling, 2015) argues that "from a geoconservation point of view, people are more likely to value geoheritage and help to manage it sustainability in case they have a deeper awareness and connection with it through more meaningful and memorable experiences." According to Pralong (2006, cited by Dowling, 2015), for people to rediscover their geoheritage through new pleasurable and memorable experiences, which helps the geoconservation community to involve with a broader audience, a move from traditional approaches to geological interpretation toward a more empirical involvement is essential. Interpretive programs could include abiotic nature (in an intrinsic manner) and its influences on life, for example, to attractiveness of abiotic influences on species distributions. Interpretive programs can consist of discerning between galaxies, stars, and planets, and revealing Earth and the origin, formation and development of Earth, and reconstructing Earth's annual orbit about the Sun and displaying the inconvenient distribution at the top of the atmosphere; describing solar altitude, solar declination, and day-length in simple words, and explaining the annual variability of each—Earth's seasonality for the public.

20.6.2 SPACE GEOTOURISM AND ENVIRONMENTAL ISSUES

Sustainable tourism is a widely used term that during the past two decades has obtained remarkable attention from researchers and policy makers (Hughes et al., 2015).

There is an essential need for the protection and conservation of abiotic nature vertically as well as horizontally in the 21st century. Also "a number of abiotic environmental factors influence species distributions, interactions, and growth" (Christopherson et al., 2016: 621) on the earth. "According to Fawkes (2009), in space tourism, the forecasts for economic revenue are $100 billion by the year 2030 and $1 trillion by 2060. Therefore, it would be a costly miscalculation not to include sustainability' (Toivonen, 2017). Astronomical tourism is a part of sustainable tourism, which has been studied less (Collison and Poe, 2013). Space tourism will intrude into completely new space, possibly leaving orbital debris and permanent infrastructure behind. Rocket launch uses much amount of natural energy and, at the current beginning stage, the economic cost is tremendous for all the parties involved. According to Bradbury (2010, cited by Toivonen, 2017), the thing that can impact on the surface temperature of the earth is the particles of black soot emitted by spacecraft and the launch of a spacecraft may leave black carbon particles that will stay in the stratosphere for 10 years. The space tourism industry should choose a long-term outlook for short-term actions to increase sustainable planning and development (Toivonen, 2017). Sustainable tourism undergoes a theoretical fragility, which not only causes its universal applicability to be under question, but also leads to a special focus on tourism resource conservation and protection (ibid: 24). In this respect, planetary geological and sociocultural sustainability connection with the formal geotourism sector is an essential matter. Developing a space tourism industry is in conflict with current attitudes in environmental protection and sustainable development aims in the tourism industry. Hence, so-called space geotourism needs to be technologically and environmentally sustainable development accompany with economically sustainable business. For instance, according to Toivonen (2017: 24) "In the Sustainable Future Planning structure, 'the Sustainability' includes operational actions, environmental assessments and indicators as well as alternative fuels (ibid: 28); 'Future scenarios' include alternative future planning, voluntary measures, and strategic global agreements as in the sustainable future planning structure" (ibid: 30).

20.7 CONCLUSIONS

As for the planetary geological content, space travels, geotourism immersion, and geopark movements in the 21st century, geotourism—celestial geotourism and space geotourism—are the proposed terms for these activities. On the other hand, for such geotourism to succeed, space and planetary geological interpretation is also needed so that social awareness increases and finally interest in the local community and interest about nature is required. As for global sustainability, recently, sustainability science has come about as a new, integrative discipline, which is broadly based on the concepts of sustainable development related to functioning systems of the Earth. Geographic concepts are fundamental to this new science, emphasis of which is on human well-being, Earth systems, and human–environment interactions (Christopherson and Birkeland, 2018). It is expected that space tourism, which targets engineers, scientists, and entrepreneurs, as well as the general public, will become a multi-billion dollar business in the near future (Toivonen, 2017: 23). When a business is not only concerned for its economic success, but also pays attention to environmental and social aspects as part of its activities, traditionally, such tourism development and tourist activities have become sustainable. UNESCO's Geoparks program was built with the basic tenet of sustainable tourism. In space tourism, there are social sustainability issues such as the industry targeting the wealthy elite, the cost of the trip acting as a "discriminating factor" (ibid: 24). However, there are still other issues and challenges:

a. Space and celestial geotourism are still unrecognized institutionally and in academic investigations.
b. The need for sustainability awareness for development in the space and celestial tourism sector by pointing out some necessary elements needed for future planning processes.
c. Formal recognition and development of celestial geotourism can be realized through the collaboration of geopark managerials with well-established and influential astronomy-related institutions.
d. There is need to involve geologists and their assessment methods to proposing the list of world-class possible geosites for amateur astronomical observation, and for interpretation of geo- and astro-archaeological remains as cultural geosites.

Geotourism promotion in new vertical destinations could be a solution for sustainability issues and it opens new windows in the tourism industry, especially from the viewpoint of knowledge-based tourism in the 21st century.

ACKNOWLEDGMENTS

This chapter is dedicated in honor of my compatriot, Iranian-born American businesswoman, Anousheh Raissyan-Ansari, the world's first female space (geo)tourist and co-founder of Prodea Systems, and to the Ansari family.

I am very grateful to Noah A. Razmara for his discussions about this chapter with me and the help of Fatemeh Rakhshi, Mahboubeh Masoudnia, and Dr. Susan Turner, Dr Samaneh Soleimani and Prof. David Bruce Weaver, which improved it.

KEYWORDS

- **space geotourism**
- **celestial geotourism**
- **meteorite geotourism**
- **space tourism and astrotourism**
- **celestial geosite**

REFERENCES

Belfiore, Michael. *Rocketeers: How a Visionary Band of Business Leaders, Engineers, and Pilots Is Boldly Privatizing Space*; Smithsonian Books: New York, *2007*.

Carter, C.; Garrod, B.; Low, T., Eds.; *The Encyclopedia of Sustainable Tourism*; CAB International, 2015.

Chen, A.; Lu, Y.; Young, C. Y. N. *The Principles of Geotourism*; Springer-Verlag GmbH: Berlin Heidelberg, Jointly published with Science Press, Beijing, 2015.

Christopherson, R. W.; Birkeland, G. H. *Geosystems: An Introduction to Physical Geography*; Tenth Edition; Pearson: NY, 2018.

Collison, Fredrick M.; Kevin, P. "Astronomical Tourism": The Astronomy and Dark Sky Program at Bryce Canyon National Park. *Tour. Manage. Perspect.* **2013**, *7*, 1–15. https://doi.org/10.1016/j.tmp.2013.01.002.

Coratza, P.; Reynard, E.; Zwoliński, Z. Geodiversity and Geoheritage: Crossing Disciplines and Approaches. *Eur. Assoc. Conserv. Geol. Heritage. Geoheritage J.* **2018**, *10*, 525–526. https://doi.org/10.1007/s12371-018-0333-9, published online: November 15, 2018.

Dowling, R. K. Geotourism's Contribution to Sustainable Tourism. In *The Practice of Sustainable Tourism: Resolving the Paradox*; Hughes, M., Weaver, D., Pforr, C., Eds.; Routledge: London, 2015; pp 207–227.

Dowling, R. K. Global Geotourism—An Emerging Form of Sustainable Tourism. *Czech J. Tour.* **2013**, *2* (2), 59–79; DOI: 10.2478/cjot-2013-0004.

Dowling, R. K.; Newsome, D., Eds.; *Geotourism*; Elsevier/Heineman: Oxford, 2006.

Dowling, R. K.; Newsome, D., Eds.; *Global Geotourism Perspectives*; Goodfellow, 2010.

Dowling, R. K.; Newsome, D., Eds.; *Handbook of Geotourism*; Edward Elgar Publishing: Cheltenham, Gloucestershire, 2018.

Farsani, N. T.; Coelho, C.; Costa, C.; Neto de Carvalho, C., Eds.; *Geoparks and Geotourism: New Approaches to Sustainability for the 21st Century*, Brown Walker Press: Boca Raton, 2012.

Fayos-Sola, E.; Marín, C.; Rashidi, M. R. Astrotourism. In *Encyclopedia of Tourism*; Jafari, J., Honggen, X., Eds.; Springer: Berlin, 2015.

Fayos-Sola, Eduardo; Marin, Cipriano; Jafari, Jafar. Astrotourism: No Requiem for Meaningful Travel. *PASOS* **2014**, *12* (4), 663–671. https://doi.org/10.25145/j.pasos.2014.12.048.

Ghiasabadi, R. M. Čārtāqihā-ye Iran (Persain Squared-Base Arcs); Tehran: IRANSHENASI Publisher, 2010.

Ghiasabadi, R. M. The Sun in Tetrapylons: A Report on Discovering the Relationship between Tetrapylons, Time and Variation in the Declination of the Sun, Traslated by Bahareh Eskandari; Persian Studies: Tehran, 2015; Available online at http://www.ghiasabadi.com/.

Helgadóttir, Guðrún; Sigurðardóttir, Ingibjörg. The Riding Trail as Geotourism Attraction: Evidence from Iceland. *Geosciences* **2018**, *8* (10), 376; DOI:10.3390/geosciences8100376.

Hamacher, D. W.; Email Personal Communication with Dr Duane W. Hamacher, 2020.

Hamacher, D. W.; Noon, K. *The First Astronomers*. Allen & Unwin Press: Sydney, 2021 (forthcoming).

Hose, T. A., Eds.; *Geoheritage and Geotourism: A European Perspective*; Boydell Press: Woodbridge, 2016.

Hose, Thomas A. Geotourism and Interpretation. In *Geotourism*; Dowling, R. K.; Newsome, D., Eds.; Elsevier/Heinemann: Oxford, 2006; pp 221–241.

Hose, T. A. Selling the Story of Britain's Stone. *Environ. Interpret.* **1995**, *10* (2), 16–17.

Howard, Peter. *Heritage: Management, Interpretation, Identity*; Continuum: London, 2003.

Hughes, M.; Weaver, D.; Pforr, C., Eds.; *The Practice of Sustainable Tourism: Resolving the Paradox*; Routledge: United States, 2015.

Jafari, J. Terrestrial Outreach: Living the Stardome on Earth (Preface). In *Starlight: A Common Heritage*; Marín, C., Jafari, J., Eds.; Tenerife: Astrophysical Institute of the Canary Islands, 2007; pp. 55–57.

James, Jane; Clark, Ian; James, Patrick. Geotourism in Australia. In *Geotourism*; Dowling, R. K., Newsome, D., Eds.; Elsevier/Heinemann: Oxford, 2006; 2006; pp 63–77.

Lemky, Kim; Jolliffe, Lee. Mining Heritage and Tourism in the Former Coal Mining Communities of Cape Breton Island, Canada. In *Mining Heritage and Tourism*; Conlin, Michael V.; Jolliffe, Lee, Eds.; Oxford: Routledge, 2011; pp 144–157.

Li, Mohan. Tourist Photography and the Tourist Gaze: An Empirical Study of Chinese Tourists in the UK. The University of Central Lancashire, Unpublished PhD Thesis, 2015; clok.uclan.ac.uk/.../Li%20Mohan%20Final%20e-Thesis (Accessed Nov. 19, 2018).

McKeever, Patrick J. Foreword 2. In *The Principles of Geotourism*; Chen, Anze, Yunting Lu, Young, C. Y. Ng; Springer-Verlag GmbH: Berlin Heidelberg, Jointly published with Science Press, Beijing, 2015; p. ix.

Newsome, D.; Dowling, R. K., Eds.; *Geotourism: The Tourism of Geology and Landscapes*; Goodfellow: Oxford, 2010.

Reddy, M.; Nica, M.; Wilkes, K. Space Tourism: Research Recommendations for the Future of the Industry and Perspectives of Potential Participants. *Tour. Manage.* **2012**, *33*, 1093–102.

Reynard, E.; Brilha, J. (Eds.). *Geoheritage. Assessment, Protection, and Management*; Elsevier, 2018.

Reynard, E.; Coratza, P.; Regolini-Bissig, G., Eds.; *Geomorphosites;* Pfeil: München, 2009.

Riganti, A.; Johnston, J. Geotourism—A Focus on the Urban Environment. In *Handbook of Geotourism*; Dowling, R. K.; Newsome, D., Eds.; Edward Elgar Publishing: Cheltenham, Gloucestershire, 2018.

Sadry, B. N. *Fundamentals of Geotourism: With Special Emphasis on Iran*, SAMT publishers, Tehran, 2009. (English summary available online at: http://physio-geo.revues.org/4873?file=1 ; Retrieved: Jan. 01, 2020) (in Persian).

Tafreshi, Babak. *The World's Best Stargazing Sites*; 2016; Available online at https://www.nationalgeographic.com/travel/top-10/worlds-best-stargazing-sites/ (Accessed on Dec 1, 2018).

Thompson, Amy. Science page, WIRED website https://www.wired.com/story/the-race-to-get-tourists-to-suborbital-space-is-heating-up/ (available online and accessed on Nov 29, 2018).

Toivonen, Annette. Sustainable Planning for Space Tourism. *MATKAILUTUTKIMUS* **2017**, *13*: 1–2; from https://journal.fi/matkailututkimus/article/download/67850/28443/.

Urry, John; Larsen, Jonas. *The Tourist Gaze*, The Third Edition; SAGE Publications, 2011.

Van Pelt, Michel. *Space Tourism: Adventures in Earth Orbit and Beyond*. Copernicus Publishing, USA, 2005.

Weaver, David. Celestial Ecotourism: New Horizons in Nature-Based Tourism. *J. Eco.* **2011**, *10* (1), 38–45; DOI: 10.1080/14724040903576116.

Weaver, David; Lawton, Laura. *Tourism Management*, 5th ed.; John Wiley & Sons Publishing, 2014.

WTO. Manila Declaration on World Tourism. World Tourism Organisation: Madrid, 1980. https://en.wikipedia.org/wiki/Planetarium

INTERNET SOURCES:

https://en.wikipedia.org/wiki/Space_tourism, (access December 01, 2018).
https://en.wikipedia.org/wiki/Planetarium. (access December 01, 2018).

CHAPTER 21

Post-Mining Objects as Geotourist Attractions: Upper Silesian Coal Basin (Poland)

KRZYSZTOF GAIDZIK[1,*] and MARTA CHMIELEWSKA[2,*]

[1]Department of Fundamental Geology, Faculty of Earth Sciences, University of Silesia, Będzińska 60, 41-200 Sosnowiec, Poland

[2]Department of Economic Geography, Faculty of Earth Sciences, University of Silesia, Będzińska 60, 41-200 Sosnowiec, Poland

*Corresponding author. E-mail: marta.chmielewska@us.edu.pl, E-mail: krzysztof.gaidzik@us.edu.pl

ABSTRACT

Postindustrial tourism and postmining in particular, is a very intensively developing branch of geotourism. It includes tourism of stone quarries, mine sites, exploitation hollows, and postexploitation areas with specific buildings, infrastructure, and culture that brings out their geo-educational, cognitive, and esthetic values. The area of the Upper Silesian Coal Basin (USCB) is a perfect spot for the development of postmining geotourism due to:

1) Hundreds of years of ore (iron, zinc, lead), and bituminous coal mining history,
2) Traditional culture developed in relation to mining activity,
3) Landscape filled with postmining buildings and infrastructure, such as shaft towers, miners' housing estates, underground passages of different levels, types, and ages, as well as coal and ore mining waste dumps, and so on.

In this chapter, we present the main attractions of the USCB region, already very popular among tourists, like:

1) Historic Silver Mine and Black Trout Adit in Tarnowskie Góry,
2) Queen Louise Coal Mine and Main Key Hereditary Adit in Zabrze, and
3) Guido Coal Mine in Zabrze.

We also aim to present some of the so-called hidden potential of this region, that is, smaller or lesser-known postmining sites, waste dumps, and objects with different functions. We believe that appropriate regional policy and promotion, long-term sustainable development, and geoprotection of the present geoheritage will lead to geotourism development and economic growth of the USCB postmining area.

21.1 INTRODUCTION

Stone quarries, mine sites, exploitation hollows and postexploitation areas with specific buildings, infrastructure and culture, present high geo-educational, cognitive, and esthetic values, being of particular interest to geotourists, as corroborated by world-wide examples (e.g., López-García et al., 2011; Garofano and Govoni, 2012; Pérez-Aguilar et al., 2013; Nita and Myga-Piątek, 2014; Prosser, 2018). As C. D. Prosser (2018) stated "quarrying and mining have played, and continue to play, a major role in 'feeding' geoconservation activity," and the resulting products can be considered as "sites of cultural or historical heritage" (López-García et al., 2011). Actually, the earliest forms of geotourism are related with touristic exploration of caves and mines (Hose, 2008). In many cases, abandoned mining sites and districts may have been given a statutory conservation designation (based on their biodiversity, cultural, esthetic, and historical or geoheritage values). Former mines, quarries and anthropogenic outcrops or even waste dumps provide key localities within UNESCO Global Geoparks and sometimes even within the UNESCO World Heritage Sites (Prosser, 2018). That is especially true when underground passages related to human activity can be accessible for visitors, such as industrial, postmining sites. Such passages related to mining activity unveil geological objects that under normal circumstances, no one would see. The underground world always has been and still is largely unknown and unexplored, and considered mysterious and dangerous. Thus, the attraction of this unusual and incomprehensible world is the main impulse for people to partake in underground geotourism (Garofano and Govoni, 2012).

The area of the Upper Silesian Coal Basin (USCB) is a perfect spot for the development of postmining geotourism due to: (1) hundreds of years of history of ore (iron, zinc, lead), and bituminous coal mining, (2) traditional culture developed in relation to mining activity, (3) landscape filled with post-mining buildings and infrastructure, such as shaft towers, miners' housing estates, underground passages of different levels, types and age, as well as coal and ore mining waste dumps, and so on. Thus, the main goal of this study is to characterize the geotourism potential of selected postmining sites in the USCB. Some of them are already well-known and very popular among tourists, like: (1) Historic Silver Mine and Black Trout Adit in Tarnowskie Góry, (2) Queen Louise Coal Mine and Main Key Hereditary Adit in Zabrze, and (3) Guido Coal Mine in Zabrze. Other sites related to mining activity that include mainly shaft towers, miners' housing estates, coal and ore mining waste dumps that are still to be discovered for wider audience. We aim to present high potential of this particular region for geotourism. We believe that appropriate regional policy and promotion, long-term sustainable development, and geoprotection of present geoheritage will lead to geotourism development and economic growth of the USCB post-mining area.

21.2 USCB—GEOLOGY AND HISTORY OF EXPLOITATION

The USCB is one of the most significant Carboniferous basins of Europe. However, well-known and commonly used, this term is not a proper tectonic unit, and should be considered an economic unit, as it is limited by the extent of the coal-bearing Carboniferous deposits in the southern Poland and north of Czech Republic (Fig. 21.1) (Buła et al., 2008). These deposits fill the 6 100 km^2 triangle Devonian-Carboniferous Upper Silesian Trough developed on the Precambrian crystalline rocks of the Upper Silesian Block (Cabała et al., 2004). The USCB represents an orogenic basin formed in the mountain foredeep of the Silesian and Moravian Variscides. The sedimentation is related to considerable subsidence and is characterized by gradual transformation from flysch to molassic coal-bearing formations. The complex of Carboniferous mudstone and sandstone, together with numerous coal seams shows considerable thickness of up to 8000 m (Cabała et al., 2004). The lower part of the complex is characterized by the paralic coal-bearing series, while the higher sections of the profile include continental lacustrine deposits. Among them the 510 seam is one of the key seams that maintains significant thickness at a vast area of the basin. In the northern part of the USCB the Carboniferous deposits are overlain by a Triassic carbonate formation (limestone and ore-bearing dolomite). In the southern and western parts only Miocene argillaceous and arenaceous facies with evaporites cover the Carboniferous sediments (ibid).

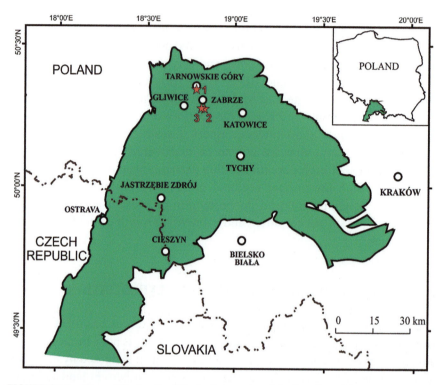

FIGURE 21.1 Upper Silesian Coal Basin (in green) and the location of the main geotourist attractions presented in this study (red stars): (1) Historic Silver Mine and Black Trout Adit in Tarnowskie Góry and its Underground Water Management System, (2) Queen Louise Coal Mine and Main Key Hereditary Adit in Zabrze, and (3) Guido Coal Mine in Zabrze.

The abundance of rich deposits of various mineral raw materials of different ages explains long history of mining activities in the area of the USCB. The most important include: coal, lead-zinc-silver (galena), iron, stowing sands, dolomite, limestone, marl, porphyry, melaphyry and so on. Over the hundreds of years of mining history, more than 13 billion tons of various mineral raw materials have been extracted in this area (Dulias, 2016). According to archaeological research, the beginning of mining activity in the area of the USCB dates back to 7000 BC. First raw materials extracted there were flints and clays. In the course of time also iron, silver, and lead ores, as well as dolomitic limestones have been exploited.

Historical sources confirming mining activities are not older than from the 12th century (Lamparska, 2017). In the middle ages in the northern part of the region (especially the area between Bytom and Tarnowskie Góry) the exploitation of non-ferrous metal ores like silver and lead was common. Silver was obtained from shallow deposits of argentinite (AgS), an admixture of galena (PbS), the source of lead. Since the middle of the 16th century also zinc in the form of galman was exploited, followed by the exploitation of zinc blend from sphalerite in the 18th century. Intensification of nonferrous metal ores mining took place between the end of the 18th century and 1960s, after that it began to decline.

The exploitation of bog iron ore (containing mostly limonite) also started in the ancient times. However, in the 19th century the Quaternary subsurface deposits were exhausted, so the exploitation was shifted to the Triassic deposits and Carboniferous sphero-siderites, which run out at the beginning of the 20th century (Piernikarczyk, 1933; Molenda, 1972; Slotta, 1985).

Hard coal exploitation started no later than in the 17th century, as confirmed by the historical source from 1657 indicating coal as the energy source in local iron forges. However, it is believed that coal was already known and used for household purposes in the 15th or 16th century.

At first coal was exploited by peasantry in small coal mines that has been converted into organized mining plants since the middle of the 18th century. The industrial revolution brought a rapid growth in mining and metallurgy and in consequence also prompt development of urban settlements. Thanks to inventions of those times the demand for hard coal increased. On the other hand, the use of steam engine for drainage in mines allowed deeper exploitation. About 100 mines have been worked in the USCB area in the second part of the 19th century. Later, the consolidation of coal mines took place along with further mechanization and electrification. Small and outdated mines were shut down or overtaken by more modern facilities. After the Second World War the intensive development of mining continued in the USCB region. In 1979, a record in coal production was obtained, reaching almost 200 million of tones exploited from 60 coal mines working in the region. In the second part of the 20th century coal mines were becoming more automated and deeper. However, starting from the 1980s the decline in coal production has been noted (Piernikarczyk, 1933; Kossuth, 1965; Jaros, 1975; Riley and Tkocz, 1998, 1999).

At the beginning of 1990s, along with the transformation of political system in Poland, it was clear that traditional industries developed in the USCB were not adjusted to the market economy of the rest of the Europe, so their restructuration was necessary. As a result, many mines, foundries, or factories had to be closed down.

As a consequence, brownfields and degraded areas appeared in urban space. These degraded areas required revitalization, that is, a complex, long-lasting, and interdisciplinary process aimed at social and economic recovery. It is mainly related to the introduction of new functions and communities into these areas, often associated with regeneration of historical buildings or technological monuments. It frequently requires expensive liquidation of industrial installations and purification of the environment from pollution (Chmielewska et al., 2016). In addition, revitalization plays an important role in redevelopment of postindustrial regions; it helps to change their character and image, and it stimulates regional economic development, by creating new workplaces (Chmielewska and Lamparska, 2011). Nowadays, after over 25 years of economic restructuration of the area of the USCB a lot has been changed. There are much fewer industrial plants and brownfields in the landscape. Most of postmining sites were converted into service facilities such as shopping and entertainment centers, cultural facilities, tourist objects, recreational places, business centers, or housing estates. However, all preserved objects are a part of postindustrial cultural heritage and remnants of hundreds of years of mining, which makes them interesting, and worth protection and promotion (Chmielewska, 2015; Chmielewska et al., 2016).

21.3 MAIN GEOTOURIST ATTRACTIONS IN THE AREA OF THE USCB

Nowadays mining in the area of the USCB goes slowly, but steadily into the past. Similar process has been observed in other European former traditional mining areas, for example, Ruhr Area in Germany, Nord-Pas-de Calais in France, South Wales, and Cornwall in Great Britain or Andalusia in Spain, where remarkable postmining objects are preserved as state protected monuments (Kostrubiec and Lamparska, 2005; Schwartz and Lorenc, 2011). Some of them were changed into underground or open-air museums based on their historical, cultural, technical, esthetic, and geo-educational values (Lamparska-Stobiecka, 2008).

Along with other postindustrial tourist attractions, they form a part of tourist trail as The Industrial Monuments Route of the Silesian Voivodeship (www.zabytkitechniki.pl/en-US). But only the most outstanding, well maintained, and recognizable objects can enter the list of anchor points of the European Route of Industrial Heritage (www.erih.net) or the UNESCO World Heritage List (Chmielewska, 2015). According to such criteria, three mines and two adits may be considered as the main geotourist attractions of the USCB area.

21.3.1 HISTORIC SILVER MINE AND BLACK TROUT ADIT IN TARNOWSKIE GÓRY AND ITS UNDERGROUND WATER MANAGEMENT SYSTEM

The history of ore mining in the USCB area dates back to the 3–4th centuries AD (Dzięgiel, 2008). First written account reporting on silver-bearing galena mines in Tarnowskie Góry area originates from the year 1136 (Długoborski, 1948). Metallic ore were exploited from middle Triassic limestone and dolomite. Since then the lead-silver-zinc excavation has been active leaving numerous remnants in landscape (waste dumps, mounds, depressions, quarries, mine objects, and infrastructure). The first period of prosperity of Tarnowskie Góry was connected mainly with lead and silver production and ended at the beginning of the 17th century with the crisis caused by difficulties with drainage and deep exploitation. In late 1700s, ore mining in the region started to thrive again, as a result of implementation of the steam technology in underground drainage in the state-owned Friedrich Mine. As a consequence, the volume of local zinc production in the 19th century covered almost half of global zinc consumption. The mining activity in Tarnowskie Góry stopped in 1912 due to resource shortage. The idea of preserving the mine as a tourist attraction occurred shortly after the Second World War. Finally, the Historic Silver Mine (www.kopalniasrebra.pl/en/) was opened for visitors on 5.09.1976. The Black Trout Adit (www.sztolniapstraga.pl/en/) is a part of a Fridrich Deep Adit, which was bored between 1821 and 1880 as an underground dewatering system for the Friedrich Mine. It was the youngest, longest, and deepest adit in the vicinity of Tarnowskie Góry, where adits have been created since 1547. Each adit was supposed to drain water by gravity from galleries, tunnels, and workings in ore deposits located above. Water was then discharged by a main adit tunnel to rivers situated below. Friedrich Deep Adit reached a length of 15 km and ended in the Drama River. Black Trout Adit was opened for tourists in 1957 after 2 years of preparations. Today, two objects (Historic Silver Mine and its underground water management system, and The Black Trout Adit) related with this bright past mining history denoting a significant contribution to the global production of lead and zinc, are available for geotourists to explore. Both were recently inscribed on the UNESCO World Heritage List,[1] confirming that these are outstanding examples illustrating a significant stage in human history. Actually, the entire object listed in the UNESCO list includes 28 postmining facilities that comprise the whole underground system of mining work: all principal drainage galleries (adits) that contribute to the underground water management system, chambers, transport galleries, corridors, and shafts, together with the water station, under/above ground structures (Figs. 21.2a and 21.2b). The Historical Mine of Silver Ores is a system of reconstructed galleries and chambers located around the Angel shaft, at depth of some 40 m below surface. The tourist trail is 1740 m long, including boat run along 270 m. There is no age limit for visitors. The Black Trout Adit on the other hand includes descent with the Eve shaft about 32 m underground, then 600-meters-long boats ride and ascent with the Sylvester shaft. This trail is available for tourists over the age of 4.

21.3.2 QUEEN LOUISE COAL MINE AND MAIN KEY HEREDITARY ADIT IN ZABRZE

The Queen Louise Coal Mine in Zabrze (Figs. 21.2c and 21.2d) was one of the biggest and most modern coal mines in the USCB area. It was founded by the Prussian government in 1791 under the name Zabrze to supply coking coal to the ironworks in Gliwice. In 1811, it was renamed Luisa in honor of the Queen of Prussia. During the 19th and 20th century it has been systematically improved and modernized. After the Second World War and the connection of this region to Poland, the coal mine was renamed again, and so the former name Zabrze was restored. In 1960, the exploitation in the western part was stopped. The eastern part in 1970 was connected to the Bielszowice Coal Mine and worked until 1998 (Jaros, 1991; Piątek, 2013).

The creation of museum from The Queen Louise Coal Mine began quite early, because already in 1960 right after the closure of the western part of the mine. The first underground tourist route of hard coal mining in Poland was opened here in the vicinity of Carnall shaft (Fig. 21.2d) in 1965 and it was created to encourage young people to work in the mining industry. The trail was in former excavations at the level 503. It was closed in 1979, because of the maintenance costs. In consequence, the Carnall shaft was buried. Shortly after, in 1980 the area around Wilhelmina shaft (in the eastern part) with shallow training adit was adapted for tourist purposes. At the turn of the 20th

---

[1] http://whc.unesco.org/en/list/1539

Post-Mining Objects as Geotourist Attractions

FIGURE 21.2 The main attractions of the historic mines in Poland: (a) and (b) Examples of corridors and galleries in the historic silver mine in Tarnowskie Góry (Photos: K. Gaidzik); (c) seam No. 510 (right side of photo) in the Queen Louise coal mine; (d) the Carnall shaft in the Queen Louise coal mine; (e) various types of mining racks at level 170 in the Guido coal mine in Zabrze; and (f) former experimental mine in the Guido coal mine in Zabrze (Photos: M. Chmielewska).

and 21st centuries both locations worked as the Queen Louise Museum. In 2011, the thorough revitalization of the facility started, along with the project of adaptation for tourists of the Main Key Hereditary Adit (Lamparska, 2017). The construction of the Main Key Hereditary Adit lasted for 64 years between 1799 and 1863. It was built to drain the mines, and as a way of transportation of coal to the ironworks. Drilling of the adit was difficult and slow due to hydrogeological conditions, and it was becoming less and less profitable over time, especially because of the use of dewatering pumps in coal mines, and the development of the railway transport. Finally, the 14.25 km long adit led from Chorzów to Zabrze connecting coal mines on its way. It was in use until 1875, then abandoned, silted up and partially collapsed (Wiśniewski, 2009; Lamparska, 2017). The idea of reconstructing the Main Key Hereditary Adit for tourist purposes appeared in 2000, whereas the revitalization started in 2009. The undertaking required digging out of the mud and unblocking the main tunnel, along with the reconstruction of Carnal shaft, and the outlet of the adit in the center of Zabrze. Even though the underground passages have been preserved in a very good conditions it was necessary to thoroughly secure the excavations from collapses and water with the use of an existing casing coming from various periods of drilling, as well as implementing safety and ventilation systems. The main attraction of the object, that is, underground boat rafting, started in September 2018 (http://sztolnialuiza.pl).

Nowadays both facilities: The Queen Luise Coal Mine and the Main Key Hereditary Adit, operate under one marketing name: The Queen Louise Adit (http://sztolnialuiza.pl), with three locations connected by underground routes. First location, around the Carnall shaft, includes monumental buildings, like chain bath and shaft tower. Tourists can admire exhibitions related to mining technology, the industrial uses of coal, and a fully functional over 100-year-old steam winding engine. In this location there is also an entrance to the Main Key Hereditary Adit. Around the Wilhelmina shaft, in the second location, tourist can visit shallow undergrounds in the oldest parts of the mine, and much younger former training adit that provide access to a 1.5 km underground route that includes displays of 1970s coal-cutting technology with operating mining machines and a ride in a mining train. Also popular among tourists is an open-air scientific park called 12C Park, with interactive educational installations presenting coal mining techniques and issues related to the four elements (earth, fire, air, and water). The most interesting installation there is the Bajtel Gruba—a miniature coal mine with abundance of moving devices, where children may play the role of miners.

Third location is the port and the outlet of the Main Key Hereditary Adit in the center of Zabrze.

Currently there are four routes in the museum. First one, "Mining journey in time," leads partly through the tunnels of the Main Key Hereditary Adit with a reconstructed transshipment port and partly through the tunnels of the Queen Luise Coal Mine including a passage drilled completely in pure carbon—in a 5-m high coal seam (seam No 510; Fig. 21.2c). It ends in the former training adit located near to the Wilhelmina shaft.

It shows mainly the history of the Main Key Hereditary Adit and the history of coal mining techniques including presentation of room-and-pillar and longwall system with shearer. It takes about 3 h and is available for tourists over 6 years old. Second route, "Underground treasures," joins regular sightseeing with treasure hunt and geocaching. This route is shorter (about 2 h), and available for tourists over 6 years old. The third trail, "Underground water tour," leads entirely through the 2 km long section of the Main Key Hereditary Adit, over the half of which is flooded and possible to cross only by boat. The route lasts for 2.5 h and may be visited by tourists over 7 years old. Finally, the fourth, "Family route," is the shortest one (1.5 h), available even for 3 years old children. The route is located only in undergrounds of the Wilhelmina shaft and educational adit. The Queen Luise Adit offers also 14 educational routes adapted to children of all ages—from kindergarten to youth.

21.3.3 GUIDO COAL MINE IN ZABRZE

The Guido Coal Mine in Zabrze was founded in 1855 by magnate and industrialist Count Guido Henckel von Donnersmarck at the mining field adjacent to the Queen Luise Coal Mine. The first exploitation level was located 80 m below the surface, but soon also deeper levels (170 m and 320 m) were in use. In 1885, the Guido Coal Mine was sold to the neighboring Queen Luise Coal Mine and started working as its southernmost part. In 1912, it was merged with the newly built Delbrück mine and coking plant. The regular exploitation stopped in 1920. The temporary recovery of Guido Coal Mine took place in 1967 when the Experimental Coal Mine M-300 was created there. It carried out research on new technologies and mining machines. In 1982, the Guido Mining Museum was opened to the public on the level 170, thanks to the efforts of the Coal Mining Museum in Zabrze. Several years later it was registered a national monument. In 2000, because of cost reduction and restructuring of the Polish coal mining industry, the Guido

Mining Museum was closed down followed by disassembling of the Mine. Fortunately, efforts of the local government of Zabrze and other institutions stopped this destructive practice. As a result, in 2007 the Historic Guido Coal Mine was founded as a cultural institution operating under the auspices of Zabrze municipality (Lamparska, 2017).

The Guido Coal Mine offers three different underground tours, all of which ends in the underground pub located in the former Pump Hall[2]:

1. Regular sightseeing of the mine takes place at two levels—170 and 320. The trail starts in the oldest available part, on level 170, where tourists can admire various types of mining racks (Fig. 21.2e), originally preserved horse stables and St. Barbara's chapel, as well as exhibitions related to the history of mining in the 19th century. The second part of the trail is located on level 320, where a lot of elements of the infrastructure of the former Experimental Mine are preserved (Fig. 21.2f). Here visitors may learn about different ways of coal exploitation, and transportation up to the surface, ride in the suspended electric train and see large scale mining machines in operation. It is possible to observe coal in its raw form—seams that are 2 m thick. The route is 3.5 km long, takes about 2.5 h, and is available for tourist over 6 years old.
2. An alternative form of visiting the mine is the trail called "The Dark of the Mine" which leads tourists by the light of headlamps through the youngest, deepest, and rawest part of the Guido Coal Mine—on level 355. The passages are tight, with steep dip-headings and raises, full of chain conveyors, pipelines used for pumping the filling, and a mass of mining steel. The main point is a tight passage through very steep and over 100 m long wall. The roof in this place is supported by thickly placed props of Valent type and the coal seam is within reach here. The route is about 1.5 km long, takes about 2.5 h, and is recommended for the physically fit people over 12 years old.
3. The most extreme tourist trail in the Guido Coal Mine, that is, the "Shift," allows visitors to feel like miners. At the beginning tourists get their full equipment and frontman's outfit. Then they go to the mining longwall no 4 at level 355, which is historically the last mining place in the Guido Mine, and there they have to shift over a pipeline weighing dozens of kilograms, erect a band conveyor, dimension a wooden prop, install a ventilation pipe, or do other

---

[2]https://kopalniaguido.pl/index.php/en/

works that they are told to perform by the guide foreman. After work the tourists go to the mining baths to take a shower. The route is only for adults in good health and physical condition. It lasts for 3.5 h.

The Guido Coal Mine organizes also cultural events at the underground stage, like concerts and theatre plays or conferences. Visitors may go also straight to the underground pub or organize a party down there. In addition, the object has its own hostel right next to the Coal Mine.

21.4 UNDISCOVERED GEOTOURISM POTENTIAL OF THE USCB

Despite many years of multidisciplinary studies, the geotourism potential of the USCB area is still not well determined. Hundreds of shafts, postmining buildings, infrastructure, and entire districts, together with thousands of kilometers of underground galleries and sidewalks and numerous waste dumps form the landscape of USCB. Various geosite inventories in this region (e.g., Staszewska and Żemła, 2013; Chybiorz and Kowalska, 2017), or just postmining sites (e.g., Gawor et al., 2011; Lamparska, 2017), show the number of sites that are already accessible or can be adopted for tourist purposes. However, these usually present only large, well known structures. Whereas the entire topography and morphology of this region can be considered as a postmining heritage as it has been modified significantly due to the hundreds of years of mining activity. The region is filled with abandoned old mine sites, waste dumps, remnants of mining infrastructure and buildings. Here we present a brief introduction to some other sites of interest in the USCB area that are still waiting to be "discovered."

Shallow mining activity in the form of quarries, apart from deep ore and coal exploitation, is also widespread in the USCB region. As shown by J. Nita (2012) abandoned quarry can be useful for geotourism if it fulfils at least one of the following functions:

(1) Scientific, (2) geo-educational, (3) evocative, (4) practical, (5) tourist interesting points, and (6) leisure.

Example of sites that combine few of the above-mentioned points are the historical abandoned mining sites between Tarnowskie Góry and Bytom that include quarries in Bobrowniki, Blachówka, Segiet nature reserve, and landscape-nature complexes Doły Piekarskie and Suchogórski rock labyrinth. Of them the most interesting is the Blachówka quarry that has been active since the 1890s, first as an iron ore mine, then as dolomite quarry, exposing over 40-meters-high walls of ore-bearing Middle Triassic dolomite. The long-lasting exploitation left the open pits together with galleries and shaft, as well as waste dumps. The underground passages are one of the largest wintering areas for bats in Poland.

Brownfields, wastelands, and waste dumps, even though considered as the most unwanted postindustrial heritage, play an important role in urban cultural landscape of mining regions and can be used as geotourist sites (Rostański, 2008; Gaidzik, 2010; Gawor et al., 2011; Chmielewska and Gaidzik, 2012; Gawor, 2014). In many cases these objects present high biological values (a place of occurrence of rare and endangered plant species, or a place of formation of new plant associations), as well as cultural, historical, geo-educational, and even esthetic ones (Rostański, 2008). In the USCB area, the waste rock output is estimated at about 2.1–4.3 billion tons (Dulias, 2016). There are ca. 220 coal mining waste dumps, covering over 4000 ha, with the largest one of over 250 ha—the Central Coal Mining Dump in Knurów (Gawor, 2014). However, this list includes only large, well-developed structures. Detailed studies by K. Gaidzik (2010), and M. Chmielewska and K. Gaidzik (2012) shown an exhausting number of small waste dumps hidden in the forest in a very small area between Zabrze and Bytom that are usually omitted during inventorying processes, sometime even unknown for the local people, but interesting from the geotourist point of view. However, without any promotion or information these places are still unknown for the general public. An example of a larger structure that shows high potential, not only for geotourism, is the waste dump in Katowice Murcki, with interestingly shaped plateau (Gawor et al., 2011).

Even though interesting and worth promotion and geoprotection, coal mining waste dumps can be a source of dangerous environmental impacts. Among them the most important include: (1) waste dump fire hazard that may affect its surroundings due to spreading the fire and ensuing air pollution, (2) emissions of toxic gases to the atmosphere, (3) surface water and groundwater pollution, and (4) danger of slope sliding (Gawor, 2014). Thus, properly produced sustainable development project is needed when adopting waste dumps into the landscape and geotourism purposes.

21.5 POSTMINING OBJECTS WITHOUT GEOTOURISTIC FUNCTIONS

Changing postmining sites into geotourist attractions is popular all over the Europe, but not every object has to be changed into mining museum. Some may become places of recreation or entertainment, others may serve as business tourism or active tourism centers, or may be converted into hotels, restaurants, and so on (Badulescu et al., 2005; Xie, 2006; Chmielewska, 2015). In the area of the USCB most of the preserved post-mining objects are not related directly to geotourism. Nevertheless, as they all commemorate previous mining activity of the area, it is worth to mention them briefly.

The most iconic parts of the mining landscape are the headframes. These structural frames above underground mine shafts along with accompanying buildings are the first choice when it comes to select a place worth redevelopment. In the area of the USCB headframe complexes have been changed into: restaurants (Maciej Schaft in Zabrze—Fig. 21.3a—and Prezydent Schaft in Chorzów), cultural centers (Ignacy Monumental Coal Mine in Rybnik, Two towers of Polska Coal Mine in Świętochłowice, Park of Tradition in Krystyn Schaft in Siemianowice Śląskie), art galleries (Wilson Schaft in Katowice), or loft (Bolko Schaft in Bytom). The headframe Warszawa II of the Katowice Coal Mine gained an interesting function—a viewing tower, with an observation deck installed at the top. It plays a key role as a central spot of the Area of Culture in Katowice (created entirely in place of the Katowice Coal Mine) amidst the headquarter of Silesian Museum, the Concert Hall of Polish Radio's National Symphonic Orchestra and the International Congress Centre (Fig. 21.3b). A headframe is also a symbol of the shopping and entertainment complex Silesia City Centre in Katowice which was created in place of former Gottwald Coal Mine. What should not be forgotten, however, is that apart from headframe complexes which are already successfully revitalized there are plenty of those that are still waiting on their turn like Andrzej winding tower in Ruda Śląska or Ewa and Krystyna towers in Bytom.

Important remnants of industrial past are workers housing estates. They were built by the industrialists who owned industrial plants to provide housing for their employees. In most cases, these settlements have survived longer than mines they were created for, but not necessarily kept their original residential function. The most outstanding workers housing estate in the area of the USCB is Nikiszowiec in Katowice, which in 2011 gained the title of Monument of History. Along with neighboring and almost equally interesting Giszowiec they were both designed by famous architects: Emil and Georg Zillmann from Berlin at the beginning of the 20th century for employees of the Giesche (later Wieczorek) Coal Mine located in Janów (nowadays a part of Katowice). As for those times they offered an extremely high standard of life, but they were significantly different from each other:

FIGURE 21.3 (a) Maciej shaft in Zabrze, (b) cultural zone in Katowice, with remnants of the Katowice Coal Mine in Poland. *Source*: M. Chmielewska.

Giszowiec was like a village, consisted of small houses with gardens, while Nikiszowiec was like a small town of three-storey tenement houses. Both are still inhabited and visited by tourist as sites on the Industrial Monuments Route of the Silesian Voivodeship. Another interesting example of the workers settlement changed into a (geo)tourist attraction is the Ficinus Colony in Ruda Śląska.

The 16 two-storey houses making up this housing estate were built in 1860s not of brick, but of sandstone extracted from the Silesian quarries. Houses were inhabited by miners working in the nearby mine Gottessegen. Nowadays, as a result of revitalization, only few are still inhabited, and the majority were transformed into service facilities such as shops, pubs and restaurants, and small enterprises.

21.6 CONCLUSIONS

Nowadays tourists, and geotourists in particular, seek for something new, different, interesting, and even exciting, with educational and historical values, but most importantly, something real that present actual life over time and space. Postindustrial heritage, especially well-preserved or still alive can fulfill that demand. The area of the USCB works ideally as a perfect region for the postmining geotourism development, due to the hundreds of years of ore and coal mining history, numerous ore and coal mines—most of them closed, but some still working—several monumental postmining objects, abandoned postmining sites, hundreds of waste dumps, abundance of postindustrial districts, and still existing mining culture, and so on. The most important objects, already open for tourists and very popular among them, include:

1) Historic Silver Mine and Black Trout Adit in Tarnowskie Góry and its Underground Water Management System
2) Queen Louise Coal Mine and Main Key Hereditary Adit in Zabrze, and
3) Guido Coal Mine in Zabrze.

However, we argue that the potential of the area is much higher and that it is not only well-developed deep mines with breathtaking corridors, shafts, beautiful buildings, and structures that can be considered tourist attractions, but also land modification, waste dumps, or subsidence trough, commonly filled with water that can be adopted and presented for a wide audience. We believe that appropriate regional policy and promotion, long-term sustainable development, and geoprotection of present geoheritage will lead to geotourism development and economic growth of this area.

KEYWORDS

- **Poland**
- **postmining geotourism**
- **coal and ore exploitation**
- **geosite**
- **mine**
- **shaft**
- **waste dump**

REFERENCES

Badulescu, A.; Bugnar, N.; Badulescu, D. Cultural Tourism in Urban Areas. Study-case: Oradea City, Romania. In *Conditions of the Foreign Tourism Development in Central and Eastern Europe*; *Urban tourism—Present State and Development Perspectives*: Wrocław, 2005; Vol. 8, pp 9–20.

Buła, Z.; Żaba, J.; Habryn, R. Tectonic Subdivision of Poland: Southern Poland (Upper Silesian Block and Małopolska Block). *Przegląd Geologiczny* **2008**, *56* (10), 912–920.

Cabala, J. M.; Ćmiel, S. R.; Idziak, A. F. Environmental Impact of Mining Activity in the Upper Silesian Coal Basin (Poland). *Geologica Belgica* **2004**, *7*, 225–229.

Chmielewska, M. Conservation of Post-Industrial Cultural Heritage in Europe in Local and Global Context. In *Geographical-Political Aspects of the Transborder Conservation of Natural and Cultural Heritage, Practice in the Field of the Transborder Heritage Conservation, Region and Regionalism*; Heffner, K., Ed.; 12, University of Łódź, Silesian Institute in Opole, Silesian Institute Society, 2015; Vol. 2, pp 133–145.

Chmielewska, M.; Gaidzik, K. The Influence of Mining on the Contemporary Cultural Landscape of Rokitnica and Miechowice (Upper Silesia). In *The History of Mining—Part of European Cultural Heritage*; Zagożdżon, P., Madziarz, M., Eds.; Oficyna Wydawnicza Politechniki Wrocławskiej: Wrocław, 2012; Vol. 4, pp 53–66.

Chmielewska, M.; Lamparska M. Post-Industrial Tourism as a Chance to Develop Cities in Traditional Industrial Regions in Europe. *Revista Sociologie Românească* **2011**, *9* (3), 67–75.

Chmielewska, M.; Sitek, S.; Zuzańska-Żyśko, E. Around the Revitalisation of Post-Industrial Urban Spaces—Case Study: Metropolitan Association of Upper Silesia. *4th International Geography Symposium—GEOMED 2016 Book of Proceedings*, 2016; pp 823–834.

Chybiorz, R.; Kowalska, M. Inventory and Assessment of the Attractiveness of Geosites in the Silesian Voivodeship (Southern Poland). *Przegląd Geologiczny* **2017,** *65* (6), 365–374.

Długoborski, W. *Rekrutacje górników w Zagłębiu Górnośląskim przed zniesieniem poddaństwa*: Warszawa, 1948.

Dulias, R. *The Impact of Mining on the Landscape*; Springer International Publishing: Switzerland, 2016.

Dzięgiel, M. The Underground Tourist Trails in Tarnowskie Góry (Upper Silesia, Poland). *Geoturystyka* **2008,** *4,* 51–62.

Gaidzik, K. Anthropogenic Landforms in the Area Between Rokitnica and Miechowice (Upper Silesia, Southern Poland). *Anthropog. Aspects Landsc. Transform.* **2010,** *6,* 29–33.

Garofano, M.; Govoni, D. Underground Geotourism: A Historic and Economic Overview of Show Caves and Show Mines in Italy. *Geoheritage* **2012,** *4* (1–2), 79–92.

Gawor, Ł. Coal Mining Waste Dumps as Secondary Deposits—Examples from the Upper Silesian Coal Basin and the Lublin Coal Basin. *Geol. Geophys. Environ.* **2014,** *40* (3).

Gawor, Ł.; Jankowski, A. T. Ruman, M. Post-mining Dumping Grounds as Geotourist Attractions in the Upper Silesian Coal Basin and the Ruhr District. *Moravian Geogr. Rep.* **2011,** *19* (4), 61–68.

Guido Coal Mine in Zabrze. https://kopalniaguido.pl/index.php/en/ (accessed Dec 03, 2018).

Historic Silver Mine in Tarnowskie Góry. www.Europekopalniasrebra.pl/en/ (accessed Dec 03, 2018).

Historic Silver Mine on the UNESCO World Heritage List. http://whc.unesco.org/en/list/1539 (accessed Dec 03, 2018).

Hose, T. A. Towards a History of Geotourism: Definitions, Antecedents and the Future. The History of Geoconservation. *Geol. Soc. London Spec. Publ.* **2008,** *300* (1), 37–60.

Industrial Monuments Route of the Silesian Voivodeship. www.zabytkitechniki.pl/en-US (accessed Dec 03, 2018).

Jaros, J. *Zarys dziejów górnictwa węglowego*: Warszawa, 1975.

Jaros, J. *Dwa wieki kopalni węgla kamiennego „Zabrze— Bielszowice"*: Zabrze, 1991.

Kossuth, S. *Górnictwo węglowe na Górnym Śląsku w połowie XIX w.*: Katowice, 1965.

Kostrubiec, B.; Lamparska, M. Mining Tourism in Hard Coal Basins in Poland and in France. In *Conditions of the Foreign Tourism Development in Central and Eastern Europe*; Wyrzykowski, J., Ed.; 2005; Vol. 8, Wroclaw; pp 97–110.

Lamparska, M. *Turystyka wśród górniczych szybów. Szlak turystyczny i przewodnik po dawnych i współczesnych kopalniach na obszarze Górnośląskiego Zagłębia Węglowego.* Wyd. „Śląsk": Katowice, 2017.

Lamparska-Stobiecka, M. Geotourism on Postindustrial Areas and in Monumental Mines. In *Teoretyczne i empiryczne zagadnienia rekreacji i turystyki*; Młynarski, W., Ed.; AWF in Katowice: Katowice, 2008; p 186–200.

López-García, J. A.; Oyarzun, R.; Andrés, S. L.; Martínez, J. I. M. Scientific, Educational, and Environmental Considerations Regarding Mine Sites and Geoheritage: a Perspective from SE Spain. *Geoheritage* **2011,** *3* (4), 267–275.

Molenda, D. *Kopalnie rud ołowiu na terenie złóż śląsko-krakowskich XVI i XVII wieku*: Warszawa, 1972.

Nita, J. Quarries in Landscape and Geotourism. *Geographia Polonica* **2012,** *85* (4), 5–12.

Nita, J.; Myga-Piątek, U. Geotourist Potential of Post-Mining Regions in Poland. Bulletin of Geography. *Phys. Geogr. Ser.* **2014,** *7* (1), 139–156.

Pérez-Aguilar, A.; Juliani, C.; de Barros, E. J.; de Andrade, M. R. M.; de Oliveira, E. S.; Braga, D. D. A.; Santos, R. O. Archaeological Gold Mining Structures from Colonial Period Present in Guarulhos and Mairiporã, São Paulo State, Brazil. *Geoheritage* **2013**, *5* (2), 87–105.

Piątek, E., Ed.; *Kopalnia Guido w Zabrzu, Fragment górnośląskiego górnictwa węglowego*: Pszczyna, 2013.

Piernikarczyk, J. *Historia górnictwa i hutnictwa na Górnym Śląsku*: Katowice, 1933.

Prosser, C. D. Geoconservation, Quarrying and Mining: Opportunities and Challenges Illustrated Through Working in Partnership with the Mineral Extraction Industry in England. *Geoheritage* **2018**, *10* (2), 259–270.

Queen Louise Coal Mine in Zabrze. http://sztolnialuiza.pl (accessed Dec 03, 2018).

Riley, R.; Tkocz, M. Coal Mining in Upper Silesia under Communism and Capitalism. *Eur. Urban Reg. Studies* **1998**, *5* (3), 217–235.

Riley, R.; Tkocz, M. Local Responses to Changed Circumstances: Coalmining in the Market Economy in Upper Silesia, Poland. *Geo. J.* **1999**, *48* (4), 279–290.

Route of Industrial Heritage. www.erih.net (accessed Dec 03, 2018).

Rostański, A. Natural Value of Post-Industrial Waste Heaps. In *Landscape Built on Coal*; IETU: Katowice, 2008; pp 149–156.

Schwarz, P.; Lorenc M. W. From Mining the Landscape to Mining Wollets: Mining Heritage Tourism in Four European Regions. In *Geotourism. A Variety of Aspects*; Słomka, T., Ed.; Kraków, 2011; pp 261–296.

Slotta, R. *Das Carnall-Service als Dokument des Oberschlesischen Metallerzbergaus*. Veröffentlichungen aus dem Deutschen Bergbau Museum: Bochum, 1985.

Staszewska, A.; Żemła, M. The Industrial Monuments Route of the Silesian Voivodeship as an Example of the Regional Tourism Product Enhancing Tourism Competitiveness of the Region. *Czech J. Tour.* **2013**, *2* (1), 37–53.

The Black Trout Adit. sztolniapstraga.pl/en/ (accessed Aug 05, 2018).

Wiśniewski, L. Główna Kluczowa Sztolnia Dziedziczna jako zabytek techniki zaliczany do dziedzictwa kulturowego. In *The History of Mining—Part of European Cultural Heritage*; Zagożdżon, P., Madziarz, M., Eds.; Oficyna Wydawnicza Politechniki Wrocławskiej: Wrocław, 2009; Vol. 4, 2, pp 311–325.

Xie, P. F. Developing Industrial Heritage Tourism: A Case Study of the Proposed Jeep Museum in Toledo, Ohio, *Tour. Manag.* **2006**, *27*, 1321–1330.

INTERNET SOURCES

www.zabytkitechniki.pl/en-US-Industrial Monuments Route of the Silesian Voivodeship (access December 03, 2018)

www.erih.net - Route of Industrial Heritage (access December 03, 2018)

www.Europekopalniasrebra.pl/en/-Historic Silver Mine in Tarnowskie Góry (access December 03, 2018) sztolniapstraga.pl/en/-The Black Trout Adit (accessed August 05, 2018)

http://whc.unesco.org/en/list/1539-Historic Silver Mine on the UNESCO World Heritage List (access December 03, 2018)

http://sztolnialuiza.pl - Queen Louise Coal Mine in Zabrze (access December 03, 2018)

https://kopalniaguido.pl/index.php/en/-Guido Coal Mine in Zabrze (access December 03, 2018)

CHAPTER 22

Geotourism vs Mass Tourism: An Overview of the Langkawi UNESCO Global Geopark

KAMARULZAMAN ABDUL GHANI

Founder President FLAG—Friends of Langkawi Geopark, Malaysia

E-mail: kzamang@gmail.com

ABSTRACT

Langkawi was endorsed as a Geopark in 2007. It was then the first global Geopark in Southeast Asia. In 2015, it officially became known as Langkawi UNESCO Global Geopark. As a Geopark, it has the opportunity to apply the concept of geotourism in what we perceive to be educational and sustainable tourism, thereby, bringing Langkawi tourism to another height with new products and interest. Langkawi in the past has always promoted itself as a nature-based tourism and ecotourism. However, as a duty-free island, it also attracted a sizeable number of visitors, mainly from the domestic front. Its popularity as an island holiday destination saw brisk physical development in terms of transport infrastructure and tourism related development such as hotels and accommodation, commercial complexes, and restaurants. The measure of progress in tourism has always been indicated by the number of visitor arrivals. Hence, mass tourism is the order of the day to a point of *overtourism*. In whatever form and terminology, the fact remains that the increasing number of visitors when uncontrolled and unregulated brings with it sustainability issues in various areas of the ecosystem related to overdevelopment, namely environment, waste disposal, electricity, and water supply. This write-up seeks to discuss the issues of conflict and balance between mass tourism, overtourism, and Geopark tourism, specifically geotourism and what the outlook of Langkawi UNESCO Global Geopark could preferably be.

22.1 INTRODUCTION

Langkawi is an archipelago of 99 islands. It has an area of 477 sq. km with a population of 155,262 inhabitants (LADA, 2017; Department of Statistics Malaysia, 2010). It lies in the northwest of Peninsular Malaysia under the State of Kedah. It was declared a Geopark by the Kedah state in 2006 and subsequently, in 2007 it became a member of the Global Geoparks Network. In 2015, it officially became Langkawi UNESCO Global Geopark (LUGG). As a 99-island Geopark, its area size is estimated to be 920 sq. km. The Langkawi Development Authority (LADA), a corporate body of the Federal Government, is charged with the development of tourism and infrastructure and socio-economy. It is named as the managers of the Geopark. It has within its organization a Geopark Division tasked with the oversight, planning, and development of the Geopark. Of course, under the Geopark, the responsibility of managing the Geopark directly or indirectly, must involve all other government departments and agencies that operate within the Geopark area. Such departments, among others, include the District Office, the Local Municipality, Department of Forestry, Department of Environment, Department of Drainage and Irrigation, Department of Fisheries, the Marine Department, the Agriculture Department, Police and Immigration. This island destination is steeped in legends. The battle of giants is connected with the Cambrian Gunung Machinchang mountain range, which is now home to Langkawi Cable Car and Skywalk, the most popular tourist product in Langkawi. The story of Mahsuri, that of a lady falsely accused of adultery and executed to death who then casted a curse that Langkawi be barren for seven generations, is one popular legend that attracts tourists to Kota Mahsuri or the Mahsuri Tomb. Tasik Dayang Bunting is a fresh water lake where it is believed that a dip in or a sip of the water will cure infertility in married couples.

Langkawi is rich with tropical forest, which are largely protected, whereby 26,266 ha (54.6%) of the island is gazetted as Permanent Forest Reserves. This includes the mangrove forests, covering 3126 ha or 6.5% of the island (Statistics by LADA, 2017). Much of the greens have been compensated for development but nature remains the mainstay of Langkawi tourism. In 2011, the government used the tagline "Naturally Langkawi" and also "the Jewel of Kedah" for promotion purposes. The endorsement of Langkawi UNESCO Global Geopark that brings on geotourism makes complete the nature-based tourism that Langkawi is known for.

22.2 GROWTH IN TOURISM

The one important initiative to further popularize Langkawi as an island destination was its declaration as a duty-free island in 1987. The local council was then named Langkawi City Tourism Municipality. Henceforth, in 2011, the government initiated the Langkawi Tourism Blueprint 2011–2015 to further chart development for Langkawi, involving the construction of more hotels and expansion of the international airport. Langkawi has a complete infrastructure of roads, jetties, and international airport to facilitate both domestic and cross-border travel. Ferry services ply to and from four major points, namely Kuala Kedah, Kuala Perlis, and Pulau Pinang; and also Satun in Thailand. The main jetty in Kuah, Langkawi handled over 2 million passengers annually. As for the airport, it served 238 flights a week and handled 1.35 million passengers in 2017 (Statistics by Malaysia Airports Sdn Bhd, 2017). Visitor arrivals in 2017 were recorded at 3.68 million, increasing over two-folds from the 1.81 million recorded in 2006. There are currently 247 hotels of various categories in LUGG, offering 11,896 rooms while a few more are coming up alongside commercial complexes to tap the tourism dollars that are coming in.

22.3 GEOTOURISM AND MASS TOURISM ISSUES

Mass tourism looks up to numbers of visitors coming in droves to a destination on planned packages, which eventually would lead to overtourism whereby local residents become outplaced. This occurs when rent prices push out local tenants to make way for holiday rentals, narrow roads become jammed with tourist vehicles, wildlife is scared away, tourists cannot view landmarks because of the crowds and when fragile environments become degraded (https://www.responsibletravel.com/copy/what-is-overtourism).

Geotourism, on the other hand, is more inclined to knowledge-based tourism and very much related to a Geopark. It is defined as the provision of interpretative and service facilities to enable tourists to acquire knowledge and understanding of the geology and geomorphology of a site beyond the level of mere esthetic appreciation (Hose, 1995). It is about people going to a place to look at and learn about one or more aspects of geology and geomorphology (Joyce, 2006).

Geotourism is a distinct subsector of natural area tourism firmly entrenched in "geological" tourism, which is a form of natural area tourism that specifically focuses on geology and landscape (Dowling and Newsome, 2006). Also, "*It is knowledge-based tourism, an interdisciplinary integration of the tourism industry with conservation and interpretation of 'abiotic nature' attributes, besides considering related cultural issues, within the geosites for the general public*" (Sadry, 2009). This definition augur well in the case of LUGG as it has placed equal importance to the cultural and socioeconomic development aspects aside from the geological knowledge and education.

It cannot be denied that surmounting pressure of mass tourism is piling on LUGG. The introduction of the Vehicular Ferry Services in 2016 allowed for holiday makers to bring in their vehicles into the island. Record shows that 125,069 vehicles entered Langkawi in 2017 compared to 66,826 just a year ago. This contributes to traffic congestion as well as increase in carbon emission, affecting air quality. The government in 2015 instituted a low carbon policy, which would augur well in arresting part of the air pollution problem.

Mass tourism creates higher demand for accommodation and commercial premises, leading to the physical development of new land areas. As land matter is under the jurisdiction of the State, some land areas previously gazetted as permanent forests were degazetted to allow for development. Although large-scale development requires that an Environment Impact Assessment be prepared and submitted to the authorities, some developers do not comply while the social impact has been neglected or not been properly addressed. Thus, we see water pollution being contributed by bad sewerage and garbage disposal affecting rivers and the beaches as well as street congestion due to lack of parking spaces. Parking problem is severe in the Pantai Chenang area, a popular mass tourist destination, showcasing the beach area as well as a vibrant shopping and eatery center.

As for solid waste disposal, it has become a very challenging task as Langkawi depends on a landfill that is fast reaching its full capacity. The incinerators have been defunct and are out of the equation in addressing waste disposal issues. Plastic wastes find its way into drains, rivers, and finally into the ocean. Authorities could not cope in clean up. On certain occasions, clean-up campaigns are conducted by volunteers and NGOs. However, in isolated beaches not frequented by tourists, shored plastic wastes are a common sight.

Mass tourism and overtourism also put pressure on utilities. Water and electricity are mainly supplied by the mainland state of Kedah via undersea cable and pipes. Even though Langkawi has never faced a severe shortage, it is a sustainability concern in the long term.

Geoparks as a tool for sustainable development must concern itself with the impending unregulated and uncontrolled influx of tourists. Geoparks must set the examples in championing sustainability through geotourism. As a tourism concept, the acceptance of geotourism is relatively slow because Langkawi in history has already been known as a tourism island. For tourism players and visitors alike, it does not really matter what brand of tourism is introduced. Mass tourism brings better income for the former while new attractions will generate visitor interest.

Geotourism can generate new interests if properly promoted and popularized. The pioneering initiative of the three Geoforest Parks (GFP) in LUGG, namely Machinchang Cambrian GFP, Kilim Karsts GFP, and Dayang Bunting Marble GFP was an excellent start in introducing geotourism. They open the windows of knowledge in understanding basic geology, biodiversity and conservation of natural assets. When properly presented and implemented, the value and quality of tourism can be effectively enhanced. Visitors will better respect and love nature as well as stay longer in order to experience a wider coverage of interesting geosites within the Geopark.

No conclusive study has been undertaken to ascertain that visitors flock into LUGG for the specific reason that it is a Geopark and that geotourism is their purpose and main interest during the visit. A proxy evidence of a specific geotourism activity would be when visitors, accompanied by a guide, would visit the Bat Cave, enjoy the mangrove boat ride tour, and probably visit the fossil island of Pulau Anak Tikusin Kilim Karsts GFP. The most convincing of proof a geotourism visit, however, would be their memorable photographic record of the most popular iconic Hollywood-type ID signage of "Kilim Geoforest Park," erected on the wall cliff of an island karst in 2016. The latest addition in LUGG in its bid to further promote geotourism is the establishment of the Geo-Bio Trail of Kubang Badak in 2018. One reason for introducing this area is to offer an alternative geotourism experience as well as to ease congestion in the already popular Kilim Karsts GFP that have similar attractions of boat rides among tropical island karsts, mangroves, caves, and a mingle with the fishing community. Just like Kilim, this area is managed by the local community through a cooperative organization, an important attribute of a successful Geopark.

22.4 CIVIL SOCIETY MOVEMENT

As far as environmental issues and care for natural assets in Langkawi, several prominent voluntary nongovernmental organizations are worth mentioning. Trash Hero Langkawi is a voluntary organization that commits itself to collect rubbish in public areas. They do it every Saturdays to create awareness among the general public about keeping the environment clean from litter. They receive support and participation from hotels and also from LADA, particularly by their Geopark Rangers. Rotary Club Langkawi, although aligned to welfare activities and assistance has embarked on producing reusable shopping bags in order to reduce the use of plastic bags on the island. The Malaysian Nature Society in Langkawi has raised issues of deforestation. This organization has also raised the issues of water pollution and degradation of mangrove forest within the Kilim Karsts Geoforest Park, a popular tourist destination within Langkawi UNESCO Global Geopark area. Together with Rotary Club, FLAG-Friends of Langkawi Geopark, LADA and the Forestry Department, they have undertaken a massive river clean-up in the Kilim River. In 2016, this group also undertook a Malaysian record-breaking feat in planting 21,000 mangrove saplings in Pulau Dayang Bunting, which is the second largest island in Langkawi.

FLAG-Friends of Langkawi Geopark, established in 2014, is a voluntary NGO designated to support the Langkawi Geopark. It arose amid concerns about the sustainability of the Geopark as revealed during the first revalidation exercise undertaken by the Global Geoparks Network in 2011 where in environmental issues and the Geopark identity were raised. Such sustainability issues if not properly addressed will put the global Geopark status of Langkawi at risk. It is believed that the success of a Geopark is driven by its people. So, FLAG was intent on popularizing the Geopark concept and show the world that we have a Geopark at work with the community playing a significant role. It conducted workshops and seminars about environmental care and sustainable development, organic farming, village guide training, including one forum in 2016 presenting the then former Prime Minister of Malaysia "Tun Dr. Mahathir" who talked about conservation of the island's natural assets. In the same year, it embarked on a marine conservation program to enhance fish population in the straits of Pulau Tuba and the riverine area of Pulau Dayang Bunting. This involves the construction and deployment of an innovative Fish Aggregating Device, locally called the APEHAL, designed to shelter juvenile fish from larger predators after they leave the mangrove sanctuaries. This device uses discarded vehicle tires, which otherwise would have been dumped in the island's waste disposal landfill or even into the sea. FLAG upholds the notion that "Geoparks are founded upon geology but the success of a Geopark is driven by its people."

It is important to note that the civil society of LUGG has raised concerns officially to the authorities about the environmental degradation in the Kilim Geoforest Park, future proposals for development in the coastal areas of

Kuah town and the Pantai Chenang Beach. Most importantly is put forward the quest for a plastic-free Langkawi for the simple reason that being an island, plastic wastes find its way faster into the sea and plastic hazards in the ocean has become a global concern.

22.5 CONCLUSION

Observing the development of tourism in LUGG, it is imperative that administrators and the community within the Geopark appreciate and understand how to move the Geopark agenda for the sustainability of natural and cultural assets, environment, and the community at large. It is still a daunting task in Langkawi UNESCO Global Geopark. Geotourism can further flourish if it encompasses a wider spectrum of attractions and interests and not confined to just geological features and appreciation but beyond that, the elements of biodiversity and culture. This would qualify "geotourism" to be "earth tourism" as what the prefix "geo," "ge" mean in Science, Latin, or Greek; and would likely be better connected to the general public or tourists alike. With better understanding and appreciation, the value of nature will be better placed in the hearts and minds and people would sincerely protect what they grow to love. Hence, the Geopark brand brings with it the hope for a sustainable large territory and perhaps the whole planet, which truly is in a sense, a Geopark area by geological concept. Each Geopark is unique in its own right and how it is presented to the public. LUGG pioneered in the concept of Geoforest Parks and together with the recent introduction of the Kubang Badak Geo-Bio Trail, it will secure the position of geotourism in allaying the negative impact of mass tourism and overtourism.

KEYWORDS

- overtourism
- mass tourism
- geotourism
- **Langkawi UNESCO Global Geopark**
- **FLAG-Friends of Langkawi Geopark**

REFERENCES

Department of Statistics Malaysia; Government of Malaysia, 2018.

Dowling, R. K.; Newsome, D. Geotourism's Issues and Challenges. In *Geotourism*; Dowling, R. K.; Newsome, D., Eds.; Elsevier/Heineman: Oxford, 2006; pp 242–254.

Hose, T. A. Selling the Story of Britain's Stone. *Environ. Interpret.* **1995,** *10,* 16–17.

Joyce, B. *Geomorphological Sites and the New Geotourism in Australia,* 2006; p 2 Available from: website: earthsci.unimelb.edu.au (accessed June 25, 2014).

Ghani, K. A. The Making of Langkawi Geopark, LESTARI Public Lecture No.13, Institute for Environment and Development (LESTARI), UniversitiKebangsaan Malaysia, 2014.

Langkawi Development Authority (LADA), Government of Malaysia. https://www.lada.gov.my.

Langkawi Tourism Blueprint 2011–2015, Prime Minister's Department Malaysia.

Langkawi UNESCO Global Geopark. https://www.lada.gov.my/mengenai-kami/produk/langkawi-unesco-global-geopark.

Sadry, B. N. *Fundamentals of Geotourism: With Special Emphasis on Iran*; SAMT Publishers: Tehran, 2009 (English summary available online at: http://physio-geo.revues.org/4873?file=1; Retrieved: Jan. 01, 2020) (in Persian).

Shafeea, L., Ed.; Langkawi Geopark, Institute for Environment and Development (LESTARI), Universiti Kebangsaan Malaysia, 2007.

CHAPTER 23

The Future of Geotourism

BAHRAM NEKOUIE SADRY

Adjunct Senior Lecturer and Geotourism Consultant

E-mail: Bahram.Sadry@gmail.com

ABSTRACT

This chapter concentrates on the key issues for future sustainable management of geotourism and the parameters which can maximize the positive impacts of abiotic nature-based tourism in the 21st century. The topics discussed in the book covered by contributors are reviewed, synthesized, and other new matters are introduced. In order to expand concepts, various theoretical and practical observations are brought together, and the editor also develops some guidelines or strategies essential for geotourism planning and development. This chapter's main effort is to provide a deep insight into abiotic nature and tourism connections, leaving a list of items for future discussion.

23.1 INTRODUCTION

According to Newsome and Dowling (2018: 481), "Globally, geotourism has come a long way and fostered a greater awareness of geodiversity which in turn has raised the profile of geoheritage and the need for geoconservation." The geotourism concept is developing dramatically. Reviewing comprehensive academic works of geotourism published since 1995 by various authors from different countries and different languages (e.g., Albanian, Chinese, English, Persian, Spanish) shows that in the last decade of the 21st century (2008–2018), there has been a considerable development of the geotourism concept in the literature alongside the geopark movement, which is an amalgam of sustainability and geoheritage conservation. Geotourism per se has emerged in the Americas not long ago and we have witnessed its systematic development in North as well as South America, in countries like Canada, Brazil, Peru, Mexico, and Uruguay; much further away, for example, Turkey has stated its geotourism and geopark development supporting its being an already successful tourism destination.

Geosites in the USA have been center of attention for a long time and there has been a significant effort in connecting people to places in the US National parks along with exclusive and progressive implications. Nevertheless, there has been a focus on the concept of modern geotourism ever since Thomas Hose (1995) introduce his definition, followed by great efforts such as by Ross Dowling and his colleague David Newsome with geotourism publications in modern era, as well as holding seminars in last 12 years. There have been many people involved in this development, some of them have cooperated in the present book such as Murray Gray, who made a great contribution in attracting academics' attention to abiotic nature conservation and studies by publishing the 2004 book "*Geodiversity*." He truly inspired other researchers and expanded their knowledge, so that the term "abiotic nature" was used to introduce a Geotourism definition in 2009 as: "*Geotourism is a knowledge-based tourism, an interdisciplinary integration of the tourism industry with conservation and interpretation of abiotic nature attributes, besides considering related cultural issues, within the geosites for the general public.*" The trend is obviously detectable, as the geopark development displays geotourism development globally. Raising awareness and interest assures a bright future for geotourism since many tourism public and private sectors, NGOs, geological institutions, environmentalists, geological and geomorphological societies, and mining engineers are increasingly addressing the current issue of geotourism around the world.

Our knowledge of geotourism is increasing dramatically all around the world, raising awareness and developing knowledge of geological heritage, geoheritage interpretation, geotourism definition, and so geopark extension in today's systematic world is not comparable to what it was in 2000. Therefore, based on recent progress, the first section below proposes some guidelines and subsequently addresses a number of issues for the development of geotourism in the 21st century. Following this, a brief conclusion will be presented at the end of the book.

23.2 GUIDELINES FOR THE DEVELOPMENT OF GEOTOURISM IN THE 21ST CENTURY

Albeit in the early stage of development, geotourism has several issues to address as part of its evolving future. They include:

23.2.1 THE NEED FOR DEVELOPMENT OF ROADSIDE GEOLOGY FOR GEOTOURISM PROMOTION

There is a huge potential for geotourism development in roadside cuttings. As mentioned by William Witherspoon and his colleague John Rimel (in chapters 12 and 13) and also based on the collection of the *Roadside Geology* book series published in the USA, promoting practical roadside geotourism would be possible through cooperation of geological organizations, tourism private sectors, and geological associations and volunteers, etc., in all countries. These efforts would gradually result in raising people's awareness and increasing the popularity of geology. It would also have a positive impact on the creation of national geoparks, geological heritage conservation, and geotourism development.

In the first stage, considering and starting the geological mapping of road cuts and providing gradually some roadside outcrops as geosites for travelers and geotourists around the world seems essential. It is worth mentioning that developing interesting outcrops and particularly those that are accessible along the roadside in order to create and develop interest among people is not a new phenomenon, already accomplished in some countries, such as Malaysia. According to Lesslar and Lee (2001 cited by Tongkul, 2006), the Miri Airport Road Outcrop is the first geological site to be set up as an outdoor museum to promote geotourism in Malaysian Borneo and this project was conceptualized in 1998 by a group of volunteers from Sarawak Shell Berhad (a gas field operator)[1], a gas field operator who started to work practically in 2000. The successful US *Roadside Geology* book and map series published and such Geosites on the other hand, as the Miri Airport Road Outcrop, besides other samples and such schemes in the UK and Australia can be models for all countries to work on, promoting and marketing their roadside geology sites, gradually as roadside geotourism sites in the 21st century.

23.2.2 THE NEED TO INTERPRET HERITAGE BY MODERN PRINCIPLES

According to Hose (Chapter 2), geotourism is a geology-focused and visitor-centered development of environmental interpretation. While there are completely distinct requirements in the modern and postmodern era, some interpreters still practice the six principles of Tilden. The American Freeman

---

[1] https://www.shell.com.my/media/2018-press-releases/fid-for-new-gas-fields-offshore-sarawak.html

Tilden was one of the first people to set down principles of interpretation in his book, *Interpreting Our Heritage* (1957), which has been mentioned in detail in this book by Dowling (Chapter 11). Therefore, practicing 15 modern principles and incorporating them into geological heritage interpretation, as Ted Cable explains with mining case studies (in Chapter 9), seems necessary in the 21st century. The importance of geological interpretation has already been noticed, but there is still a long way to go in doing research and designing geological interpretation for different age groups [using Beck and Cable's (1998) 15 principles of interpretation] and for the disabled. Regarding the new geointerpretation and non-formal education strategies, "It is noted that geoheritage is about understanding and conserving geological formations and their processes due to their esthetic, artistic, cultural, ecological, economic, educational, recreational, and scientific values" (Newsome and Dowling, 2018a). According to the argument discussed by Ross Dowling (Chapter 11), the ABC approach toward nature could be addressed in interpretation centers for the public through cooperation of cultural, biological, and geological specialists in geosites. This would greatly add to the diversity and attraction of the geopark and could contribute to the environment and raise public awareness. As a result, combining the "ABC" approach to the interpretation with Larry Beck and Ted Cable's 15 principles can play an effective role as a new strategy for geointerpretation within geoparks in the 21st century.

23.2.3 THE NEED FOR PROMOTION OF GEOTRAILS

According to Hose (Chapter 10), "Geotrails are a significant element in geotourism provision and are the commonest and most geographically widespread form of modern geotourism provision;" as the Geological Society of Australia (2018) has formally defined: *"A Geotrail delivers geotourism experiences through a journey linked by an area's geology and landscape as the basis for providing visitor engagement, learning and enjoyment."* And added to that, "Geotrails are relatively easy to establish and represent a very cost-effective means of enhancing regional development. Geotrails are best constructed around routes currently used by tourists; in essence, geotrails should form logical journeys linking accommodation destinations." Various geotrail project proposals by researchers and executives and work to expand geotrails in geoparks and geotourism sites in the 21st century could be taken into consideration as a strategy to develop geotourism. Geotrails have a positive role in reducing negative impacts of tourism. In order to mitigate these negative impacts, developing geotrails and ecotrails as a strategic solution in

Langkawi Geopark in Malaysia as a successful tourism destination case can be considered, which is suggested by Kamarulzaman Abdul Ghani in Chapter 22 of the present book.

As Hose discussed earlier (Chapter 10), new geotrails might better appeal and be accessible to wider and younger audiences and some of the interpretive panels will be gradually replaced by smartphone, tablets, and computers in geotrails by geotrail providers in the 21st century, such as some works already done in Portuguese Geoparks in 2007.

According to Macadam (pers. comm., Dec. 17, 2018) "Guided walks are expensive – on staff time – but incredibly valuable if used intelligently to find out what people want to know, what they already know, and what they think of the present provision for visitors." (see also Macadam, 2015). Macadam continues his statement about apps, "I have used a good app for Rokua Geopark's Ice Age landscape, and another (TOURinSTONE) for the use of stone in the historic centre of Turin, but it is early days." Making use of three mobile phone apps employed in the context of geotourism for the development of urban geotourism is a good example in Italy; Pica et al.(2017, cited by Newsome and Dowling, 2018: 477) mentioned one of these as, "the GeoGuide Rome mobile application provides the opportunity for tourists to access geological features and information about Rome in 18 stops." Therefore, digital transformation including development of smart phone applications, website, and visitor center enhancements gradually will be quite popular in the geotourism market and it will make heritage interpretation and designing geotrails grow in the 21st century.

23.2.4 THE NEED FOR DEVELOPMENT OF URBAN GEOTOURISM

Urban geotourism is a sub-sector that could be studied and developed under urban tourism courses in various countries [e.g., see Gaidzik (Chapter 3 in this book)]. Many cities with outcrops of an abiotic nature, as its background, present geological phenomena that can be studied as urban geology and geomorphology (see "Start talking about geoheritage where people are" in Macadam, 2018b). These could also play a significant role in raising people's awareness on how to confront natural—and geological—hazards, cities most likely to be hit by earthquake, for instance. After the Tohoku earthquake and tsunami in 2011, a new theme "natural hazards" was adopted in Japanese geoparks as an example, which has also been mentioned in this book by Vafadari and Cooper (Chapter 14). Secondary geosites like geoscience museums and meteorite museums, dinosaur museums, etc., are also capable of being equipped with some educational tools such as geological heritage interpretation, which is compatible with geotourism development objectives. Using a "natural hazards" theme is useful for other major cities (for development of urban geotourism) and geoparks in other countries.

23.2.5 THE NEED FOR DEVELOPMENT OF NEW TRAINING COURSES AND ACADEMIC MAJORS

According to Farsani et al. (2012), "Tourists have changed in the 21st century compared to the past: their expectations of travel and sightseeing experience and also destinations' attraction is quite different today. Geotourism development mainly depends on an efficient approach for destination and product management. Educated travelers are looking for enriched travel experiences, therefore destinations should enhance their proficiency to compete with other destinations. Also, according to Dowling and Newsome (2017), "It is suggested that while many techniques (e.g., electronic devices and media) are available to deliver interpretive content, it is direct face-to-face contact with guides that can deliver the most responsive, directed, and varied content to visitors." To develop geotourism, designing special sub-sectors for geotour guide training courses for geoparks and for any individual primary and secondary geosites seems essential. To illustrate, the official training courses of geotour guides (Sadry, 2014), which have been held since 2014 in Iran and since 2008 in Fernando de Noronha archipelago of Brazil (Moreira and Bigarella, 2010), or specific georafting-tour guide training courses recently developed and held in Austria (GGN Newsletter, 2018) are successful educational examples that seem crucial to develop the geotourism industry. Here some specific training courses are proposed to develop, for example, urban geotourguides, celestial geotour guide training courses, cave geotour guide training courses, volcano geotour guide training courses, mining heritage geotour guide training courses or "training geotour guides for deaf/mute visitors" (Farsani et al., 2012: 156) and so on; all could be practical options.

Allan (Chapter 19) states that "Providing accessible geotourism facilities and services for people with special needs is an ethical practice and a primary responsibility for geotourism authorities and organizations to help visitors with special needs to practice their basic rights and enhance their wellbeing'. As Dowling and Newsome (2017) argue, "there is a real need for supervision in the form of guided tours and/or ranger presence." Developing specific geotour guide training courses can play an important role for geotourism promotion and be a management strategy for geotourism authorities and organizations in the 21st century. Also, hiring specific well-trained accredited geotour guides in geosites is economically beneficial for individual geotourism sites and geoparks since, according to Cheung (2016, cited by Dowling and Newsome, 2017), the findings from a survey of visitors to Hong Kong Global Geopark imply that geotourists are willing to pay extra for high-quality geo-guided tour services. As for higher education, although some programs relating to geotourism, geopark, and geoconservation management and their related publications have been introduced at higher education level in Portugal, Poland, Australia, Iran, and Romania (Dowling and Newsome, 2010; Sadry, 2012a; J. Macadam, pers. comm., Jan 31, 2020), the development of such academic programs or publishing of national textbooks seems logical in all corners of the world to service the increasing geopark creations around the world.

23.2.6   THE NEED FOR DEVELOPMENT OF DINOSAUR GEOTOURISM

Many dinosaur and related fossils have been found in different countries during the last two centuries. These sites have been or have the potential to be developed as dinosaur or other fossil geotourism sites, which are especially popular, thanks to movies and TV programs and educational children's books. Dinosaur geotourism sites are among the segments capable of being expanded in the 21st century. This great potential for geotourism development in the 21st century has been explored in this book by Cayla (Chapter 18). "There are successful examples of such sites around the world, such as Fukui geopark in Japan. There are also some sites discovered in recent months, for instance, in Germany, where a site is being investigated for potential geotourism activities with a museum or cultural differences in presentations and in displays" (N. Cayla, pers. comm., Dec. 24, 2018). Also, Dinosaurierpark Munchenhagen, the Colorado Dinosaur Monument wall, Ghost Ranch, Zhigong Geopark, Drumheller, Winton Lark Quarry, and the Australian Age of Dinosaurs are other successful examples. The topic could also be studied among interdisciplinary researchers in various countries.

23.2.7   THE NEED TO DEVELOP AN ACCESSIBLE GEOTOURISM

It is important for geotourism management to improve facilities for disabled and senior adults. Disabled "geo-visitors" need new and special services and facilities in terms of physical facilities, comfort and styles of heritage interpretation. In the 21st century, the geotourism sub-sector of the sustainable tourism industry should pay particular attention to the accessible geotourism market and elder geotourism market. Retired elder people are known to have more spare time and more money to travel—the "gray nomads." On the other hand, there have been limited investigations on the geological interpretation for the disabled, yet, according to Turner (pers. comm., Sep. 24, 2018), this is not the case in the museums world where it has been done for over 100 years—using sound, tactile displays, etc. (e.g., Mather, 1986); for example, she opened up the Hancock Museum geology displays to visitors from the Blind School, Newcastle upon Tyne in the 1970s.

Accessing the geotourism market for everyone including people with special needs, seniors, and those with temporary illnesses seems essential. For example, in Naturtejo da Meseta Meridional UNESCO Global Geopark, Portugal, prepared a Braille book for blind children for the development of an accessible geotourism market, and, in southern Spain there is an outstanding example in the Casa de los Volcane in the Cabo de Gata-Níjar UNESCO Global Geopark, where the managerial body decided to design wheelchair access for handicapped visitors into lava tunnels and also "they provided a tactile diorama of a volcano for blind people" (Farsani et al., 2012: 156); geology trails for the blind have been developed in the Dinosaur Diamond Byway, Utah, USA (see Walden and Pearce, 2010).

In terms of holding geotour and interpretive provisions for people with special needs, treating equally all divergences from some notional "ideal" target audience, making geotourism infrastructure and/or writing interpretation is an issue (J. Macadam, pers. comm., Dec. 17, 2018). As a solution, preparing geotrails and interpretative centers and bringing non-formal educational tools into existence for disabled geotourists can help to promote interpretive activities in geosites that sometimes would be easily achievable. Also, applying modern media for ordinary people, as well as those with special needs, such as outdoor apps, apps in museums, AR, VR, MR, drones, virtual (and real) mine models, 3D printing, virtual and real reconstructions, QR codes, real and virtual timelines, etc., relevant to differently abled visitors and VR for use in accessible places (e.g., caves) is essential (J. Macadam, pers. comm. Dec. 17, 2018). This could be used as a modern tool to geological interpretation and mining heritage for ordinary people and those with special needs (e.g., see "New, and not so new, media" in Macadam, 2018b, 2018). For example, there are some successful uses of mining heritage stories in Cornwall and the Copper Coast geopark, Ireland.

23.2.8 THE NEED FOR DEVELOPMENT OF CELESTIAL GEOTOURISM

The potential of geotourism development is still neglected in 21st century. As I discussed in detail in Chapter 20 about "space and celestial geotourism," a new categorization has been proposed, geotourism development would be possible from astronomical observation, expansion of secondary geosites to the planetariums and lands similar to other planetary perspectives (as geotourism sites). There are also geoparks with observation capabilities that could be used with the purpose of merging these activities with planetary (space) geological interpretation. The development of celestial geotourism in the form of celestial geosites has the potential to create a new market in geotourism. It would also raise social and environmental awareness, which is aligned with the geotourism paradigm and sustainable development. Geotourism development through celestial geosites, and as a branch of sustainable tourism, can be a solution for rural development. This is particularly a strategy to alleviate poverty in the developing countries, preventing migration from rural areas and creating employment in remote areas with magnificent, free of pollution, starry skies, ideal for celestial observation, which could be taken into consideration in the 21st century to develop the geotourism industry.

23.2.9 THE NEED TO CREATE MORE NATIONAL AND GLOBAL GEOPARKS

The more comprehensive approach to geoconservation aims to protect all natural resources on Earth including the physical and abiotic; we should do our best to take care of nature, not rule over it (Gray, 2004). According to Hose (2006), geoconservation measures and tourism promotion are two fundamental aspects of geotourism. Establishing geoparks is a kind of strategy toward removing poverty and providing geoconservation for management in the 21st century. According to UNESCO (2016), a geopark is a geographical site with amazing geological features with preservation, educational, and sustainable development purposes. Farsani et al. (2012) identifies geoparks as a basic factor in the expansion of geotourism, which is both a unique strategy to preserve the Earth as well as developing sustainable tourism in rural areas. Geotourism contributes to the development of geoparks, which has led to regional investment, provided career opportunities (especially for women: e.g., Turner, 2013) and increases income generation among local people.

UNESCO started to cooperate with the already established European geoparks since 2001. By 2004, 17 European and eight Chinese geoparks gathered together in Paris to establish Global Geoparks Networks (GGNs). All members of the Network can benefit from the actions that take place to preserve geological heritage (UNESCO, 2018).

The program within the Earth Sciences Division began as a UNESCO-assisted program and started immediately after the First International Geopark conference held in Beijing, China, in June–July 2004 but was not an official program until 2018. UNESCO announced that there are already 147 Global geoparks in 41 countries by the end of 2019. There is hardly any available evidence on the economic impacts and the effects of geoparks on local communities, except for some reports by China and Europe. The UK National Commission for UNESCO estimated that the UK economy had been assisted with about £18.8M from the GGN in 2013. Investigating the eight Chinese global geoparks shows that income generation in 4 years after the creation of these geoparks had tripled in those regions compared to the period before the creation (Ng, 2015, cited by Dowling and Newsome, 2017). Another successful example is Yuntaishan Global Geopark, where the economy of the region was revolutionized with income generation from the geotourism sector; it was significantly increased about 50 times during 2001 ($1.189M) to 2012 ($5B). Newsome and Dowling (2018) point out that the Chinese government has developed a strategy to reduce poverty in rural areas and local people involved in tourism activities are enjoying a considerable income. Ng (2017) stated that each geopark in China is worth annually around US$26 million and about 1.42 billion tourists visit these geoparks, which bring about a revenue of US$90 billion. Accordingly, Newsome and Dowling (2018) believed that the considerable economic impact represents a great opportunity for managers, staffs, and even indirect service providers. Geotourism is still in its infancy. However, it is expected that geoparks will become a primary tourism destination for those who are looking for the earth's amazing landscapes (Farsani et al., 2012; Turner, 2013). About 12 Chinese national parks have been established annually since 2000 (Ng, 2017), although as noted, except for some Chinese reports that are quite valuable, there are limited reports and publications available on the economic effects of geoparks, the European network and the GGN. According to Newsome and Dowling (2018), geotourism can be considered as a force driving geoparks that ultimately leads to UNESCO's purpose for all geoparks, which is sustainable development. The GGN that was created originally by the UNESCO Earth Sciences Division (ESD) under Wolfgang Eder now has spread over five continents except for Australia (which did have a global geopark "Kanawinka" 2007–2013). Although the environmental and economic impacts of geotourism have hardly yet been recognized, it is expected that geopark expansion will continue in the future as more governments discern the value.

There has also been scarce record of analytic reports to prove poverty alleviation as an impact of geopark development in the different regions of the world, except for Young Ng's reports (e.g., 2015, 2017) on China, and a few relating to the European network, which were published in recent years. Earlier attempts to get economic figures in 2001–2007during promotion of the geopark concept in Australia were equally fraught although individual geoparks did provide some (S. Turner, then UNESCO International Advisory Group for Geoparks, pers. comm., Dec. 15, 2018). Promoting publications of such records would be beneficial to persuade chief decision makers, perhaps other than Europeans and Chinese, and could contribute significantly to the growth of geoparks, conservation of geological identity, and geological heritage all over the world.

Also, regarding the importance of volunteer management programs in Geotourism Development, as discussed by Cristian Ciobanu and Alexandru Andrăşanu (Chapter 15), Youth is a reservoir of creativity and energy for any community and, if properly managed, engaged and empowered, it can have a significant impact on the growth of a geopark (there is a successful story about a Volunteer Group in the Haţeg UNESCO Global Geopark in Romania). However, there are some challenges regarding the development of geoparks in developing countries. For example, Iran's primary geopark (Qeshm, accepted in 2005: Turner, 2005) was de-listed in 2012 from the network by the GGN, despite adding to infrastructure, and upholding the objectives and functions of a geopark but the cultural differences in this region may have been at the root of the problem (their neglect of western-style geotourism and poor participation in the GGN may have been underlying factors: Turner pers. comm., Dec.15, 2018). However, Qeshm was listed as a global geopark once more and has regained the opportunity to prove itself. One of the challenges of the development of geoparks in developing countries is lack of true recognition that, firstly, this type of tourism is knowledge oriented, and, secondly, the monopolization of the sector's activists is in relation to public sector budgets.

Interpretation of heritage and specially (geo)heritage is both a science and an art that is less well known in developing countries. Their tourism development is not familiar with such facilities, and, mostly, at the national level, initiation with interpretation of heritage and generalization of science are serious issues that both academics and executive directors are less familiar with.

Basically, one of the main differences between developing and developed countries (other than economic differences) is the educational issues in the former countries where, indeed, non-formal edutainment is usually less considered and is often neglected. In fact, the lack of proper formal education in general and lack of non-formal education in tourism are major problems to overcome across the board. Sadry (2009) recalled how around 140 years ago the Iranian Hajj Sayyah travelled for18 years to more than 30 countries, including the USA, and concluded that the main difference between west and east is in education. Of course, we don't all have the opportunity to do this and so sharing information within the geoparks movement, for instance, is an important factor in promoting education.

The second issue hindering the establishment and development of geoparks is preferring personal interests to national interests by those engaged in geoparks projects. Not allowing the presence and cooperation of other informed persons and even specialists by those influential and close to governmental authorities leads to failure of geoparks. We need many good and enthusiastic people to foster the geopark movement and not a narrow clique. These patterns might be repeated in developing countries, such as the Middle East, countries of which really require rural and local economic growth, and sometimes slow down the development of geoparks in developing countries in the 21st century. To address the issue, developing some monitoring mechanisms by the governments and also by the UNESCO appears to be inevitable in these countries.

Actually, there is a need to formulate a strategy and to define the role of geotourism in directing tourism in different countries. Considering strategy development and defining the role of geotourism, like the 2030 or 2050 vision, should be addressed in various countries around the world by top-bottom management and through recognizing geotourism and all its subsectors proper for each country. This should be considered by governments to alleviate poverty in developing countries and to enhance economic and social development. The issue of cooperation between organizations and local society facilitation and interpretation of heritage and designing it in developing geoparks are issues that are weak in developing countries and sometimes the abovementioned issues are not even believed.

Therefore, for developing geoparks as leverage for poverty alleviation especially in developing countries, removing the weakness of these abovementioned three issues in Middle Eastern geoparks, and developing countries, the problems will be solved. One of these mechanisms is for

UNESCO to establish codes and standards and more clear instructions for developing countries.

Founding a World Forum of Geotourism and establishing a global data base from developed and developing countries regarding specialists of Geotourism, and referring consultation works to them are possible approaches for recognizing activists and specialists in developed and developing countries. Political support is a substantial factor both in developing and developed countries, as the only proclaimed Australian global geopark, Kanawinka, was de-listed by UNESCO due to "a lack of political support!" (Newsome and Dowling, 2018) at the federal level. Perhaps also Kanawinka was just too big and remote and its members did not participate enough in the GGN; it is still a viable national geopark (S. Turner, pers. comm. Dec. 15, 2018).

According to Newsome and Dowling (2018), "UNESCO performs an important role in stimulating geopark development and in fostering optimal geopark management. Of course, all of this depends on cooperation, support, and funding." Lack of political support in some developed countries and the issue of cooperation between organizations, local community facilitation, interpretation of heritage and its use in the design for establishing geoparks are among the shortcomings of developing countries; these factors are sometimes highly ignored and should be addressed through developing macro-strategies at a national level. It is worth mentioning that the "Halal Geopark," which was introduced by Sadry (2012b), can be incorporated into a geopark that would attract a huge number of Muslims to expand the geopark market in all countries. This market expansion needs further research and consideration in geopark strategy development in the future.

23.3 CONCLUSION

I have re-explored the encompassing topics of "geotourism concepts in the 21st century," "geo-assessments: geoheritage assessments for geotourism," "geo-interpretation—interpreting geological and mining heritage," "geoparks & community developments—a base for geotourism promotion," "globalization & the future of geological attraction destinations," through all the chapters of this book. Also, considering the previous educational and consulting experiences of the Editor and reviewing the previous texts (e.g., the latest comprehensive one, Dowling and Newsome, 2018), it is concluded that geopark territories are essential for cultural and natural heritages of any given region and "Geopark tourism" in a holistic manner, including geotourism as an essential (see Chapter1), in addition to any other types/subsectors of sustainable tourism, with using all geographical resources within a geopark territory (namely: all existing things around us, based on Chinese attitudes and thoughts about the environment (see Chapter 2 and also Chen et al., 2015) for sustainable tourism promotion such as ecotourism, nature-based activities, wildlife tourism, all sub-segments of cultural tourism, and so on, in a sustainable package—the so-called national/global geopark can be considered to be the new "wave" of the sustainable tourism industry's movement in the 21st century.

I would suggest that geoparks have the potential to be developed in all countries. The expansion of global geoparks in the last 15 years is proving evidence that this should be a focus.

My prospect is that geotourism and its potential for alleviating poverty would be increasingly noticed by governments of various countries. However, there are constraints in developing geoparks in such countries, mainly due to lack of knowledge, recognition, and sympathy.

There has been extensive research done on the dimensions and elements of geotourism development in the 21st century, from interpretation for different age groups to incorporating emerging technologies, applying social media and social networks to interpret heritage and environmental conservation are obvious examples, which lead to the promotion of non-formal education of people.

The development of geotourism in "vertical" (meaning deep ocean to space) destinations is one of the main issues that will be remarkably addressed under abiotic nature in the 21st century.

As I mentioned previously in Chapter 1, it is unnecessary to extend the geotourism definition to cover all biotic and abiotic aspects of nature, and all cultural elements (without considering its relationship to geoscience) in the geotourism definition. Actually, it would be the geopark mission to keep all this diversity alongside its main function, which is promoting and conserving geological heritage. Therefore, geotourism is considered both as geological tourism (as a type of tourism) or better to say abiotic nature-based tourism, which is an approach to abiotic nature conservation.

Furthermore, the "ABC approach" to our environment for interpretation, proposed by Dowling (Chapter 11) is a key strategy for interpretive provision within all geopark interpretive centers around the globe. Finally, to rewrite the British geologist Michel Walland's (1946–2017) phrase, as I mentioned in the Preface of this book—in his book "Sand: the never ending story"—From individual grains to desert dunes, from the bottom of the sea to the landscapes of Mars, and from billions of years in the past to the future 'just in a grain of sand, there are worlds to see!'

And, considering the third neglected and hitherto unrecognized component of our environment, namely abiotic nature, for conservation and tourism promotion (as a neglected resource) can broaden our horizons and insights and enable us to define from Earth to the universe and from geointerpretation on Earth to the planetary geological interpretive provision in space for sustainable tourism promotion in the 21st and in future centuries.

So, to observe our planet Earth from space, as a geosite, and listen to the space geological interpretation while in orbit within a shuttle or on a space station is not unrealizable in the 21st century.

ACKNOWLEDGMENTS

I am very grateful to Dr. Susan Turner, Fatemeh Fehrest, and John Macadam for their help, which has improved this chapter.

KEYWORDS

- **future of geotourism**
- **geotourism industry**
- **Halal geoparks**
- **Chinese geoparks**
- **global geopark network**

REFERENCES

Allan, M. Accessible Geotourism: Constraints and Implications. In *The Geotourism Industry in the 21st Century*; Sadry, B. N. Ed.; The Apple Academic Publishers Inc., 2020; pp 471–477.

Beck, L.; Cable, T. T. *Interpretation for the 21st Century: Fifteen Guiding Principles for Interpreting Nature and Culture*; Sagamore Publishing: USA, 1998, p. 242.

Cable, T. T. Interpreting Mining: A Case Study of a Coal Mine Exhibit. In *The Geotourism Industry in the 21st Century*; Sadry, B. N. Ed.; The Apple Academic Publishers Inc., 2020; pp 229–245.

Cayla, N. Email personal communication with Nathalie Cayla, 2018.

Cayla, N. Dinosaur Geotourism, A World Wide Growing Tourism Niche. In *The Geotourism Industry in the 21st Century*; Sadry, B. N. Ed.; The Apple Academic Publishers Inc., 2020; pp 447–470.

Chen, A.; Yunting, L.; Young, C. Y. Ng. *The Principles of Geotourism*; Springer Geography, Springer-Verlag Berlin Heidelberg and Science Press Ltd., 2015.

Cheung, L. T. O. The Effect of Geopark Visitors' Travel Motivations on Their Willingness to Pay for Accredited Geo-Guided Tours. *Geoheritage* **2016,** *8*, 201–209. DOI: 10.1007/s12371-015-0154-z.

Ciobanu, C.; Andrășanu, A. The Role of Volunteer Management Programs in Geotourism Development. In *The Geotourism Industry in the 21st Century*; Sadry, B. N., Ed.; The Apple Academic Publishers Inc., 2020; pp 375–386.

Dowling, R.; Newsome, D. The Future of Geotourism: Where to From Here? In *Geotourism: The Tourism of Geology and Landscape*; Newsome, D.; Dowling, R., Eds.; Goodfellow Publishers Limited: Oxford, UK, 2010, pp. 231–244.

Dowling, R. K.; Newsome, D. Geotourism Destinations—Visitor Impacts and Site Management Considerations. *Czech J. Tour.* **2017,** *6* (2), 111–129. DOI: 10.1515/cjot-2017-0006.

Dowling, R. K.; Newsome, D., Eds.; *Handbook of Geotourism*; Edward Elgar Publishing: Cheltenham, Gloucestershire, 2018.

Dowling, R. K. Interpreting Geological and Mining Heritage. In *The Geotourism Industry in the 21st Century*; Sadry, B. N. Ed.; The Apple Academic Publishers Inc., 2020; pp 277–298.

Farsani, N. T; Coelho, C.; Costa, C.; Neto de Carvalho, C., Eds.; *Geoparks and Geotourism: New Approaches to Sustainability for the 21st Century*. Brown Walker Press: Boca Raton, 2012.

Gaidzik, K. Urban Geotourism in Poland. In *The Geotourism Industry in the 21st Century*; Sadry, B. N. Ed.; The Apple Academic Publishers Inc., 2020; pp 93–116.

Geological Society of Australia. *Geotourism and Geotrails*, 2018. Available at: https://www.gsa.org.au/Public/Geotourism/Geotourism%20and%20Geotrails/Public/Geotourism/Geotourism%20and%20Geotrails.aspx?hkey=f3ee82ea-de9a-44eb-82f7-377d2d28d2ba, retrieved Dec.15 2018.

GGN Newsletter. Introducing a New GeoProduct GeoRafting—Experiencing Geology; ISSUE2-3M; available (Dec.1, 2018). at: www.globalgeopark.org/.../2018.../GGN%20Newsletter%202018%20ISSUE2-3M.pdf

Gray, M. *Geodiversity. Valuing and Conserving Abiotic Nature*. J. Wiley & Sons: Chichester, 2004.

Hose, T. A. Geotourism and Interpretation. In *Geotourism*; Dowling, R. K.; Newsome, D., Eds.; Oxford-Burlington (Elsevier Butterworth-Heinemann), 2006; 221–241.

Hose, T. A. Selling the story of Britain's stone. *Environmental Interpretation* **1995,** *10* (2), 16–17.

Hose, T. A. Historical Viewpoints on the Geotourism Concept in the 21st Century. In *The Geotourism Industry in the 21st Century*; Sadry, B. N., Ed.; The Apple Academic Publishers Inc., 2020; pp 23–92.

Hose, T. A. Geotrails. In *The Geotourism Industry in the 21st Century*; Sadry, B. N., Ed.; The Apple Academic Publishers Inc., 2020; pp 247–275.

Lesslar, P.; Lee, C. Preserving Our Key Geological Exposures-Exploring the Realm of Geotourism. In *Geological Heritage of Malaysia*, I. Komoo; H. D. Tjia; S. Leman, Eds; LESTARI, Universiti Kebangsaan Malaysia, 2001; pp 417–425.

Mather, P., Ed. *A Time for a Museum. History of the Queensland Museum 1862–1986.* Memoirs of the Queensland Museum, S. Brisbane, 1986.

Macadam, J. *Geoheritage: what do we want to tell them?* Keynote talk to Norwegian Geological Survey, 2015. Available at: https://www.youtube.com/watch?v=XJKe5YYScBI

Macadam, J. Geoheritage: Getting the Message Across. What Message and to Whom? In *Geoheritage: Assessment, Protection and Management*; Reynard, E., Brilha, J., Eds.; Elsevier Inc.: Amsterdam, 2018; pp 267–288.

Macadam, J. Email personal communication with John Macadam, 2018.

Macadam, J. Email personal communication with John Macadam, 2020.

Moreira J. C.; Bigarella, J. J. Geotourism and Geoparks in Brazil. In *Global Geotourism Prespectives*; Dowling, R.; Newsome, D., Eds.; Goodfellow Publishers Limited: Oxford, UK, 2010; pp 137–152.

Newsome, D.; Dowling, R. K. Geoheritage and Geotourism. In *Geoheritage: Assessment, Protection and Management*; E. Reynard; J. Brilha, Eds. Elsevier Inc.: Amsterdam, 2018a; pp 305–322.

Newsome, D.; Dowling R. K. Conclusions: Thinking About the Future. In *Handbook of Geotourism*; Dowling, R. K.; Newsome, D., Eds.; Edward Elgar Publishing: Cheltenham, Gloucestershire, 2018, pp. 475–482.

Ng, Y. Geoparks: *A Global Approach to Promote Sustainable Tourism for Local Communities.* Paper Presented at the Global Eco Asia Pacific Conference, *Expanding Ecotourism Horizons.* Rottnest Island, Australia, 2015, Nov. 17–19.

Ng, Y. Economic Impacts of Geotourism and Geoparks in China; Global Eco Asia-Pacific Conference, Adelaide, Australia, 2017, Nov. 27–29. Available at: https://2017.globaleco.com.au/perch/resources/Gallery/dr-young-ng.pdf.

Pica, A.; Reynard, E.; Grangier, L.; Kaiser, C.; Ghiraldi, L.; Perotti, L.; Del Monte, M. *GeoGuides, Urban Geotourism Offer Powered by Mobile Application Technology. Geoheritage J.* **2017.** DOI: 10.1007/s12371-017-0237-0

Reynard, E.; Brilha, J., Eds.; *Geoheritage: Assessment, Protection and Management*; Elsevier Inc.: Amsterdam, 2018.

Sadry B. N. *Inauguration Ceremony of the First Official Geotour Guide Training Courses*, 2014. Available at: http://global-geotourism.blogspot.com/2014/06/inauguration-of-training-course-on.html (Dec. 15, 2018).

Sadry, B. N. *Geodiversity & Geoheritage Education through Geotourism Textbooks in Iran*, Geophysical Research Abstracts, EGU General Assembly: Vienna, Austria, April 22–27, 2012a, Vol 14, EGU2012-14008.

Sadry, B. N. *Islamic Geopark Brand in the Middle East: A Potential Lever for Socio-Economic Growth.* Proceedings of the First Geological Conference of Kurdistan—GEOKURDISTAN 2012, Sulaimani, Kurdistan Region, Iraq, Nov. 14–16, 2012b; pp 31–32.

Sadry, B. N. *Fundamentals of Geotourism: With Special Emphasis on Iran*; SAMT Publishers: Tehran, 2009 (English summary available online at: http://physio-geo.revues.org/4873?file=1 ; Retrieved: Jan. 01, 2020) (in Persian).

Sadry, B. N. Space and Celestial Geotourism. In *The Geotourism Industry in the 21st Century*; Sadry, B. N., Ed.; The Apple Academic Publishers Inc., 2020; pp 479–503.

Sadry, B. N. The Scope and Nature of Geotourism in the 21st Century. In *The Geotourism Industry in the 21st Century;* Sadry, B. N., Ed.; The Apple Academic Publishers Inc., 2020; pp 3–21.

Tongkul, F. Geotourism in Malaysian Borneo. In *Geotourism*; Dowling R. K., Newsome, D., Eds.; Elsevier/Heineman: Oxford, 2006.

Tilden, F. *Interpreting Our Heritage: Principles and Practices for Visitor Services in Parks, Museums, and Historic Places*; The University of North Carolina Press, 1957.

Turner, S. "Geotourism": Where Wise Birds Stay Awhile. *GeoExpro Mag.*, **2005**, 56–58, 60 (an online version published in 2008, available at: https://www.geoexpro.com/articles/2016/08/where-wise-birds-stay-awhile) (retrieved: Jan. 01, 2020).

Turner, S. Geoheritage and Geoparks: One (Australian) Woman's Point of View. *Geoheritage* **2013**, *5*, 249–264.

Turner, S. Email personal communication with Susan Turner, 2018.

UNESCO. *Wider value of UNESCO to the UK*. UK National Commission for UNESCO: London, 2014.

UNESCO. *UNESCO Global Geoparks*; UNESCO: Paris, 2016.

UNESCO. *UNESCO Global Geoparks*; UNESCO: Paris, 2018. Available at: http://www.unesco.org/new/en/natural-sciences/environment/earth-sciences/unesco-global-geoparks/.

Vafadari, K.; Cooper, M. J. M. Community Engagement in Japanese Geoparks. In *The Geotourism Industry in the 21st Century*; Sadry, B. N., Ed.; The Apple Academic Publishers Inc., 2020; pp 357–373.

Walden, J.; Pearce, S. USA Scenic Byways—Connecting People to Places. In *Geotourism: The Tourism of Geology and Landscape*; Newsome, D.; Dowling, R. K., Eds.; Goodfellow: Oxford, UK, 2010; pp 211–220.

Walland, M. *Sand: The Never Ending Story*; University of California Press, 2010.

Witherspoon, W.; Rimel, J. Commercially Successful Books for Place-Based Geology: Roadside Geology Covers the US. In *The Geotourism Industry in the 21st Century*; Sadry, B. N., Ed.; The Apple Academic Publishers Inc., 2020; pp 325–353.

Witherspoon, W. Evolving Geological Interpretation Writings about a Well-Traveled Part of California, 1878–2016. In *The Geotourism Industry in the 21st Century*; Sadry, B. N., Ed.; The Apple Academic Publishers Inc., 2020; pp 209–324.

Index

A

Accessible geotourism
  constraints for
    challenge, 475
    International Association for Geoscience Diversity (IAGD)
    visions, 475–476
    special needs, 474
Albania
  geoheritage and geotourism in
    Alpine tectogenesis, 180
    development of tourism and, 182
    earth science sites, criteria for selection of, 176
    European geology, nature and, 174–175
    features of geology, 179
    fundamentals of, 171
    geomorphologic cycles, 180–181
    groups of sites, 177
    mantel harzburgite–dunite, 179
    metamorphic rocks, 179
    Mirdita ophiolite, 182
    natural monuments, categorization of, 177
    northeastern geotour, 182
    northern geotour, 182
    picturesque landscapes, 181
    ProGEO Symposium, 177
    pupils and students, formative education of, 181
    sites, classification and categorization of, 175–177
    southeastern geotour, 182, 184
    southern geotour, 182
    studies on, 173
    Terra Rossa, 180
    Tethys Ocean, 179
    Third ProGEO Symposium in Madrid, 170
    tourism, 173
    UNESCO/IUGS Geosites and Geoparks, 170
American Geological Institute sponsored *Roadside Geology of US Interstate 80 Between Salt Lake City and San Francisco: The Meaning Behind the Landscape,* 306
APEHAL, 532
Area of Outstanding Natural Beauty (AONB), 72
Australian National Landscape (ANL), 74
Azores archipelago, 190
  geotourism and sustainable development, strategy for, 197–198
    courses, 200
    geopark accessibility, 201, 203
    info wood post at site, 202
    interactive brochure, 199
    interpretative geological panels, 202
    Queijo do Morro, 201
    touristic fairs, 203
    UGG, targets from, 198, 200
  UNESCO global geopark, 190
    culture man and volcanoes, 191–192
    geoconservation strategy, 192–193
    geotourism in, 193–196

B

Brazil, 119
  geological masterpieces, 121
  map of, 120
  mining heritage and geotourism
    Bahia, diamonds and carbonados in, 126–128
    challenges to development of, 138–140
    concepts and knowledge, 123
    cultural landscapes, 124
    definition of, 122
    Minas Gerais, gold and gemstones in, 124–126
    mobile elements, 122
    Piauí, opals in, 131–134
    regional and local levels, 121–122

Rio Grande Do Norte, scheelite in, 134–135
Rio Grande Do Sul, amethyst in, 129–131
São Paulo, Varvito park, 136–138
selective gathering, 123
technical information, 123
tourism in industrial heritage, 123
valorization, 122
tourism in, 120
Bryce Canyon National Park (BCNP), 494

C

Case studies
Eldheimar Volcano Museum, 287–288
former coal mine tour, 289–291
geosite attraction, 285–286
Iceland, 287–288
Jeju Island, 285–286
Jordan, 294–295
museum of science and industry coal mine exhibit, 233, 236–239
New Zealand, 289–291
proposed phosphate mine museum and experience, 294–295
rock salt mine and underground chapel, 292–294
Ruseifa, 294–295
South Korea, 285–286
Teton Geotourism Center, 289
Vestmannaeyjar, 287–288
Westland, 289–291
Wieliczka Poland, 292–294

D

Dinosaur geotourism
destinations relying on existing network
Haţeg Declaration, 462–465
Isle of Wight, 461–462
education and play go hand in hand, 466–468
scientific discoveries
paleontological destinations evolving with, 455–459
scope and typology
paleontological geotourism, 451
in situ dinosaurs geosites, 451–453
territorial relays, 453–454

E

European Geoparks Network (EGN), 73–74

F

FLAG-Friends of Langkawi Geopark, 532

G

Geo-Bio Trail of Kubang Badak in 2018, 531
Geoforest Parks (GFP), 531
Geoheritage, 278
geoconservation
China, 37
geosites program, 37–39
Germany and geotopes, 33–34
local and national campaign, 28
nature, 27–28
Ottoman Empire, 34–37
soft-rock quarries, 29
SSSI, 29
UK recognition and definitions, 31–33
journal, 4
minor threats, 29
hard-rock exposures, 30
Isle of Arran, 31
Isle of Wight, 30
Malvern Hills, 30
modern museum collections, 30
National Geological Natural Reserve (NGNR), 37
Geo-interpretation
Abiotic, Biotic, and Cultural (ABC) features, 282, 284
case studies
Eldheimar Volcano Museum, 287–288
former coal mine tour, 289–291
geosite attraction, 285–286
Iceland, 287–288
Jeju Island, 285–286
Jordan, 294–295
New Zealand, 289–291
proposed phosphate mine museum and experience, 294–295
rock salt mine and underground chapel, 292–294
Ruseifa, 294–295
South Korea, 285–286
Teton Geotourism Center, 289

Vestmannaeyjar, 287–288
Westland, 289–291
Wieliczka Poland, 292–294
communication, 284
Eldheimar—World of Fire, 287
face-to-face methods, 282
form, 284
hand, 284
head, 284
heart, 284
process, 284
public engagement and appreciation, 282
The Rule of Thirds, 285
storytelling, 285
time, 284
Geological interpretation writings
California geotourism trip with Google maps, 318, 320–321
road guide map, 319
full-color guide to rocks along I-80
AAPG Geological Highway Map along I-80, 306
introduction, part, 307
Geological Highway Map series, 304
John McPhee articles, 311
Annals of the Former World, 311
assemblage of rock types, 312
Assembling California, 311, 312
Basin and Range, 311, 312
Cretaceous Period, 312
geologic fieldwork and interpretation, 312
New Yorker, 311
Rising from the Plains, 312
Roadside Geology of Northern and Central California, 313
In Suspect Terrain, 312
literary approach and updated geology
gold-bearing riverbeds, 317
Rough-Hewn Land, 316–317
path across the Sierra Nevada, 301
plate tectonics revolution
Alpine geology, 305
Cold War military research, 305
comprehensive theory, 305
Cretaceous age granitic rocks of Sierra Nevada, 305
debate, 306
Geological Society of America, 305
Great Valley and Coast Ranges, 306
subduction, 305
plate tectonics up front and labeled geologic maps
Cretaceous-age granite, 309
elements, 308
The Great Collision, *Roadside Geology* series, 308
Jurassic-age sediments, 308
Roadside Geology of Northern California, 308
12-page Sierra Nevada-Klamath introduction, 311
railroad era (1878, 1915)
An American Geological Railway Guide, 301
fault movement, 304
foldout map, portion of, 303
Guidebook of the Western United States, 302
list of stations in, 302
Omaha-to-San Francisco, 303
Sierra Nevada Geologic history, panorama of, 308
Donner Lake, 307
hill builds geologic knowledge, 307
Plate tectonic theory, 307
Third Pass of *Roadside Geology* Series
full color, 318
trips with clarity and simplicity
Ted Konigsmark, 313–315
USGS *Geologic Map of the US*
traveler's appetite, 304
Geomorphosites, 147
aesthetic value of, 149–151
natural monuments
Buzau valley, 154
concept of, 151–153
cultural and geotourism, 157–162
Euseigne Pyramids, Swiss Alps, 154
geocultural sites on Matmata-Dahar plateau, 161
glacier landscapes in Swiss Alps, 158
glacier tourism, 153–157
Monthey, 154
Pâclele Mari mud volcanoes, 154
Pierre à Dzo erratic boulder, 154

Romania, 154
  Sphynx, Bucegi Mountains, Romania, 154
  Switzerland, 154
Geoparks
  Area of Outstanding Natural Beauty (AONB), 72
  Australasian geoparks, 74–76
  Australian National Landscape (ANL), 74
  Canadian geoparks, 74–76
  Chinese geoparks, 74–76
  European Geoparks Network (EGN), 73–74
  Luochuan Loess National Geopark (LLNG), 76
  Site of Special Scientific Interest (SSSI), 73
  sustainable geotourism, 72
  UNESCO-recognized, 71–72
Geotourism
  clarifying, situation
    conceptual and practical definitions, 11–12
    correspondence with, 10
    geopark, core elements and key components of, 11
  defined, 5
  definitions and approaches
    Asia, 41–43
    Australasian and European, 40–41
    discrete term, 39
    Europeans, 44–45
    intellectual stimulation, 40
    National Geographic, 45–48
    outside of Europe, 45
    rockhounding and rockhounds, 39
    UK and first ever, 43–44
  elements of, 5–6
  First International Geoscience Education Conference, 61
  geo-interpretation, 280
    geology and mass media, 70–71
    geology visitor centres, 67–68
    historical aspects of, 62–65
    Ludford corner, 66–67
    UK country parks, 65–66
    UK geology museums, 69–70
    UK geotrails, 68–69
    Wren's Nest National Nature Reserve, 66–67
  guidelines for development of, 536
    accessible geotourism, 541–542
    celestial geotourism, 543
    development of roadside geology, 537
    dinosaur geotourism, 541
    interpret heritage by modern principles, 537–538
    national and global geoparks, creation, 543–547
    new training courses and academic majors, 540–541
    promotion of geotrails, 538–539
    urban geotourism, 539–540
  influences on growth
    geology education and universities, 62
    international school geology education, 61–62
    natural history societies, 58–59
    philosophical and naturalist societies, 57–58
    school curriculum, geology in, 59–61
  mining district, 280
  natural resources, 279
  niche tourism, 48–49
  pilgrims and travelers, 25–26
  precursors of
    Association, 56
    Circular notices, 56
    Grand Tour, 52–54
    manufactures, 56–57
    seven wonders, 54–55
    socio-cultural construct, landscapes, 50
    three aesthetic movements, 50–52
  Quechua miners guide tourists, 280
  recreational geology, 48
  socio-economic activity, 148
  tourism industry, 26–27
    and tourist, 25
  tourtellotic geotourism, 7
    sustainable tourism principles, 8
  true geotourism
    basic principles, 7
    modern era, 6
      Feasibility Study, 7
  US National Park Service, 279
Geotourism *vs* mass tourism

Index

civil society movement, 531–533
growth in tourism, 529
issues, 529–530
Geotrails, 247
  attracting and holding
    actual number of persons viewing panel, 256
    effort required, 256–257
    holding power, 255
    panel placed, 256
    trailside panel, 255
  Barnack Holes and Hills NNR, 260
  Bradwell Station site, 272
  Bradwell to Newport Pagnell Geotrail, 271
  Brown End Quarry, 270
  environmental interpretation, 249
  geo-interpretative provision
    conceptual model for, 250
    geosite cube plot, 251
    quick response (QR) labeling, use of, 250–251
    sophisticated trails, 250
  geointerpretative provision, models, 249
  geotourists and recreationalists
    actual number of persons viewing panel, 256
    effort required, 256–257
    holding power, 255
    panel placed, 256
    trailside panel, 255
  interpretative media, 251
    leaflets, 252
    Nerja cave geoarchaeological site, 254
    traditional geological themes, 254
    trailside panels, 252–253
  North East Pingo Trail, 271
  perceived best practice, 257–260
  types of users, 248
  United Kingdom (UK), historical and modern geotrails in, 260
    Crystal Palace Park dinosaur reconstructions, 261
    geological and geomorphological interest, 267
    Geological and Geomorphological Sites (RIGS), 266
    hard-rock geotrails, 268

Lower Carboniferous limestone, 268
marine promenade in, 267
Mortimer Forest Geology Trail, 264–265
multi-panel provision, 266
National Nature Week, 263
North-East Wales RIGS (NEWRIGS) group, 266
original stone plinths of, 263
proactive educational geoconservation approach, 262
start and finish stones, 262
Wildlife Trusts, 267
Geoturystyka, 44

I

Interpretation
  benefits for, 232–233
  case study
    museum of science and industry coal mine exhibit, 233, 236–239
  coal mine
    approaches, 239
    curators, 240
    Lohr's leadership labels, 241
    personal interpretation and, 242–244
    principles of, 241
  definitions, 229–230
  information, 281
  institutional commitment, 242
  Last Child in the Woods, 281
  mine sites
    Old Ben No. 17 Coal Mine exhibit, 236
    selection of some excellent, 234–235
  mining
    National Mining Association (nma.org), 233
  principles, 230–232
  techniques and philosophy, 241

J

Japanese geoparks
  geoconservation, 358
  geodiversity, defined, 358
  geoheritage conservation, 358
  geotourism
    conservation and community involvement, 363–365

defined, 361
heritage and landscapes, 362
Kunasaki Peninsula and GIAHS, 365, 367–370
Japan as in 2020, 362
threats and tourism
   fragmentation, 359
   geopark concept, 360–361
   human impact, 359

K

Kilim Geoforest Park, 531

L

LADA, 532
Langkawi UNESCO Global Geopark (LUGG)
   Geoforest Parks (GFP), 531
Last Child in the Woods, 281
Latin American and Caribbean Geopark Network (GeoLAC), 422
Latin American UNESCO
   distinctive geoproducts, 423
   geological and geomorphological diversity, 424
   global geoparks, 425
      Araripe UNESCO global geopark, Brazil, 426–428
      AUTOCOLCA, 441
      Cerro San Cristobal, 436
      Colca Canyon, 441
      Colca River and Andamayo River, 439–442
      COLCA Y Volcanes de Andagua aspiring geopark, 439–442
      Comarca Minera, 436, 437
      Cruz del Condor site, 439
      El Chico National Park, 436
      Grutas del Palacio Geopark, 428–432
      Hidalgo's *Comarca Minera,* 438
      INGEMMET, 442
      landscapes and characteristics, 440
      Mixteca Alta UNESCO, 433–435
      Nazca Plate, 439
      Omitlan of Juarez, 439
      Pachuca and *Las Navajas,* 435
      Peñas Cargadas, 436
      Pueblos Mágicos, 437
      Santa María Regla, 436
      Scientific Expedition to Peru (PSEP), 442
      Sierra Madre Oriental, 436
      State of Hidalgo, 435
      Trans-Mexican Volcanic Belt, 436
      Universidad Nacional Autónoma de México (UNAM), 437
   global geoparks, aspiring geoparks, and projects, 424
   promotion of geotourism
      regional initiatives for, 442
Luochuan Loess National Geopark (LLNG), 76

M

Malatya Photography and Cinema Art Association (MAFSAD), 403
Mountain Press Publishing (MPP)
   covering continent with geology for the masses
   series won, 335

N

Narman, Oltu, and Tortum (NOT), 398–400
National Geological Natural Reserve (NGNR), 37
National Nature Reserve (NNR), 73
National Park (NP), 72
The Northern Anatolian Mountains, 389

P

Post-mining objects as geotourist attractions
   Upper Silesian Coal Basin (USCB)
      archaeological research, 510
      bog iron ore, 511
      Carboniferous deposits, 509–510
      consequences, 512
      geoturistic functions, objects without, 520–523
      Guido Coal Mine in Zabrze, 517–520
      hard coal exploitation, 511
      historic Silver Mine and Black Trout Adit, 513–514
      Queen Louise Coal Mine in Zabrze, 514–517
      transformation of political system in Poland, 511

Index

Proposed geopark projects
  Turkey and geotourism
    BITLIS volcanic geopark project, 405–407
    geological and geographical evolution, 389
    geological structures, 390
    geotourism and geoheritage, history of, 390–393
    glass terrace in Levent Valley, 403
    Hasanağa Stream Valley, 403
    Küçükkürne Village in Dipsiz Valley, 402
    Kula Geopark Area, 395
    Kula Volcanic Geopark, 393–394
    Kızılcahamam-Çamlıdere, 396–398
    Levent Valley, 400–401, 403–404
    Malatya Photography and Cinema Art Association (MAFSAD), 403
    Mount Ağrı (Ararat), 413–414
    Narman, Oltu, and Tortum (NOT), 398–400
    natural and cultural landscapes, 388
    The Northern Anatolian Mountains, 389
    process of, 388
    Sandal volcanic cone, 395
    The Taurus Mountains, 389
    UNESCO and the EGN geopark network, 404
    Upper Kızılırmak proposed geopark area, 407–413

R

Roadside Geology (RG) series, 326
  authorship and use of first-person narrative
    Don Hyndman's, 327
    MPP's work for geoscience education, 327
  covering continent with geology for the masses, MPP
    The Psychology of Cornet and Trumpet Playing, 328
    general format set in, 333
    Petrology of Igneous and Metamorphic Rocks, 328
    road guide map, 331
    Roadside Geology of Northern California, 333
    Roadside Geology of the Northern Rockies, 330, 332
    Roadside Geology of the Yellowstone Country, 330
    series won, 335
    1955 U.S. Geological Survey (USGS) geological map of Montana, 329
  geological interpretation books since 2008, MPP
    California Rocks!, 341
    color illustration of the changes in the Farallon slab, 340
    Geology Underfoot in Yosemite National Park, 341
    Geology Underfoot series, 341
    good color photography, 338
    Roadside Geology of Connecticut and Rhode Island, 338
    Roadside Geology of Nevada, 339
    Roadside Geology of Oregon, 339
    Roadside Geology of Washington, 339
    Roadside Geology of Yellowstone Country, 339
    Rocks! series, 341
  geological tourism
    Decatur Book Festival bills, 345
    Geographic Information System (GIS) database, 350
    Georgia Geological Society, 347
    Georgia Geo-Travelogue, 342
    Golddiggers, Generals, and Tightrope Walkers, program, 347
    Google Earth®, 342
    Hyndman notes, 344
    key Civil War locations, 348
    Kids Rock!, game, 349
    matching game, 345
    Natural Communities of Georgia, 347
    Natural Environments of Georgia, 347
    rail line and, 348
    Roadside Geology of Georgia, 350
    sock rock, 345
    Two Google Earth tools, 345
    USGS, 347
  Georgia
    Geologic Map of Georgia, 336
    Georgia Geological Society, 337
    Google Maps®, 337

Historical Geology Lab Manual, 336
  Pamela Gore of Georgia Perimeter College, 335
  US and part of Canada, coverage, 334

S

Site of Special Scientific Interest (SSSI), 73
Space and celestial geotourism
  boundaries
    astronomical tourism, 486
    celestial geotourism, 488
    classification and synopsis of, 489
    defined, 484–485
    space tourism, 485–486
    space tourism, typology of, 487
  celestial geotourism
    astronomical observations, 496
    calendar structure, 496
    celestial geosite (daytime-scapes), 493–494
    geosites (nightscapes), 490, 492–493
    meteorite geotourism, 496–497
    nocturnal activities within geoparks, 494
    planetariums, 497
    secondary celestial geosites, 495–497
    terrestrial space geo-landscapes, 494
    urban celestial geotourism, 495
  destinations, resources, and meanings, 497
    primary space geosite, 498–499
    space geotour guides, 499–500
  man in the moon, iconic image, 484
  management considerations
    and environmental issues, 502
    mindful visit for, 500–502

T

Threats and protection, geoheritage geoconservation
  China, 37
  geosites program, 37–39
  Germany and geotopes, 33–34
  local and national campaign, 28
  nature, 27–28
  Ottoman Empire, 34–37
  soft-rock quarries, 29
  SSSI, 29
  UK recognition and definitions, 31–33
  minor threats, 29
    hard-rock exposures, 30
    Isle of Arran, 31
    Isle of Wight, 30
    Malvern Hills, 30
    modern museum collections, 30
    National Geological Natural Reserve (NGNR), 37
Trash Hero Langkawi, 531

U

Upper Silesian Coal Basin (USCB)
  archaeological research, 510
  bog iron ore, 511
  Carboniferous deposits, 509–510
  consequences, 512
  geoturistic functions, postmining objects without, 520–523
  Guido Coal Mine in Zabrze, 517–520
  hard coal exploitation, 511
  historic Silver Mine and Black Trout Adit, 513–514
  Queen Louise Coal Mine in Zabrze, 514–517
  transformation of political system in Poland, 511
Urban geotourism
  in large cities
    Between Bridges, 103
    Gdañsk, 105–106
    Kielce, 104–105
    Kraków, 98–102
    local materials, 99–100
    location of, 99
    OstrówTumski, 103–104
    Poznañ, 105–106
    The Venice of the North, 102
    Warszawa, 105–106
    Wrocławia, 102–104
  Poland, geotouristic potential
    geological structure, 95
    Global UNESCO Geoparks Network, 97
    Kraków, 98
    Muskau Arch Geopark, 96–97
    Precambrian, 96
    Siedliska and The Dragon's Cave, 97
    tectonic provinces and main cities, 96

Index

postindustrial geotourism
  Silesia region, 108–109
  small towns, 107, 109
  attractions, 108

V

Volunteer management programs
  geopark ambassadors
    defined, 379
    Haţeg Country Geoparks Ambassadors, 380
  geotourism development and promotion of territory, 381
    contributions and community, 383
    volunteering results, 382
  House of Science and Art, 384
  need and functionality
    captive target group, 378
    Foundation for Partnership, 378
    geopark team, 367
    key elements, 377
    MOL Romania, 378
    recruitment, 378
    Romanian volunteering law (78/2014), 378
    VOLUM Federation and ProVobis, 378
    youth of, 377
  volunteer engagement, structure of, 379

W

World's top geotourism destinations
  Brazilian City of Rio de Janeiro, 222
  Central Swiss Alps, 220
  Golden Circle, 221
  Grand Canyon (USA), 216
  Great Barrier Reef, 218
  mountain site, 220
  palaeontological site, 221
  South China Karst, 218
  Sugar Loaf Mountain, 223
  Table Mountain, 222
  Table Mountain National Park, 222
  top geological sites, lists
    Africa's Top Geological Sites, 213
    13 Geologic Wonders of the Natural World, 210–211
    20 Geological Wonders of the World in 2016, 211–212
    100 Great Geosites, 213
    10 most spectacular geologic sites in the United States, 212–213
    Niagara Falls, 215
    Scottish Geodiversity Forum, 213
    Top 10 Geological Wonders of the World, 212
  Uluru, 220
  Victoria Falls, 215
  volcanic site, 219
  Yellowstone National Park Act (1872), 219
World's top geotourism destinations
  top tourism/nature sites, lists, 208
    7 Natural Wonders of the World, 209
    CNN, 210

Y

Yellowstone National Park Act (1872), 219